# Starch-based Blends, Composites and Nanocomposites

**RSC Green Chemistry**



*Titles in the Series:*

1: The Future of Glycerol: New Uses of a Versatile Raw Material
2: Alternative Solvents for Green Chemistry
3: Eco-Friendly Synthesis of Fine Chemicals
4: Sustainable Solutions for Modern Economies
5: Chemical Reactions and Processes under Flow Conditions
6: Radical Reactions in Aqueous Media
7: Aqueous Microwave Chemistry
8: The Future of Glycerol: 2nd Edition
9: Transportation Biofuels: Novel Pathways for the Production of Ethanol, Biogas and Biodiesel
10: Alternatives to Conventional Food Processing
11: Green Trends in Insect Control
12: A Handbook of Applied Biopolymer Technology: Synthesis, Degradation and Applications
13: Challenges in Green Analytical Chemistry
14: Advanced Oil Crop Biorefineries
15: Enantioselective Homogeneous Supported Catalysis
16: Natural Polymers Volume 1: Composites
17: Natural Polymers Volume 2: Nanocomposites
18: Integrated Forest Biorefineries
19: Sustainable Preparation of Metal Nanoparticles: Methods and Applications
20: Alternative Solvents for Green Chemistry: 2nd Edition
21: Natural Product Extraction: Principles and Applications
22: Element Recovery and Sustainability
23: Green Materials for Sustainable Water Remediation and Treatment
24: The Economic Utilisation of Food Co-Products
25: Biomass for Sustainable Applications: Pollution Remediation and Energy
26: From C–H to C–C Bonds: Cross-Dehydrogenative Coupling
27: Renewable Resources for Biorefineries
28: Transition Metal Catalysis in Aerobic Alcohol Oxidation
29: Green Materials from Plant Oils
30: Polyhydroxyalkanoates (PHAs) Based Blends, Composites and Nanocomposites

31: Ball Milling Towards Green Synthesis: Applications, Projects, Challenges
32: Porous Carbon Materials from Sustainable Precursors
33: Heterogeneous Catalysis for Today's Challenges: Synthesis, Characterization and Applications
34: Chemical Biotechnology and Bioengineering
35: Microwave-assisted Polymerization
36: Ionic Liquids in the Biorefinery Concept: Challenges and Perspectives
37: Starch-based Blends, Composites and Nanocomposites

*How to obtain future titles on publication:*
A standing order plan is available for this series. A standing order will bring delivery of each new volume immediately on publication.

*For further information please contact:*
Book Sales Department, Royal Society of Chemistry, Thomas Graham House, Science Park, Milton Road, Cambridge CB4 0WF, UK
Telephone: +44 (0)1223 420066, Fax: +44 (0)1223 420247
Email: booksales@rsc.org
Visit our website at www.rsc.org/books

# *Starch-based Blends, Composites and Nanocomposites*

Edited by

**Visakh P. M.**
*Tomsk Polytechnic University, Russia*
*Email: visagam143@gmail.com*

**Long Yu**
*CSIRO, Clayton, Australia*
*Email: Long.yu@csiro.au*

ROYAL SOCIETY OF CHEMISTRY

THE QUEEN'S AWARDS FOR ENTERPRISE: INTERNATIONAL TRADE 2013

RSC Green Chemistry No. 37

Print ISBN: 978-1-84973-979-5
PDF eISBN: 978-1-78262-279-6
ISSN: 1757-7039

A catalogue record for this book is available from the British Library



Published by The Royal Society of Chemistry,
Thomas Graham House, Science Park, Milton Road,
Cambridge CB4 0WF, UK

Registered Charity Number 207890

For further information see our web site at www.rsc.org

Printed in the United Kingdom by CPI Group (UK) Ltd, Croydon, CR0 4YY, UK

# Preface

This book summarizes many of the recent research accomplishments in the area of starch-based polymer blends, composites and nanocomposites such as starch: structure–property relationships, preparation and characterization of starch nanocrystals, natural fibre-reinforced thermoplastic starch composites, applications of starch nanocrystal-based blends, composites and nanocomposites, chemically modified thermoplastic starches, outstanding features of starch-based hydrogel nanocomposites, starch-based blends, fracture and failure of starch-based composites, application of starch nanocomposites in the food industry and effect of additives on the properties of starch. As the title indicates, the book emphasizes the various aspects of starch-based blends, composites and nanocomposites, and it is intended to serve as a "one-stop" reference resource for important research accomplishments in this area. This book should be a very valuable reference source for university and college faculties, professionals, postdoctoral research fellows, senior graduate students and researchers in R&D laboratories working in the area of starch-based blends, composites and nanocomposites. The various chapters in this book have been contributed by prominent researchers in industry, academia and government/private research laboratories across the globe, and provide an up-to-date record of the major findings and observations in the this subject area.

The first chapter gives an overview of the state-of-the-art, new challenges and opportunities in starch-based studies and research, preparation and characterization and applications of starch-based blends, composites and nanocomposites and also future trends in this area.

Chapter 2 provides up-to-date information about starch and its structure–property relationships. The complexes of starch with other components, starch hydrolysis, starch modification methods (chemical, physical, enzymatic and genetic/biotechnological modification), industrial production

RSC Green Chemistry No. 37
Starch-based Blends, Composites and Nanocomposites
Edited by Visakh P. M. and Long Yu

Published by the Royal Society of Chemistry, www.rsc.org

of starches and their uses are also discussed. Chapter 3, on the preparation and characterization of starch nanocrystals, critically reviews recent advances in the preparation of starch nanocrystals and reports the critical steps needed to modify them chemically for new applications. Novel and improved techniques are outlined in order to optimize the preparation of starch nanocrystals from a wide range of starch sources. Advanced methods used to characterize starch nanocrystals and to understand their interactions with polymer matrices are also discussed. In addition, the physicochemical properties, including mechanical, barrier, morphological, thermal, and swelling and emulsification, are presented with potentially new insights into the development of starch nanocrystals for industrial applications. Chapter 4, on natural fibre-reinforced thermoplastic starch composites, covers aspects such as natural fibre, starch, thermoplastic starch and natural fibre-reinforced thermoplastic starch composites. Of interest here is that the authors prepared an environmentally friendly composite where the matrix (sugar palm starch) and fibre (sugar palm fibre) were derived from one source, *i.e.* the sugar palm tree. The resulting materials are termed "biocomposites" or "green" composites, which are considered to be totally biodegradable.

Chapter 5 concentrates mainly on applications of starch nanocrystal-based blends, composites and nanocomposites and highlights recent research on the preparation, characterization and properties of polymeric matrix–starch nanocrystal nanocomposites. The results discussed indicate that starch nanocrystals were able to improve the properties of polymers because strong interactions between the matrix and the nanofillers were formed. The particular characteristics, such as the polymer used as the matrix and the amount of nanofillers, that also affected the properties of the nanocomposites are discussed. Chapter 6 considers the chemical modification of thermoplastic starch. Topics covered include starch as an industrial raw material, destructuring of starch by extrusion, chemical modification of starch by reactive extrusion, depolymerization of starch by reactive extrusion of thermoplastic starch and starch modification with citric acid. The aim of this chapter is to review the chemical modification of starch as thermoplastic starch by extrusion processing or other similar techniques in which the material is in the melt state. Generally, this process is conducted in the presence of plasticizers such as glycerol, glycols and urea. Chapter 7 discusses the outstanding features of starch-based hydrogel nanocomposites and is focused on the formulation of functional materials with enhanced properties for more suitable applications in different fields. Relevant aspects of the current knowledge of starch-based hydrogel nanocomposites are discussed, mainly those based on structure–property relationships, and also specific functionalities for a given application. Further comprehensive studies will broaden the understanding of structure–property relationships. Chapter 8, on starch-based blends, includes topics such as the preparation, modification and applications of starch-based blends. Chapter 9 covers the fracture and failure of starch-based composites with topics such as the fracture and failure of starch reinforced with natural fibers and the fracture and failure of starch-based nanocomposites (starch reinforced with carbon nanotubes and with nanoclays), and summarizes future trends. Chapter 10 surveys the applications and potential of starch nanocomposites in the food industry. The major topics include nanotechnology in foods, starch as the matrix with different nanofillers and starch nanoparticles. The final chapter, on the effects of additives on the properties of starch, provides an overview of recent progress in the interaction between starch and its additives. It includes the effect of additives on the properties of starch, the mechanism of the interactions between starch and additives and recent applications of additives in starch-based products.

The editors would like to express their sincere gratitude to all the contributors to this book, whose excellent support ensured the successful completion of this venture. We are grateful to them for the commitment and the sincerity they have shown towards their contributions. Without their enthusiasm and support, the compilation of this book would have not been possible. We also thank all the reviewers who devoted their valuable time to make critical comments on each chapter. Finally, we thank the publisher, the Royal Society of Chemistry, for recognizing the demand for such a book, for realizing the increasing importance of the area of starch-based blends, composites and nanocomposites and for initiating this new project.

Visakh P. M.

# Contents



RSC Green Chemistry No. 37
Starch-based Blends, Composites and Nanocomposites
Edited by Visakh P. M. and Long Yu

Published by the Royal Society of Chemistry, www.rsc.org

CHAPTER 1

# *Starch: State-of-the-Art, New Challenges and Opportunities*

VISAKH P. M.

Tomsk Polytechnic University, Lenin Av. 30, 634050 Tomsk, Russia
Email: visagam143@gmail.com

## 1.1 Starch: Introduction and Structure–Property Relationships

Starch is a polysaccharide consisting of D-glucose units, referred to as homoglucan or glucopyranose, and has two major biomacromolecules – amylose and amylopectin. Amylopectin is a much larger molecule than amylose, with a molecular weight of $1\times10^7$–$1\times10^9$ and a heavily branched structure built from about 95% α-(1→4) and 5% α-(1→6) linkages. Amylopectin unit chains are relatively short compared with amylose molecules, with a broad distribution profile. Starch varieties contain primarily two different types of anhydroglucose polymers, amylase and amylopectin.

Both amylose chains and the exterior chains of amylopectin can form double helices, which in turn may associate to form crystalline domains. In most starches these are confined to the amylopectin component. Double helices form more or less ordered arrays where the ordered structures are crystalline entities. The starch granule is a very complex structure, the complexity being built around variations in the composition (α-glucan, moisture, lipid, protein and phosphorylated residues) and structure of the components. In wheat, the starch surface protein friabilin has attracted much attention because of its proposed association with grain hardness.[1–4]

RSC Green Chemistry No. 37
Starch-based Blends, Composites and Nanocomposites
Edited by Visakh P. M. and Long Yu

Published by the Royal Society of Chemistry, www.rsc.org

Integral proteins have a higher molecular weight than surface proteins (~50–150 and ~15–30 kDa, respectively) and include residues of enzymes involved in starch synthesis, especially starch synthase. Starches also contain relatively small quantities (<0.4%) of minerals (calcium, magnesium, phosphorus, potassium and sodium), which are, with the exception of phosphorus, of little functional significance. As the starch paste cools, the viscosity increases due to the formation of a gel held together by intermolecular interactions involving amylose and amylopectin molecules. The retrogradation of amylose in processed foods is considered to be important for properties relating to stickiness, ability to absorb water and digestibility, whereas retrogradation of amylopectin is probably a more important determinant in the staling of bread and cakes.

Most starches contain a portion that digests rapidly [rapidly digesting starch (RDS)], a portion that digests slowly [slowly digesting starch (SDS)] and a portion that is resistant to digestion [resistant starch (RS)].[5] Starch modification not only decreases retrogradation, gelling tendencies of pastes and gel syneresis but also improves paste clarity and sheen, paste and gel texture, film formation and adhesion.[6] These highly functional derivatives have been tailored to create competitive advantages in new products, improve product aesthetics, lower recipe/production costs, eliminate batch rejects, ensure product consistency and extend shelf-life while clearly making starch relevant in all stages of a food product's life cycle.[7] Modification of starch is an ongoing process as there are numerous possibilities. There is a huge market for the many new functional and added-value properties resulting from these modifications.

## 1.2 Preparation and Characterization of Starch Nanocrystals

Acid hydrolysis is possibly the most common and optimized method to produce starch nanocrystals.[8] Acid treatment dissolves the regions of low lateral order to reveal the concentric lamellar structure of starch granules. By this approach, water-insoluble and highly crystalline residues may be converted into stable suspensions by a subsequent vigorous mechanical shearing action.[9] During acid hydrolysis, regions of low lateral order and also amorphous phases in the starch granules start to dissolve, while the highly crystalline water-insoluble lamellae remain undissolved.[10] Le Corre *et al.*[11] conducted an experiment to determine whether starch from many different sources could be used to prepare starch nanocrystals and if the amylose content and/or botanic origin of the starch influenced their final properties. Starch nanocrystals are reported to be derived from starch granule crystallites[12] and result from the disruption of the semicrystalline structure of native starch granules at temperatures below the gelatinization temperature. Under these conditions, the amorphous regions in starch granules are hydrolysed, which allows the separation of nanoscale crystalline residues.[13] Starch nanocrystals of different sizes and shapes can be obtained depending on the origin of the starch and the isolation process. Xu *et al.*[14] prepared starch nanocrystals from corn, barley, potato, tapioca, chickpea and mung bean starches using an acid hydrolysis method.

Mélé *et al.*[15] studied the processing of nanocomposite materials consisting of natural rubber filled with waxy maize starch nanocrystals. Angellier *et al.*[16] employed starch nanocrystals in natural rubber composites and found a remarkable enhancing effect, but when the starch nanocrystal content exceeded 20%, the enhancement decreased. Bouthegourd *et al.*[17] reported the extraction and characterization of potato starch nanocrystals and their nanocomposites with a natural rubber latex matrix with the preparation performed using sulfuric acid at 40 °C. Kim *et al.*[18] claimed that obtaining individual nanoparticles from starch was almost impossible regardless of the origin of the starch. In another study by the same group,[19] a hydrolysis process combined with a physical treatment such as ultrasonication for the formation of a uniform dispersion of starch nanocrystal was investigated.

Li *et al.* reported three stages corresponding to the stepwise hydrolysis of the amorphous, semicrystalline and crystalline layers of the starch structure.[20] Some authors have suggested that high-amylose starches are more susceptible to acid hydrolysis than those with lower amylose contents, which are more easily hydrolysed.[21–23] This can be explained by either the greater extent of starch inter-chain associations in the amorphous regions, which are more compactly organized,[24] or by the slower penetration of hydrogen ions into the granules due to the limited swelling of high-amylose starch.

Large-scale starch nanocrystals (10–50 nm) obtained from the acid hydrolysis of amylopectin-rich waxy maize starch have been employed to prepare nanobiocomposites with natural rubber using a mastication technique.[25] Habibi and Dufresne[26] found that the mechanical characteristics of nanocomposite materials were improved by using chemically modified starch nanocrystals, which resulted in better dispersion of the filler within the matrix. Chen *et al.*[27] reported a reduction in the moisture uptake of a poly(vinyl alcohol) (PVA) matrix from 78 to 62% for the unfilled matrix and 40% w/w starch nanocrystal-reinforced composites, respectively. Angellier *et al.*[28] modified starch nanocrystals with alkenyl succinic anhydride or isocyanates and observed that the toluene uptake of the composite was higher than that of unmodified starch nanocrystals.

Thielemans *et al.*[29] observed that the thermal behaviour of starch nanocrystals was improved by grafting to alkyl polymer chains, which they suggested may be due to the protective crystalline layer formed by the oxygen-poor stearate surface. Namazi and Dadkhah[30] found similar results in relation to hydrophobically modified starch nanocrystals using octanoyl, nonanoyl and decanoyl chloride in an aqueous medium under mild conditions. They evaluated the thermal properties of the starch nanocrystals using thermogravimetric analysis (TGA) and observed that the decomposition onset temperature increased for modified starch nanocrystals, revealing their higher thermal stability than the unmodified form. The synthesis of starch nanocrystals with various sizes and shapes has been widely reported,[31] with the industrially important platelet starch nanocrystals obtained from hydrolysis of native starch granules being relatively easy to obtain with thicknesses of 6–8 nm, lengths of 40–60 nm and widths of 15–30 nm.[32]

## 1.3 Natural Fibre-reinforced Thermoplastic Starch Composites

Starch can be processed into a mouldable thermoplastic known as thermoplastic starch (TPS). TPS is plasticized starch that has been processed (typically using heat and pressure) to destroy completely the crystalline structure of the starch to form an amorphous thermoplastic material. Water contained in the starch and the added plasticizers play an indispensable role because the plasticizers can form hydrogen bonds with the starch, replacing the strong interactions between the hydroxyl groups of the starch molecules, thus making the starch thermoplastic. McHugh *et al.* suggested that, owing to its small size, glycerol was more effective than sorbitol in plasticizing the starch.[33] Many studies have been carried out on the preparation of TPS using glycerol as plasticizer. Park *et al.* developed biodegradable thermoplastic potato starch by using 30% glycerol as plasticizer.[34] Averous and Boquillon studied the thermal and mechanical behaviour of composites made from TPS reinforced with agro-materials (cellulose and lignocellulose fibres).[35] The TPS composite modulus displays a regular behaviour where the reinforcement effect increases with increase in the fibre length from short-length fibre (SF) to medium-length fibre (MF) and fibre content whereas the elongation at break decreases with increase in fibre content and length.

## 1.4 Applications of Starch Nanocrystal-based Blends, Composites and Nanocomposites

The main advantages of starch nanocrystals (SNCs) are their renewable nature, low cost, high barrier properties, availability, compatibilization with biopolymers, high specific strength, non-abrasive and non-toxic nature that allows easier processing even at high filling levels, biodegradability and a relatively reactive surface. They are edible, versatile and light weight and have a high aspect ratio, high specific strength and high modulus. Starch nanoparticles and nanocrystals have many potential applications, such as plastic fillers, food additives, drug carriers, implant materials, vehicles for carrying bioactive substances and nutraceuticals, fillers in biodegradable composites, coating binders, adhesives and a source of energy at the end of their life cycle.[36] The starch nanocrystals can also be used in biomedical, biochemical and technological applications and as vehicles for carrying bioactive substances and nutraceuticals. However, they tend to aggregate and settle down in aqueous solutions, which is a limitation to their application in most biological and food systems.[37] Wei *et al.*[38] mentioned that SNCs are crystalline platelets originating from the breakdown of the semicrystalline structure of starch granules by acid hydrolysis of amorphous parts. Because of their unique properties, SNCs have been widely used as particle emulsifiers to prepare Pickering emulsions[39] and as a reinforcement to prepare nanocomposites, such as biodegradable films and natural polymers.[40–42]

## 1.5 Chemically Modified Thermoplastic Starches

Grafting with maleic anhydride (starch-*g*-MA) is one of the most traditional and relevant techniques used to compatibilize non-polar polymers, such as polyolefins [*e.g.* low- and high-density polyethylene (LDPE and HDPE) and polypropylene (PP)] and polar materials, such as starch. Diisocyanates, such as 4,4′-diphenylmethane diisocyanate (MDI) and hexamethylene diisocyanate (HDI), have been employed for the reactive compatibilization of starch with polyesters such as poly(lactic acid) (PLA).[43–47]

Starch can be esterified by processing TPS with acid anhydrides and catalysts. The main anhydrides used are acetic, propionic, maleic and succinic anhydride.[48] Other reagents such as formic acid,[49,50] acid halides and vinyl acetate have also been reported.[51] Other reactions may occur during the esterification process, such as hydrolysis and glycosylation with glycerol.[52]

Propylene oxide is the most commonly used reagents in the preparation of hydroxypropyl starch.[53,54] Reaction with 3-chloro-2-hydroxypropyltrimethylamine chloride with reactive extrusion (REX), in the presence of 15 wt% glycerol, has been used to prepare cationic starch.[55] Reactions with vinyl acetate,[56] styrene[57] and acrylamide[58,59] have been described in processes for the grafting of starch in the molten state. Reduction of the molar mass of starch by glycolysis, catalysed by inorganic or organic acids, has been used to prepare modified TPS with lower melt viscosity.[60] Citric acid was used to improve the compatibility of TPS and other polymers, including LDPE.[61,62] Blends with linear LDPE were described by Ning *et al.*[63] In their studies, it was shown that citric acid can indeed improve the compatibility of the system. A recent study by our group also investigated blends of starch modified with citric acid and polyethylene in a two-stage process. Citric acid has also been used in nanocomposites of TPS with clays such as montmorillonite (MMT), for the purpose of modifying MMT to increase its rate of exfoliation[64–66] and to alter the properties of TPS so as to increase the wettability of the clay. The future of the reactive extrusion of TPS lies in its use in an intensive form so that starch can be radically modified, generating new materials that can be tuned to a wide assortment of uses. Reactive extrusion of TPS has proved to be a green process in that native starch is used without previous modification and the reactions take place at the same time as TPS is produced.

## 1.6 Outstanding Features of Starch-based Hydrogel Nanocomposites

The highly branched structure renders high molecular weight amylopectin that is in general 1000 times higher than that of amylose. Indeed, amylopectin is a "titan" natural molecule: one of the largest with a molecular weight of up to $400 \times 10^6$.[1] The ratio of amylose to amylopectin in any native starch is dependent on the source. In addition to these two main components, starch granules could also present other minor components in their composition such as some particulate material (*e.g.* cell-wall fragments) or surface and internal components (*e.g.* proteins, enzymes, lipids, amino and nucleic acids).[67,68]

Another important feature of starch is the large number of hydroxyl groups in their backbone. These groups have great affinity for other hydroxyl groups (hydrogen bonding), which can act as a driving force to hold the starch chains together in a regular pattern. Where such ordering occurs, crystalline regions are deposited in the starch granules. Many research groups have explored in depth the potential of starch-based hydrogels and their derivatives (such as starch-based hydrogel composites) in a wide variety of technological and biological fields. As a result, outstanding advances in starch-based hydrogels have published in recent decades. The application of hydrogels is no longer focused only on liquid absorption/retaining and the advantages of this very promising class of materials are now being exploited in the most varied industrial, technological and biotechnological sectors.[69–72]

Hydrogels prepared from biopolymers (mainly from polysaccharides) have found great applicability as biomaterials.[73–75] The interesting properties of polysaccharides derive from their structure, which, in general, contains a large number of functional groups (–COOH, –OH, $-NH_2$, $-NHOCCH_3$ and $-OSO_3H$) that can be crosslinked by reaction with a coupling agent or that allow the insertion of crosslinkable groups or polymeric chains on the polysaccharide backbone.[76,77] Most of the hydrogels prepared by this methodology show semi-interpenetrating network (semi-IPN) characteristics. IUPAC defines a semi-IPN as a polymeric material comprising at least one network and at least one linear or branched polymer characterized by the penetration of both on a molecular scale.[78] Starch-based hydrogels prepared from hydrophilic polymers/monomers, in a general way, have the capacity to absorb and retain large amounts of liquid, which classifies these hydrogels as superabsorbent. Raw starch is not so hydrophilic owing to its granular structure, and for this reason the association of starch with more hydrophilic polymers is required in order to prepare materials with a high liquid uptake capacity. In polymer science, hydrogels have evolved into materials with outstanding features and many potential applications, from soil conditioners and hygienic products to tissue engineering, drug delivery systems and imprinted polymers.[79,80] Al *et al.*[81] prepared superabsorbent hydrogel composites by grafting acrylic acid onto a starch backbone at a monomer to starch weight ratio of 1.5 (*ca.* 40 wt% maize starch) using

*N*,*N*′-methylenebisacrylamide as crosslinker, cerium ammonium nitrate as initiator and Na/MMT as the reinforcing phase. Eid[82] reported the preparation of starch polyelectrolyte hydrogels [starch/polyvinylpyrrolidone, starch/polyacrylamide and starch/poly(acrylic acid)] polymerized *via* gamma irradiation loaded with silver nanoparticles prepared by *in situ* reduction of silver nitrate with sodium borohydride, at room temperature.

The addition of active fillers can confer new properties on hydrogels, broadening their range of applications. In this respect, silver nanoparticles have been incorporated into hydrogels owing to their antibacterial features. Magnetic nanoparticles have also been added to starch hydrogels, allowing the modulation of release and absorption features by the use of an external magnetic field and also magnetic separation from aqueous media. Hydrogels are also potential candidates for the transport and storage of acid-responsive drugs. Mauricio *et al.* also developed a novel synthetic strategy for the synthesis of magnetite-nanostructured amylose microspheres.[83]

## 1.7 Starch-based Blends

Sabetzadeh *et al.*[84] prepared LDPE–thermoplastic corn starch (TPCS) blends containing different amounts of TPCS (0–40 wt%) and a constant amount of LDPE-*g*-MA (3 wt%) by using a single-screw extruder. Tanrattanakul and Panwiriyarat[85] enhanced the compatibility of LDPE–cassava starch blends by the addition of potassium persulfate (PPS) and benzoyl peroxide (BPO) to increase interfacial adhesion between the LDPE matrix and starch granules. Wang *et al.*[86] prepared compatible TPS–PE blends by reactive extrusion and reported that in the presence of dicumyl peroxide (DCP), the thermal plasticization of starch and its compatibilizing modification with polyethylene (PE) were accomplished by one-step reactive extrusion in a single-screw extruder. Thipmanee and Sane[87] reported the compatibilizing effect of zeolite 5A on linear low-density polyethylene (LLDPE)–TPS blends. The mechanical properties of the starch–poly(ε-caprolactone) (PCL) blends become poorer with increasing starch content in the blend[88] This may be ascribed to the incompatibility between the hydrophobic PCL and the hydrophilic starch.[89,90] Tan *et al.*[91] used synthesized starrch-modified polyurethane (St-PCL) as compatibilizer to compatilibilize the starch-PCL blend and found that a smaller amount of St-PCL can effectively improve the compatibility of the blends. Starch-*g*-PCL has been used as a compatibilizer to enhance the interfacial adhesion between PCL and starch phases, and the properties of the blend were obviously improved.[92,93]

Lai *et al.*[94] used three types of TPS, namely potato starch, corn starch and soluble potato starch, to blend with polyhydroxybutyrate (PHB), and the TPS (soluble starch)–PHB blend gave the highest level of water absorption and weight loss. PVA was also one of the first synthetic polymers to be blended with starch and shows excellent mechanical properties and compatibility with starch.[95–97] Moreover, the PVA–starch blend is assumed to be biodegradable since both components are biodegradable in various microbial environments.[98] Dean *et al.*[99] investigated ungelatinized and gelatinized starch–PCL blends and light microscopy indicated that blends with gelatinized starch had better interfacial adhesion.

## 1.8 Fracture and Failure of Starch-based Composites

Wang *et al.*[100] characterized the fracture behaviour of poly(propylene carbonate) (PPC) modified with starch. During EWF (essential work of fracture) testing, they found that a significant amount of plastic deformation took place around the initial ligament region (*i.e.* outer plastic zone) for pure PPC and the starch–PPC blends containing 10 and 20 wt% of starch. Studies of most starch-based materials have not been specifically focused on the mechanical performance and the fracture behaviour is usually disregarded.[101] Discontinuous fibres have been widely used as reinforcement for thermoplastic matrices[102] in order to improve their properties and also to allow better processing. Various cellulosic products and wastes were preferred as reinforcement mainly to achieve cost savings. Natural fibres also have many advantages compared with artificial fibres,[103] *e.g.* they are inexpensive, abundant and renewable, light weight, biodegradable and non-abrasive to processing equipment. In the case of biodegradable matrices, the adoption of these fibres is primarily due to environmental concerns.[104–108] Processing conditions have an important effect on the mechanical properties of fibre-reinforced composites due not only to the fibre orientation, but also to fibre breakage.[109] The combined effect of fibre content and microstructure on the fracture toughness of short-fibre-reinforced composites has been widely investigated in the case of different polymers reinforced with inorganic fibres.[110]

Muller *et al.*[111] reported a comprehensive investigation of the interfacial interactions in TPS reinforced with wood particles and a study of the mechanical properties including the failure mechanisms. They observed that modification of TPS with wood led to a very strong reinforcement, especially for fibres with larger aspect ratios. Romhány *et al.*[112] studied TPS-based composites containing flax fibres in unidirectional and crossed-ply arrangements. They determined the tensile behaviour of the composites with different fibre contents and arrangements and also identified the failure mechanisms by using the acoustic emission technique. Intermediate values were found by Alvarez *et al.*[113] for randomly oriented fibre composites. In the case of natural fibre composites, additional factors contribute to the low reinforcement in the transverse direction: the boundary layer between elementary sisal fibres is easily split and fibre–matrix and elementary fibre interfaces are thus created. Famá and co-workers reported significant improvements in the mechanical properties of cassava starch-based nanocomposites with extremely small amounts of carbon nanotubes previously wrapped with a starch–iodine complex.[114,115]

## 1.9 Application of Starch Nanocomposites in the Food Industry

Starch nanocomposites are mainly used as biopolymers or sources of packaging materials in the food industry. Recently, fossil energy depletion and growing environmental concerns have led to increasing interest in bio-based eco-efficient and high-technology materials. Starches used in food or industrial applications are first modified: the starches are chemically, enzymatically and/or physically modified to accentuate their positive characteristics, reduce their undesirable qualities (such as high viscosity, susceptibility to retrogradation and lack of process tolerance) or add new attributes (retention, film formation, digestibility, solubility, *etc.*). Indeed, mild acid hydrolysis has been used for many years to modify starch and its properties. Studies revealed that the starch granules consist of numerous nano-sized semicrystalline blocklets. Through mild hydrolysis using acids and/or enzymes, the nano-blocklets could be isolated from starch.[116–121] Physical treatments may also disintegrate the starch granules and thus release the nano-blocklets.[122,123]

These starch nanoparticles (SNPs) have a crystalline moiety with advantages derived inherently from starch granules, which include renewability and biodegradability.[124] Small starch particles ($\sim$500 nm) could be prepared by hydrolysing waxy rice starch using $\alpha$-amylase. However, ultrasonication after this enzymatic treatment increased the mean diameter of the starch hydrolysates, which indicates that the ultrasonication induced aggregation of the starch hydrolysates. Under the size distribution profile, the population density of the two main fractions (peaks of 0.5 and 3.6 μm) was decreased by the ultrasonication. Additionally, when the degree of enzymatic hydrolysis was greater, the size increase from the ultrasonication was more significant. This finding indicates that the starch hydrolysates became more susceptible to the sonication treatment when the hydrolysis level was higher. Therefore, to use the present process (combination of enzymatic hydrolysis and ultrasonication) for producing SNPs, more precise control of the hydrolysis and ultrasonication are needed.

## 1.10 Effect of Additives on the Properties of Starch

Food hydrocolloids have been widely applied in the food industry as a thickener, gelling agent, coating agent, stabilizer and emulsifier. The amount of the food hydrocolloids that are added to food usually represents a very small proportion compared with other ingredients, but they can effectively improve the performance of the food system.[125,126] The main hydrocolloids include cellulose derivatives, such as carboxymethylcellulose. Since both starch and hydrocolloids are polysaccharide molecules with a certain degree of similarity, hydrocolloids can affect the starch gelatinization and retrogradation behaviour and improve the textural properties and freeze–thaw stability of starchy foods.[127]

The addition of hydrocolloids could make such a trend more complex. A decrease in the peak viscosity of native potato starch or anionic modified potato starch with negatively charged xanthan gum but an increase in that of cationic modified potato starch with xanthan gum have been demonstrated.[128] Heyman *et al.* also reported a decreasing trend in the viscosity of waxy potato, waxy maize and pea starches by xanthan gum,[129] because xanthan gum inhibited granule swelling in most cases. Guar and xanthan gums induced an increase in the viscosity of rice starch, while such an increase in the viscosity of guar or xanthan–rice starch mixtures started much earlier than with starch alone.[130]

Hydrocolloids are often added to starch to improve the taste of food owing to their advantageous gel performance and transparency. However, as polysaccharide macromolecules, mixtures of starch and hydrocolloids can show thermodynamic incompatibility and phase separation phenomena, especially in the gel system. Guar and xanthan gums resulted in decreased elasticity and lowered resistance to deformation of gels. The incorporation of hydroxypropylmethylcellulose (HPMC) resulted in a more elastic network but a low resistance to deformation. These hydrocolloids led to a weaker structure with less gel-like character than with rice starch alone. There was an increase in plasticity on addition of either κ-carrageenan, guar gum or xanthan gum to potato starch gel.[131] However, the plasticity of such mixed gels was lost to a significant extent within the first day of storage, resulting in an elastically characteristic gel, which then stabilized rheologically and texturally for subsequent 30-day storage.

The addition of xanthan gum appeared to contribute to the elasticity of the weak network of starch paste. The addition of three non-starch polysaccharides [agar, xanthan gum and konjac glucomannan (KGM)] suppressed rice starch hydrolysis in the gels compared with starch alone. At the same concentration, agar had a lower suppressing effect on starch hydrolysis than xanthan gum and KGM.[132] Hydrocolloids could interact with the amylose leaching from the starch granules to produce a more complex matrix surrounding the gelatinized granules, thus leading to further restricted swelling and pasting.[133] In starchy foods, such as noodle or jelly foods, starch with hydrocolloids usually has a better texture and appearance. The gel is prepared by combination of wheat starch and hsin-tsao leaf gum, called grass jelly, which is consumed in a beverage. Onion pasta with HPMC can have improved quality and sensory characteristics.[134] Hydrocolloids are widely used in starch noodle processing.[135] The addition of a small amount of gum can improve the rehydration characteristics of noodles during cooking and modify the texture and overall "mouth-feel" of the finished product.[136] Hydrocolloids also contribute to reduced cooking losses and the swelling index of starch noodles.[137,138]

## References

1. P. M. Baldwin, *Starch Stärke*, 2001, **53**, 475.
2. P. Greenwell and J. D. Schofield, *Cereal Chem.*, 1986, **63**, 379.

3. S. Oda and J. D. Schofield, *J. Cereal Sci.*, 1997, **26**, 29.
4. C. F. Morris, *Plant Mol. Biol.*, 2002, **48**, 633.
5. K. N. Englyst, H. N. Englyst, G. J. Hudson, T. J. Cole and J. H. Cummings, *Am. J. Clin. Nutr.*, 1999, **69**, 448.
6. J. N. BeMiller, *Starch Stärke*, 1997, **49**, 127.
7. O. B. Wurzburg, *Nutr. Rev.*, 1986, **44**, 74.
8. H. Angellier, L. Choisnard, S. Molina-Boisseau, P. Ozil and A. Dufresne, *Biomacromolecules*, 2004, **5**, 1545–1551.
9. H. Angellier, S. Molina-Boisseau, M. N. Belgacem and A. Dufresne, *Langmuir*, 2005, **21**, 2425–2433.
10. S. Song, C. Wang, Z. Pan and X. Wang, *J. Appl. Polym. Sci.*, 2008, **107**, 418–422.
11. D. Le Corre, J. Bras and A. Dufresne, *J. Nanopart. Res.*, 2011, **13**, 7193–7208.
12. A. Dufresne, *Molecules*, 2010, **15**, 4111–4128.
13. J.-L. Putaux, S. Molina-Boisseau, T. Momaur and A. Dufresne, *Biomacromolecules*, 2003, **4**, 1198–1202.
14. Y. Xu, E. Sismour, C. Grizzard, M. Thomas, D. Pestov, Z. Huba, T. Wang and H. L. Bhardwaj, *Cereal Chem.*, 2014, **19**(4) 383–389.
15. P. Mélé, H. Angellier-Coussy, S. Molina-Boisseau and A. Dufresne, *Biomacromolecules*, 2011, **12**, 1487–1493.
16. H. Angellier, S. Molina-Boisseau, M. N. Belgacem and A. Dufresne, *Langmuir*, 2005, **21**, 2425–2433.
17. E. Bouthegourd, K. Rajisha, N. Kalarical, J. M. Saiter and S. Thomas, *Mater. Lett.*, 2011, **65**, 3615–3617.
18. H. Y. Kim, J. H. Lee, J. Y. Kim, W. J. Lim and S. T. Lim, *Starch Stärke*, 2012, **64**, 367–373.
19. H.-Y. Kim, J.-A. Han, D.-K. Kweon, J.-D. Park and S.-T. Lim, *Carbohydr. Polym.*, 2013, **93**, 582–588.
20. W. Li, H. Corke and T. Beta, *Carbohydr. Polym.*, 2007, **69**, 398–405.
21. Y. Nakazawa and Y.-J. Wang, *Carbohydr. Res.*, 2003, **338**, 2871–2882.
22. T. Vasanthan and R. Bhatty, *Cereal Chem.*, 1996, **73**, 199–207.
23. L. Jayakody and R. Hoover, *Food Res. Int.*, 2002, **35**, 665–680.
24. R. Hoover and H. Manuel, *J. Cereal Sci.*, 1996, **23**, 153–162.
25. M. Valodkar and S. Thakore, *Int. J. Polym. Anal. Charact.*, 2010, **15**, 387–395.
26. Y. Habibi and A. Dufresne, *Biomacromolecules*, 2008, **9**, 1974–1980.
27. G. Chen, M. Wei, J. Chen, J. Huang, A. Dufresne and P. R. Chang, *Polymer*, 2008, **49**, 1860–1870.
28. H. Angellier, S. Molina-Boisseau, M. N. Belgacem and A. Dufresne, *Langmuir*, 2005, **21**, 2425–2433.
29. W. Thielemans, M. N. Belgacem and A. Dufresne, *Langmuir*, 2006, **22**, 4804–4810.
30. H. Namazi and A. Dadkhah, *Carbohydr. Polym.*, 2010, **79**, 731–737.
31. Y. Xu, W. Ding, J. Liu, Y. Li, J. F. Kennedy, Q. Gu and S. Shao, *Carbohydr. Polym.*, 2010, **80**, 1078–1084.

32. C. Li, P. Sun and C. Yang, *Starch Stärke*, 2012, **64**, 497–502.
33. T. H. McHugh, J. F. Aujard and J. M. Krochta, *J. Food Sci.*, 1994, **59**, 416.
34. H. M. Park, X. Li, C. Z. Jin, C. Y. Park, W. J. Cho and C. S. Ha, *Macromol. Mater. Eng.*, 2002, **287**, 553.
35. L. Averous and N. Boquillon, *Carbohydr. Polym.*, 2004, **56**, 111.
36. N. L. García, L. Ribba, A. Dufresne, M. Aranguren and S. Goyanes, *Carbohydr. Polym.*, 2011, **84**, 203–210.
37. M. J. Jivan, A. Madadlou and M. Yarmand, *Food Chem.*, 2013, **141**, 1661–1666.
38. B. Wei, X. Hu, H. Li, C. Wu, X. Xu, Z. Jin and Y. Tian, *Food Hydrocolloids*, 2014, **36**, 369–373.
39. V. P. Cyras, L. B. Manfredi, M.-T. Ton-That and A. Vázquez, *Carbohydr. Polym.*, 2008, **73**, 55–63.
40. N. Lin, J. Huang and A. Dufresne, *Nanoscale*, 2012, **4**, 3274–3294.
41. H. Zheng, F. Ai, P. R. Chang, J. Huang and A. Dufresne, *Polym. Compos*, 2009, **30**, 474–480.
42. H. Angellier, S. Molina-Boisseau, P. Dole and A. Dufresne, *Biomacromolecules*, 2006, **7**, 531–539.
43. L. Yu, K. Dean, Q. Yuan, L. Cheng and X. Zhang, Effect of compatibilizer distribution on the blends of starch/biodegradable polyesters, *J. Appl. Polym. Sci.*, 2007, **103**, 812–818.
44. F. Xie, T. Xue, L. Yu, L. Chen, X. Li and X. Zhang, Rheological properties of starch-based materials and starch/poly(lactic acid) blends, *Macromol. Symp.*, 2007, **249**, 529–534.
45. H. Wang, X. Sun and P. Seib, Strengthening blends of poly(lactic acid) and starch with methylenediphenyl diisocyanate, *J. Appl.Polym. Sci.*, 2001, **82**, 1761–1767.
46. C. L. Jun, Reactive blending of biodegradable polymers: PLA and starch, *J. Polym. Environ.*, 2000, **8**, 33–37.
47. E. Schwach, J. L. Six and L. Averous, Biodegradable blends based on starch and poly(lactic acid): comparison of different strategies and estimate of compatibilization, *J. Polym. Environ.*, 2008, **16**, 286–297.
48. V. D. Miladinov and M. A. Hanna, Starch esterification by reactive extrusion, *Ind. Crops Prod.*, 2000, **11**, 51–57.
49. T. Divers, I. Pillin, J. F. Feller, G. Levesque and Y. Grohens, Starch modification, destructuration and hydrolysis during O-formylation, *Starch Stärke*, 2004, **56**, 389–398.
50. F. Bossard, I. Pillin, T. Aubry and Y. Grohens, Rheological characterization of starch derivatives/polycaprolactone blends processed by reactive extrusion, *Polym. Eng. Sci.*, 2008, **48**, 1862–1870.
51. G. Moad, Chemical modification of starch by reactive extrusion, *Progr. Polym. Sci.*, 2011, **36**, 218–237.
52. J. M. Raquez, R. Narayan and P. Dubois, Recent advances in reactive extrusion processing of biodegradable polymer-based compositions, *Macromol. Mater. Eng.*, 2008, **293**, 447–470.

53. G. Lammers, E. J. Stamhuis and A. A. C. M. Beenackers, Continuous production of hydroxypropyl starch in a static mixer reactor, *Starch Starke*, 1993, **45**, 227–232.
54. R. A. De Graaf and L. P. B. M. Janssen, The hydroxypropylation of starch in a self-wiping twin screw extruder, *Adv. Polym. Technol.*, 2003, **22**, 56–68.
55. R. Bhosale and R. Singhal, Process optimization for the synthesis of octenyl succinyl derivative of waxy corn and amaranth starches, *Carbohydr. Polym.*, 2006, **66**, 521–527.
56. R. A. De Graaf, A. Broekroelofs and L. P. B. M. Janssen, The Acetylation of Starch by Reactive Extrusion, *Starch Stärke*, 1998, **50**, 198–205.
57. R. A. De Graaf and L. P. B. M. Janssen, The production of a new partially biodegradable starch plastic by reactive extrusion, *Polym. Eng. Sci.*, 2000, **40**, 2086–2094.
58. J. L. Willett and V. L. Finkenstadt, Reactive extrusion of starch-polyacrylamide graft copolymers using various starches, *J. Polym. Environ.*, 2006, **14**, 125–129.
59. V. L. Finkenstadt and J. F. Willett, Reactive extrusion of starch–polyacrylamide graft copolymers: effects of monomer/starch ratio and moisture content, *Macromol. Chem. Phys.*, 2005, **206**, 1648–1652.
60. A. J. F. Carvalho, M. D. Zambon, A. A. S. Curvelo and A. Gandini, Thermoplastic starch modification during melting processing: hydrolysis catalyzed by carboxylic acids, *Carbohydr. Polym.*, 2005, **62**, 387–390.
61. V. R. Miranda and A. J. F. Carvalho, Blendas de polietileno e amido termoplástico modificado, *Polímeros*, 2011, **21**, 347–352.
62. A. Taguet, M. A. Huneault and B. D. Favis, Interface/morphology relationships in polymer blends with thermoplastic starch, *Polymer*, 2009, **50**, 5733–5743.
63. W. Ning, Y. Jiugao, M. Xiaofei and W. Ying, The influence of citric acid on the properties of thermoplastic starch/linear low-density polyethylene blends, *Carbohydr. Polym.*, 2007, **67**, 446–453.
64. M. F. Huang, J. G. Yu and X. F. Ma, High mechanical performance MMT–urea and formamide-plasticized thermoplastic cornstarch biodegradable nanocomposites, *Carbohydr. Polym.*, 2006, **63**, 393–399.
65. K. Majdzadeh-Ardakani, A. H. Navarchian and F. Sadeghi, Optimization of mechanical properties of thermoplastic starch/clay nanocomposites, *Carbohydr. Polym.*, 2010, **79**, 547–554.
66. K. Majdzadeh-Ardakani and B. Nazari, Improving the mechanical properties of thermoplastic starch/poly(vinyl alcohol)/clay nanocomposites, *Compos. Sci. Technol.*, 2010, **70**, 1557–1563.
67. A. Buleon, P. Colonna, V. Planchot and S. Ball, *Int. J. Biol. Macromol.*, 1998, **23**, 85–112.
68. M. A. Araujo, A. M. Cunha and M. Mota, *Biomaterials*, 2004, **25**, 2687–2693.
69. A. Jamshidi, F. Ahmad Khan Beigi, K. Kabiri and M. J. Zohuriaan-Mehr, *Polymer Testing*, 2005, **24**, 825–828.

70. D. Buenger, F. Topuz and J. Groll, *Progr Polym. Sci.*, 2012, **37**, 1678–1719.
71. Y. Qiu and K. Park, *Adv. Drug Delivery Rev.*, 2001, **53**, 321–339.
72. Y. Aizawa, S. C. Owen and M. S. Shoichet, *Progr. Polym. Sci.*, 2012, **37**, 645–658.
73. A. R. Fajardo, J. F. Piai, A. F. Rubira and E. C. Muniz, *Carbohydr. Polym.*, 2010, **80**, 934–943.
74. E. G. Crispim, J. F. Piai, A. R. Fajardo, E. R. F. Nakamura, T. U. Nakamura, A. F. Rubira and E. C. Muniz, *Express Polym. Lett.*, 2012, **6**, 383–395.
75. S. Van Vlierberghe, P. Dubruel and E. Schacht, *Biomacromolecules*, 2011, **12**, 1387–1408.
76. J. T. Oliveira and R. L. Reis, *J. Tissue Eng. Regener. Med.*, 2011, **5**, 421–436.
77. M. R. Guilherme, A. V. Reis, A. T. Paulino, A. R. Fajardo, E. C. Muniz and E. B. Tambourgi, *J. Appl. Polym. Sci.*, 2007, **105**, 2903–2909.
78. A. D. Jenkins, P. Kratochvíl, R. F. T. Stepto and U. W. Suter, *Int. J. Pure Appl. Chem.*, 1996, **68**, 2287–2311.
79. M. J. Zohuriaan-Mehr, H. Omidian, S. Doroudiani and K. Kabiri, *J. Mater. Sci.*, 2010, **45**, 5711–5735.
80. K. Deligkaris, T. S. Tadele, W. Olthuis and A. van den Berg, *Sens. Actuators, B*, 2010, **147**, 765–774.
81. E. Al, G. Güçlü, T. B. İyim, S. Emik and S. Özgümüş, *J. Appl. Polym. Sci.*, 2008, **109**, 16–22.
82. M. Eid, *J. Inorg. Organomet. Polym.*, 2011, **21**, 297–305.
83. M. R. Mauricio, M. R. Guilherme, M. H. Kunita, E. C. Muniz and A. F. Rubira, *Chem. Eng. J.*, 2012, **189–190**, 456–463.
84. M. Sabetzadeh, R. Bagheri and M. Masoomi, *J. Appl. Polym. Sci.*, 2012, **126**, E63.
85. V. Tanrattanakul and W. Panwiriyarat, *J. Appl. Polym. Sci.*, 2009, **114**, 742.
86. S. Wang, J. Yu and J. Yu, *Polym. Int.*, 2005, **54**, 279.
87. R. Thipmanee and A. Sane, *J. Appl. Polym. Sci.*, 2012, **126**, E252.
88. A. Mahieu, C. Terrié, A. Agoulon, N. Leblanc and B. Youssef, *J. Polym. Res.*, 2013, **20**, 1.
89. P. Sarazin, G. Lia, W. J. Ortsb and B. D. Favis, *Polymer*, 2008, **49**, 599.
90. L. L. Averous, M. P. Doleb and C. Fringant, *Polymer*, 2000, **41**, 4157.
91. Y. Tan, P. X. Wang, K. Xu, H. Y. An and L. Dong, *Chem. Res. Chin. Univ*, 2010, **26**, 483.
92. E. J. Choi, C. H. Kim and J. K. Park, *J. Polym. Sci., Part B: Polym.*, 1999, **37**, 2430.
93. R. Mani, J. Tang and M. Bhattacharya, *Macromol. Rapid Commun.*, 1998, **19**, 283.
94. S. M. Lai, T. M. Don and Y. C. Huang, *J. Appl. Polym. Sci.*, 2006, **100**, 2371.
95. L. T. Sin, W. A. W. A. Rahmana, A. R. Rahmata and A. A. Samadb, *Polymer*, 2010, **51**, 1206.

96. S. Y. Yang, I. Liu Chia, J. Y. Wu, J. C. Kuo and C. Y. Huang, *Macromol. Symp.*, 2008, **272**, 150.
97. M. Khan, S. K. Bhattachariaa, M. A. Kaderb and K. Baharic, *Carbohydr. Polym.*, 2006, **63**, 500.
98. W. L. Chai, J. D. Chow and C. C. Chen, *J. Polym. Environ.*, 2012, **20**, 550.
99. K. Dean, L. Yu, S. Bateman and D. Y. Wu, *J. Appl. Polym. Sci.*, 2007, **103**, 802.
100. X. L. Wang, R. K. Y. Li, Y. X. Cao and Y. Z. Meng, Short communication, Essential work of fracture analysis for starch filled poly(propylene carbonate) composites, *Mater. Des.*, 2007, **28**, 1934–1939.
101. F. Tuba, L. Oláh and P. Nagy, Characterization of reactively compatibilized poly(D,L-lactide)/poly(ε-caprolactone) biodegradable blends by essential work of fracture method, *Eng. Fract. Mech.*, 2011, **78**, 3123–3133.
102. D. E. Mouzakis, T. Harmia and J. Karger-Kocsis, Fracture behaviour of discontinuous long glass fiber reinforced injection molded polypropylene, *Polym. Polym. Compos.*, 2000, **8**, 167–175.
103. H. D. Rozman, G. B. Peng and Z. A. Mohd Ishak, The effect of compounding techniques on the mechanical properties of oil palm empty fruit bunch–polypropylene composites, *J. Appl. Polym. Sci.*, 1998, **70**, 2647–2655.
104. S. Iannace, L. Nocilla and L. Nicolais, Biocomposites based on sea algae fibers and biodegradable thermoplastic matrices, *J. Appl. Polym. Sci.*, 1999, **73**, 583–592.
105. H. Hanselka, A. S. Herrman, Fiber properties and characterization. in *7th Internationales Techtexil Symposium*, 1994, pp. 20–22.
106. A. Dufresne and M. R. Vignon, improvement of starch film performances using cellulose microfibrils, *Macromolecules*, 1998, **31**, 2693.
107. J. Kuruvilla and L. H. C. Mattoso, Sisal fibre reinforced polymer composites: status and future, *Nat. Polym. Compos.*, 1998, 333.
108. C. Biastioli, Principles and Applications: Starch Polymer Composites, in *Degradable Polymers*, ed. G. Scott and D. Gilead, Chapman & Hall, London, 1995.
109. P. V. Joseph, K. Joseph and S. Thomas, Effect of processing variables on the mechanical properties of sisal-fiber-reinforced polypropylene composites, *Compos. Sci. Technol.*, 1999, **59**, 1625–1640.
110. S.-Y. Fu, Y.-W. Mai, B. Lauke, G. Xu and C.-Y. Yue, Combined effect of fiber content and microstructure on the fracture toughness of SGF and SCF reinforced polypropylene composites, *J. Mat. Sci.*, 2002, **37**, 3067–3074.
111. P. Muller, K. Renner, J. Móczó, E. Fekete and B. Pukánszky, Thermoplastic starch/wood composites: interfacial interactions and functional properties, *Carbohydr. Polym.*, 2014, **102**, 821–829.
112. G. Romhány, J. Karger-Kocsis and T. Czigány, Tensile fracture and failure behavior of thermoplastic starch with unidirectional and cross-ply flax fiber reinforcements, *Macromol. Mater. Eng.*, 2003, **288**, 699–707.

113. V. Alvarez, A. Vázquez and C. Bernal, Effect of microstructure on the tensile and fracture properties of sisal fiber/starch based composites, *J. Compos. Mater.*, 2006, **40**, 21–35.
114. L. Famá, P. G. Gañán Rojo, C. Bernal and S. Goyanes, Biodegradable starch based nanocomposites with low water vapor permeability and high storage, *Carbohydr. Polym.*, 2012, **87**, 1989.
115. L. M. Famá, V. Pettarin, S. N. Goyanes and C. R. Bernal, Starch/multi walled carbon nanotubes composites with improved mechanical properties, *Carbohydr. Polym.*, 2011, **83**, 1226.
116. D. J. Gallant, B. Bouchet and P. M. Baldwin, *Carbohydr. Polym*, 1997, **32**, 177.
117. H. Tang, T. Mitsunaga and Y. Kawamura, *Carbohydr. Polym*, 2006, **63**, 555.
118. N. L. García, L. Ribba, A. Dufresne, M. Aranguren and S. Goyanes, *Carbohydr. Polym.*, 2011, **84**, 203.
119. H. Namazi and A. Dadkhah, *Carbohydr. Polym*, 2010, **79**, 731.
120. J.-L. Putaux, S. Molina-Boisseau, T. Momaur and A. Dufresne, *Biomacromolecules*, 2003, **4**, 1198.
121. J.-Y. Kim, D.-J. Park and S.-T. Lim, *Cereal Chem.*, 2008, **85**, 182.
122. D. Liu, Q. Wu, H. Chen and P. R. Chang, *J. Colloid Interface Sci.*, 2009, **339**, 117.
123. S. Bel Haaj, A. Magnin, C. Pétrier and S. Boufi, *Carbohydr. Polym.*, 2013, **92**, 1625.
124. H.-Y. Kim, S. S. Park and S.-T. Lim, *Colloids Surf., B.*, 2015, **126**, 607–620.
125. J. A. Ramírez, R. M. Uresti, G. Velazquez and M. Vázquez, *Food Hydrocolloids*, 2011, **25**, 1842–1852.
126. P. Varela and S. M. Fiszman, *Food Hydrocolloids*, 2013, **30**, 477–484.
127. J. N. BeMiller, *Carbohydr. Polym.*, 2011, **86**, 386–423.
128. X. Cai, Y. Hong, Z. Gu and Y. Zhang, *Food Res. Int.*, 2011, **25**, 5.
129. B. Heyman, D. De Hertogh, P. Van der Meeren, F. Depypere and K. Dewettinck, *Carbohydr. Polym.*, 2013, **96**, 568–577.
130. C. M. Rosell, W. Yokoyama and C. Shoemaker, *Carbohydr. Polym.*, 2011, **84**, 373–382.
131. M. Krystyjan, G. Adamczyk, M. Sikora and P. Tomasik, *Food Hydrocolloids*, 2013, **31**, 270–276.
132. T. Sasaki and K. Kohyama, *Food Chem.*, 2011, **127**, 541–546.
133. M. S. Alamri, A. A. Mohamed and S. Hussain, *Food Hydrocolloids*, 2013, **30**, 541–551.
134. P. Chantaro and R. Pongsawatmanit, *J. Food Eng.*, 2010, **98**, 44–50.
135. C. Gamonpilas, W. Pongjaruvat, A. Fuongfuchat, P. Methacanon, N. Seetapan and N. Thamjedsada, *J. Food Eng.*, 2011, **105**, 233–240.
136. B. X. Fu, *Food Res. Int.*, 2008, **41**, 888–902.
137. S. Heo, S. Jeon and S. Lee, *LWT—Food Sci. Technol.*, 2014, **55**, 627–631.
138. E. Silva, M. Birkenhake, E. Scholten, L. M. C. Sagis and E. van der Linden, *Food Hydrocolloids*, 2013, **30**, 42–52.

CHAPTER 2

# *Starch: Introduction and Structure–Property Relationships*

KHONGSAK SRIKAEO

Faculty of Food and Agricultural Technology, Pibulsongkram Rajabhat University, Muang, Phitsanulok 65000, Thailand
Email: khongsak@live.psru.ac.th

## 2.1 Introduction

Starch is present as a macro-constituent in many foods and its properties and interactions with other constituents are of interest to the food industry and for human nutrition. Starch contributes 50–70% of the energy in the human diet, providing a direct source of glucose, which is an essential substrate in brain and red blood cells for generating metabolic energy. Indeed, the availability of a reliable source of starch from agriculture is considered to have been an important factor in human development.[1] Starch is also an important industrial material as it is a versatile biopolymer with great potential and of low price for both the food and non-food industries. Approximately, 60 million tonnes are extracted annually worldwide from various cereal, tuber and root crops, of which roughly 60% is used in foods (for example, bakery products, sauces, soups, confectionery, sugar syrups, ice cream, snack foods, meat products, baby foods, fat replacers, coffee whitener, beer, soft drinks) and 40% in pharmaceuticals and for non-edible purposes, such as fertilizers, seed coatings, paper, cardboard, packing materials, adhesives, textiles, fabrics, diapers, bioplastics, building materials,

RSC Green Chemistry No. 37
Starch-based Blends, Composites and Nanocomposites
Edited by Visakh P. M. and Long Yu

Published by the Royal Society of Chemistry, www.rsc.org

cement and oil drilling.[2,3] Starch varies greatly in form and functionality between and within botanical species, and even from the same plant cultivar grown under different conditions. Starches with a wide range of functional properties are needed to ensure fitness-for-purpose for such a diverse range of end uses. Starch can also be converted into starch nanocomposites to enhance its properties and serve as an ideal "green" biopolymer that comes from Nature and readily goes back to Nature. Various research activities are ongoing worldwide in different areas of starch nanocomposites, opening up new avenues for its application.[4]

## 2.2 Starch Structure

Starch has been the subject of intensive research over many decades, resulting in a vast body of published literature on preparative and analytical methods, molecular structure, physical, chemical and biochemical properties, functionality and uses.

Starch is a polysaccharide consisting of D-glucose units, referred to as homoglucan or glucopyranose, and has two major biomacromolecules – amylose and amylopectin (Figure 2.1). Starch consists primarily of D-glucopyranose polymers linked together by $\alpha$-(1$\rightarrow$4) and $\alpha$-(1$\rightarrow$6) glycosidic bonds. In forming these bonds, C1 on a D-glucopyranose molecule reacts with C4 or C6 from an adjacent D-glucopyranose molecule. Starch polymers always have one reducing end because the aldehyde group on one end is

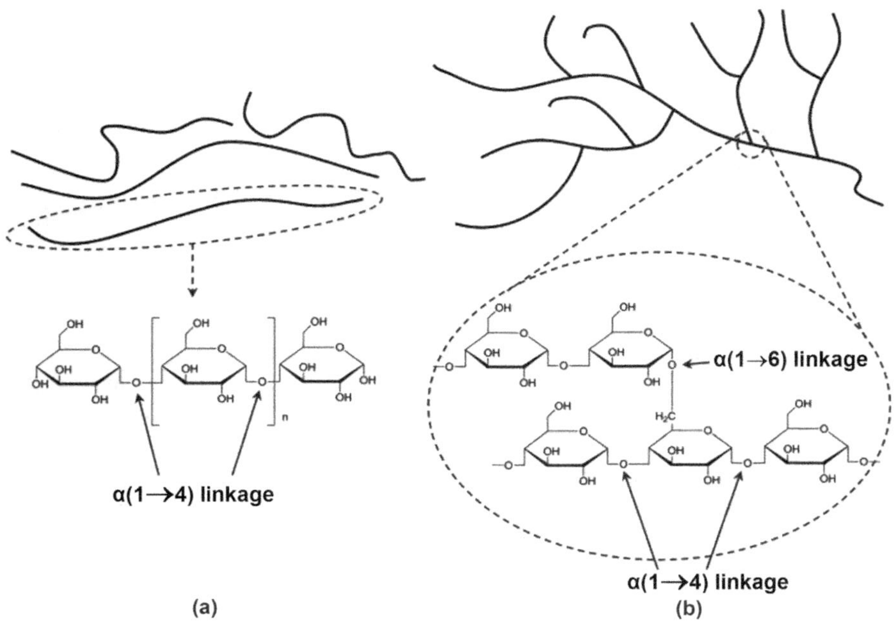


**Figure 2.1** Structures of (a) amylose and (b) amylopectin.
Reproduced from Xie *et al.*[11] with permission of Elsevier.

free. The other end of the polymer is called the non-reducing end. There could be a large number of non-reducing ends in a starch molecule depending on the number of polymeric branches. The glycosidic linkages in starch are in the $\alpha$ configuration. Formation of the $\alpha$ linkages is determined by the orientation of the hydroxyl group (–OH) on C1 of the pyranose ring. The $\alpha$ linkage allows some starch polymers to form helical structures, which contribute to its unique properties and enzyme digestibility.[5,6]

Regardless of their botanical origins, starch varieties contain primarily two different types of anhydroglucose polymers, amylose and amylopectin. These polymers have different structures and properties, which have been discussed and reviewed by many authors.[5–10]

Amylose is a relatively long, linear $\alpha$-glucan containing around 99% $\alpha$-(1$\rightarrow$4) and $\alpha$-(1$\rightarrow$6) linkages and differs in size and structure depending on botanical origin. Amylose has a molecular weight of approximately $1\times10^5-1\times10^6$ and a degree of polymerization (DP) by number ($DP_n$) of 324–4920, with around 9–20 branch points equivalent to 3–11 chains per molecule.[8,12–17] Each chain contains approximately 200–700 glucose residues, equivalent to a molecular weight of 32 400–113 400. Because of the low degree of branching, dissolved amylose has a tendency to form insoluble semicrystalline aggregates, depending on the location of the branches in the structure.[6,7]

Amylopectin is a much larger molecule than amylose, with a molecular weight of $1\times10^7-1\times10^9$ and a heavily branched structure built from about 95% $\alpha$-(1$\rightarrow$4) and 5% $\alpha$-(1$\rightarrow$6) linkages. The $DP_n$ is typically within the range 9600–15 900 but comprises three major species with $DP_n$ 13 400–26 500, 4400–8400, and 700–2100.[8,12,13,18] In common with amylose, the size, shape, structure and polydispersity of the molecule vary with botanical origin. Unlike amylose, however, there is great additional variation with respect to the unit chain lengths and branching patterns. Amylopectin unit chains are relatively short compared with amylose molecules, with a broad distribution profile. They are typically $\sim$18–25 units long on average, although the range is extended (19–31) if high-amylose starches are also included.[13,15,18–21] The individual chains can be specifically classified in terms of their lengths (chain lengths) and positions.[19,20]

### 2.2.1 Starch Molecular Structure

The structure of starch in a grain can be categorized into six levels (Figure 2.2), ranging in scale from nanometres to millimetres (six orders of magnitude).[22,23]

*Level 1: Individual branches.* This is the distribution of the chain lengths or DP of the branches in a sample, often termed the chain-length distribution.

*Level 2: Whole starch molecules.* This is the structure of the branched molecules, which is exceptionally hard to characterize fully, both theoretically and experimentally, but which can usefully be characterized by averages such as the number- and weight-average molecular weight or the

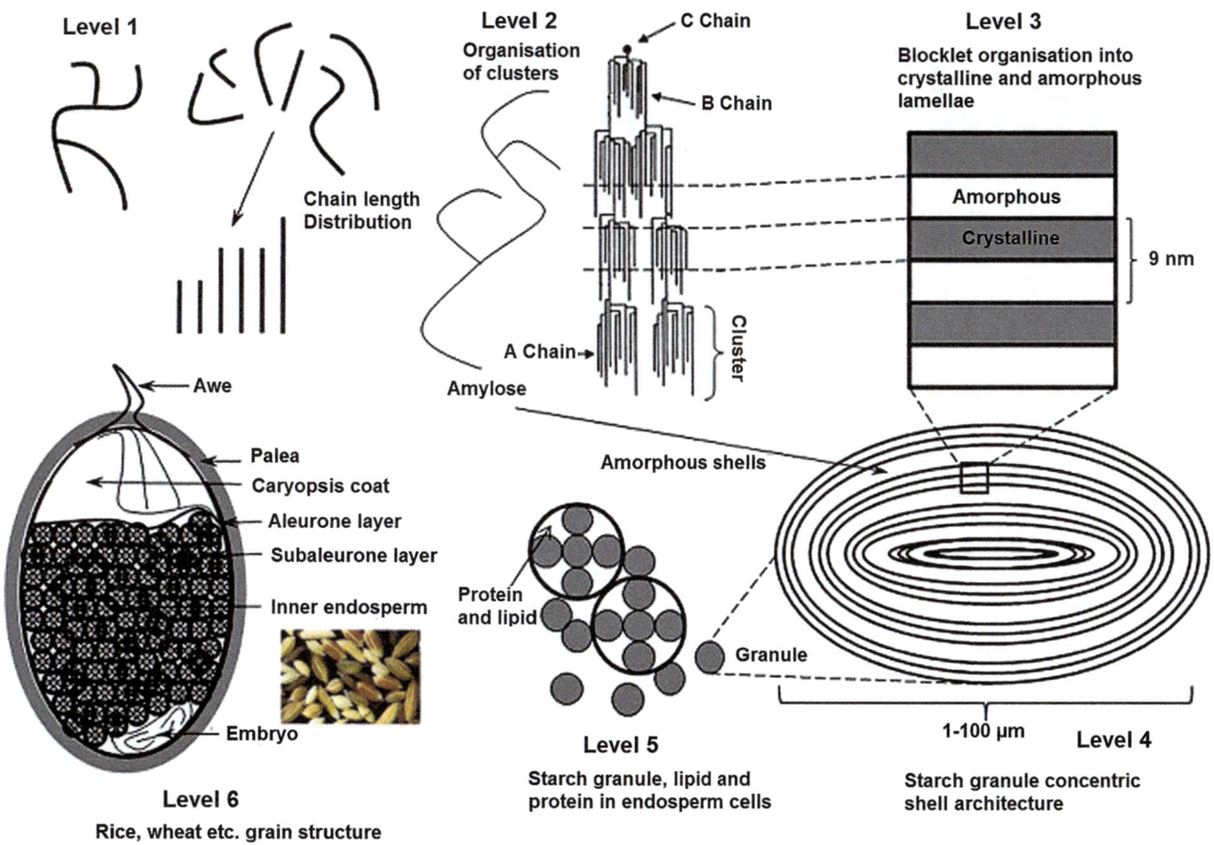


**Figure 2.2** Six supramolecular levels of the grains, highlighting the microscopic structural contribution of starch. Reproduced from Dona *et al.*[23] with permission of Elsevier.

weight-average size distribution.[24,25] The branching properties of each amylopectin anhydroglucose chain can be separated into three distinct groups: (1) A chains, which have no consequential branches and are joined to a B or C chain through an α-(1→6) bond at their reducing end; (2) B chains, which differ from A chains by possessing subsequent branches that are connected to them *via* an α-(1→6) bond at the reducing end of branches; and (3) C chains, which are the backbone of the amylopectin molecule, having a free reducing end.[20,26] Numerous branches are attached to a C chain, thus laying the foundation for the complex structure of amylopectin. A schematic representation of an amylopectin model is shown in Figure 2.3.

*Level 3: Lamellar structure.* In native starch, the tight packing of glucan chains enables them to intertwine to form double helices, which in turn create regions of crystalline order, with the crystalline component largely comprising clusters of shorter portions (typically ∼17 monomer units) of amylopectin branches; the amylose is thought largely to reside in the amorphous layers. This structure can be characterized by small-angle X-ray and neutron scattering and by scanning electron microscopy (SEM). Isolated amylose molecules can form into a sixfold left-handed double-helical structure, which also can exist in whole starch after gelatinization.

Both amylose chains and exterior chains of amylopectin can form double helices, which may in turn associate to form crystalline domains. In most starches these are confined to the amylopectin component. Double helices form more or less ordered arrays where the ordered structures are crystalline entities. The traditional view was that cereal starches generate A-type X-ray diffraction (XRD) patterns, tuber and high-amylose starches generate B-type and legume, root and some fruit and stem starches generate C-type, although it is now believed that C-type starch diffraction patterns represent a combination of A- and B-type polymorphs.[8,10,27–36] The A- and B-type polymorphic structures are represented schematically in Figure 2.4.

Moreover, recent evidence showed that a structure level termed the "blocklet" exists between the macromolecules and the organization of starch granules. The blocklet has a very asymmetric structure and is similar in shape but differs in size with the plant, and it is continuous throughout

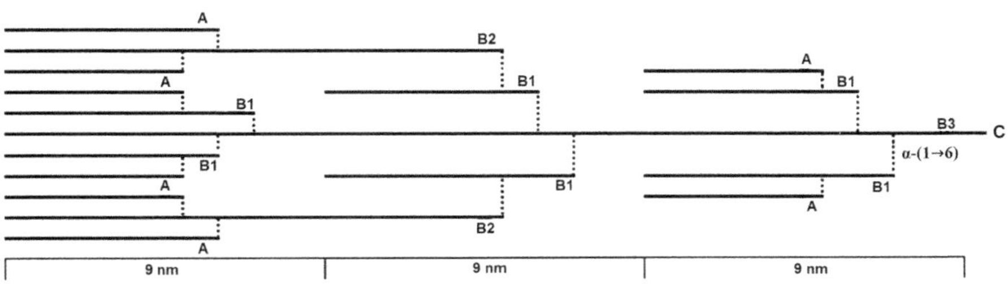


**Figure 2.3** Schematic representation of a section of amylopectin indicating the branching pattern of unit α-(1→4) chains (A, B1–B3) joined together by α-(1→6) linkages (branch points).
Reproduced from Tester *et al.*[7] with permission of Elsevier.

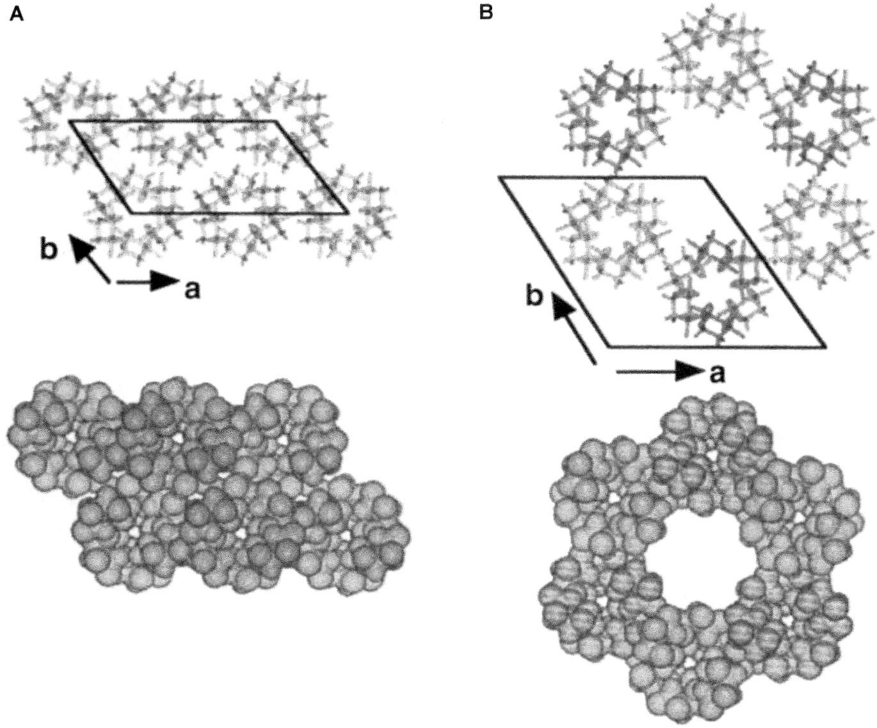


**Figure 2.4** Crystalline packing of double helices in A-type (A) and B-type (B) amylose. Projection of the structure onto the (*a*, *b*) plane.
Reproduced from Buléon *et al.*[8] with permission of Elsevier.

the granule.[37,38] A schematic representation of normal blocklet structure is shown in Figure 2.5.[38] Figure 2.6 illustrates SEM images of the blocklet model of potato starch granules, showing the smaller blocklets that are compacted densely to form the stiffer outer shell, and the larger blocklets that are packed loosely to form the inner shells.[39]

*Level 4: Starch granules.* The lamellar structures occupy the inner architecture of native starch granules in concentric growth ring shells of thickness 100–400 nm; these are separated by regions of amorphous structure. The crystalline and amorphous lamellae alternate radially, creating concentric shells. The amorphous regions of the granule contain the branch points of amylopectin molecules, as they tend not to align, and amylose, as it also does not form tightly packed crystalline regions.[40] These semicrystalline and amorphous "growth rings" that make up the granules are ~15–30 μm in size. Characterization is mostly performed by SEM. This level of structure is greatly changed by the process of gelatinization.

A conceptual model of starch granule structure called the "hairy billiard ball" illustrates both the single-helical organization of amylose and the double-helical organization and clustering of amylopectin.[41–43] The model

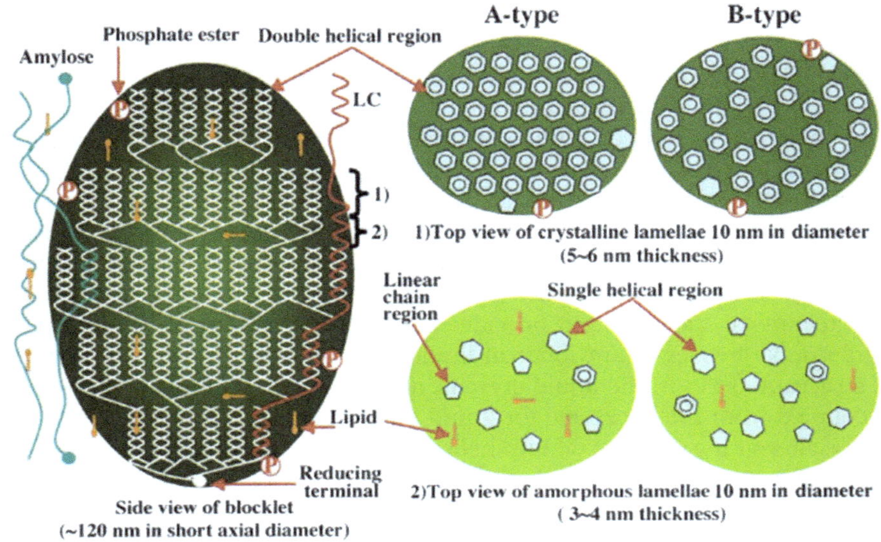


**Figure 2.5** Schematic of normal blocklet structure.
Reproduced from Tang *et al.*[38] with permission of Elsevier.

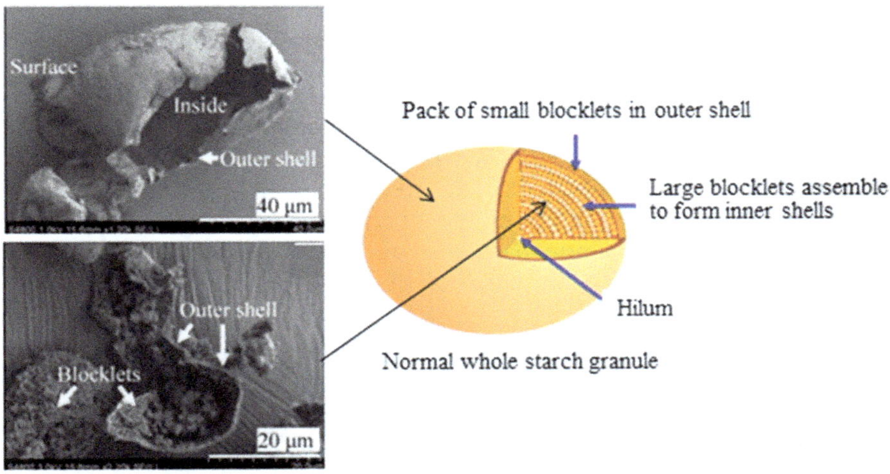


**Figure 2.6** The refined blocklet model of a potato starch granule and SEM images showing the blocklets in outer and inner shells.
Reproduced from Huang *et al.*[39] with permission of Elsevier.

describes the outside of the granule with protruding branches of amylopectin (hair) rather than a smooth exterior. The outside of the billiard ball, inside the hairy layer, defines a boundary between initially accessible, or surface, substrate sites and initially inaccessible, or interior, substrate sites.[43] The hairy billiard ball model was developed to explain short-range surface order in XRD patterns. These patterns are unrelated to the X-ray patterns observed due to the semicrystalline lamellae of granules.[36] The digestion time courses of the action of glucohydrolase on starch granules are also best explained by variations in accessibility of the enzyme, seen in the hairy billiard ball model. Unlike synthetic polymers, such as polyethylene, starch amylopectin crystallites do not assemble into semicrystalline rings with chain-folded lamellae. The repeat distance of starch crystallites upon annealing also does not appear to change as with synthetic polymers.[44] Instead, the double helices appear to arrange side-by-side in a smectic- or nematic-type structure.[45] These properties led to another conceptual model of the granular architecture of starch, based on side-chain liquid crystalline polymers. The rigid double helices of amylopectin branches provide the fundamental units of structural order in the liquid crystals, although they need to be adequately decoupled from the backbone for liquid crystalline packing to occur.[23,46]

The alternating semicrystalline shells, elliptical in shape, regularly alternate with a repeat distance of ~9 nm.[47,48] Regardless of their botanical origins, amylopectin molecules exhibit the same repeat distance for crystalline and amorphous regions, as measured by XRD.[45,47] The molecular order of starch granules is also confirmed by their birefringent nature when viewed in polarized light, displaying a dark birefringence cross called a Maltese cross.[6]

The intensity of birefringence depends on the granule size, relative crystallinity and microcrystalline orientation. Polarized light microscopic images of A- and B-type granules of wheat starch are shown in Figure 2.7. It is seen that the A-type granules are generally much larger than the B-type granules and the birefringence intensity of the former was also higher than that of the latter. This can be explained by its greater number of B2 chains and fewer short A and B1 chains, which have better fits to disk-shaped granules.[49] In contrast, B-type granules were spherical and irregular.[50]

Within the lamellae, the crystalline layers are considered to be formed mainly by amylopectin chains packed into a crystalline lattice, whereas the amorphous layers contain the amylopectin branching points and amylose

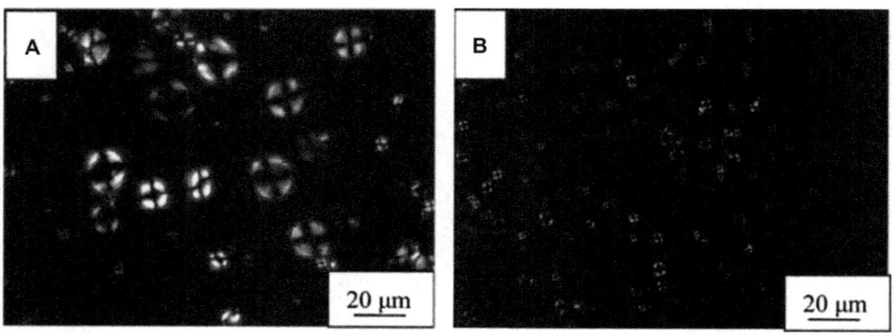


**Figure 2.7** Polarized light microscopic images of the A-type (A) and B-type (B) granules of wheat starch.
Reproduced from Zhang *et al.*[50] with permission of Elsevier.

and amylopectin molecules in a disordered conformation. Longer amylopectin chains are considered to pass from the crystalline region into the amorphous region of the lamellae.[50,51] Amylopectin molecules near the surface of granules may have a different structure to those closer to the centre of the molecule. The ratio of long- to short-branch chains affects the shape of amylopectin, which influences their packing into granules.[52] In this way, genetic and environmental factors that affect amylopectin biosynthesis can influence its molecular architecture and, in turn, granule morphology.[53]

*Level 5: Endosperm.* This comprises the starch granules, together with protein and lipids. This level of structure is not usually important for digestion, with the exception of grains such as sorghum that lack a hull and instead have a dense protein–starch matrix that is not readily broken down.

*Level 6: Whole grain.* This final level, ~1 mm in size, includes all outer layers of the grain and the highest level structures such as the hull.

The role of the granular structure in nature is as an energy-storage medium for the germinating plant, to release glucose slowly given the right external stimuli. The compact crystalline structure of amylopectin is the most important structural component for this purpose. Although amylose generally makes up a significant fraction (20–30%) of starch varieties, its role in the starch granule as a source of glucose for energizing an embryo is yet to be completely understood. While amylopectin alone is sufficient for the formation of starch granules, amylose also plays a central role in the initial stages of granule crystallization.[54] Interestingly, details such as the amylopectin to amylose ratio, size, morphology and size distribution of the starch granule all determine the physicochemical characteristics of the granule.[55]

The composition and structure of starch granules vary considerably between different plants (Table 2.1), affecting the properties and functions of starches from different crops. Tuber or root starches have both larger granules and lower protein and lipid contents than cereal starches.

The starch granule is a very complex structure. The complexity is built around variations in the composition (α-glucan, moisture, lipid, protein and

**Table 2.1** Characteristics of starch from various sources. Reproduced from Jobling[56] with permission of Elsevier.

| Characteristic | Source | | | |
|---|---|---|---|---|
| | Maize | Wheat | Potato | Cassava |
| Type | Cereal | Cereal | Tuber | Root |
| Granule shape | Round, polygonal | Round, bimodal | Oval, spherical | Oval, truncated |
| Granule size/μm | 2–30 | 1–45 | 5–100 | 4–35 |
| Phosphate/g per 100 g | 0.02 | 0.06 | 0.08 | 0.01 |
| Protein/g per 100 g | 0.35 | 0.4 | 0.06 | 0.1 |
| Lipid/g per 100 g | 0.7 | 0.8 | 0.05 | 0.1 |
| Size of world market/$10^6$ tons per year | 39.4 | 4.1 | 2.6 | 2.5[a] |

[a]Includes other minor starch sources, such as sweet potato.

phosphorylated residues) and structure of the components. In addition, there is an additional level of variation, which reflects the way in which amorphous and crystalline elements are associated. There is still much to learn about the complexities of starch granules, although considerable knowledge gained over the last century led to new views regarding this important component of many food and non-food products.[7]

### 2.2.2 Minor Components in Starch Granules

Amylose and amylopectin represent approximately 98–99% of the dry weight of starch granules. However, they also contain other components. Cereal starches contain integral lipids in the form of lysophospholipids (LPLs) and free fatty acids (FFAs), which are positively correlated with the amylose fraction, and the LPLs may account for up to ~2% of the starch weight (in high-amylose cereal starches). However, the granules may also be contaminated with surface lipids.[57,58] The contaminants comprise triglycerides, glycolipids, phospholipids and FFAs derived from the amyloplast membrane and non-starch sources. These differ from integral (internal) lipids, which are composed exclusively of the FFAs and LPLs.[57,58] Starches from the Triticeae contain almost exclusively LPLs (-choline, -ethanolamine and -glycerol) whereas other cereal starches are rich in FFAs. Part of the amylose fraction within lipid-containing granules exists as an amylose inclusion complex where the fatty acid chains occupy a hydrophobic core located within the single amylose helix. The presence of these complexes in native starch granules is apparent from carbon-13 cross-polarization/magic-angle spinning nuclear magnetic resonance ($^{13}$C CP/MAS NMR) spectra.[59–61] The presence of this fraction is also evident from iodine binding studies, where non-defatted amylose from cereal starches has a lower iodine binding capacity than the corresponding lipid-extracted material. The amount of lipid-complexed amylose ranges from <15% to >55% of the amylose fraction in cereal starches, with oat starches being especially rich in lipids and correspondingly complexed amylose.[7,58,59]

Purified starches contain <0.6% protein. In common with starch lipids, proteins occur on the surface (and include non-starch-derived proteins) and are embedded within the matrix of granules regardless of origin. Both the starch lipids and proteins have the potential to moderate starch functionality.[62] Collectively, the proteins are referred to as starch granule-associated proteins and may be associated with lipids on granule surfaces.[63] In wheat, the starch surface protein friabilin has attracted much attention because of its proposed association with grain hardness.[63–66] Integral proteins have a higher molecular weight than surface proteins (~50–150 and ~15–30 kDa, respectively) and include residues of enzymes involved in starch synthesis, especially starch synthase.[63]

Starches also contain relatively small quantities (<0.4%) of minerals (calcium, magnesium, phosphorus, potassium and sodium), which are, with the exception of phosphorus, of little functional significance. The phosphorus is found in three major forms: phosphate monoesters, phospholipids and inorganic phosphates.[67–71] Phosphate monoesters are selectively bound to specific regions within the amylopectin molecule.[67–71] In the Triticeae starches, the phosphorus content is very close to the LPL content and there is a relatively small amount of α-glucan phosphate monoesters. Other starches generally contain little phosphate, with the exception of potato starch, where there is essentially no lipid. Here, the phosphate monoester content may exceed 0.1%.[67]

Starch composition and structure have been a major topic of interest for many decades. More details regarding starch composition may be found in many excellent publications.[7–9,42,72]

## 2.3 Starch Functionalities

Most starches consumed by humans and processed for non-food uses have undergone some form of processing, which usually involves heating in the presence of moisture under shear, and then cooling. The granular structural changes that take place during the phase transition, including starch gelatinization, pasting and retrogradation, are important aspects of starch functionality.

### 2.3.1 Phase Transition of Starch

Gelatinization has been defined as the phase transition of the starch granules from an ordered state to a disordered state, which takes place during heating in excess water. However, more recent studies using various techniques including light microscopy, SEM, differential scanning calorimetry (DSC), the Rapid Visco Analyser (RVA), XRD and high-performance size-exclusion chromatography (HPSEC) suggested that starch gelatinization process is more complex than a simple granular order-to-disorder transition.[73] In terms of pasting, it is important to note that this process is not completely separated from gelatinization but rather is an overlapping occurrence, probably best described as a continuation of gelatinization. Although there is no definite point at which pasting begins and gelatinization ends, pasting is usually related to the development of viscosity. However, the link between gelatinization and pasting often results in the use of these two terms interchangeably.[5]

When starch is heated progressively in an excess of water, a honeycomb-like structure is observed within the granules at the gelatinization temperature as observed by SEM. It was believed that the formation of this structure caused irreversible, profound swelling and loss of the polarization crosses. Higher temperatures resulted in further swelling and melting, which led to strong deformation of the granules, and ultimately the molten granules connected with each other to form a sponge-like structure.[74] Figure 2.8 compares the SEM images of normal maize starch and high-amylose maize starch under heat treatment. High-amylose maize starch did

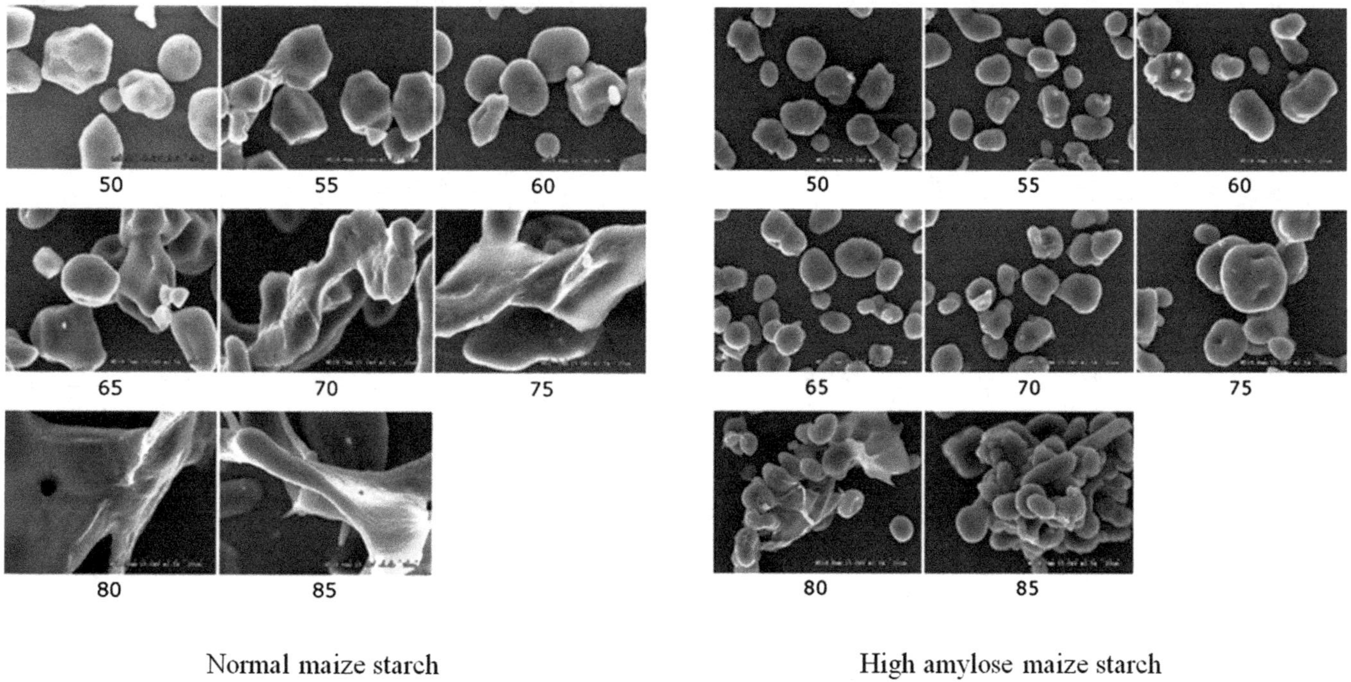


**Figure 2.8** SEM images of normal maize starch and high-amylose maize starch under heat treatment (50–85 °C). Reproduced from Ratnayake and Jackson[73] with permission of Elsevier.

not show granular swelling at the same temperatures as normal maize starch, indicating the necessity for higher amounts of amylopectin for the swelling process to occur.[73]

During heat treatment, the starch granules are gelatinized, losing their crystallinity and structural organization. On cooling, the disaggregated starch molecules first form a gel and then retrograde gradually into semi-crystalline aggregates that differ in form from the native granules. Starch-rich foods that have been cooked and cooled often contain substantial amounts of retrograded starch. Understanding the steps that occur during gelatinization and retrogradation of a particular starch are key steps to predicting better the functional properties of processed starch from knowledge of the structure of native granules.

Gelatinization occurs when native starch is heated in the presence of sufficient moisture. The granules absorb water and swell, and the crystalline organization is irreversibly disrupted. According to the theory of Jenkins and Donald,[75] water first enters the amorphous growth rings and, at a certain degree of swelling, disruptive stress is transmitted through connecting molecules from the amorphous to the crystalline regions. Amylose molecules begin to leach from the granules as they are disrupted under shear and the viscosity of the resulting paste increases to a maximum, which corresponds to the point when the number of swollen but still intact starch granules is at a maximum. The maximum is followed by a decrease in paste viscosity, as the granules rupture and starch molecules are dispersed in the aqueous phase. The rate and extent of swelling and breakdown are dependent on the type and amount of starch, the temperature gradient, shear force and the composition of the mixture, for example, the presence of lipids and proteins.[76] The gelatinization temperature of most starches is between 60 and 80 °C. In general, there is a negative relationship between the amylose content of starch and the gelatinization temperature and peak viscosity.[53]

As the starch paste cools, the viscosity increases owing to the formation of a gel held together by intermolecular interactions involving amylose and amylopectin molecules. In gels that contain about 25% amylose, the starch molecules form a network, resulting in a firm gel, in contrast to waxy starch gels, which are soft and contain aggregates but no network.[77] On standing, starch gels retrograde and form insoluble B-type crystallites due to association of linear regions of $\alpha$-(1$\rightarrow$4)-linked glucose units in the polymers.[78] Retrogradation is an ongoing process occurring over an extended period. Amylose retrogrades over minutes to hours and amylopectin over hours to days, depending on the ability of the branched chains to form associations. The retrogradation of amylose in processed foods is considered to be important for properties relating to stickiness, ability to absorb water and digestibility, whereas retrogradation of amylopectin is probably a more important determinant in the staling of bread and cakes. In rice, amylose contributes texture and stickiness, whereas gelatinization temperature and cooking and pasting properties are more closely related to amylopectin.[53,79]

The effects of retrogradation in starch-based products can be desirable or, more usually, undesirable. There is a general consensus that starch retrogradation contributes significantly to staling or undesirable firming of bread and other starch-based products.[80] Similarly, the susceptibility of legume starch gels to retrogradation and syneresis makes these types of starches unsuitable for products requiring low-temperature storage. However, retrogradation is sometimes promoted to modify the structural, mechanical or organoleptic properties of certain starch-based products. This is true, for example, in the production of breakfast cereals and parboiled rice, since retrogradation results in hardening and reduced stickiness.[81] Freezing–thawing, which accelerates retrogradation, is applied to cooked potato mash in the production of dehydrated mashed potatoes to decrease the amount of soluble starch and to improve the consistency of the reconstituted product.[82] The production of Japanese harusame noodles also involves a freeze–thaw cycle to reduce stickiness and to obtain a characteristic chewiness. Similarly, Chinese rice vermicelli (a type of rice noodle) strands are conditioned after complete steam gelatinization of starch in order to attain the desired textural characteristics.[83,84]

Because of its industrial significance, many methods for the study of starch retrogradation have been developed. Changes in physical and chemical properties, attributable to changes in the starch component of model starch systems or actual starch-based products during aging, form the usual bases for these methods. These time-dependent changes may directly contribute to or correlate with the sensory perception or digestibility of starchy foods. However, it is important to emphasize that, in most cases, following changes in a single parameter with time may not provide an adequate description of retrogradation. Furthermore, retrogradation kinetics, determined using different methods, may also not be in total agreement.[85] Correct interpretations of results, therefore, would depend on an exact knowledge of the physical and/or chemical basis, and also an appreciation of the limitations, of any given method. The adoption of proper procedures is, of course, critical in ensuring validity of results.

Methods to study starch retrogradation can be conveniently classified as (i) macroscopic techniques, *e.g.* those methods which monitor alterations in certain physical properties as manifestations of retrogradation, for example, mechanical or textural changes, and (ii) molecular techniques, *e.g.* those methods which study changes in starch polymer conformation or water mobility in starch gels at molecular levels. Thus, rheological techniques, sensory evaluation of texture, DSC, light scattering, turbidimetry and measurement of syneresis may be used to study the macroscopic manifestations of retrogradation. On the other hand, XRD, NMR, vibrational spectroscopy (*e.g.* Raman spectroscopy) and Fourier transform infrared (FTIR) spectroscopy may be classified as molecular techniques. In all cases, the inclusion of microcomputer technology of ever-increasing sophistication in the designs of instruments will continue to increase the precision, resolution, speed of analysis and range of capability of any technique.[84]

### 2.3.2 Starch in Limited Water Systems

It should be noted that the mentioned starch functionalities only work at dilute starch-in-water concentrations. These conditions are completely different from the concentrated starch-in-water systems. It has always been, and still is, difficult to study starch in limited water systems. DSC has been beneficial for studying starch in such concentrated systems.[86] Several views on the impact of moisture content on DSC enthalpic transitions have been formulated. One of these maintains that, initially, water enters the more-accessible, amorphous regions of the starch granules. Swelling of the amorphous regions induces stress on the crystallites, and starch chains are stripped from the surface of crystallites, thereby reducing the crystallinity and causing the loss of birefringence of the starch granules. In excess water, the degree of swelling is sufficient to gelatinize the granule completely. At lower water levels, swelling is insufficient to disrupt the starch granule completely. The remaining order is then disrupted at higher temperatures by a conventional melting transition. At very low moisture contents (<30%, based on starch), only the latter process occurs.[87]

### 2.3.3 Glass Transition in Starch

The role of the glass transition in starch has been found to be an important issue. As mentioned earlier, starch is a semicrystalline polymer composed of partially crystalline amylopectin molecules and amorphous amylose molecules. The amorphous form in starch can exhibit glass transition behaviour.[88]

Glass transition refers to the phase transition when a glass is changed into a supercooled melt or the reverse.[89] Rapid changes in the physical, mechanical, electrical, thermal and other properties of a material can be observed through the glass transition; these changes are described by mathematical expressions such as Vogel–Tamman–Fulcher (VTF) and Williams–Landel–Ferry (WLF) equations. Through the measurement of these rapidly changed properties, the glass transition temperature ($T_g$) can be determined. Mathematical models, described by the Gordon–Taylor and Couchman–Karasz equations, are able to predict the $T_g$ of multicomponent mixtures. Although $T_g$ has been proven to be an effective indicator for food quality changes during storage, there is evidence that physicochemical changes also take place below $T_g$.[90,91]

$T_g$ helps explain the physical and chemical behaviour of starch systems in both foods and non-foods,[92,93] *e.g.* changes of the textural properties of bread.[94] When a glassy material is stored below its $T_g$, it spontaneously approaches a more stable state viscosity.[95] This phenomenon is called enthalpy relaxation, which is due to the local molecular motion of certain molecules or certain parts of some polymer molecules. Enthalpy relaxation is important for food materials stored below the $T_g$, in consideration of the stability of the physicochemical properties of the materials.[96] In the glassy state, molecule mobility and diffusion are virtually non-existent. Above $T_g$, glassy materials soften or become rubbery and demonstrate decreased viscosity and increased mobility.[97]

Both the glass and the supercooled melt are in the non-crystalline state. In contrast to the glass, which is a rigid solid, the supercooled melt, between $T_g$ and $T_m$ (melting temperature), can be a viscoelastic "rubber" in the case of polymer materials, or a mainly viscous liquid for low molecular weight materials. Unlike $T_m$, $T_g$ is a kinetic parameter, depending on the temperature scanning rate and the sample's thermal history. Nonetheless, $T_g$ is a useful material descriptor owing to its good correlation with the structural and thermodynamic properties of the material.[96]

The glass transition is accompanied by a change in heat capacity, which enables $T_g$ to be determined calorimetrically. However, difficulties often arise in the accurate determination of $T_g$ owing to the heterogeneity of starch systems, especially in foods, and the low energy associated with the transition. A frequently used technique for measuring $T_g$ is DSC.[98] All published work appears to agree that $T_g$ decreases as the moisture content (plasticizer) increases. Extrapolation from experimental results for starch–water systems indicates that the $T_g$ for pure starch is about 226.85 °C and decreases by about 6 °C per 1% water for the first 10% water addition. Once the water content reaches about 20%, the starch is plasticized at room temperature. As the water content increases further, $T_g$ continues to decline and is eventually below the freezing point of the aqueous system.[99,100] High-amylose starch was found to exhibit high $T_g$ for the same kind of starch.[101]

## 2.4 Starch–Lipid Complexes

Complexes between amylose and lipids, such as fatty acids, LPLs and monoacylglycerides, can significantly modify the properties and functionality of starch in ways that are of interest to the food industry and for human nutrition. For example, complexation with lipids reduces the solubility of starch in water, alters the rheological properties of pastes, decreases swelling capacity, increases gelatinization temperature, reduces gel rigidity, retards retrogradation and reduces the susceptibility to enzymic hydrolysis.[102–108]

Complexation of starch with endogenous or added lipids may occur during gelatinization, or when starch and lipids are heated together in the RVA. A variety of cooking and extrusion processes have been used to form complexes under conditions that mimic processing.[109,110] Starch–lipid complexes have been studied extensively using the same techniques as for the analysis of starch granules and starch molecules, namely DSC, XRD, iodine-binding capacity, viscometry and image analysis. The complexes may be quantified using competitive iodine-binding studies that measure the reduction in iodine-binding colour when complexes are formed.[105,111]

Inclusion complexes with lipids form mainly with the amylose component of the starch, hence the amylose to amylopectin ratio is an important factor that produces variability in the ability of natural starches to bind lipids.

Lipid-complexed amylose has been confirmed to occur in native cereal starch granules, but as cereal starches contain about 1% lipid, only 15–55% of the amylose fraction is complexed with lipid.[57,58] Owing to its high degree of branching, the lipid binding capability of amylopectin is considered to be much weaker than that of amylose. There is little direct evidence to suggest that amylopectin forms true inclusion complexes with lipids, although some studies have led to proposals that favour interaction of some lipids and surfactants with outer amylopectin branches.[112–114]

The extent of crystallinity and the thermal stability of complexes between amylose and complex-forming lipids increase with increasing amylose chain length.[115,116] The effect of the lipid component on complex formation is less well defined. The heat stability of complexes between fatty acid anions and amylose increases with fatty acid chain length[107,108,117] and decreases with unsaturation.[113] Monoglycerides of saturated fatty acids with chain lengths from $C_{12}$ to $C_{18}$ are very effective amylose complexing agents in aqueous solution at 60 °C, as are *trans*-unsaturated analogues, whereas the *cis*-unsaturated analogues form thermally less stable complexes.[57,106,118–120] Maximum complex formation occurs at a different concentration for different lipids, depending on the water solubility and critical micellar concentration of the lipid.[111] Above a certain concentration, some lipids (for example, palmitic and stearic acids) tend to self-associate in preference to forming starch–lipid complexes, which helps to explain why a particular lipid has been described as forming strong complexes with amylose in some studies and weak or no complexes in others. There is no evidence to indicate that diglycerides or triglycerides form inclusion complexes with amylose.[57,77,111]

Studies with NMR and molecular modelling of amylose–lipid complexes indicate that the hydrophobic fatty acid chain is included inside the amylose helix, whereas the polar groups of fatty acids and monoacylglycerols are too bulky to enter the helical tube.[8,121,122]

## 2.5 Starch Hydrolysis

Many processing operations are dependent on native starch hydrolysis or could be improved by adopting or optimizing this process. The ability to manipulate and understand the hydrolysis of starch granules offers several advantages, particularly industrial and nutritional implications.[123]

### 2.5.1 Acid Hydrolysis

Acid hydrolysis is widely used in industry for the chemical treatment of starch particles.[6] The underlying idea regarding acid hydrolysis is to exploit the susceptibility differences of semicrystalline and amorphous starch lamellae to acid. Crystalline lamellae were more resistant to hydrolysis than amorphous lamellae, tending to display negligibly slow hydrolysis, whereas starch amorphous zones were prone to fast acid hydrolysis.[124,125] The degree of crystallinity increased gradually with the time of acid hydrolysis.[125,126]

For relatively short hydrolysis times (up to 72 h), acid thinning did not cause disruption of the granular crystalline structure.[127,128] Other studies indicated that the hydrogen ion primarily attacked first the interior and then the exterior of C-type starch granules. B-type starch granules started to crack after about 4 days of hydrolysis, whereas C-type starch granules cracked until the hydrolysis progressed up to 16 days.[129] The degradation of amylose and amylopectin by a high acid concentration resulted in a decrease in storage modulus ($G'$), loss modulus ($G''$), gelling temperature and gel strength of acid-thinned starches.[125] The microstructure of starch granules was strongly disrupted by the action of hydrogen ions, depending on acid concentration and temperature. From an industrial applications standpoint, close monitoring of the effects of acid hydrolysis on the functionality and microstructure of starch granules should provide important guidelines for hydrolysis process design oriented to the production of starches with desired functional properties (*e.g.* crystallinity, shear thinning).[130] Acid hydrolysis of starch has had widespread use in the past but it is now largely replaced by enzymatic processes.

## 2.5.2 Starch Digestive Enzymes

Phosphorylase is a starch-degrading enzyme produced by many plants. It is an *exo*-acting enzyme that removes single glucosyl units from the non-reducing ends of starch chains by reaction with inorganic phosphate ($P_i$) to give α-D-glucopyranose 1-phosphate (α-Glc 1-P).[131,132] More details about starch metabolism in plants may be obtained elsewhere.[133,134]

Enzymes involved in the breakdown of starch chains are primarily of four types: [6,135] (1) those that hydrolyse α-(1→4) D-glucosidic bonds (amylases), (2) those that hydrolyse α-(1→6) D-glucosidic bonds (isoamylases), (3) those that transfer α-(1→4) D-glucosidic bonds (glucanosyltransferases) and (4) branching enzymes [α-(1→4)-α-(1→6) transferases]. The amylases can be divided into three classes: (1) the *endo*-acting α-amylases, (2) the *exo*-acting β-amylases and (3) isoamylases.

*endo*-Hydrolases or *endo*-acting enzymes are generally excreted by cells and so operate outside them. Carbohydrate *endo*-hydrolases cleave large carbohydrates that are unable to diffuse into cells, to give smaller products that can traverse cell membranes. *endo*-Hydrolases randomly cleave a hydrated starch molecule into two smaller molecules; in the specific case of α-amylase this is done by cleaving any accessible α-(1→4) bond. On the other hand, *exo*-hydrolases or *exo*-acting enzymes release a monomer or dimer from the non-reducing end of the substrate molecule. Enzymes such as amyloglucosidase and β-amylase release glucose or maltose units, respectively, from the non-reducing end of starch and oligosaccharides.[23,136]

The surfaces of starch granules are affected in different ways by enzyme action, as seen by light microscopy, confocal laser scanning microscopy[137] and SEM.[138,139] Many different modes of enzyme attack have been identified, including pin-hole, sponge-like erosion, medium-sized holes, single holes in individual granules, and surface erosion.[140] Depending on the enzyme and variety of starch, enzymes can erode the entire granule surface (*exo*-corrosion) or digest channels at specific locations on the surface towards the centre (*endo*-corrosion). Starches from wheat, barley and rye have specific "susceptible zones" that become pitted as a result of *endo*-corrosion. Pits become enlarged and form numerous channels within the granule, thus weakening its structure. The granules eventually fragment, leaving behind residual starch. Although the physiochemical properties of the crystalline and amorphous regions of the granule are different, the residual starch after digestion has a virtually unchanged debranched distribution, as measured by SEC with multiple-angle laser light scattering and refractive index detection. A similar profile after debranching implies a progression of digestion through the crystalline and amorphous lamellae; this is also seen with SEM. Residues left after digestion also show comparable properties, such as gelatinization temperature and enthalpy, with native granules as analysed by DSC.[139]

High-amylose starches characterized at different stages of *in vitro* digestion display properties that suggest that amorphous areas of the granule are hydrolysed by α-amylase, leaving the helical crystalline region intact. Analysis with SEM, small-angle X-ray scattering (SAXS), infrared spectroscopy, solid-state $^{13}$C NMR spectroscopy and XRD show that an increase in the molecular order is favoured by the hydrolytic action of the enzyme.[141] The changes suggest that resistant starch (RS) does not have a specific structure in predigested samples, but may be formed during digestion by a rearrangement of amylose chains into resistant structures of higher crystallinity.[123,142] Consequently, the resistance to enzyme digestion is a result of competition between the kinetics of enzyme hydrolysis and the kinetics of amylose retrogradation. Furthermore, when hydrolysis is stopped after different times, the yield of high-amylose RS fragments diminishes, but the size of the fragments remains constant. This is attributed to the amorphous regions being digested more readily, leaving the chains in the helical crystalline regions.[23] More detail is discussed in Section 2.5.4.

### 2.5.3 Human Digestive System

Starch is met by salivary amylase in the mouth, which is the first enzyme to act on carbohydrates during digestion.[143] In a relatively short time, the bolus of food is carried by oesophageal peristalsis into the stomach. One of the key cells in the stomach for starch digestion is the parietal cell, which secretes HCl. The pH of the gastric juice is ~2.6, retarding the action of α-amylase but increasing the acid hydrolysis of starch. Also in the upper gastrointestinal tract, lipids bound to starch are hydrolysed by lipases secreted by various exocrine glands. A key step in lipid digestion is the creation of an emulsion that increases the area of the oil–water interface, allowing more efficient action of the enzymes. The emulsion is produced first through the mechanical process of mastication and then peristaltic movements of the digestive tract. The emulsion is stabilized by its droplets being coated with membrane lipids, denatured proteins and fatty acids, thus preventing them from coalescing.[144]

From the stomach, the ingested food proceeds to the duodenum, where it encounters the pancreatic secretion whose rate of release is controlled by "satiety" hormones. Pancreatic fluid contains two important components for starch digestion. Sodium hydrogencarbonate (bicarbonate) neutralizes the acidity of the fluid arriving from the stomach to a pH of ~8. Pancreatic fluid also contains α-amylase that continues the hydrolysis of starch into glucose and oligosaccharides. The latter include linear and branched structures that are not absorbed into the bloodstream without further hydrolysis to glucose. Glucose is only a very minor product of *endo*-hydrolases, such as α-amylase which operates during the initial stages of digestion, so further enzymic processes are definitely required.

Substrates not digested by α-amylase, such as α-limit dextrins, and small linear oligomers, along with larger α-glucans, are later degraded into single glucose units. This conversion occurs *via* enzymes that are integral to the plasma membrane of enterocytes in the small intestine; these include mucosal maltase–glucoamylase and sucrase–isomaltase. These are *exo*-glucosidases that act on the non-reducing end of glucose oligomers and catalyse not only the hydrolysis of α-(1→4) bonds but also to a lesser extent that of α-(1→6) branch bonds, ensuring further degradation of non-linear oligosaccharides. The resulting monosaccharides, such as glucose and galactose, are absorbed by secondary active transport across the apical membrane of enterocytes and subsequently exit the gastrointestinal tract across the basolateral membrane, into the bloodstream.[144]

The large intestine (colon) contains a very large population of microorganisms (showing considerable variation between individuals) that are beneficial to the digestion of polysaccharides in humans who might be enzyme deficient, or in normal subjects with carbohydrates that have not been broken down in the jejunum and ileum. The bacteria inhabiting this segment of the digestive tract metabolize, through fermentation, undigested polysaccharides such as RS and soluble fibre. Fermentation creates short-chain fatty acids, which stabilize blood glucose levels and suppress cholesterol synthesis in the liver.[23,145,146]

### 2.5.4 Starch Digestibility

Starch and starchy food products can be classified according to their digestibility, which is generally characterized by the rate and duration of the glycaemic response. Predicting and controlling the glucose absorption due to ingestion of starchy foods are of great interest in the context of worldwide health concerns. Most starches contain a portion that digests rapidly [rapidly digesting starch (RDS)], a portion that digests slowly [slowly digesting starch (SDS)] and a portion that is resistant to digestion [resistant starch (RS)].[147] RS has been defined as the portion of starch that is not hydrolysed by the enzymes in the small intestine and passes to the large intestine. One of the most widely used methods to classify starches based on the kinetics of *in vitro* digestion was suggested by Englyst *et al.*[148] The method is based on the *in vitro* digestion of starch by simulating stomach and intestinal conditions and measuring glucose release at different times. Based on the method, different starch fractions are defined as follows:

- RDS: amount of glucose released after 20 min.
- SDS: amount of glucose released between 20 and 120 min of *in vitro* digestion.
- RS: total starch minus amount of glucose released within 120 min of *in vitro* digestion.

RS has attracted much attention owing to both its potential health benefits (similar to those of soluble fibre) and its functional properties. RS positively influences the functioning of the digestive tract, microbial flora, blood cholesterol level and glycaemic response, and assists in the control of diabetes. Apart from the potential health benefits of RS, another positive advantage is its lower impact on the sensory properties of food compared with traditional sources of fibre, such as whole grains, fruits or bran. Among its desirable physicochemical properties are its swelling capacity, viscosity, gel formation and water-binding capacity, which make it useful in a variety of foods.[149–152]

The four distinct classes of RS in foods are (1) RS1 – physically inaccessible starch, which is entrapped within whole or partly milled grains or seeds; (2) RS2 – some types of raw starch granules (such as banana and potato) and high-amylose starches (*e.g.* high-amylose maize); (3) RS3 – retrograded starch (either processed from unmodified starch or resulting from food processing applications); and (4) RS4 – starches that are chemically modified to obtain resistance to enzymatic digestion (such as some starch ethers, starch esters and crosslinked starches).[153] Therefore, RS may not be digested for four reasons:

1. The compact molecular structure limits the accessibility of digestive enzymes and various amylases and explains the resistant nature of raw starch granules.[154] The starch may not be physically bioaccessible to the digestive enzymes such as in grains, seeds or tubers.
2. The starch granules themselves are structured in a way that prevents the digestive enzymes from breaking them down (*e.g.* raw potatoes, unripe bananas and high-amylose maize starch).[155]
3. Starch granules are disrupted by heating in an excess of water in a process commonly known as gelatinization, which renders the molecules fully accessible to digestive enzymes. Some sort of hydrated cooking operation is typical in the preparation of starchy foods for consumption, rendering the starch rapidly digestible.[154] However, if these starch gels are then cooled, they form starch crystals that are

resistant to enzyme digestion. This form of "retrograded" starch is found in small quantities (~5%) in foods such as "corn-flakes" or cooked and cooled potatoes, as used in a potato salad.

4. Selected starches that have been chemically modified by etherization, esterization or cross-bonding cannot be broken down by digestive enzymes.[156]

## 2.6 Modification of Starch

Modification of starch was carried out to overcome the shortcomings of native starches and increase the usefulness of starch for industrial applications. Native starches when cooked can easily retrograde and pastes have a gelling tendency in addition to easily undergoing syneresis. Therefore, starch modification not only decreases retrogradation, gelling tendencies of pastes and gel syneresis but also improves paste clarity and sheen, paste and gel texture, film formation and adhesion.[157] Modification of starches has brought about an evolution of new processing technologies and market trends. These highly functional derivatives have been tailored to create competitive advantages in a new product, improve product aesthetics, lower recipe/production costs, eliminate batch rejects, ensure product consistency and extend shelf-life, while clearly making starch relevant in all stages of a food product's life cycle.[158] Modification of starch is an ongoing process as there are numerous possibilities. There is a huge market for the many new functional and added-value properties resulting from these modifications.

Over the last few decades, starch has been modified by various methods to achieve functionalities suitable for various industrial applications. There are four broad-based kinds of modifications: chemical, physical, enzymatic and genetic. Review articles on starch modification are available.[56,157,159] However, within the last decade there has been intense interest among researchers in developing novel methods of starch modification with more emphasis on enzymatic, physical and genetic modifications.[160]

### 2.6.1 Chemical Modification

Chemical modification of starch involves the polymer molecules of the starch granule in its native form. Modification is generally achieved through derivatization methods such as etherification, esterification and cross-linking, oxidation, cationization and grafting of starch. However, there has been a dearth of new methods for chemical modification as this kind of modification gives rise to issues concerning consumers and the environment. There has been a trend to combine different kinds of chemical treatments to create new kinds of modifications. Similarly, chemical methods have been combined with physical modifications such as microwave radiation and extrusion methods to produce modified starch with specific functional properties. The overall advantages of these modifications were a shortened time of modification and increased production.[160]

New methods to produce hydrophobic starch have been developed in the last decade. Modification of maize and cassava starch with microwave radiation was used to esterify FFAs with lipase as catalyst. The FFAs were obtained from recovered coconut oil, which was hydrolysed with lipase. A degree of substitution (DS) of 1.1 was obtainable using microwave radiation for esterification with hydrolysed coconut oil and a DS of 0.33 when palmitic acid was the acyl donor. These modified starches have potential uses in surface coating materials and biomedical materials applications.[161,162] Hydrophobically modified starch was prepared by an enzyme-catalysed reaction of starch and alkenyl ketene dimer (AKD), which is a fatty acid residue. The enzymes used were lipases from *Pseudomonas* sp. and *Pseudomonas fluorescens*. This modified starch can be used as an ingredient in applications where high solution viscosity and hydrophobic interactions are required, for example as paint thickeners, construction materials, emulsion stabilizers and emulsifiers in cosmetics.[163]

Selective esterification of starch nanoparticles was performed using *Candida antarctica* lipase B (CAL-B) in its immobilized and free forms as catalyst. Acylation reactions were made accessible for the starch nanoparticles by formation of Aerosol-OT (AOT) [bis(2-ethylhexyl) sodium sulfosuccinate] to stabilize the microemulsions. Acylations of ε-caprolactone (CL), vinyl stearate and maleic anhydride were achieved when AOT-coated dispersions of starch nanoparticles in toluene were exposed to physically immobilized CAL-B (Novozym 435). The modified starch nanoparticles were found to retain their nanodimensions upon removal of the surfactant when dispersed in dimethyl sulfoxide or water.[164]

Starch modification using a combination of chemical and physical or chemical and enzymatic methods has grown rapidly. A combined method of modification of rice starch using crosslinking and phosphorylation provided a product with good freeze–thaw stability.[165]

Crosslinking of tapioca starch with sodium trimetaphosphate in the presence of osmotic pressure-enhancing salts caused an increase in the peak and final viscosity with a decrease in breakdown. Enhancement of osmotic pressure increases the activity of the crosslinking agent.[166] Starch-based hydrogels were prepared by UV-induced polymerization of acryloylated starch with the zwitterionic monomer 3-dimethyl(methacryloyloxyethyl)-ammonium propanesulfonate (DMAPS). A unique salt-tolerant swelling behaviour was observed for this modified starch.[167] Starch modified through esterification with ferulic acid giving rise to starch ferulate showed lower viscosity, higher water-holding capacity and much less retrogradation during low-temperature storage compared with native starch.[168]

Microwave-assisted esterification to produce starch maleate using the dry method had a reaction efficiency of up to 98% and a reaction time of under 5 min. This is considered to be an efficient method for esterifying starch.[169] A high efficiency in producing succinylated cassava starch with microwave assistance was also observed.[170] This is a good method to decrease the use of chemicals to enhance production.

Microwave and ultrasound irradiation were used for the esterification of carboxymethyl cold water-soluble potato starch with octenylsuccinic anhydride. The esterification time was shortened from several hours to a few minutes. The derivatives displayed excellent emulsifying and surfactant performance properties.[171]

Maize and mung bean starches have been modified using a dual modification process, whereby the native starch was subjected to partial enzymatic hydrolysis using a mixture of fungal α-amylase and glucoamylase followed by hydroxypropylation with propylene oxide. The resultant modified starch proved to have significantly different functional properties compared with hydroxypropyl starch prepared with untreated native starch.[172]

The process of ozonation oxidizes starch and ozone is a powerful oxidant as it has an extra O atom compared with oxygen. During the ozonation process, the carboxyl and carbonyl contents were found to increase with the time of exposure to ozone. There was a difference in the extent of starch oxidation among starches from different sources.[173] Ozonated starches were found to be useful as thickening agents whereas those treated in the presence of amino acids were suitable alternatives to highly chemically oxidized starch.[174] Ozone is a clean and powerful oxidant and leaves no residues behind, unlike the hypochlorite oxidation process where large amounts of salts are produced.[175]

### 2.6.2 Physical Modification

Physical modification can be safely used as a modification process in food products as it does not involve the presence of any chemicals. There has been a surge in new methods for the physical modification of starches including, but are not limited to, osmotic pressure treatment,[176] deep freezing,[177] multiple deep freezing and thawing,[178] instantaneous controlled pressure drop (DIC) process,[179,180] mechanical activation with a stirring ball mill,[181] micronization in a vacuum ball mill,[182] pulsed electric field treatment,[183] corona electrical discharges,[184] thermally inhibited treatment (dry heating),[185] iterated syneresis[186] and superheated starch.[187]

Osmotic pressure treatment (OPT) was carried out in the presence of high-salt solutions.[176] Starch solution was suspended in sodium sulfate to obtain a uniform starch suspension and heat distribution. The potato starch studied changed from a B- to an A-type after treatment with the gelatinization temperatures increasing significantly. A uniform heat distribution is provided for using this method as compared with heat–moisture treatment and modified starch can be produced on a large scale.

Deep freezing and thawing of moistened starch increased the crystallinity of the granules,[177] but multiple deep freezing and thawing[178] caused irreversible disruption of the crystalline order. Iterated syneresis was similar to multiple deep freezing and thawing. The freezing and thawing were repeated until the moisture content in the solid phase was less than 20%. All modified starches showed a change towards a B-type X-ray diffraction pattern, indicating disruption of the crystalline properties.[186] Again, as there are no chemicals involved, there is no concern about the effect on the environment or safety issues to be addressed.

The instantaneous controlled pressure drop (DIC) process involves a short pressurization by injection of saturated steam at a fixed pressure and predetermined time before it drops towards vacuum.[179,180] Gelatinization transition temperatures and enzymatic hydrolysis increased whereas gelatinization enthalpy decreased after treatment.

The use of friction, collision, impingement, shear and other mechanical actions to alter the crystalline structures and properties of the starch granule relates to mechanical activation or micronization. This process causes large particles to be crushed to form smaller particles whereas tiny particles agglomerate and form large particles. The gelatinization temperature and viscosity are decreased for the treated sample.[181,182]

A non-thermal food preservation method, pulsed electric field (PEF) technology, has been used to study the effect of the treatment on starch. Rearrangement and destruction of starch molecules were observed in addition to a decrease in gelatinization properties, viscosity and crystallinity.[183] The solubility, gel consistency and clarity of starches decreased with increase in exposure time to corona electrical discharges.[184]

Thermal inhibition of starch is carried out by dehydrating starch until it is anhydrous ($<$1% moisture) and heating it at a temperature of 100 °C or greater for a sufficient time to inhibit starch. Alkaline conditions enhanced the effect of heating. Pastes formed from these starches had increased resistance to viscosity breakdown and a non-cohesive texture. Thermal inhibition with ionic gums used sodium alginate, carboxymethylcellulose and xanthan as crosslinking agents and formed graft copolymers through ester formation.[185]

Superheated starches were prepared by heating a starch solution to a temperature between 180 and 220 °C to produce spreadable particle gels with spherulite morphology and a cream-like texture upon cooling. Dry superheated starches mixed with cold water are able to give an immediate gel-like texture.[187]

Extrusion heating (EH) and fluidized bed heating (FBH) was used on an amaranth starch-rich fraction. EH caused a high degree of granule disruption and almost complete loss of crystallinity whereas FBH gave some loss of crystallinity but the granule integrity was preserved.[188]

### 2.6.3 Enzymatic Modification

Enzymatic modification has mainly used hydrolysing enzymes and one of its products is a syrup, either glucose syrup or high-fructose maize syrup. In research studies, more enzymes are being identified for use in the modification of starch.

The use of amylomaltases (EC 2.4.1.25) to modify starches is expected to find applications in the food industry as a plant and chemical-free alternative to gelatine. Starch treated with α-(1→4)-α-(1→4) glucosyl transferases, also known as amylomaltases, is used in forming a thermoreversible gel. This is done by breaking an α-(1→4) bond between two glucose units to make subsequently a novel α-(1→4) bond. These enzymes are found in the Eukarya, Bacteria and Archaea. The enzyme used has to be free of enzymatic components that can cause undesirable damage to the starch molecule. The starches that can be used for modification should contain amylose, such as potato, maize, wheat, rice and tapioca starch. The average molecular weight, reducing power and branching percentage remain unchanged from the starting material. It is believed a mutual rearrangement between the starch molecules occurs without an increase in oxidation-sensitive places or parts having reducing activity. Little or no retrogradation occurs. This modified starch can be used in foodstuffs, cosmetics, pharmaceutics, detergents, adhesives and drilling fluids. It is also a good source of plant-derived substitutes for gelatine, except that it forms a turbid gel whereas gelatine gels are transparent.[189,190] Similar work has also been carried out on granular maize starch.[191] The thermodynamics of this gelatine-like starch-based system have been studied.[192] An amylomaltase-modified potato starch has been used as a fat replacer and enhancer of creaminess in yoghurt.[193] In the study on gel texture formed in the modification of potato, high-amylose potato, maize and pea starch with amylomaltase (AM) isolated from the hyperthermophilic bacterium *Thermus thermophilus*, there was an improvement in gel texture compared with the parent starch.[192] All the modified starches showed broadened amylopectin chain-length profiles.[160]

A putative glycogen branching enzyme (GBE) was cloned and expressed from *Streptococcus mutans*, thereafter known as SmGBE. This enzyme differs from the first bacterial GBE group in that it is encoded with four highly conserved regions of the α-(1→4) GBE family and does not have the extra N-terminal stretch. SmGBE is able to modify starches and produce shorter branches with a greater branched structure compared with native starch. Starch retrogradation is also retarded with enzyme reactions due to low-temperature storage.[194] Cyclomaltodextrinase (CDase; EC 3.2.1.54), isolated from alkalophilic *Bacillus* sp. I-5 (CDase I-5), was used to modify rice starch to produce low-amylose starch products. The amylose content was found to have decreased significantly from 28.5 to 9%, whereas there was no significant change in the side-chain length distribution of the amylopectin. Storage of the modified rice starch at 4 °C for 7 days showed that the retrogradation rate had significantly decreased compared with the control sample.[195]

As glycaemic response in foods is related to diabetes, prediabetes, cardiovascular disease and obesity, it is becoming important to have more low glycaemic index foods to control these diseases.[196] One of the methods is to produce foods containing starch that is digested slowly. There was a significant reduction of rapidly digested starch by 14.5, 29.0, 19.8, and 31.0%

for maize starch modified with β-amylase, β-amylase and transglucosidase, maltogenic α-amylase, and maltogenic α-amylase and transglucosidase, respectively. Increases in the starch branch density and crystalline structure in the modified starches were thought to contribute to the slow digestion.[197]

In the reaction of glucoamylase (1,4-α-D-glucan glucohydrolase, EC 3.2.1.3) with starch, 8–32% of the D-glucose remains inside the granule during the early stages of the reaction, but over 80% of the D-glucose is found in the reaction supernatant outside the granule during the later stages of the reaction. Glucoamylase was reacted with waxy maize, maize and amylomaize-7 starches to retain 100% of the D-glucose produced inside the granule by controlling the length of time of the reaction and the type of starch. The amount of water used in the reaction was also decreased to an amount equal to the weight of starch which gave a solid-granule reaction system. The amount of water used was sufficient for hydrolysis to take place, but the solid granules did not allow diffusion of the D-glucose. The starch granules containing D-glucose rapidly diffuse out the D-glucose when suspended in water. In addition, cyclomaltodextrin glucanosyl transferase (CGTase, EC 2.4.1.19) in the presence of isoamylase was used to produce cyclodextrins (CDs) with a maximum yield of 3.4% and 100% retention inside waxy maize starch granules. Cyclomaltodextrins are also formed *in situ*, with retention of CDs in the granule, and this leads to the production of a new material that has the properties of starch granules and cyclomaltodextrins. Formation of complexes of organic molecules with cyclomaltodextrins provides stabilization of light-, heat- and oxygen-sensitive materials in the starch granules and also a mechanism for their slow release, in addition to conferring special tastes, odours and flavours on the starch granules.[198,199]

### 2.6.4 Genetic/Biotechnological Modification

Advances in genetic engineering technologies have made the genetic modification of starch in plants possible by targeting the enzymes of the starch biosynthetic pathway. This transgene technology has the potential to produce novel starches which can reduce or eliminate the use of environmentally hazardous post-harvest chemical and enzymatic modification.[200] The activity of these enzymes affects the reactivity, functionality and food processing applications of these modified starches, thereby providing a market for "niche" products. Genetic modification can be carried out by the traditional plant-breeding techniques or through biotechnology. The genetic modification that is carried out uses the methods[160] shown in Figure 2.9.

Repression of glucan water dikinase (GWD) (starch phosphorylating enzyme R1) was used in the alteration of specific structural motifs of potato starch.[202] An *Escherichia coli* glg B encoding a glycogen branching enzyme from a patatin promoter onto potato lines was also studied.[203] Starch containing amylopectin molecules with a relatively higher number of amylopectin branches and a higher amount of short amylopectin chains with a

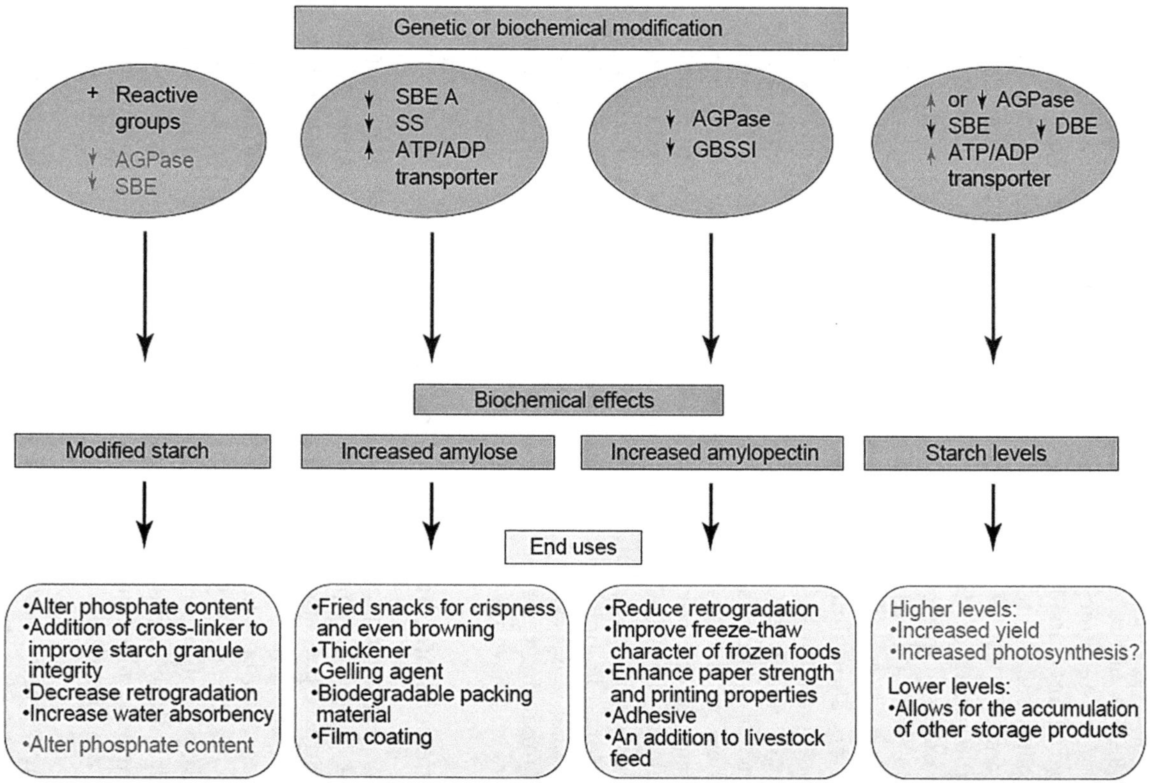


**Figure 2.9** Modification of starch and end uses. An arrow pointing upwards indicates an increase and an arrow pointing downwards indicates a decrease in the level of an enzyme; a plus symbol indicates the addition of reactive groups. DBE = debranching enzyme; GBSSI = granular-bound starch synthase I; SBE = starch branching enzyme; SBE A = class A SBE; SS = starch synthase. The text colour used in the modification portion corresponds to the points of the same colour under end uses. Reproduced from Slattery *et al.*[201] with permission of Elsevier.

lower content of phosphate was obtained. This starch also gave rise to hard and adhesive gels. AGPase (ADP-glucose pyrophosphorylase) was used as a catalyst to increase the total cassava root biomass by 2.6-fold.[204] When a full-length cDNA encoding a second starch branching enzyme (SBE A) isoform was isolated and an antisense SBE A RNA was generated on transgenic potato plants, a complete reduction in SBE A was observed. The composition and structure of the potato starch were completely altered in that the average chain length of amylopectin was greater. This caused an increase in the apparent amylose content and higher levels of phosphorus were also observed.[205] A similar observation had been reported.[206] Potato starches with low SBE values showed an increase of up to 5 °C in DSC peak temperature and viscosity onset temperature. This was speculated to be due to an alteration in amylopectin branch patterns.

Three mutagenized grains of the diploid oat *Avena strigosa* Schreb were identified: mutants lam-1, lam-2 and sga-1. The lam-1 and lam-2 lacked GBSS activity and an amylose component, therefore representing mutations of the waxy type. The nature of the mutation in sga-1 is yet to be discovered.[207]

## 2.7 Industrial Production of Starch and its Uses

The major sources of starch for human use are the cereals, but roots and tubers are also important. Maize or maize starch makes up more than 80% of the world market for starch, and most of this is produced in the USA. Europe is the major producer of wheat and potato starches, whereas cassava and tapioca starch are produced mainly in Asia. Other starches, such as those from rice and sweet potato, make up only a minor proportion of the total.[208]

Starch has been used in both food and non-food products for centuries. The ancient Egyptians and later the Romans used it as an adhesive, and it is still used as such today. The Greeks used it for medicinal purposes. In the developed world, starch provides at least 35% of a human's daily calorific intake, but this can be much higher. The main crops that are harvested for this purpose are maize, rice, wheat and potatoes. In many areas, especially Africa and the Far East, starch can provide 80% of human daily calorific intake, and this may be only from a single source such as rice.[209]

### 2.7.1 Industrial Production of Starch

The starch manufacturing process differs with the raw materials. Starch can be produced by wet or dry milling or a combination of both. Wet milling is commonly used for industrial manufacturing as it provides better quality products. The manufacture of maize starch including nutritive sweeteners and dextrin by wet milling is shown in Figure 2.10.

The wet-milling process starts with the softening of the grain by steeping in sulfur dioxide solution. Proper steeping is essential for high yields

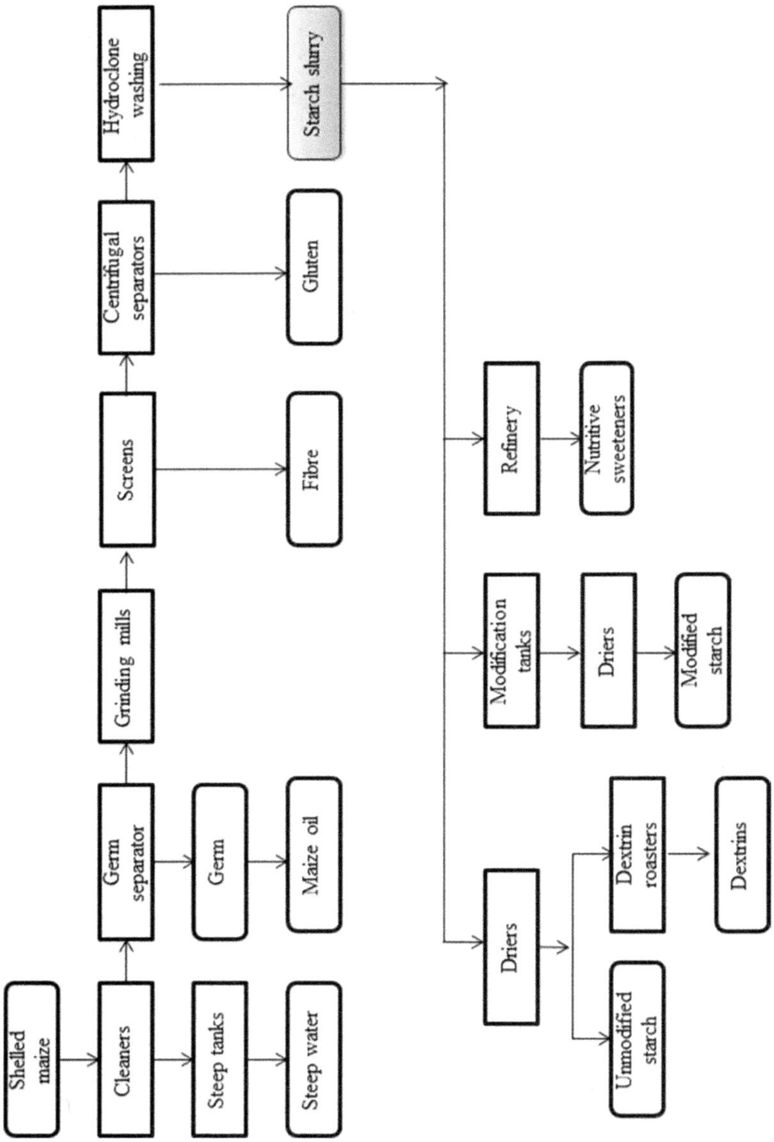


**Figure 2.10** The wet milling process of maize starch. Adapted from Corn Refiners Association.[210]

and high starch quality. The steeping is carried out in a continuous counter-current process. The cleaned maize is filled into a battery of large steeping tanks, where the maize is soaked in hot water (52–54 °C) for 30–48 h to begin breaking the starch and protein bonds. The gluten bonds within the maize begin to loosen and release the starch. The steeping is actually a controlled fermentation. Sulfur dioxide improves the fermentation by enhancing the growth of favourable microorganisms, preferably *Lactobacillus* sp., while suppressing detrimental bacteria, moulds, fungi and yeast. The kernel swells to more than double its size and increases its moisture content from about 15% to 45%. The soften grains are then ground, coarsely, in mills having one stationary and one rotating disk. The separated germs are collected in floatation cells or hydroclones. The ground maize from which some of the germs have been separated is ground through a second grind mill to release germs not released in the first pass. After germ separation, the maize slurry is passed through Entoleter mills, which sling the material against pins at high speed or counter-rotating disc mills to free the starch with minimum damage to the fibre. The fibre is recovered on a series of screens from which the adhering starch and gluten are washed off and then dewatered in screw presses. After the germs and fibre have been removed, the starch and gluten are separated in a disc-nozzle type of centrifuge. Advantage is taken of the density difference: 1.06 for gluten and 1.6 for starch. The gluten is dewatered on rotary vacuum filters and then dried to give gluten meal with a protein concentration of >60%. The starch stream from the centrifugal separator still contains a considerable amount of protein and this is removed by washing in a counter-current fashion in a series of hydroclones. In the wet milling of maize, the only entry point for fresh water is in the last stage of the starch washing system. The water from one operation is used in another. Efficient water use is an important factor in a successful wet-milling process that could lead to over 99.5% of the dry matter being recovered. The recovered starch and co-products generally consist of starch 67.5%, fibre 12.5%, germ 7.3%, maize extractives 6.5% and gluten meal 5.4%.

For wheat (Figure 2.11), starch is the co-product from the manufacture of gluten, a protein of wheat that gives dough its unique properties. In contrast to maize, wheat starch is produced by first dry milling to flour. The flour is then made into either dough or batter with water. The starch is subsequently separated from the gluten. The starch fraction also contains fibres, pentosans and bran. The gluten contains about 75–85% protein. The starch fraction is refined through centrifugal separators to give two fractions, A-starch (60–65% purity, 15–50 μm particle size) and B- starch (15–20% purity, 2–15 μm particle size).

For tubers such as tapioca and potato (Figure 2.12), although there are some differences in details, the production processes of potato and tapioca starch are essentially the same.

Starch is a versatile and useful polymer not only because it is a cheap, natural material but also because of the ease with which its physicochemical

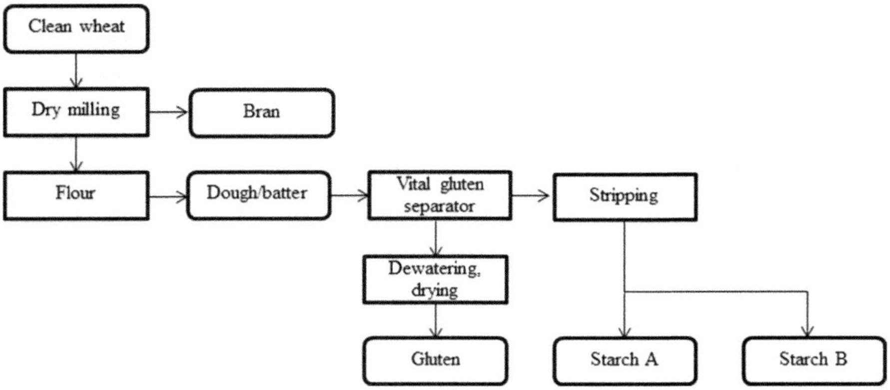


**Figure 2.11** Wheat starch manufacturing process.

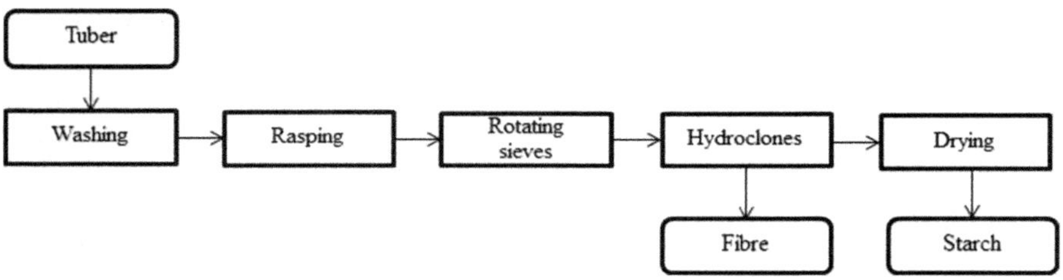


**Figure 2.12** Tuber starch manufacturing process.

properties can be altered through chemical or enzyme modification and/or physical treatment. It is found in many different plant organs, including seeds, fruits, tubers and roots, where it is used as a source of energy during periods of dormancy and regrowth. Many of these starch-storing organs are staple foodstuffs in the human diet. Increasingly, starch is also used as a renewable raw material, as a source of energy after conversion to ethanol, and for many different industrial applications.

## 2.7.2 Utilization of Starches

For food uses, starch affects texture, viscosity, gel formation, adhesion, binding, moisture retention, film formation and product homogeneity. It is used mainly in soups, sauces and gravies, bakery products, dairy confectionery, snacks, batters and coatings and meat products. Non-food applications of starch include pharmaceuticals, textiles, alcohol-based fuels and adhesives. New uses of starch include low-calorie substitutes, biodegradable packaging materials, thin films and thermoplastic materials with improved thermal and mechanical properties. Table 2.2 summarizes the products in which maize starch is used.

**Table 2.2** Products in the manufacture of which maize starch is used. Adapted from Corn Refiners Association.[210]

| Unmodified starch | Modified starch | Dextrins |
|---|---|---|
| • Beverages, non-alcoholic (powdered mixes)<br>• Baking, snack foods (baking powder, biscuits, breads and rolls, cakes, cookies, crackers, doughnuts, frosting, icing, glazes, pies, powdered sugar, pretzels, yeast)<br>• Canned fruits and vegetables (fruits and berries, fruit fillings, soups, tomato sauces, vegetables)<br>• Cereals (cereals, cereal bars)<br>• Condiments (ketchup, gravies, mayonnaise, mustard, oriental sauces, salad dressings, sauce mixes, vinegar)<br>• Confectionary and gum (chewing gum, chocolates, confectionery, marshmallows, nougats)<br>• Fats and oils (margarine, pan coatings)<br>• Formulated dairy products (cheese spreads and foods, coffee whitener, frozen cream)<br>• Ice cream and frozen desserts (frozen puddings/custards, ice cream or milk, powdered mixes, sherbets, water ices)<br>• Jams, jellies, preserves (fruit butters)<br>• Meat products (bologna, fish, seafood, mincemeat, sausages, surimi)<br>• Mixes, prepared (cake mixes, cookie mixes, dessert mixes, dried foods, frozen or dried eggs, frosting, icing mixes, gravy mixes, instant breakfast, pancake or waffle mixes, bread mixes, seasoning mixes, dried soups) | • Beverages, alcoholic (beer, spirits)<br>• Beverages, non-alcoholic (carbonated, fruit drinks and juices, powdered mixes)<br>• Baking, snack foods (baking powder, bars, energy snacks, biscuits, breads and rolls, cakes, cookies, crackers, doughnuts, flavours, frosting, icing, glazes, pies, potato chips, powdered sugar, pretzels)<br>• Canned fruits and vegetables (fruits and berries, fruit fillings, soups, tomato sauces, vegetables)<br>• Cereals (cereals, cereal bars)<br>• Condiments (ketchup, gravies, mayonnaise, mustard, oriental sauces, pickles, relishes, salad dressings, sauce mixes)<br>• Confectionary and gum (chewing gum, confectionery, marshmallows, nougats)<br>• Fats and oils (margarine)<br>• Formulated dairy products (cheese spreads and foods, coffee whitener, frozen cream, yoghurt)<br>• Ice cream and frozen desserts (frozen puddings/custards, ice cream or milk, powdered mixes, sherbets, water ices)<br>• Jams, jellies, preserves (fruit butters, jams, jellies, marmalade)<br>• Meat products (bologna, breakfast meats, chicken products, dried meats, fish, seafood, hotdog, mincemeat, sausages, surimi) | • Beverages, non-alcoholic (protein drinks, fruit drinks and juices, powdered mixes)<br>• Baking, snack foods (biscuits, breads and rolls, cakes, cookies, crackers, doughnuts, flavours, food colouring, frosting, icing, glazes, pies, potato chips, pretzels, spices)<br>• Canned fruits and vegetables (soups, tomato sauces)<br>• Cereals (cereals, cereal bars)<br>• Condiments (gravies, oriental sauces, salad dressings, sauce mixes)<br>• Confectionary and gum (chewing gum, confectionery, marshmallows, nougats)<br>• Fats and oils (margarine)<br>• Formulated dairy products (cheese spreads and foods, coffee whitener, condensed milk, frozen cream)<br>• Ice cream and frozen desserts (frozen puddings/custards, ice cream or milk, powdered mixes, sherbets, water ices)<br>• Jams, jellies, preserves (fruit butters, jams, jellies, marmalade, preserves)<br>• Meat products (bologna, mincemeat)<br>• Mixes, prepared (cake mixes, cookie mixes, dessert mixes, dried foods, frozen or dried eggs, frosting, icing mixes, gelatin mixes, gravy mixes, instant breakfast, instant tea, pancake or waffle mixes, bread mixes, seasoning mixes, dried soups) |

**Table 2.2** (*Continued*)

| Unmodified starch | Modified starch | Dextrins |
|---|---|---|
| • Syrups and sweeteners (chocolate, cocoa, dessert toppings, fruit and table sugars, soda fountain)<br>• Miscellaneous foods (baby foods, desserts, dietetic preparations, invalid feedings, precooked frozen meats, rice and coffee polish)<br>• Household needs (briquettes, diapers, metal cleaner, trash bags, twine, cord, string)<br>• Personal care (cosmetics, surgical dressings)<br>• Pharmaceuticals (antibiotics, aspirin, drug coatings, cough drops, drugs, medicine syrups)<br>• Animal feed (cat, cattle, dog, fish, swine)<br>• Chemicals (acetic acid, agrochemicals, dispersing agents, enzymes, fermentation processes, food acids, alcohol, insecticides, organic solvents)<br>• Paper and related products (abrasive paper and cloth, bookbinding, envelopes, glassine, labels, parchment, printing inks, straws, wallpaper)<br>• Paste and adhesives (binders, glues, gums, mucilage, pastes) | • Mixes, prepared (cake mixes, cookie mixes, dessert mixes, dried foods, frozen or dried eggs, frosting, icing mixes, gelatin mixes, gravy mixes, instant breakfast, instant tea, pancake or waffle mixes, bread mixes, seasoning mixes, dried soups)<br>• Syrups and sweeteners (chocolate, cocoa, dessert toppings, fruit and table sugars, low-calorie sweeteners, soda fountain)<br>• Miscellaneous foods (baby foods, desserts, dietetic preparations, invalid feedings, precooked frozen meats)<br>• Household needs (air freshener, batteries, briquettes, cleaners, crayons, chalk, diapers, laundry care, matches, trash bags, twine, cord, string)<br>• Personal care (cosmetics, deodorants, hair styling products, surgical dressings)<br>• Pharmaceuticals (aspirin, drug coatings, cough drops, drugs, medicine syrups)<br>• Tobacco (tobacco ingredient)<br>• Animal feed (cat, cattle, dog, fish)<br>• Chemicals (agrochemicals, dispersing agents, enzymes, fermentation | • Syrups and sweeteners (chocolate, cocoa, dessert toppings, fruit and table sugars, low-calorie sweeteners, soda fountain)<br>• Miscellaneous foods (baby foods, desserts, dietetic preparations, invalid feedings, peanut butter, rice and coffee polish)<br>• Household needs (briquettes, crayons, chalk, matches)<br>• Personal care (cosmetics)<br>• Pharmaceuticals (antibiotics, drug coatings, cough drops, drugs, medicine syrups)<br>• Tobacco (tobacco ingredient)<br>• Animal feed (cat, dog, fish)<br>• Chemicals (fermentation processes, organic solvents)<br>• Paper and related products (abrasive paper and cloth, bookbinding, envelopes, glassine, labels, parchment, printing inks, straws, wallpaper)<br>• Paste and adhesives (binders, glues, gums, mucilage, pastes)<br>• Textile (dyes, printing, sizing materials, window shade, shade cloth)<br>• Building materials (ceramics, fibreboard, laminated, tile, ceiling, wallboard) |

| | | |
|---|---|---|
| • Textile (cord polishing, dyes, oilcloth, printing, sizing materials, window shade, shade cloth)<br>• Building materials (cardboard, ceramics, coatings, cork products, fibreglass, fibreboard, glass or rock wool, laminated, linoleum, paints and vanishes, tile, ceiling, wallboard)<br>• Mining/metallurgy (ore refining, separation)<br>• Miscellaneous industrial (explosives, filters, fireworks, oil-well drilling, plastics, rubber (cold process), tyres) | processes, food acids, alcohol, insecticides, organic solvents)<br>• Paper and related products (abrasive paper and cloth, bookbinding, envelopes, glassine, labels, parchment, printing inks, straws, wallpaper)<br>• Paste and adhesives (binders, glues, gums, mucilage, pastes)<br>• Textile (cord polishing, dyes, oilcloth, printing, sizing materials, window shade, shade cloth)<br>• Building materials (cardboard, ceramics, coatings, cork products, fibreglass, fibreboard, glass or rock wool, laminated, linoleum, paints and vanishes, tile, ceiling, wall treatment compounds, wallboard)<br>• Mining/metallurgy (electroplating, galvanizing, metal plating, ore refining, separation)<br>• Miscellaneous industrial (boiler compounds, explosives, filters, fireworks, leather tanning, lubricating agents, oil-well drilling, plastics, protective colloids, refractories, shoes, water recovery) | • Mining/metallurgy (ore refining, separation)<br>• Miscellaneous industrial (explosives, fireworks, protective colloids) |

## References


1. G. H. Perry, N. J. Dominy, K. G. Claw, A. S. Lee, H. Fiegler, R. Redon, J. Werner, F. A. Villanea, J. L. Mountain, R. Misra, N. P. Carter, C. Lee and A. C. Stone, *Nat. Genet.*, 2007, **39**, 1256.
2. M. M. Burrell, *J. Exp. Bot.*, 2003, **54**, 451.
3. L. Copeland, J. Blazek, H. Salman and M. C. M. Tang, *Food Hydrocolloids*, 2009, **23**, 1527.
4. D. Ray and S. Maiti, in Natural Polymers, *Nanocomposites*, ed. M. J. John and T. Sabu, Royal Society of Chemistry, Cambridge, 2012, vol. 2, p. 185.
5. D. J. Thomas and W. A. Atwell, *Starches*, Eagan Press, St Paul, MN, 1999.
6. J. N. BeMiller and R. L. Whistler, *Starch: Chemistry and Technology*, Academic Press, Burlington, MA, 3rd edn, 2009.
7. R. F. Tester, J. Karkalas and X. Qi, *J. Cereal Sci.*, 2004, **39**, 151.
8. A. Buléon, P. Colonna, V. Planchot and S. Ball, *Int. J. Biol. Macromol.*, 1998, **23**, 85.
9. J. Swinkels, *Starch Stärke*, 1985, **37**, 1.
10. H. Zobel, *Starch Stärke*, 1988, **40**, 44.
11. F. Xie, E. Pollet, P. J. Halley and L. Avérous, *Prog. Polym. Sci.*, 2013, **38**, 1590.
12. C. Biliaderis, *ChemInform*, 1998, DOI: 10.1002/chin.199847350.
13. J. Mua and D. Jackson, *J. Agric. Food Chem.*, 1997, **45**, 3840.
14. Y. Takeda, S. Hizukuri, C. Takeda and A. Suzuki, *Carbohydr. Res.*, 1987, **165**, 139.
15. L. Wang and P. White, *Cereal Chem.*, 1994, **71**, 263.
16. Y. Yoshimoto, J. Tashiro, T. Takenouchi and Y. Takeda, *Cereal Chem.*, 2000, **77**, 279.
17. Y. Yoshimoto, T. Takenouchi and Y. Takeda, *Carbohydr. Polym.*, 2002, **47**, 159.
18. Y. Takeda, S. Shibahara and I. Hanashiro, *Carbohydr. Res.*, 2003, **338**, 471.
19. S. Hizukuri, *Carbohydr. Res.*, 1985, **141**, 295.
20. S. Hizukuri, *Carbohydr. Res.*, 1986, **147**, 342.
21. J. Jane, Y. Chen, L. Lee, A. McPherson, K. Wong, M. Radosavljevic and T. Kasemsuwan, *Cereal Chem.*, 1999, **76**, 629.
22. S. Ball, H.-P. Guan, M. James, A. Myers, P. Keeling, G. Mouille, A. Buléon, P. Colonna and J. Preiss, *Cell*, 1996, **86**, 349.
23. A. C. Dona, G. Pages, R. G. Gilbert and P. W. Kuchel, *Carbohydr. Polym.*, 2010, **80**, 599.
24. A. Gray-Weale and R. G. Gilbert, *J. Polym. Sci., Part A: Polym. Chem.*, 2009, **47**, 3914.
25. M. J. Gidley, I. Hanashiro, N. M. Hani, S. E. Hill, A. Huber, J.-L. Jane, Q. Liu, G. A. Morris, A. Rolland-Sabaté and A. M. Striegel, *Carbohydr. Polym.*, 2010, **79**, 255.
26. S. Peat, W. Whelan and G. Thomas, *J. Chem. Soc.*, 1952, 4546.

27. A. Sarko and H. C. Wu, *Starch Stärke*, 1978, **30**, 73.
28. R. Veregin, C. Fyfe, R. Marchessault and M. Taylor, *Macromolecules*, 1986, **19**, 1030.
29. M. J. Gidley and P. V. Bulpin, *Carbohydr. Res.*, 1987, **161**, 291.
30. H. Zobel, *Starch Stärke*, 1988, **40**, 1.
31. T. A. Waigh, M. J. Gidley, B. U. Komanshek and A. M. Donald, *Carbohydr. Res.*, 2000, **328**, 165.
32. A. Imberty, A. Buléon, V. Tran and S. Péerez, *Starch Stärke*, 1991, **43**, 375.
33. P. Cairns, T. Y. Bogracheva, S. Ring, C. Hedley and V. Morris, *Carbohydr. Polym.*, 1997, **32**, 275.
34. A. Buléon, C. Gérard, C. Riekel, R. Vuong and H. Chanzy, *Macromolecules*, 1998, **31**, 6605.
35. A. McPherson and J. Jane, *Carbohydr. Polym.*, 1999, **40**, 57.
36. O. Sevenou, S. Hill, I. Farhat and J. Mitchell, *Int. J. Biol. Macromol.*, 2002, **31**, 79.
37. D. J. Gallant, B. Bouchet and P. M. Baldwin, *Carbohydr. Polym.*, 1997, **32**, 177.
38. H. Tang, T. Mitsunaga and Y. Kawamura, *Carbohydr. Polym.*, 2006, **63**, 555.
39. J. Huang, N. Wei, H. Li, S. Liu and D. Yang, *Carbohydr. Polym.*, 2014, **103**, 355.
40. A. M. Donald, in *Starch in Food: Structure, Function and Applications*, ed. A. C. Eliasson, Woodhead Publishing, Abington, 2004, p. 156.
41. J. I. Jane, K. S. Wong and A. E. McPherson, *Carbohydr. Res.*, 1997, **300**, 219.
42. D. R. Lineback, *Baker's Dig.*, 1984, **58**, 17.
43. G. H. Robertson, D. W. Wong, C. C. Lee, K. Wagschal, M. R. Smith and W. J. Orts, *J. Agric. Food Chem.*, 2006, **54**, 353.
44. P. Jenkins, R. Comerson, A. Donald, W. Bras, G. Derbyshire, G. Mant and A. Ryan, *J. Polym. Sci., Part B: Polym. Phys.*, 1994, **32**, 1579.
45. T. A. Waigh, I. Hopkinson, A. M. Donald, M. F. Butler, F. Heidelbach and C. Riekel, *Macromolecules*, 1997, **30**, 3813.
46. V. Shibaev, *Mol. Cryst. Liq. Cryst.*, 1994, **243**, 201.
47. T. A. Waigh, A. M. Donald, F. Heidelbach, C. Riekel and M. J. Gidley, *Biopolymers*, 1999, **49**, 91.
48. T. A. Waigh, K. L. Kato, A. M. Donald, M. J. Gidley, C. J. Clarke and C. Riekel, *Starch Stärke*, 2000, **52**, 450.
49. J.-l. Jane, *J. Appl. Glycosci.*, 2006, **53**, 205.
50. B. Zhang, X. Li, J. Liu, F. Xie and L. Chen, *Food Hydrocolloids*, 2013, **31**, 68.
51. X. Qi, R. Tester, C. Snape and R. Ansell, *J. Cereal Sci.*, 2003, **37**, 363.
52. J. I. Jane, *J. Appl. Glycosci.*, 2007, **54**, 31.
53. L. Copeland, J. Blazek, H. Salman and M. C. Tang, *Food Hydrocolloids*, 2009, **23**, 1527.
54. G. R. Ziegler, J. A. Creek and J. Runt, *Biomacromolecules*, 2005, **6**, 1547.

55. R. P. Ellis, M. P. Cochrane, M. F. B. Dale, C. M. Duffus, A. Lynn, I. M. Morrison, R. D. M. Prentice, J. S. Swanston and S. A. Tiller, *J. Sci. Food Agric.*, 1998, 77, 289.
56. S. Jobling, *Curr. Opin. Plant Biol.*, 2004, 7, 210.
57. W. R. Morrison, *J. Cereal Sci.*, 1988, **8**, 1.
58. W. Morrison, *Cereal Foods World*, 1995, **40**, 437.
59. W. Morrison, R. Law and C. Snape, *J. Cereal Sci.*, 1993, **18**, 107.
60. W. R. Morrison, R. F. Tester, M. J. Gidley and J. Karkalas, *Carbohydr. Res.*, 1993, **245**, 289.
61. W. R. Morrison, R. F. Tester, C. E. Snape, R. Law and M. J. Gidley, *Cereal Chem.*, 1993, **70**, 385.
62. I. A. Appelqvist and M. R. Debet, *Food Rev. Int.*, 1997, **13**, 163.
63. P. M. Baldwin, *Starch Stärke*, 2001, **53**, 475.
64. P. Greenwell and J. D. Schofield, *Cereal Chem.*, 1986, **63**, 379.
65. S. Oda and J. D. Schofield, *J. Cereal Sci.*, 1997, **26**, 29.
66. C. F. Morris, *Plant Mol. Biol.*, 2002, **48**, 633.
67. T. Kasemsuwan and J.-L. Jane, *Cereal Chem.*, 1996, **73**, 702.
68. A. Blennow, A. M. Bay-Smidt, B. Wischmann, C. E. Olsen and B. L. Møller, *Carbohydr. Res.*, 1998, **307**, 45.
69. A. Blennow, A. M. Bay-Smidt, C. E. Olsen and B. L. Møller, *Int. J. Biol. Macromol.*, 2000, **27**, 211.
70. A. Blennow, S. Engelsen, L. Munck and B. Møller, *Carbohydr. Polym.*, 2000, **41**, 163.
71. A. Blennow, T. H. Nielsen, L. Baunsgaard, R. Mikkelsen and S. B. Engelsen, *Trends Plant Sci.*, 2002, 7, 445.
72. H. Fredriksson, J. Silverio, R. Andersson, A.-C. Eliasson and P. Åman, *Carbohydr. Polym.*, 1998, **35**, 119.
73. W. S. Ratnayake and D. S. Jackson, *Carbohydr. Polym.*, 2007, **67**, 511.
74. J.-m. Liu and S.-l. Zhao, *Starch Stärke*, 1990, **42**, 96.
75. P. J. Jenkins and A. M. Donald, *Carbohydr. Res.*, 1998, **308**, 133.
76. M. R. Debet and M. J. Gidley, *J. Agric. Food Chem.*, 2007, **55**, 4752.
77. M. C. Tang and L. Copeland, *Carbohydr. Polym.*, 2007, **70**, 1.
78. M. J. Gidley, *Macromolecules*, 1989, **22**, 351.
79. U. T. Tran, H. Okadome, M. Murata, S. Homma and K. I. Ohtsubo, *Food Sci. Technol. Res.*, 2001, 7, 323.
80. K. Kulp, J. Ponte Jr and B. L. D'Appolonia, *Crit. Rev. Food Sci. Nutr.*, 1981, **15**, 1.
81. P. Colonna, V. Leloup and A. Buleon, *Eur. J. Clin. Nutr.*, 1992, **46**, S17.
82. B. Ooraikul, G. Packer and D. Hadziyev, *J. Food Sci.*, 1974, **39**, 358.
83. C. Seow and C. Teo, *Starch Stärke*, 1996, **48**, 90.
84. A. A. Karim, M. H. Norziah and C. C. Seow, *Food Chem.*, 2000, **71**, 9.
85. P. Roulet, W. MacInnes, P. Würsch, R. Sanchez and A. Raemy, *Food Hydrocolloids*, 1988, **2**, 381.
86. M. Fukuoka, K.-I. Ohta and H. Watanabe, *J. Food Eng.*, 2002, **53**, 39.
87. J. W. Donovan, *Biopolymers*, 1979, **18**, 263.
88. K. Zeleznak and R. Hoseney, *Cereal Chem.*, 1987, **64**, 121.

89. M. Le Meste, D. Champion, G. Roudaut, G. Blond and D. Simatos, *J. Food Sci.*, 2002, **67**, 2444.
90. D. Champion, M. Le Meste and D. Simatos, *Trends Food Sci. Technol.*, 2000, **11**, 41.
91. S. Rahman, *Food Properties Handbook*, CRC Press, Boca Raton, FL, 2nd edn, 2009.
92. L. Slade, H. Levine and D. S. Reid, *Crit. Rev. Food Sci. Nutr.*, 1991, **30**, 115.
93. L. Slade, H. Levine, J. Ievolella and M. Wang, *J. Sci. Food Agric.*, 1993, **63**, 133.
94. M. Le Meste, V. Huang, J. Panama, G. Anderson and R. Lentz, *Cereal Foods World*, 1992, **37**, 264.
95. T. Noel, S. Ring and M. Whittam, *Trends Food Sci. Technol.*, 1990, **1**, 62.
96. Y. Liu, B. Bhandari and W. Zhou, *J. Agric. Food Chem.*, 2006, **54**, 5701.
97. L. N. Bell and D. E. Touma, *J. Food Sci.*, 1996, **61**, 807.
98. C. G. Biliaderis, C. M. Page, T. J. Maurice and B. O. Juliano, *J. Agric. Food Chem.*, 1986, **34**, 6.
99. P. Orford, R. Parker, S. Ring and A. Smith, *Int. J. Biol. Macromol.*, 1989, **11**, 91.
100. M. Kalichevsky, E. Jaroszkiewicz and J. Blanshard, *Int. J. Biol. Macromol.*, 1992, **14**, 257.
101. P. Liu, L. Yu, X. Wang, D. Li, L. Chen and X. Li, *J. Cereal Sci.*, 2010, **51**, 388.
102. C. Biliaderis and H. Seneviratne, *Carbohydr. Polym.*, 1990, **13**, 185.
103. T. C. Crowe, S. A. Seligman and L. Copeland, *J. Nutr.*, 2000, **130**, 2006.
104. H. S. Guraya, R. S. Kadan and E. T. Champagne, *Cereal Chem.*, 1997, **74**, 561.
105. K. Kaur and N. Singh, *Food Chem.*, 2000, **71**, 511.
106. S. Ozcan and D. S. Jackson, *Starch Stärke*, 2002, **54**, 593.
107. F. Tufvesson, M. Wahlgren and A. C. Eliasson, *Starch Stärke*, 2003, **55**, 61.
108. F. Tufvesson, M. Wahlgren and A. C. Eliasson, *Starch Stärke*, 2003, **55**, 138.
109. S. Bhatnagar and M. A. Hanna, *Cereal Chem.*, 1994, **71**, 582.
110. G. Fanta, R. Shogren and J. Salch, *Carbohydr. Polym.*, 1999, **38**, 1.
111. M. C. Tang and L. Copeland, *Carbohydr. Polym.*, 2007, **67**, 80.
112. A. C. Eliasson, *Thermochim. Acta*, 1994, **246**, 343.
113. D. E. Hahn and L. H. Hood, *Cereal Chem.*, 1987, **64**, 81.
114. V. K. Villwock, A. C. Eliasson, J. Silverio and J. N. BeMiller, *Cereal Chem.*, 1999, **76**, 292.
115. G. G. Gelders, J. P. Duyck, H. Goesaert and J. A. Delcour, *Carbohydr. Polym.*, 2005, **60**, 379.
116. M. Godet, H. Bizot and A. Buléon, *Carbohydr. Polym.*, 1995, **27**, 47.
117. M. Godet, V. Tran, P. Colonna, A. Buleon and M. Pezolet, *Int. J. Biol. Macromol.*, 1995, **17**, 405.
118. A. C. Eliasson and N. Krog, *J. Cereal Sci.*, 1985, **3**, 239.

119. S. Raphaelides and J. Karkalas, *Carbohydr. Res.*, 1988, **172**, 65.
120. T. Riisom, N. Krog and J. Eriksen, *J. Cereal Sci.*, 1984, **2**, 105.
121. O. Nimz, K. Gessler, I. Usón, G. M. Sheldrick and W. Saenger, *Carbohydr. Res.*, 2004, **339**, 1427.
122. C. E. Snape, W. R. Morrison, M. M. Maroto-Valer, J. Karkalas and R. A. Pethrick, *Carbohydr. Polym.*, 1998, **36**, 225.
123. C. G. Oates, *Trends Food Sci. Technol.*, 1997, **8**, 375.
124. R. Hoover, *Food Rev. Int.*, 2000, **16**, 369.
125. Y.-J. Wang, V.-D. Truong and L. Wang, *Carbohydr. Polym.*, 2003, **52**, 327.
126. S.-J. Wang, W.-Y. Gao, J.-L. Yu and P.-G. Xiao, *Chin. Chem. Lett.*, 2006, **17**, 1255.
127. H. Singh, N. S. Sodhi and N. Singh, *Int. J. Food Prop.*, 2009, **12**, 713.
128. K. Singh Sandhu, N. Singh and S.-T. Lim, *LWT-Food Sci. Technol.*, 2007, **40**, 1527.
129. J.-p. Pang, S.-j. Wang, J.-l. Yu, H.-y. Liu, J.-g. Yu and W.-y. Gao, *Food Chem.*, 2007, **105**, 989.
130. R. G. Utrilla-Coello, C. Hernández-Jaimes, H. Carrillo-Navas, F. González, E. Rodríguez, L. A. Bello-Pérez, E. J. Vernon-Carter and J. Alvarez-Ramirez, *Carbohydr. Polym.*, 2014, **103**, 596.
131. J. Bae, D. Lee, D. Kim, S.-J. Cho, J. E. Park, S. Koh, J. Kim, B.-H. Park, Y. Choi and H.-J. Shin, *Process Biochem.*, 2005, **40**, 3707.
132. R. Schinzel and B. Nidetzky, *FEMS Microbiol. Lett.*, 1999, **171**, 73.
133. S. C. Zeeman, J. Kossmann and A. M. Smith, *Annu. Rev. Plant Biol.*, 2010, **61**, 209.
134. O. Kötting, J. Kossmann, S. C. Zeeman and J. R. Lloyd, *Curr. Opin. Plant Biol.*, 2010, **13**, 320.
135. J. F. Robyt, in *Starch: Chemistry and Technology*, ed. J. N. BeMiller and R. L. Whistler, Academic Press, Burlington, MA, 3rd edn, 2009, p. 237.
136. C. Åkerberg, G. Zacchi, N. Torto and L. Gorton, *J. Chem. Technol. Biotechnol.*, 2000, **75**, 306.
137. S. Apinan, I. Yujiro, Y. Hidefumi, F. Takeshi, P. Myllärinen, P. Forssell and K. Poutanen, *Starch Stärke*, 2007, **59**, 543.
138. R. Tester, W. Morrison, M. Gidley, M. Kirkland and J. Karkalas, *J. Cereal Sci.*, 1994, **20**, 59.
139. G. Zhang, Z. Ao and B. R. Hamaker, *Biomacromolecules*, 2006, **7**, 3252.
140. M. Sujka and J. Jamroz, *Int. Agrophys.*, 2007, **21**, 107.
141. A. Lopez-Rubio, B. M. Flanagan, A. K. Shrestha, M. J. Gidley and E. P. Gilbert, *Biomacromolecules*, 2008, **9**, 1951.
142. V. Leloup, P. Colonna and S. Ring, *Biotechnol. Bioeng.*, 1991, **38**, 127.
143. M.-F. D. Hoebler, A. Karinthi, C. Belleville and J.-L. Barry, *Int. J. Food Sci. Nutr.*, 2000, **51**, 353.
144. W. F. Boron and E. L. Boulpaep, *Medical Physiology*, Elsevier Saunders, Philadelphia, PA, 2nd edn, 2012.
145. S. Macfarlane and G. T. Macfarlane, *Appl. Environ. Microbiol.*, 2006, **72**, 6204.

146. J. M. Wong, R. de Souza, C. W. Kendall, A. Emam and D. J. Jenkins, *J. Clin. Gastroenterol.*, 2006, **40**, 235.
147. K. N. Englyst, H. N. Englyst, G. J. Hudson, T. J. Cole and J. H. Cummings, *Am. J. Clin. Nutr.*, 1999, **69**, 448.
148. H. N. Englyst, S. Kingman and J. Cummings, *Eur. J. Clin. Nutr.*, 1992, **46**, S33.
149. E. Fuentes-Zaragoza, M. Riquelme-Navarrete, E. Sánchez-Zapata and J. Pérez-Álvarez, *Food Res. Int.*, 2010, **43**, 931.
150. M. Sajilata, R. S. Singhal and P. R. Kulkarni, *Compr. Rev. Food Sci. Food Saf.*, 2006, **5**, 1.
151. K. Srikaeo, S. Mingyai and P. A. Sopade, *Int. J. Food Sci. Technol.*, 2011, **46**, 2111.
152. K. Srikaeo and J. Sangkhiaw, *LWT-Food Sci. Technol.*, 2014, **59**, 1129.
153. W. Ratnayake and D. Jackson, *J. Food Sci.*, 2008, **73**, C356.
154. S. Haralampu, *Carbohydr. Polym.*, 2000, **41**, 285.
155. A. Nugent, *BNF Nutr. Bull.*, 2005, **30**, 27.
156. J. Lunn and J. Buttriss, *BNF Nutr. Bull.*, 2007, **32**, 21.
157. J. N. BeMiller, *Starch Stärke*, 1997, **49**, 127.
158. O. B. Wurzburg, *Nutr. Rev.*, 1986, **44**, 74.
159. R. N. Tharanathan, *Crit. Rev. Food Sci. Nutr.*, 2005, **45**, 371.
160. B. Kaur, F. Ariffin, R. Bhat and A. A. Karim, *Food Hydrocolloids*, 2012, **26**, 398.
161. A. Rajan, V. Prasad and T. Emilia Abraham, *Int. J. Biol. Macromol.*, 2006, **39**, 265.
162. A. Rajan, J. Sudha and T. E. Abraham, *Ind. Crops Prod.*, 2008, **27**, 50.
163. L. Qiao, Q.-M. Gu and H. Cheng, *Carbohydr. Polym*, 2006, **66**, 135.
164. S. Chakraborty, B. Sahoo, I. Teraoka, L. M. Miller and R. A. Gross, *Macromolecules*, 2005, **38**, 61.
165. P. Deetae, S. Shobsngob, W. Varanyanond, P. Chinachoti, O. Naivikul and S. Varavinit, *Carbohydr. Polym.*, 2008, **73**, 351.
166. S. Varavinit, W. Paisanjit, T. Tukomane and C. Pukkahuta, *Starch Stärke*, 2007, **59**, 290.
167. J. M. Li and L. M. Zhang, *Starch Stärke*, 2007, **59**, 418.
168. S. Ou and A. L. Yang, *Food Chem.*, 2001, **74**, 91.
169. G. X. Xing, S. F. Zhang, B. Z. Ju and J. Z. Yang, *Starch Stärke*, 2006, **58**, 464.
170. A. N. Jyothi, K. N. Rajasekharan, S. N. Moorthy and J. Sreekumar, *Starch Stärke*, 2005, **57**, 556.
171. A. Čížová, I. Srokova, V. Sasinková, A. Malovikova and A. Ebringerova, *Starch Stärke*, 2008, **60**, 389.
172. A. Karim, E. Sufha and I. Zaidul, *J. Agric. Food Chem.*, 2008, **56**, 10901.
173. H. T. Chan, R. Bhat and A. A. Karim, *J. Agric. Food Chem.*, 2009, **57**, 5965.
174. H. An and J. King, *J. Food Sci.*, 2009, **74**, C278.
175. H. T. Chan, C. P. Leh, R. Bhat, C. Senan, P. A. Williams and A. A. Karim, *Food Chem.*, 2011, **126**, 1019.

176. C. Pukkahuta, S. Shobsngob and S. Varavinit, *Starch Stärke*, 2007, **59**, 78.
177. J. Szymońska, F. Krok and P. Tomasik, *Int. J. Biol. Macromol.*, 2000, **27**, 307.
178. J. Szymońska, F. Krok, E. Komorowska-Czepirska and K. Rebilas, *Carbohydr. Polym.*, 2003, **52**, 1.
179. I. Zarguili, Z. Maache-Rezzoug, C. Loisel and J.-L. Doublier, *J. Food Eng.*, 2006, **77**, 454.
180. Z. Maache-Rezzoug, T. Maugard, I. Zarguili, E. Bezzine, M.-N. El Marzouki and C. Loisel, *J. Cereal Sci.*, 2009, **49**, 346.
181. Z.-Q. Huang, J.-P. Lu, X.-H. Li and Z.-F. Tong, *Carbohydr. Polym.*, 2007, **68**, 128.
182. L.-M. Che, D. Li, L.-J. Wang, X. Dong Chen and Z.-H. Mao, *Int. J. Food Prop.*, 2007, **10**, 527.
183. Z. Han, X.-a. Zeng, B.-s. Zhang and S.-j. Yu, *J. Food Eng.*, 2009, **93**, 318.
184. M. R. Nemtanu and R. Minea, *Macromol. Symp.*, 2006, **245-246**, 525.
185. S.-T. Lim, J.-A. Han, H. Lim and J. BeMiller, *Cereal Chem.*, 2002, **79**, 601.
186. G. Lewandowicz and M. Soral-Śmietana, *Carbohydr. Polym.*, 2004, **56**, 403.
187. P. A. Steeneken and A. J. Woortman, *Food Hydrocolloids*, 2009, **23**, 394.
188. R. González, C. Carrara, E. Tosi, M. Añón and A. Pilosof, *LWT-Food Sci. Technol.*, 2007, **40**, 136.
189. T. Kaper, M. van der Maarel, G. Euverink and L. Dijkhuizen, *Biochem. Soc. Trans.*, 2004, **32**, 279.
190. M. J. van der Maarel, I. Capron, G. J. W. Euverink, H. T. Bos, T. Kaper, D. J. Binnema and P. A. Steeneken, *Starch Stärke*, 2005, **57**, 465.
191. E. Oh, S. Choi, S. Lee, C. Kim and T. Moon, *J. Food Sci.*, 2008, **73**, C158.
192. M. R. Hansen, A. Blennow, S. Pedersen, L. Nørgaard and S. B. Engelsen, *Food Hydrocolloids*, 2008, **22**, 1551.
193. A. C. Alting, F. Van de Velde, M. W. Kanning, M. Burgering, L. Mulleners, A. Sein and P. Buwalda, *Food Hydrocolloids*, 2009, **23**, 980.
194. E.-J. Kim, S.-I. Ryu, H.-A. Bae, N. T. Huong and S.-B. Lee, *Food Chem.*, 2008, **110**, 979.
195. J.-H. Auh, H. Y. Chae, Y.-R. Kim, K.-H. Shim, S.-H. Yoo and K.-H. Park, *J. Agric. Food Chem.*, 2006, **54**, 2314.
196. D. S. Ludwig, *JAMAAP*, 2002, **287**, 2414.
197. Z. Ao, S. Simsek, G. Zhang, M. Venkatachalam, B. L. Reuhs and B. R. Hamaker, *J. Agric. Food Chem.*, 2007, **55**, 4540.
198. Y.-K. Kim and J. F. Robyt, *Carbohydr. Res.*, 1999, **318**, 129.
199. Y.-K. Kim and J. F. Robyt, *Carbohydr. Res.*, 2000, **328**, 509.
200. J. P. Davis, N. Supatcharee, R. L. Khandelwal and R. N. Chibbar, *Starch Stärke*, 2003, **55**, 107.
201. C. J. Slattery, I. H. Kavakli and T. W. Okita, *Trends Plant Sci.*, 2000, **5**, 291–298.
202. A. Viksø-Nielsen, A. Blennow, K. Jørgensen, K. H. Kristensen, A. Jensen and B. L. Møller, *Biomacromolecules*, 2001, **2**, 836.

203. B. Wischmann, A. Blennow, F. Madsen, K. Jørgensen, P. Poulsen and O. Bandsholm, *Food Hydrocolloids*, 2005, **19**, 1016.
204. U. Ihemere, D. Arias-Garzon, S. Lawrence and R. Sayre, *Plant Biotechnol. J.*, 2006, **4**, 453.
205. S. A. Jobling, G. P. Schwall, R. J. Westcott, C. M. Sidebottom, M. Debet, M. J. Gidley, R. Jeffcoat and R. Safford, *Plant J.*, 1999, **18**, 163.
206. R. Safford, S. A. Jobling, C. M. Sidebottom, R. J. Westcott, D. Cooke, K. J. Tober, B. H. Strongitharm, A. L. Russell and M. J. Gidley, *Carbohydr. Polym.*, 1998, **35**, 155.
207. T. Verhoeven, B. Fahy, M. Leggett, G. Moates and K. Denyer, *J. Cereal Sci.*, 2004, **40**, 69.
208. F. F. Dias, *J. Sci. Ind. Res.*, 1999, **58**, 403.
209. M. Burrell, *J. Exp. Bot.*, 2003, **54**, 451.
210. Corn Refiners Association, *Corn Starch*, Corn Refiners Association, Washington, DC, 12th edn, 2013.

CHAPTER 3

# *Preparation and Characterization of Starch Nanocrystals*

MEHRAN GHASEMLOU,*[a]
SEYED MOHAMMAD TAGHI GHARIBZAHEDI[b] AND
MARLENE J. CRAN[c]

[a] School of Packaging, Michigan State University, East Lansing, MI 48824, USA; [b] Department of Food Science, Engineering and Technology, Faculty of Agricultural Engineering and Technology, University of Tehran, Karaj, Iran; [c] Institute for Sustainability and Innovation, Victoria University, PO Box 14428, Melbourne, Victoria 8001, Australia
*Email: ghasemlo@msu.edu

## 3.1 Introduction

Packaging derived from petroleum-based plastic materials is extensively used in various applications because of the low density, high mechanical and barrier properties and low cost. Although many types of petroleum-based packaging plastics can be recovered and recycled, large volumes of these materials end up in landfill or the environment. As a result, increasing environmental concerns have led to the development of new flexible barrier biobased packaging products as potential replacements for these materials. In addition, the depletion of fossil fuel energy sources and subsequent increasing costs of crude oil have enhanced the global interest in biobased materials.[1] It is estimated that if petroleum continues to be consumed at the

RSC Green Chemistry No. 37
Starch-based Blends, Composites and Nanocomposites
Edited by Visakh P. M. and Long Yu

Published by the Royal Society of Chemistry, www.rsc.org

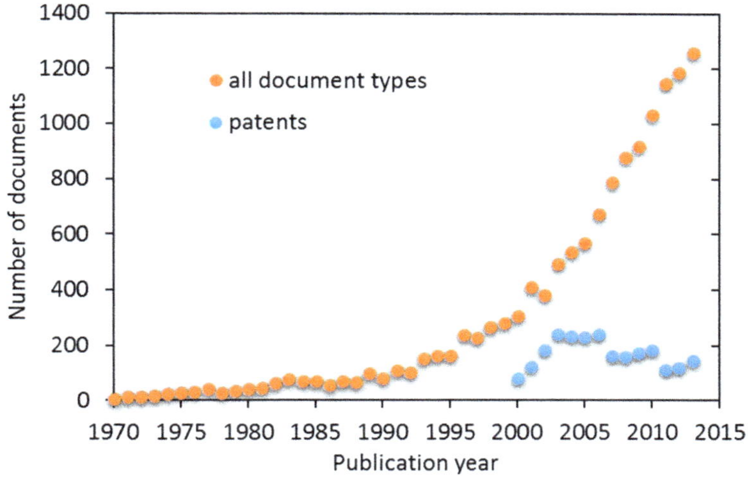


**Figure 3.1** Number of publications containing the term "biopolymer" in all documents and patents.
Source: www.scopus.com.

current rate, it will potentially be depleted within the next 50–60 years.[2] To address the challenges arising from plastic waste and diminishing resources, the development of new and sustainable materials has been extensively investigated in recent decades. Figure 3.1 shows the rising scientific interest in this field based on the total number of publications with the term "biopolymer" in the title, keywords or abstract. It is clear that a steady rise in the number of publications occurred in the last century with an upward trend that began in the mid to late 1990s. The current rate of published research in this field is now almost exponential. The number of patents with the same term in the title is also shown with what appears to be a peak around 10 years ago, with a more stepped profile suggesting more variable interest in the patenting of new biopolymer materials.

Considerable efforts are also being directed towards developing biodegradable composites using renewable resources in an attempt to replace petroleum-based plastics with more sustainable materials.[2] Edible coatings and films comprise a relatively new category of packaging materials that offer alternatives to conventional packaging owing to their excellent biodegradability, biocompatibility and edibility and the extensive range of their potential applications.[3] Although a wide range of naturally derived materials have been investigated for use as packaging plastics, researchers have identified some commonly occurring factors that restrict their broader industrial application. For instance, some biopolymer materials are known to have relatively poor mechanical properties compared with many petroleum-based plastics, owing mainly to the inherent lower stiffness and strength.[4] Modification of these materials can often improve these properties, but in some cases modified materials can be sensitive to water with a significant decline in mechanical properties especially in moist environments.[5] Low production levels, competition with food crops and high costs are factors that can also limit the wider industrial application of biopolymer packaging materials. Therefore, researchers are trying to find solutions to improving the mechanical and/or barrier properties of biobased films to compete with synthetic plastics.

One of the more promising options to improve the barrier and mechanical properties of packaging materials is by the use of nanocomposites based on materials of either synthetic or natural origins.[6] In recent years, research into polymer nanocomposites has seen significant growth in both industrial and academic laboratories because they can possess many desirable properties, including remarkable improvements in mechanical properties, enhanced barrier properties and reduced flammability.[7] With the development of biodegradable materials, polysaccharides and nanocrystals/nanoparticles in particular are attracting more scientific interest. These materials are among the more promising sources for the production of nanoparticles owing to their semicrystalline structure, among other properties. Cellulose nanocrystals are perhaps the most studied polysaccharide for nanoparticles production, with a number of published reviews on the synthesis and properties of these materials.[8,9] In addition to cellulose, starch is another promising candidate for the production of nanoparticles because it also has a semicrystalline structure that confers some interesting properties.

Cellulose and starch are increasingly viewed as potential sustainable alternatives to many petroleum-based polymers, as reported recently.[1,10] However, given the broad interest and increasing number of publications in this field, there are potentially new developments in methods of polysaccharide-based nanocrystal preparation and characterization. This chapter reviews these recent advances and discusses potential applications of these materials and also suggestions regarding the direction of future research. Although a brief summary of starch nanocrystal structure is given, it is not intended to represent a complete review of the literature on this topic and readers are referred to other recent comprehensive reviews and book chapters published on this aspect of starch nanomaterials.

## 3.2 Starch

Starch is an abundant, inexpensive, renewable and biodegradable polymer. It is the second most abundant natural polymer after cellulose and can be obtained mainly from the roots, stalks and seeds of staple crops such as rice, corn, wheat, tapioca and potato.[11] In all of these plants, starch is produced in the form of granules whose size, shape, morphology and composition vary from plant to plant.[12] It has been reported that the United States, followed by Europe, are the two largest producers of starch in the world and contribute about half of the globally produced commercial starch.[13] Worldwide, the main sources of starch are maize (82%), wheat (8%), potatoes (5%) and cassava (5%), from which tapioca starch is derived.[14] Native starch is naturally composed of nano-sized blocklets that have semicrystalline arrangements of starch chains.[15] It is now well established that the crystalline regions are created as thin lamellar domains by the intertwining of amylopectin side chains above 10 glucose units to form double helices. These double helices are packed together to form the basis of crystalline domains.[16] The amorphous regions (starch molecules in a single-chain state) and ordered regions (starch molecules in a double-helix conformation) arrange to form a growth ring structure that encompasses the point of initiation of the granule, and they are easily visible by optical and electron microscopy.[17] Morphologically, starch consists of mostly microscopic spherical granules having diameters ranging from 2 to 100 μm, depending on the botanical origin, and with a density of 1.5.[18] Chemically, starch is composed of two main macromolecular components as presented in Figure 3.2. These are amylose, an essentially linear or slightly branched (1→4)-α-D-glucan, which can have a molecular mass as high as $10^6$ g mol$^{-1}$, and amylopectin, with molecular weights between $10^6$ and $10^7$ g mol$^{-1}$, which is a highly branched structure of short (1→4)-α-D-glucan chains linked through α-(1→6) bonds.[19] In most common types of starch, the amylose content is between 72 and 82% of the granules, with the amylopectin content ranging from 18 to 28%.[20] Morphologically, the branched amylopectin component consists of crystalline areas and the linear amylose is mostly amorphous or semicrystalline. Amylose is therefore soluble in hot water whereas amylopectin is not.[21] In addition, small starch granules can mimic lipid micelles, thereby providing a fat-like texture in some products.[22]

Industrially, starch is extracted and refined from seeds, roots and tubers by wet grinding, washing, sieving and drying. After its extraction from plants, starch is in the form of a flour-like white powder and is called "native starch," or it can be subjected to one or more chemical modifications to meet some specific properties, after which it is called "modified starch."[1,10] Native starches are categorized by three crystalline types called A, B and C according to their X-ray diffraction (XRD) pattern, reflecting long-range ordering in the granule.[23] Hizukuri *et al.*[24] suggested that amylopectin chain length was a main factor for crystalline polymorphism. To explain the difference between A and B types starches, Imberty *et al.*[25] proposed a model for the double-helix packing configuration. A transition from B-type to A-type was observed by rearrangement of the pairs of double helices. It is known that the A-type adopts a close-packed arrangement with water molecules between each double-helix structure, whereas the B-type is more open and water molecules are located in the central cavity formed by six double helices.[26] The C-type is a mixture of A- and B-types since its XRD pattern can be resolved as a combination of the other two, and it can be found in bean starch.[27] Gallant *et al.*[16] reported that the blocklets in B- and C-type crystalline starches are generally larger (400–500 nm) than those in the A-type (25–100 nm). Bogracheva *et al.*[28] studied the structure and properties of C-type starch from pea seeds. They suggested that C-type starch granules include both types of polymorph, with the B-type at the centre of the granule

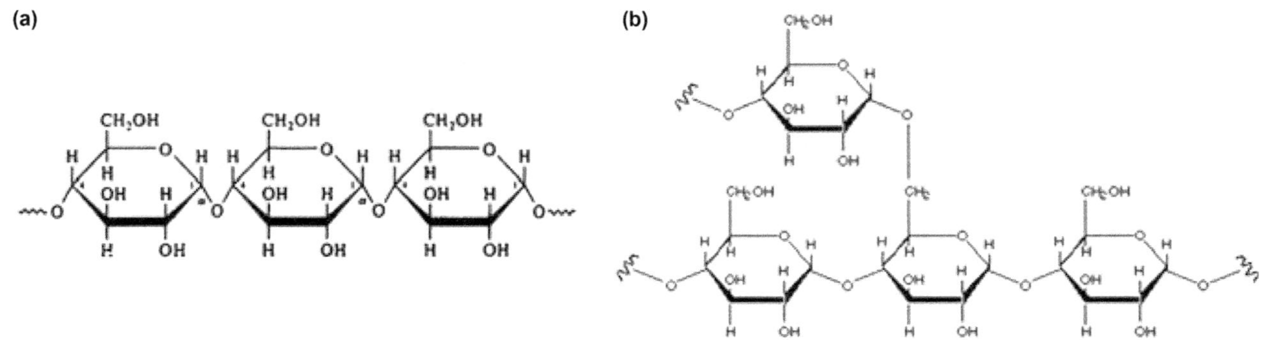

**Figure 3.2** Structures of (a) amylose and (b) amylopectin. Reprinted from Vazquez *et al.*[32] with permission of Springer-Verlag London.

and the A-type at the surrounding. Kim *et al.*[29] reported that A- or C-type crystals in native starches are more resistant to acid hydrolysis than B-type. In general, amylose and the branching chains of amylopectin form amorphous regions whereas the short-branched chains in amylopectin are the main crystalline component in granular starch. It is now well known that the partial crystallinity of native starch granules is due to a clustered organization of amylopectin side chains. The amylose to amylopectin ratio depends on the source and age of the starch and it can also be monitored by the extraction process used.[30] In recent decades, considerable efforts were made to explain the fine structure of the molecules that make up starch granules, but many details still remain unknown.[25,31,32] Starch has a very complex hierarchical structure, which can be better described if divided into the different levels of organization as illustrated in Figure 3.3. Starch granules can be gelatinized in water at lower temperatures in alkaline solutions and can be employed as a thickening, stiffening and gluing agent, giving wheat paste. However, they are not always suitable for many industrial applications owing to functional limitations. These natural limitations can be substantially solved using a variety of modifications, including physical (heat–moisture treatment, annealing, pregelatinization, high-pressure treatment, radiation, sonication), chemical (crosslinking, substitution, acid hydrolysis, oxidation/bleaching) or enzymatic processes.[26]

Many papers have reported the development of starch-based polymers for reducing the environmental impact and in order to broaden the range of

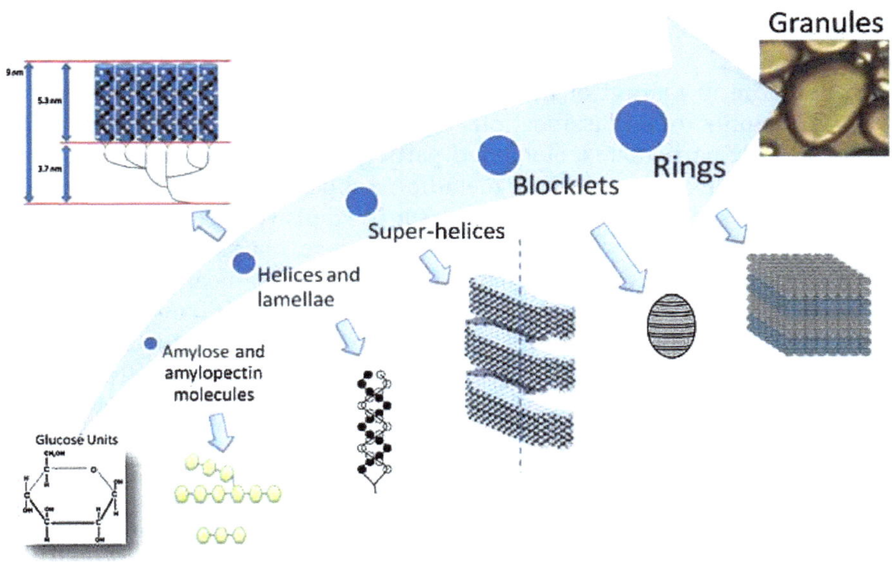


**Figure 3.3** Starch structure at different levels of organization.
Reprinted from Vazquez *et al.*[32] with permission of Springer-Verlag London.

applications for this type of biopolymer.[33–35] There are two important features in starch molecules that make it a unique and promising material. First, starch is naturally biodegradable in soil and water and degrades into sugars and organic acids, which is of great advantage from an environmental point of view.[36] Second, owing to its unique chemical characteristics, starches are amenable to a variety of chemical and enzymatic modifications that allow the development of new functionalities in starch polymers. However, the strongly hydrophilic character of starch makes it susceptible to moisture attack, resulting in changes in dimensional stability and poor mechanical properties. In addition, retrogradation and crystallization of the mobile starch chains can lead to undesirable changes in mechanical and thermo mechanical properties,[37] which can render the material unsatisfactory for many packaging applications. Although some of these problems can potentially be solved with the addition of plasticizers, they cannot meet all the requirements of packaging applications. In a recent study, we prepared corn starch films incorporated with two types of essential oils, from *Zataria multiflora* Boiss. and *Mentha pulegium*. The film properties were improved markedly through emulsification and the water-vapour barrier was further modified.[38] However, even after this modification, the physical and mechanical properties of these films were still not suitable for general packaging applications. To compete with petroleum-based plastics, starch films should have comparable mechanical and/or barrier properties. Therefore, further improvement is needed to improve these physical and mechanical properties.

Currently, one of the most effective alternatives for improving the barrier and mechanical properties of packaging materials, either synthetic or natural, is by the formation of nanocomposites.[7] Apart from size, one of the more commonly used classifications of nanoparticles is based on particle shape, namely particulates, elongated particles and layered structures. Particulates include polymeric and metallic nanoparticles with a defined or regular shape such as carbon black, which is extensively used in reinforced composites. Elongated particles include cellulose nanofibrils (also called whiskers or nanocrystals) that impart acceptable mechanical properties due to their high aspect ratio. Layered particles include starch nanocrystals and nanoclays, with the latter belonging to the most common industrial type of nanoparticle used to form layered polymer nanocomposites. Nanoclays undergo three different degrees of dispersion or filling in nanocomposite materials: intercalated nanocomposites (intercalated polymer chains between layered nanocomposites), exfoliated nanocomposites (separation of individual layers) and flocculated or phase-separated nanocomposites, which are also called microcomposites, and consequently impart poorer mechanical properties. The packaging industry has focused its attention mainly on layered inorganic solids such as clays and silicates owing to their commercial availability, low cost, significant enhancements and relatively simple processability.

## 3.3 Synthesis of Starch Nanocrystals

A range of terms are used to describe the various forms of starch that are referred to in the literature, including starch crystallites, starch nanocrystals, microcrystalline starch and hydrolysed starch. These all refer to the crystalline part of starch obtained by hydrolysis although to different extents, and it is worth emphasizing that they are different from starch nanoparticles, which can be amorphous and are not included in this chapter. In native starch granules, the crystalline and amorphous regions coexist and form an onion-like structure. During the clustering organization process, the amylopectin chains spiral due to the stacking of nanometric subunits forming the crystalline regions and, since the starch granule is inherently composed of nanoscale crystalline blocklets, there is the possibility to form starch nanocrystals.[16] By subjecting native starch to acid hydrolysis, regions of low lateral order and also amorphous phases in the starch granules starting to dissolve, whereas the highly crystalline water-insoluble lamellae remain undissolved.[39] Le Corre *et al.*[40] conducted an experiment to determine whether starch from many sources could be used to prepare starch nanocrystals and if the amylose content and/or botanical origin of the starch influenced their final properties. The results indicated that for the same amylose content but different sources, maize, potato and wheat starches showed similar crystal sizes. However, differences in size were more pronounced when sourced from the same botanical origin but with different amylose contents, suggesting the strong influence of the composition and molecular structure on the resulting crystallite dimensions.

Starch nanocrystals are reported to be derived from starch granule crystallites[41] and result from the disruption of the semicrystalline structure of native starch granules at temperatures below the gelatinization temperature. Under these conditions, the amorphous regions in starch granules are hydrolysed, which allows the separation of nanoscale crystalline residues.[14,42] This can be attributed to the higher resistance to hydrolysis of the crystalline lamellae than the amorphous lamellae by either chemicals or enzymes. Owing to the difference in acid susceptibility, crystalline dextrans in starch granules are able to be formed by the mild acid hydrolysis of the amorphous regions.[43] There appears to be little or no variation in the reported hydrolysis temperatures used, which are usually between 35 and 45 °C. The main reason for this low temperature range may be to prevent starch gelatinization and destruction of the starch crystalline structure.[44]

Starch nanocrystals of different sizes and shapes can be obtained depending on the starch origin and isolation process. Xu *et al.*[45] prepared starch nanocrystals from corn, barley, potato, tapioca, chickpea and mung bean starches using an acid hydrolysis method. Although the shape and size of the granules varied between starches, there was no obvious difference in shape among different types of starch nanocrystals. They found that the structure and morphology of starch nanocrystals depends on many factors, such as the botanical origin, including the crystalline type, the relative proportion of amylose and amylopectin and the granule morphology.[45] Putaux *et al.*[42] reported that the morphology of nanocrystals from waxy maize starch results from the acid hydrolysis of the granules. Their study suggested that they have a platelet-like shape from transmission electron microscopy observations, with a thickness of 5–7 nm and diameters ranging from 15 to 40 nm. However, in the case of other sources of starch, nanocrystals of different sizes and shapes were obtained. For example, Chen *et al.*[46] observed round and grape-like nanocrystals obtained from potato starch granules with sizes ranging from 40 to 100 nm. Liu *et al.*[47] carried out an experiment to reduce the size of starch granules from the micro- to the nanoscale using a high-pressure homogenization method and found that the starch particle size could be reduced from 3–6 μm to 10–20 nm after 20 passes.

Because of its unique properties, such as the nanoscale platelet morphology, intrinsic rigidity, high crystallinity and low permeability, starch nanocrystals have been compounded as reinforcements with many different kinds of polymer matrices to prepare nanocomposites.[48] Mélé *et al.*[49] studied the process to obtain nanocomposite materials consisting of natural rubber filled with waxy maize starch nanocrystals. Angellier *et al.*[50] employed starch nanocrystals in natural rubber composites and found a remarkable enhancing effect, but when the starch nanocrystal content exceeded 20%, the enhancement decreased. Sorbitol-plasticized pullulan films containing waxy maize starch nanocrystals showed increased mechanical and water barrier properties.[51] This effect was attributed to the establishment of strong interactions between the starch nanocrystals and between the filler and the matrix.[51] Interesting findings were also reported for starch nanocrystal-reinforced/plasticized starch films using glycerol[52] and sorbitol plasticizers.[37] Zheng *et al.*[53] reported an increase in strength and Young's modulus, and also a decrease in elongation at break, when the starch nanocrystal loading levels were lower than 2 wt% in a soy protein isolate (SPI) matrix. These results can be explained by the strong interfacial interaction between the starch nanocrystal filler and the SPI matrix. Bouthegourd *et al.*[54] reported the extraction and characterization of potato starch nanocrystals and their nanocomposites with a natural rubber latex matrix, the preparation being performed using sulfuric acid at 40 °C. The structure and the performance of the nanocomposites were investigated by scanning electron microscopy, atomic force microscopy and Fourier transform infrared (FTIR) spectroscopy and indicated a relatively uniform dispersion of starch nanocrystals when incorporated in the rubber latex.

### 3.3.1 Preparation Protocols

Perhaps one of the first studies on starch nanocrystals was conducted using cellulose whiskers with the aim of using these as reinforcing fillers in a matrix. Dufresne *et al.*[55] reported a procedure for producing what they termed "microcrystalline starch," which was actually agglomerated particles of a few tens of nanometres in diameter. The method consisted of hydrolysing starch (5 wt%) in a 2.2 M HCl suspension for 15 days. In the following years, two main challenges for the development and use of starch nanocrystals were identified. The first was related to its limited large-scale production, with the aforementioned protocol not capable of producing large quantities of starch nanocrystals, with a limited yield (10–15%) after a long production time (5 days). The second challenge was related to the hydrolysis of starch with respect to its onion-like structure.

Le Corre *et al.*[56] showed that starch nanocrystals are mixed with other microparticles and that some nanocrystals were converted to sugar by the end of the batch production process, resulting in the observed low yields. They proposed that the extent of hydrolysis may reflect the amount of starch nanocrystals in the final suspension. To evaluate the kinetics of hydrolysis, they explored the formation of nanocrystals for up to 5 days to optimize the protocol. The hydrolysis was terminated after 1, 3 and 5 days, as shown in Figure 3.4, after which the yield was assessed. It was concluded that at any time, both micro- and nanoscale particles are present and that differential centrifugation is not suitable for fractionation due to hydrogen bonding and different densities within starch granules. This study recommended that an isolation process should be adopted in the preparation method; however, differential centrifugation as the washing step and isolation technique was unsuccessful.[56] It was concluded that a continuous extraction technique such as microfiltration is needed for separation.

In a recent study, however, Kim *et al.*[29] claimed that obtaining individual nanoparticles from starch was almost impossible regardless of the origin of the starch. In another study by the same group,[57] a hydrolysis process combined with a physical treatment such as ultrasonication for the formation of a uniform dispersion of starch nanocrystals was investigated. They concluded that ultrasonic treatment at 60% vibration amplitude for 3 min could be effective in breaking up the aggregates of nanoparticles formed through hydrogen bonds. Although ultrasonication can effectively prevent the aggregation of nanoparticles occurring during the hydrolysis, starch nanocrystals treated by ultrasonication may have decreased crystallinity.

Le Corre *et al.*[58] developed an innovative solution to address the limitations of current processes for generating starch nanocrystals. They considered filtration of hydrolysates obtained from wheat starch using a microfiltration unit equipped with ceramic membranes. As shown in Figure 3.5, microfiltration was a superior process for limiting the conversion of starch nanocrystals into oligo- or monosaccharides. They reported that not only were the collected nanoparticles more crystalline, but also that mostly B-type particles were produced during the first day of hydrolysis. The critical factor in the preparation of starch nanocrystals is to maintain the starch crystalline structure while completely removing the amorphous region of starch granules.

Putaux *et al.*[42] reported the optimization of hydrolysis conditions for the extraction of starch nanocrystals with HCl and $H_2SO_4$ and investigated its

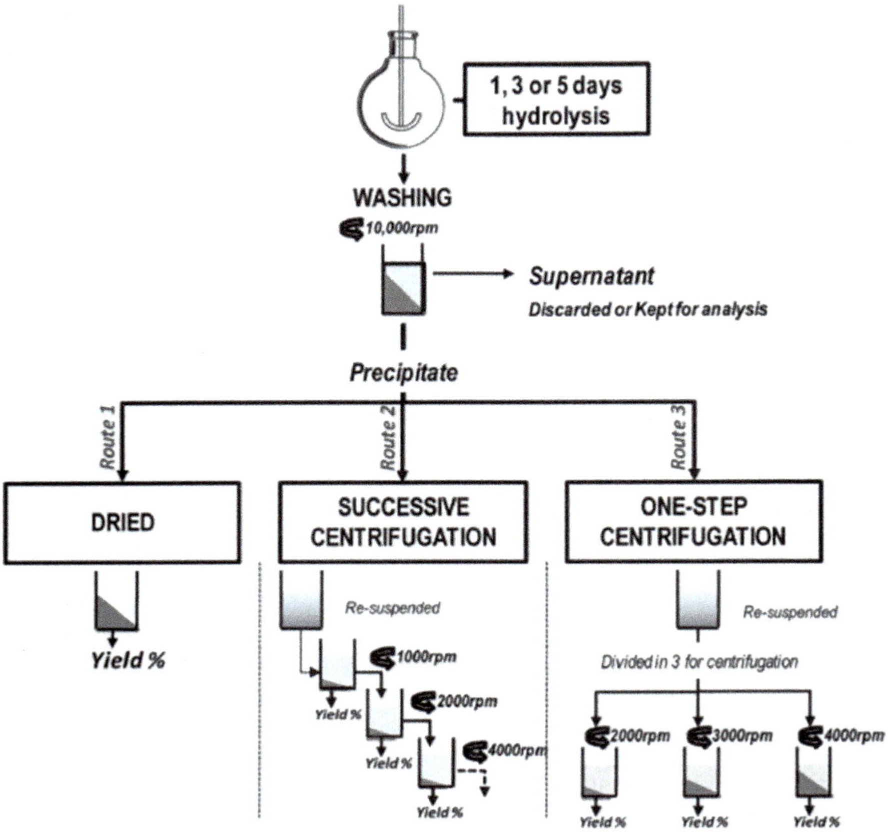


**Figure 3.4** Schematic representation of the analysis of starch nanocrystals obtained from hydrolysis for 1, 3 or 5 days *via* three different routes. Adapted from Le Corre *et al.*[56] with permission of the American Chemical Society.

effects on morphology and yield. The extraction process for starch nanocrystals by acid hydrolysis is illustrated in Figure 3.6. They concluded that although HCl hydrolysis gave individual starch nanoplatelets, the long treatment time and low yield limit its application as a nanofiller in nanomaterials. Angellier *et al.*[14] developed an optimized process using $H_2SO_4$ by response surface methodology (RSM). A given weight of native waxy maize with a concentration of 14.69% in 3.16 M $H_2SO_4$ was hydrolysed at 40 °C under constant stirring at 100 rpm. When the hydrolysis processes was finished, the insoluble residue was washed by successive centrifugations with distilled water for 10 min at 10 000 rpm until neutral pH was achieved. The suspension was then processed using a batch disperser for 2 min at 13 000 rpm to redisperse the aggregates and obtain a "stable" suspension. It should be noted that using the $H_2SO_4$ hydrolysis method introduces sulfate groups on the surface of the starch nanocrystals, which contribute to

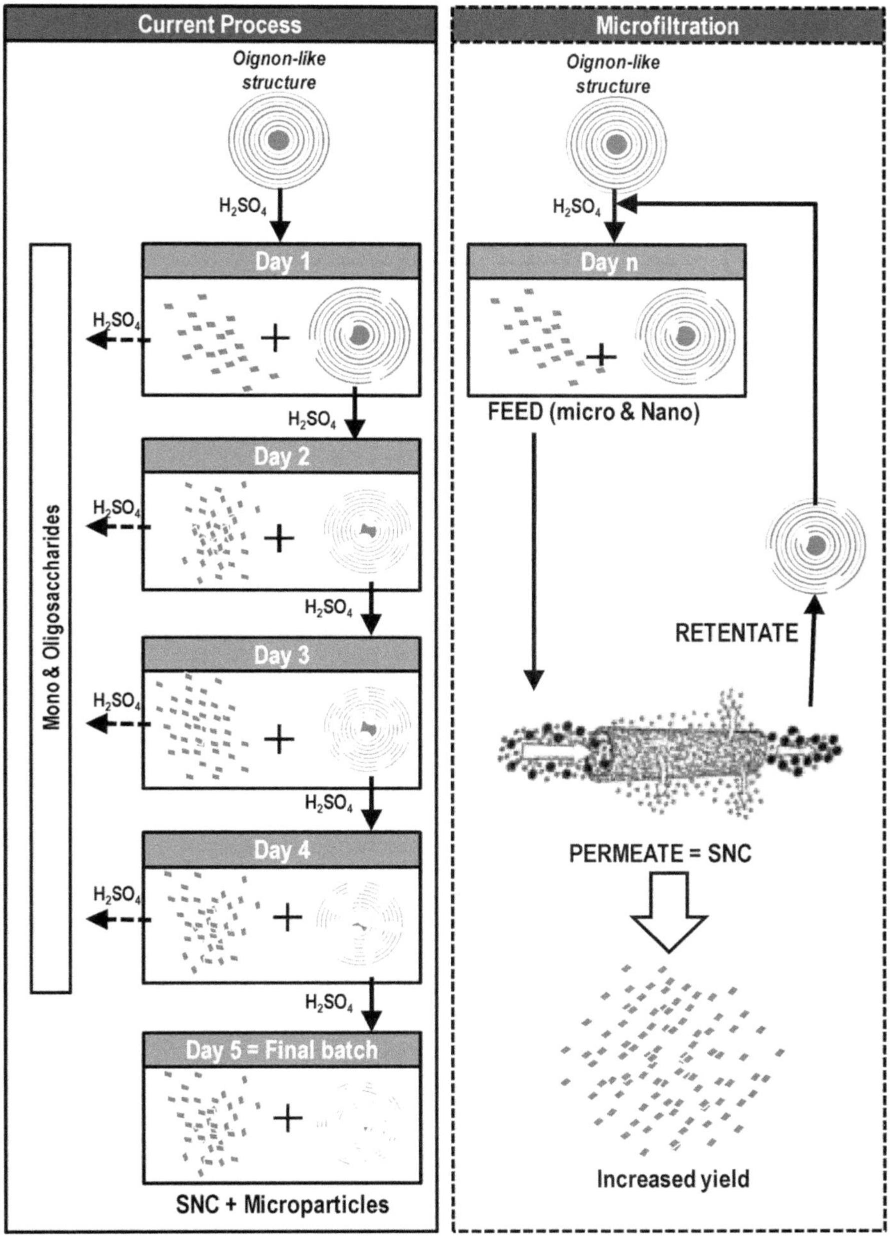


**Figure 3.5** Schematic representation of comparison between the current preparation process involving the progressive production of starch nanocrystals and the proposed microfiltration process.
Reproduced from Le Corre *et al.*[58] with permission of Elsevier.

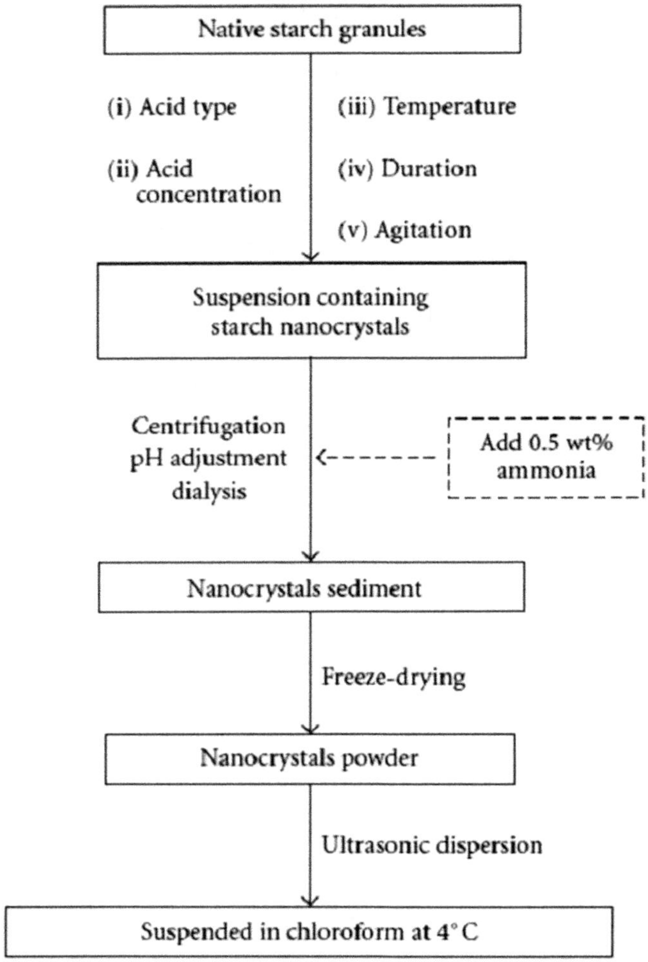


**Figure 3.6** Schematic diagram of procedure for extraction of starch nanocrystals. Adapted from Lin *et al.*[44]

dispersion and stability in aqueous solution; however, the thermal stability of the nanocrystals decreases. To solve this problem, a small amount of ammonia (0.5 wt%) was added to the $H_2SO_4$ nanocrystal suspension before centrifugation. It was concluded that it was possible to obtain starch nanocrystals after only 5 days of $H_2SO_4$ hydrolysis with a yield of 15 wt% and that they have the same shape as those obtained from the classical procedure after 40 days of HCl treatment at a much lower yield of 0.5 wt%.

Despite some improvements to reduce the production time significantly, the main drawbacks to the widespread use of starch nanocrystals continue to be the preparation time and the low yields. In other attempts to shorten the preparation time further and enhance the production yield after filtration, RSM analysis was employed in an effort to obtain starch nanocrystals after 1 day.[59] In another example, Jivan *et al.*[60] hydrolysed potato starch with 2.2 or 3.7 M HCl to obtain nanocrystals and revealed that when hydrolysis was carried out with 2.2 M HCl, the smallest sized crystals were obtained after 50 days whereas this occurred just after 24 days using 3.7 M HCl. Moreover, the stronger acidity not only decreased the time required to obtain the minimum size of particles but also led to much smaller crystals, the mean size of particles obtained using 2.2 M HCl at day 50 being 89 nm whereas the size of those obtained using 3.7 M hydrochloric acid at day 24 was 48 nm.[60] Wang *et al.*[61] observed that both amylose and amylopectin were hydrolysed to a greater extent on increasing the concentration of acid. In the few reports on starch nanocrystal production published over the past decade, only two main processes were used: that of Dufresne *et al.*[55] using HCl and the optimized process of Angellier *et al.*[14] using $H_2SO_4$.

### 3.3.2 Acid Hydrolysis

Acid hydrolysis is possibly the most common and optimized method to produce starch nanocrystals.[14] Acid treatment dissolves the regions of low lateral order to reveal the concentric lamellar structure of starch granules. By this approach, water-insoluble and highly crystalline residues may be converted into stable suspensions by a subsequent vigorous mechanical shearing action.[50] Acid hydrolysis has been known for many years to modify starch and its properties. In 1874, Nägeli[62] obtained a low molecular weight acid-resistant fraction after the hydrolysis of potato starch at room temperature for 30 days in a 15% $H_2SO_4$ suspension and named this fraction "amylo dextrin". In 1886, Lintner[63] carried out the hydrolysis of potato starch at 30–40 °C using 7.5% w/v HCl to generate a high molecular weight starch suspension which at that time was named "lintnerized starch." For all types of starches, a two-stage hydrolysis mechanism can be observed, with an initial fast hydrolysis step that is thought to correspond to the hydrolysis of the amorphous regions of starch granules, followed by a second, slower step due to the hydrolysis of the crystalline regions.[64] Other workers reported three stages, corresponding to the stepwise hydrolysis of the amorphous, semicrystalline and crystalline layers of the starch structure.[65]

There are two possible reasons to account for the slower hydrolysis rate of the crystalline regions. The first is that crystalline regions of starches are denser than the amorphous regions, which do not readily allow the rapid penetration of hydrogen ions into these regions. Second, the transformation of glucopyranose rings from the chair to half-chair conformation (required for hydrolysis of the glucosidic bonds) occurs slowly owing to immobilization of the sugars in the starch crystallites.[26] Starch acid hydrolysis has also been explained based on the molecular composition. Some workers suggested that high-amylose starches are more susceptible to acid hydrolysis than those with lower amylose contents, which are more easily hydrolysed.[66–68] This can be explained by either the greater extent of starch interchain associations in the amorphous regions, which are more compactly organized,[69] or by the slower penetration of hydrogen ions into the granules due to the limited swelling of high-amylose starch.[66] On the atomic scale, it was reported that α-1,6 bonds (found in amylopectin) are more resistant to hydrolysis than α-1,4 bonds.[57] When observed on the microscopic scale, using an optical or scanning electron microscope, acid hydrolysis occurs in three steps: (1) erosion of the surface, (2) collapse of the granule's wall due to radial diffusion of the acid and (3) fragmentation of the growth rings.[1,56]

In practice, the application of an acidic method is sometimes difficult to implement owing to the low yield, long treatment time and the use of concentrated acids.[70] Recent developments have therefore focused on the use of enzymes such as α-amylase as a pretreatment to enhance the efficiency of preparation. Kim *et al.*[71] prepared waxy rice starch nanocrystals with average diameters of 500 nm using selective enzymatic hydrolysis. In another study, they investigated starch nanocrystals prepared by the complex formation of high-amylose maize starch and *n*-butanol and subsequent enzymatic hydrolysis. Although they obtained spherical and oval nanocrystals with diameters of 10–20 nm, the significant loss by hydrolysis (around 85–90%) of the initial starch complex resulted in a poor total yield that was not enough for large-scale production.[72] In another attempt, Sun *et al.*[73] reported the preparation of starch nanocrystals by pullulanase debranching and recrystallization of native waxy maize starch. In this method, the use of chemical reagents was avoided and the yield was subsequently increased.

### 3.3.3 Chemical Modification

The unique properties of starch nanocrystals, such as biodegradability, outstanding mechanical properties, low density and low permeability, make it an ideal candidate to prepare nanocomposites within the matrices of natural polymers. However, the polar surface and hydrophilicity of starch nanocrystals limits their dispersibility in non-polar solvents and results in poor compatibility with hydrophobic polymers. Fortunately, they have reactive surfaces suitable for chemical derivatization and grafting reactions that can manipulate the surface hydrophobicity and facilitate their dispersion in water.[74] At first it was reported that it was difficult to decrease the surface energy of starch nanocrystals to improve their level of dispersion in a polymer matrix. However, it is now accepted that the nature of the hydroxyl groups in starch nanocrystals offers the possibility of modification *via* a chemical reaction strategy and accordingly the surface chemical modification of starch nanocrystals has been reviewed extensively in several publications.[75] The most common surface chemical modifications of starch nanocrystals are shown schematically in Figure 3.7.

There are several strategies that can be used to convert starch particles into nanocrystals, including the following: extraction by acid hydrolysis,[42,76,77] enzymatic hydrolysis[78,79] or co-crystallization during the regeneration process; *via* regeneration–precipitation[72,80,81] or extrusion,[70] where crosslinking can contribute to the stability of the nanoparticle structure;[9] the formation of nanocolloids by microfluidization;[47,82] and

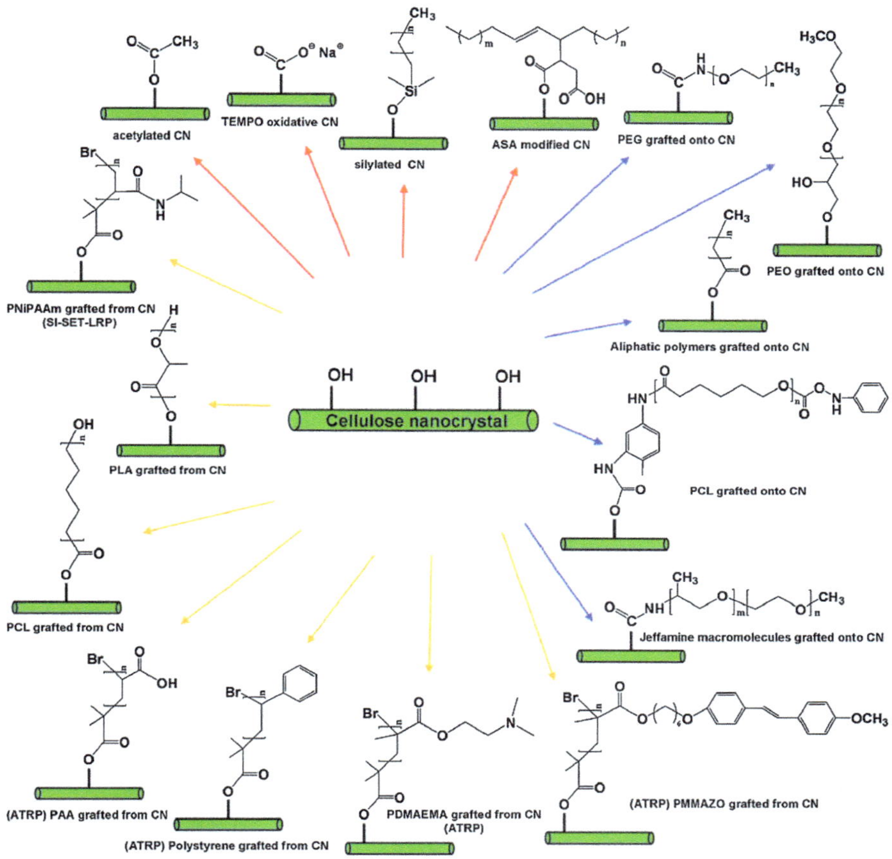


**Figure 3.7** Common chemical modifications of starch nanocrystals. PTHF = poly-tetrahydrofuran; PPGBE = poly(propylene glycol) monobutyl ether. Reproduced from Lin *et al.*[48] with permission of the Royal Society of Chemistry.

self-assembly of nanomicelles from chemically modified starch.[83,84] A summary of the different methods for the preparation of starch nanocrystals is shown in Figure 3.8.

Although recent progress in understanding the chemistry of starch and starch nanocrystals has motivated studies into reactions under heterogeneous conditions, some researchers have focused on the modification of starch nanocrystals in aqueous media rather than organic solvents such as toluene.[85–87] For example, Namazi and Dadkhah[77] modified starch nanocrystals hydrophobically with octanoic, nonanoic and decanoic acid by mild esterification using acyl chlorides in aqueous media *via* the reaction pathway shown in Figure 3.9. This method is promising as it offers a very convenient modification technique using water as an alternative to potentially toxic and expensive solvents. Following the reaction, chemical modification was confirmed by FTIR spectroscopy and wettability tests showed that modified

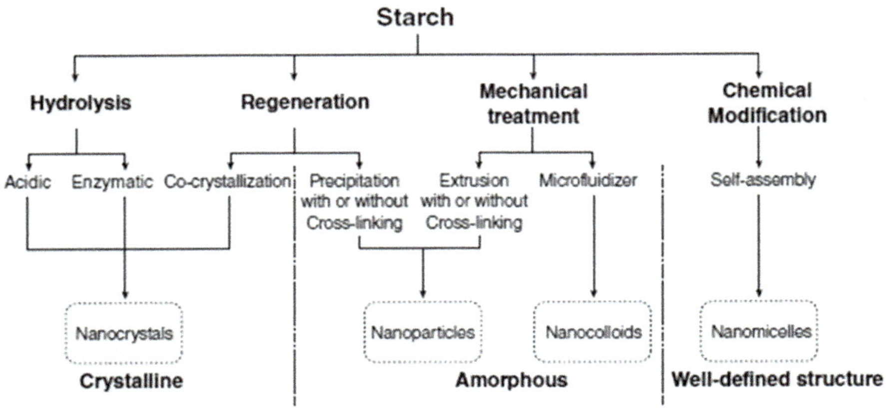


**Figure 3.8** Different methods for the preparation of starch nanoparticles. Reprinted from Huang *et al.*[75] with permission of Wiley-VCH.

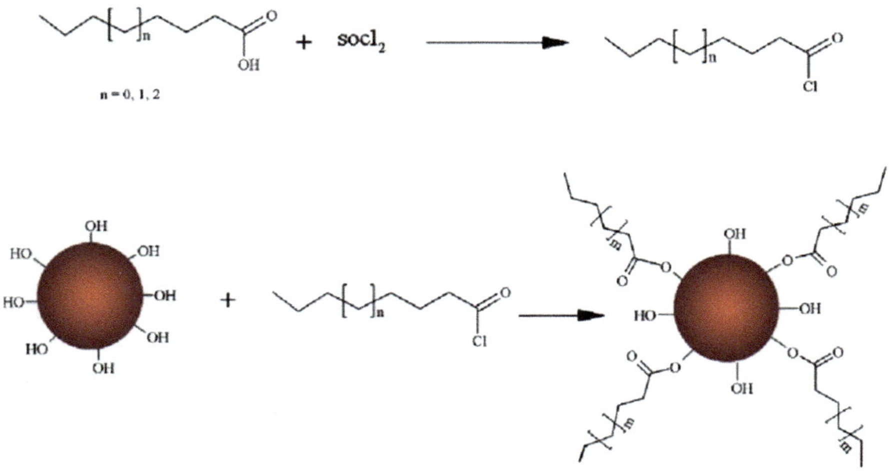


**Figure 3.9** Typical process for esterification of starch nanocrystals. Reproduced from Namazi and Dadkhah[77] with permission of Elsevier.

starch nanocrystals had a higher affinity towards organic solvents than aqueous media. As shown in Figure 3.10, native starch nanoparticles were observed after shaking the suspension in a chloroform solution, which was due to the higher affinity for the aqueous medium. However, after adding modified starch nanoparticles to the mixture and shaking, a visible migration of the particles towards the chloroform, occurred suggesting a reduced polar nature after chemical modification of starch nanoparticles by fatty acids.

In another study, Ren *et al.*[88] improved the dispersibility of starch nanocrystals obtained from the acid hydrolysis of waxy maize starch. This study suggested that modification of starch nanocrystals was achieved through

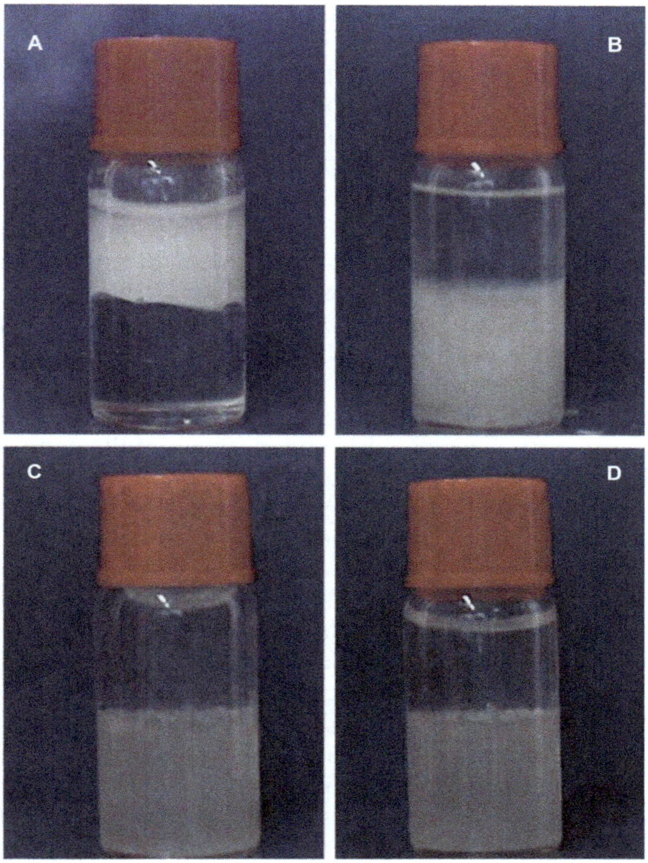


**Figure 3.10** Starch nanocrystals: (A) unmodified starch nanocrystals in distilled water; (B) octanoyl starch nanoparticles; (C) nonanoyl starch nanoparticles; (D) decanoyl starch nanoparticles in chloroform.
Reproduced from Namazi and Dadkhah[77] with permission of Elsevier.

crosslinking in an aqueous medium using sodium hexametaphosphate as the crosslinker. It was reported that the crosslinking modification did not disrupt the crystalline structure of starch nanocrystals and resulted in a stable and uniform aqueous suspension of starch nanocrystals. Other studies have been devoted to modifying starch nanocrystals with various reagents such as isocyanate, anhydrides and fatty acids in organic solvents. However, a severe loss of mechanical performance was reported for nanocomposites processed from these modified nanocrystals due to the coating of the nanoparticles and resulting damage to the percolating network of nanoparticles.[89] Regardless of the modification technique used, it is important to preserve the integrity of the starch nanocrystal crystalline domains and to minimize the potential for damaging the micro- and macromolecular structures.

Modification of starch nanocrystals has also been reported *via* graft polymerization of caprolactone[90] and using poly(ethylene oxide)[91] and also esterification with alkenyl succinic anhydrides (ASAs) and acetic anhydride (AA),[50,92,93] with varying results. Over recent years, researchers have proposed and tested new and promising strategies for the chemical modification of starch nanocrystals, one of the more effective methods being *via* the grafting of polymer chains from the matrix directly onto the surface of reinforcing nanoparticles before composite preparation.[94] Using this method, a highly effective transfer of stress between the matrix and the filler was achieved when the polymer was grafted onto the surface of starch nanocrystals. A continuous interface between the polysaccharide phase and the polymer matrix was formed subsequently improving the interfacial adhesion. In addition, entanglements can occur, reinforcing the matrix with the ungrafted polymeric chains if the length of the grafted chains is high enough.[95] This approach would allow enhancement of the non-polar nature of the original nanoparticles and dispersion in organic media.[95]

In practice, the "graft onto" strategy is relatively ineffective, so a novel "graft from" strategy has been proposed in which the polymerization of monomers and chain propagation of polymers are induced and finally are coupled with the starch nanocrystals. Various polymers have been grafted from the surface of starch nanocrystals based on this strategy, together with ring-opening polymerization for poly(ε-caprolactone) (PCL) and free radical polymerization for polystyrene. A schematic illustration of grafting polymer chains onto a starch nanocrystal based on the "graft onto" and "graft from" strategies is shown in Figure 3.11. Song *et al.*[39] prepared and characterized amphiphilic starch nanocrystals by graft copolymerization with styrene in an aqueous emulsion system. They found that the size of the amphiphilic starch nanocrystals was between 80 and 100 nm and, since there was no amorphous region, they concluded that the polystyrene side chains were grafted on the surface of starch nanocrystals. The surface chemical modification of starch nanocrystals with short grafting agents was also reported;[50] however, the mechanical properties of the nanocomposite materials obtained from these modified nanocrystals and natural rubber were found to be insufficient.

Thielemans *et al.*[91] reported the grafting of starch nanocrystals with poly(ethylene glycol) methyl ether and stearic acid chloride. Although successful surface modification was confirmed using infrared and X-ray photoelectron spectroscopy, thermal analysis and contact angle measurements, the length of the grafted chains was insufficient to allow the formation of a continuous film by hot pressing the modified nanoparticles. The same group conducted another study using polytetrahydrofuran, poly(propylene glycol) monobutyl ether and PCL. The resulting surface modifications had a layer thickness ranging between 0.5 and 20 times the thickness of the starch nanoparticles.[96] They concluded that these co-continuous materials are able to be hot pressed and, if cohesive enough, the resulting films can be tested mechanically. Habibi and Dufresne[89] processed

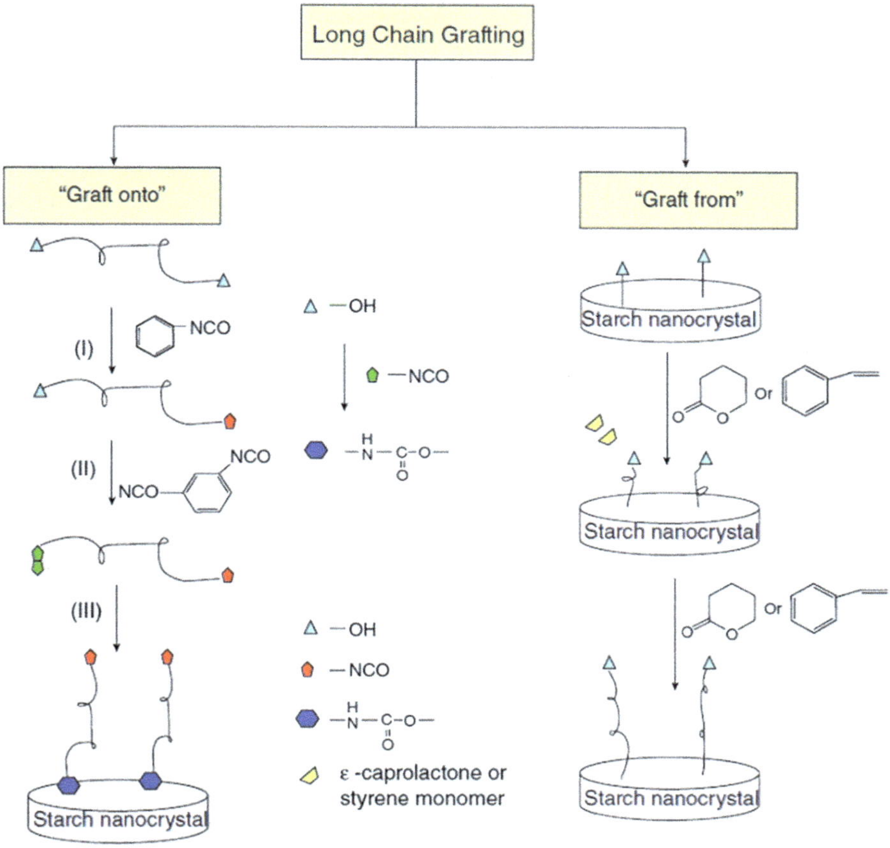


**Figure 3.11** Schematic illustration of two strategies used for grafting polymer chains onto a starch nanocrystal. For example, 2,4-toluene diisocyanate (2,4-TDI) is used as a coupling agent for premodification in the "graft onto" strategy; and ring-opening polymerization uses poly(ε-caprolactone) in the "graft from" strategy.
Reproduced from Huang *et al.*[75] with permission of Wiley-VCH.

nanocomposite films from both unmodified and PCL-grafted nanoparticles and showed that their mechanical properties were notably different.

## 3.4 Characterization of Starch Nanocrystals

### 3.4.1 Mechanical Characteristics

Depending on the polysaccharide source, the particles formed have various shapes such as rod-like particles from cellulose and chitin and platelet-like particles from starch. Incorporating the nanofillers derived from these polysaccharides in crystalline forms with uniform structures can thus be an ideal choice for forming bionanocomposites with high rigidity.[97] For example, the positive reinforcement of non-vulcanized polymeric blends has been widely reported using crab shell chitin whiskers,[98] waxy maize starch nanocrystals[99,19] and bagasse[100] or date palm tree cellulose nanocrystals.[101] Significant improvements in the mechanical performance of water-borne polyurethane (WPU) matrices by incorporating rod-like cellulose whiskers and starch nanocrystals have also been reported.[102,46] The results in these reports indicated that composites incorporated with waxy maize starch nanocrystals had similar mechanical behaviour and better barrier properties and composites incorporated with bagasse cellulose nanocrystals were more readily biodegradable, mainly due to their platelet morphology.[99,100]

Starch with a higher content of essentially linear amylose molecules results in stronger films, whereas starch comprised mainly of the highly branched amylopectin typically forms films with poorer mechanical characteristics, including decreased tensile strength.[2] Le Corre *et al.*[40] studied the effect of amylose content of starch and its botanical origin on the functional attributes of the resulting starch nanocrystals. This study demonstrated that the amylose content and crystalline type of starch are the most significant factors for nanocrystal morphology, crystallinity level and rheological behaviour. Interestingly, it was concluded that nanocrystals generated from higher amylose content starch may release loosely bonded amylose chains during preparation and/or soaking in water, thus resulting in fewer nanocrystals available to participate in the formation of a reinforcing and diffusing network. It was also reported that a higher content of the initial amylopectin can lead to more square-shaped platelets. In other studies, it was demonstrated that the incorporation of starch nanocrystals obtained from amylopectin-rich waxy maize starch granules into in poly(styrene-*co*-butyl acrylate) resulted in advantageous mechanical reinforcement of the polymer matrix.[99,103] It was also demonstrated that nanocrystals derived from waxy maize starch can significantly improve the mechanical properties of natural rubber composites.[99,31] Large-scale starch nanocrystals (10–50 nm) obtained from the acid hydrolysis of amylopectin-rich waxy maize starch have been employed to prepare nanobiocomposites with natural rubber using a mastication technique.[104] The results of this research showed that the tensile strength and elongation at break were higher than those of conventional composites with untreated waxy maize starch and amylose-rich starch and also carbon black at all loadings, suggesting that starch nanocrystals can clearly be considered as a good substitute for carbon black.[104] Angellier *et al.*[19] claimed that the formation of a continuous network within a natural rubber matrix through hydrogen-bonding forces in starch nanocrystals at a critical volume fraction of 10% w/w. Rajisha *et al.*[94] characterized nanocrystals obtained from potato starch reinforced into natural rubber and, as they expected, the tensile strength and modulus of their composites were significantly improved on increasing the nanocrystal loading owing to the formation of a nanocrystal–polymer network and subsequent immobilization of the polymer chains.

The development and characterization of starch nanocrystal composites derived from rice (5–30%) have been studied using a casting film–solution method.[106] It was demonstrated that as the concentration of nanocrystals is increased, the tensile strength and elongation at break increased and decreased, respectively. They observed that the films containing a 20% loading of nanocrystals had the best mechanical characteristics and provided a high tensile strength and elongation at break of about 16.43 MPa and 5.76%, respectively.[106] Starch nanocrystals physically similar to cellulose whiskers reportedly have higher reactive surface areas occupied with hydroxyl groups, which provide the possibility of modification through chemical and grafting reactions.[18,37,96] Modification *via* chemical reaction with small molecules, grafting of polymer chains using coupling agents (*i.e.*, the "grafting onto" strategy) and polymer chain grafting by polymerization of a monomer ("grafting from" strategy) are three strategies for the chemical modification of starch nanocrystals.[107] Application of these strategies can facilitate the formation of nanocrystal dispersions and improve the miscibility or adhesion between them and the polymer matrix. This is achieved *via* manipulation of the surface hydrophobicity and transformation of the surface polar hydroxyl groups of the starch nanocrystals, leading to the formation of an interfacial structure and thus enhancing the mechanical performance.[39,50,74,96]

The grafting approach mostly produces long tails on the starch nanocrystal surface and some reports have shown that these tails can penetrate more easily into the polymer matrix and then provide stronger interfacial interactions.[74] Some studies have suggested that modification of waxy maize starch nanocrystals formulated into a natural rubber matrix using chemical reaction leads to a remarkable decrease in mechanical performance and thermal stability as a result of the partial or total destruction of the three-dimensional network structure in the nanocomposites.[19,99] This decrease can be explained by hindrance of the interactions between chemically modified nanocrystals resulting from their coating with the grafting agent.[99] Angellier *et al.*[50] reported that the surface of starch nanocrystals could be modified using the reagents alkenyl succinic anhydride and phenyl isocyanate. PCL was grafted onto the surface of starch nanocrystals using the "grafting from" strategy *via* ring-opening polymerization under microwave irradiation.[108] The synthesized nanoparticles were then formulated into a poly(lactic acid) (PLA) matrix in order to produce the biodegradable nanocomposites with ideal mechanical characteristics. A significant improvement in mechanical properties, including strength and elongation at break, was observed in the resulting nanocomposites containing 5% w/w PLC-grafted starch nanocrystals. However, this increase was inhibited when the loading of nanocrystals was increased, mainly due to aggregation of the nanocrystals with PCL. An important finding in this study was that the rubbery PCL imparted sufficient flexibility to increase the strain of PLA plastics, but the modulus was not increased. It was proposed that this strategy for achieving enhanced mechanical properties is due to the construction of a miscible interfacial layer mediated with long chains grafted onto natural nanoparticles.[108]

Chang *et al.*[74] also modified the surface of pea starch nanocrystals *via* grafting with PCL using a microwave-assisted technique. The grafted materials was then incorporated into a WPU matrix and the results showed the concurrent improvement of tensile strength and elongation at break in neat WPU was much lower than that in WPU loaded with 5% w/w nanocrystal-*g*-PCL. The simultaneous enhancement of strength and elongation at break was attributed to the uniform dispersion of the nanocrystals, the increased entanglements mediated with grafted PCL chains and the reinforcing function of rigid starch nanocrystals. However, an increase in the nanofiller content induced self-aggregation of the crystalline domains, which impeded improvements in tensile strength and elongation at break, but significantly enhanced Young's modulus.[74] Wang *et al.*[107] considered reinforcing WPU with both cotton linter pulp–cellulose nanocrystals and waxy maize starch nanocrystals. They found that there was a synergistic effect on adding 1% w/w starch nanocrystals and 0.4% w/w cellulose nanocrystals, with significant improvements in Young's modulus, tensile strength and tensile energy at break without a considerable reduction in elongation at break.[107] In a study by Habibi and Dufresne,[89] the mechanical characteristics of nanocomposite materials were improved using chemically modified starch nanocrystals, which resulted in better dispersion of the filler within the matrix. This established a strong adhesion between the filler and matrix, which noticeably enhanced the Young's modulus and tensile stress at break.[89]

### 3.4.2 Barrier Properties

The shelf life of perishable packaged products can be significantly extended by improving the barrier properties of packaging materials. It is known that a polymeric material with excellent barrier to water vapour typically has higher oxygen permeability and *vice versa*. Therefore, protection of packaged products from deterioration due to oxidation, high temperatures, moisture and microorganisms can successfully be achieved through the use of multilayer structures consisting of different polymers, each contributing certain specific functions. The permeation rate of a gas or vapour through a polymeric material depends on its chemical nature and the physicochemical properties of the permeating molecules.[109] Optimizing the resistance to water vapour permeation (WVP) and oxygen diffusion is therefore a necessary requirement in composites/nanocomposites for the packaging of many food and drug products.[2]

The use of cellulose and starch nanocrystals with high crystallinity, their morphology and the inherent tortuosity that they impart can considerably limit the migration of water vapour and oxygen.[99] Indeed, platelet-shaped starch nanocrystals can potentially change the diffusion path of penetrative molecules more than rod-like cellulose nanocrystals and thus improve the barrier properties of materials containing these materials.[110] The WVP properties of rice starch-based nanocomposites were improved when the nanocrystal concentration was increased from 5 to 20%.[105] The reported slow decrease in WVP of the rice starch-based films with high levels of starch nanocrystal in this concentration range was related to their better water resistance imparted to the rice starch matrix. Thus, there were possibly fewer pathways for diffusing water molecules to be transported through in the composite matrix.[111] In another study, the WVP properties of carboxymethyl chitosan nanocomposites also were significantly improved by adding starch nanocrystals up to 20% w/w, which then remained almost unchanged with increasing nanocrystal content.[110] It was suggested that the starch nanocrystals at concentrations greater than 30% w/w would tend to create aggregates that could not contribute to the prevention of the diffusion of water molecules, so the WVP is minimized at starch nanocrystal contents below 20% w/w.

Chang *et al.*[111] observed a reduction in the WVP of potato starch-based films reinforced by chitin nanoparticles and García *et al.*[106] reported a 40% decrease in the WVP of cassava starch nanocomposites with increasing content of waxy maize starch nanocrystals. In the latter study, X-ray spectra revealed that the composites formed were more amorphous than the neat matrix, which was due to a higher equilibrium water content in the structure of the composites.[106] However, the same group in another study claimed that the WVP of nanocomposites formed using thermoplastic starch and waxy maize starch nanocrystal can be increased to 79% by adding only 2.5% starch nanocrystals.[76] These findings were in contrast to those reported by Angellier *et al.*,[19] who studied natural rubber composite films reinforced with starch nanocrystals. However, a similar trend was achieved for natural rubber nanocomposites filled with starch nanocrystals extracted from various starch sources.[112] In this study, it was postulated that the starch nanocrystals were glue-glycerol bonded, forming threads with a high concentration of OH groups and hence formed a preferential path for water vapour diffusion *via* the nanothreads, which clearly enhanced the permeability rate of the films formed. Kristo and Biliaderis[51] reported an unexpected result in their study of unfilled pullulan films and those containing up to 20% w/w starch nanocrystals. They observed that there were no significant differences in WVP between the two materials and, although it was not discussed, this result may be attributed to the need for a minimum content of starch nanocrystals necessary to form a tortuous pathway which may vary significantly between systems. This was confirmed with a significant decrease in WVP following the addition of more than 30% w/w starch nanocrystals to the pullulan films.

In addition to polymeric films, starch nanocrystals can considerably improve the barrier characteristics of bio-based coated papers. For example, a 5% w/w addition to a starch based-coating can generate lower WVP than with glycerol-plasticized starch-coated papers.[52] From a technological perspective, starch nanocrystals can resist a drying processes without melting due to a higher viscosity and better processability.[4] Although the barrier properties of starch nanocrystals are not as notable as those of organically modified montmorillonite (OMMT) nanoclay, it has recently been shown that the optimization of the isolation/classification process of starch nanocrystals can lead to equivalent or complementary barrier properties compared with OMMT nanoclays. Under these optimized conditions, the production of bio-based biodegradable packaging using starch nanocrystals is very promising.[113]

### 3.4.3 Swelling Characteristics

Swelling is a useful technique to determine the presence of specific interactions between the filler and the polymeric matrix by creating additional crosslinks.[114] From a technological perspective, the interaction of polymeric materials with various solvents can be very important owing to changes in the material dimensions and physical characteristics imparted by the penetrating solvent molecules into the polymer.[115] It is well known that crosslinking modification inhibits the swelling of starch, the inhibition being related to the degree of crosslinking, and this technique has been applied to characterize the crosslinking modification of starch nanocrystals.[88] Many researchers have investigated the effects of variables, including the nature of the polymeric matrix (polar or apolar), the starch nanocrystal content, chemical modification of the nanocrystals, different swelling liquids (water, toluene) and reaction times on the swelling behaviour of different nanocomposites.[114,115] It has been demonstrated that the water uptake rate for most of the compositions containing starch nanocrystals increased rapidly during the early stages of immersion and then, after reaching a maximum point, decreased until an equilibrium was reached.[94] Therefore, the absorption kinetics in the initial stages are faster, followed by an absorption plateau.[51] The observed decrease in the water uptake during this process can be ascribed to the leaching or partial release of starch nanocrystals in water, even though starch is insoluble at low temperatures. The release of starch nanocrystals into the aqueous phase during swelling is a possible reason for the increase in the water uptake rate in the initial stages. It has been suggested that the interface between starch nanocrystals and a polymeric matrix such as natural rubber is weakened by the swelling of the starch domains during the exposure time and thus enhances the swelling rate by the phenomenon known as the "overshooting effect."[94] Recently, Rajisha *et al.*[94] studied the swelling behaviour of natural rubber composites containing 5–20% w/w starch nanocrystals and they found that the water diffusion coefficient can be enhanced by increasing the nanocrystal content. Thus, further additions of starch nanocrystals resulted in an enhancement of water diffusivity compared with the unfilled natural rubber matrix. These results may be due to the formation of a continuous polar network by these nanocrystals within the natural rubber matrix and the presence of pathways created by these networks and also hydrogen bonding can explain the increase in the water diffusion coefficient.

There appears to be a critical loading level or volume fraction of starch nanocrystals in order to change the swelling behaviour of a nanocomposite by creating low/high interactions between nanoparticle clusters and also between the polymeric matrix chains and nanocrystals. Nevertheless, the determination of the critical volume fraction of starch nanocrystals during percolation is difficult owing to their heterogeneity and ill-defined geometry. The swelling properties of potato and waxy maize starch nanocrystals have been reported to be 20 and 6.7% by volume, respectively, and this difference may be explained by the potentially higher surface area of waxy maize starch nanocrystals.[10] A strong relationship between starch composition (amylose content) and the water uptake rate for natural rubber-based nanocomposite films has recently been reported.[112] An increase in the content of the amorphous, less organized and poorly bonded amylose chains in the starch nanocrystal structure can lead to a considerable decrease in the water diffusion coefficient.[19] It was postulated that the amorphous segments make no contribution to the creation of diffusing networks but these segments participate in the water sorption, explaining why similar equilibrium values are reached.[14,112]

Kristo and Biliaderis[51] found a reduction in the moisture uptake of sorbitol-plasticized pullulan films by incorporating starch nanocrystals, especially at high loadings (20–40% w/w), with no significant difference in the water uptake rate of films with a 6–15% w/w nanocrystal content. Since pullulan polysaccharide polymers adsorb more water than starch nanocrystals, especially at high water activities ($a_w > 0.8$), composites with starch nanocrystals can decrease the water sensitivity of the composite films.[116] This phenomenon can be attributed to the formation of a rigid three-dimensional starch nanocrystal network due to the strong hydrogen bonding between the starch particles[117,118] and also to strong interactions between the filler and polymer chains, thus decreasing the swelling and water absorbance of the polymeric chains located in the interfacial regions.[119] The crystallinity of starch nanocrystals and the strong interfacial interactions have been reported to lead to a decrease in the swelling properties of carboxymethyl chitosan.[110] Chen *et al.*[46] reported a reduction in the moisture uptake of a poly(vinyl alcohol) (PVA) matrix from 78 to 62% for the unfilled matrix and 40% w/w starch nanocrystal-reinforced composites, respectively.

Kristo and Biliaderis[51] reported that variations in the compatibility between the filler and the matrix could be a reason for observed differences in the swelling behaviour with different filler concentrations. The similar hydrophilic and polar nature of starch or soy protein biopolymers and nanoparticles (cellulose microfibrils, tunicin or chitin whiskers, starch nanocrystals) can result in strong adhesion between the polymer matrix and the filler as a result of hydrogen bonding. This adhesion reinforces the matrix by increasing the filler loading level and consequently contributing to the observed lower water sensitivity of highly hydrophilic composites.[51] In contrast, a reduced interfacial compatibility of hydrophilic particles with hydrophobic polymeric matrices such as natural rubber can lead to weak adhesion between the components and subsequent poor dispersion of filler within the polymer matrix.[28] Moreover, although these polymers have a low affinity to water, the inclusion of polar and hydrophilic particles can possibly enhance the moisture uptake rate of the composites compared with an unfilled hydrophobic matrix.[51] Conversely, Zheng *et al.*[53] found no noticeable changes in moisture uptake between an unfilled soy protein matrix and one filled with pea starch nanocrystals, which they claimed was due to the strong interfacial interaction between the filler and the soy protein matrix.

The swelling of packaging biopolymers is typically performed in the presence of water, but in some cases organic solvents are used to simulate fatty foods. The diffusion of toluene into a natural rubber matrix, for example, is significantly decreased when a low content of starch nanocrystals is added.[19] This reduction can be explained by the favourable interactions between the matrix and starch nanocrystals and by the formation of a three-dimensional network *via* hydrogen linkages.[19] It was shown that at a lower content of starch nanocrystals, there was a direct correlation between the toluene uptake and specific surface area of nanocrystals of different botanical origins. At low filler contents, the nanoparticles can disperse homogeneously and interact more efficiently with the matrix by forming a network *via* strong hydrogen bonds, thus preventing the swelling of polymeric chains located at the interfacial zone.[120] However, this correlation cannot be achieved at high nanocrystal contents owing to the aggregation phenomenon and the prevention of the formation of a denser network.[112] The compatibility between PVA and starch molecules can be improved by grafting and crosslinking reactions, resulting in an improvement of the water resistivity and swelling properties.[121,122] A similar modification can be applied to improve the swelling behaviour in toluene, as reported by Angellier *et al.*,[50] who modified starch nanocrystals with alkenyl succinic anhydride or isocyanates and observed that the toluene uptake of the composite was higher than those of unmodified starch nanocrystals. It was hypothesized that the modification reduced interactions between modified starch nanocrystals and improved interactions between the filler and the matrix, thus resulting in a higher affinity between modified starch nanocrystals and toluene.[50] Perhaps the most promising feature of chemically modifying starch nanocrystals used in rubber composites is that the cohesion and thermal stability of the polymer matrix can be significantly increased by adding the modified filler.[98]

### 3.4.4 Thermal Properties

The development of biodegradable packaging materials with improved thermal properties is central to improving the processability of polymeric composites. The thermal behaviour, including the determination of the glass transition temperature ($T_g$), is typically determined by dynamic mechanical analysis (DMA) and differential scanning calorimetry (DSC). The DMA technique is more widely used to study polymer chain mobility *via* $\alpha$-relaxation at the molecular level and $T_g$ is typically determined at the point of inflection of the curve of loss tangent ($\tan \delta$) as a function of temperature.

The thermal stability of polymeric materials can also be measured using thermogravimetric analysis (TGA).[75] It has been demonstrated that the nanocomposites reinforced with nanocrystals usually display enhanced thermal characteristics when nanofillers are well dispersed into the matrix.[123,124] Interestingly, the thermal properties of polysaccharide nanocrystals depend on the origin, preparation method and type of surface modification.[23] For example, during sulfuric acid hydrolysis, acidic sulfate ester groups are formed on the surface of polysaccharide nanocrystals, and this would result in a significant decrease in the degradation temperature.[125] In recent years, nanocrystals produced from starch have shown improved thermal properties and unique temperature resistance.[76,10,126]

Valodkar and Thakore[104] investigated the thermal characteristics of natural rubber and starch nanobiocomposites and showed that $T_g$ increased with the incorporation of relatively large nanocrystals (10–50 nm) or with higher loadings. In a series of natural rubber composites, the highest $T_g$ value was reported in composites with starch nanocrystals followed by those from amylose-rich starch nanocrystals and finally waxy starch nanocrystals. The use of nanoscale starch particles increases the surface area exposed to the natural rubber matrix, leading to improved stiffness and higher $T_g$ values in their nanocomposites. It has been suggested that there is a critical loading level for the incorporation of starch nanocrystals within the various polymeric matrixes to optimize the thermal properties.[104] The thermal characteristics of soy protein plastics reinforced with starch nanocrystals were evaluated using DSC, XRD and DMA techniques.[53] The DSC data revealed that an initial increase in $T_g$ was observed by increasing the starch nanocrystal content to 8% w/w into the soy protein matrix. This was confirmed using XRD, with the results showing that an increase in starch nanocrystal content induced the formation of large domains that can restrict the motion of soy protein segments *via* steric hindrance. They also observed that the ordered structure in the soy protein matrix was disrupted with an increase in starch nanocrystal loading to 16% w/w, which resulted in the lowest $T_g$ value. The DMA curves showed similar results including changes in the α-relaxation temperature ($T_{max}$), which gradually increased with up to an 8% w/w starch nanocrystal content resulting in nanocrystal domains of sufficient active area and size. However, higher concentrations of starch nanocrystals (16% w/w) produced negative results, mainly due to reduction of the nanocrystal domains, which can decrease the effective active area and destroy the original ordered structure of the soy protein matrix, resulting in a decreased $T_{max}$.[53] Piyada *et al.*[105] investigated the thermal behaviour of rice starch films reinforced with starch nanocrystals using DMA, DSC and TGA. Although no endothermic peaks were observed in the DSC thermogram, an increase in the endothermic heat flow on addition of the starch nanocrystals to the films was clearly observed. This can be explained by the limitations on the mobility of the rice starch chains due to the establishment of strong interactions between the rice starch and the nanocrystals. In general, the addition of starch nanocrystals to rice starch films can improve the thermal stability of the films produced.

Chen *et al.*[46] reported that the melting temperature ($T_m$) can be considerably improved in composites between starch nanocrystals and a WPU matrix and this result is in accordance with that claimed by Wang *et al.*[107] The latter group reported that the inclusion of starch nanocrystals in WPU matrices resulted in a significantly better reinforcing effect than in other polymer matrices. They related the synergistic reinforcement of polysaccharide nanocrystals and whiskers to the formation of strong hydrogen bonding networks.[107] The corresponding DMA data revealed that the composite film was in the glassy state at temperatures below $T_g$ and thus the storage modulus decreased on increasing the temperature.

Kristo and Biliaderis[51] and Chang *et al.*[111] reported similar explanations for improving the storage modulus of a polymer matrix by increasing the volume of starch nanocrystals. For example, Kristo and Biliaderis[51] observed a shift in the $T_g$ value towards higher temperatures with increasing starch nanocrystal content as a result of the restriction in polymer chain mobility. Angellier *et al.*[19] reported that the storage modulus and thermal stability of natural rubber-based composites was increased with higher loadings of starch nanocrystals compared with the material without the filler. They presented a detailed study of the thermal behaviour of freeze-dried waxy maize starch nanocrystals using DSC and stated that there were two endothermic peaks in the presence of excess water over the studied temperature range. For example, in the case of an 83.78% water content, they observed one endothermic peak at 80 °C and another one at 178 °C. In a similar study, Jivan *et al.*[60] also reported two endothermic peaks but at much lower temperatures of 26 and 125 °C. They suggested that the former peak could be attributed to the easy destruction of starch nanocrystals arranged in a regular lattice and the latter peak to the mobilization of the firm and well-ordered structures of starch nanocrystals. It could therefore be suggested that higher temperatures are needed to unravel the crosslinked lattice of starch nanocrystals. Piyada *et al.*[105] observed that the major DMA tan $\delta$ peak shifted to higher temperature with increasing starch nanocrystal content in rice starch films. Smith *et al.*[127] suggested that this observation may be due to the strong hydrogen bonding interactions between the amorphous polymeric matrixes and starch nanocrystal which results in strong absorption of polymer chains on the starch nanocrystal surface. Angellier *et al.*[52] also reported an enhancement in the relaxation temperature of thermoplastic starch films with the incorporation of starch nanocrystals as a result of direct contact between the nanocrystals and the amylopectin-rich domains.

Several studies have been conducted to investigate whether starch nanocrystals derived from different botanical sources have any effect on the thermal behaviour of composites using the nanocrystals. Le Corre *et al.*[23] evaluated the thermal characteristics of five different types of starch (waxy maize, normal maize, high-amylose maize, potato and wheat) and their starch nanocrystals using DSC and TGA. They reported evidence of one and two thermal transitions for all of the native starch materials and their nanocrystals, respectively. The first and second transitions were ascribed to the first and second stages of the melting of crystallites as a result of the unpacking and unwinding of the double helices. They concluded that thermal stability of B-type crystalline starch nanocrystals is higher than that of the A-type crystalline forms due to the presence of more rigid crystallites. They finally proposed that starch nanocrystals can be applied *via* wet processing (*i.e.* coating) if the temperature remains lower than 80–100 °C and processed dry at temperatures below 150–200 °C. Xu *et al.*[45] investigated the functional characteristics of starch nanocrystals prepared from corn, barley, potato, tapioca, chickpea and mung bean sources. The reported that the acid hydrolysis method for producing starch nanocrystals contributed to an increase in the crystallinity, melting temperature and enthalpy of fusion, with a reduced thermal decomposition temperature. They observed that among the different types of starch nanocrystals studied, those obtained from potato starch had the lowest degree of crystallinity and onset and melting temperature. The lower melting temperature of the potato starch was attributed to the existence of B-type crystalline structure, which melt more readily than the other two crystalline structures.[52]

The increased melting transition of starch nanocrystals compared with native starch has been ascribed to the higher crystallinity, tighter packing and increased intermolecular bonding as a result of freeze-drying, and/or to the removal of amorphous regions during the hydrolysis process.[23] Le Corre *et al.*[23] evaluated the thermal characteristics of starch nanocrystals prepared from various starch sources and reported a decomposition temperature ($T_d$) close to 250 °C. The highest (33.8 °C) and lowest (5.7 °C) $T_d$ values were reported for the starch nanocrystals originating from tapioca and potato, respectively.[128,129] The observed differences in $T_d$ values between the various starch nanocrystals can be attributed to the sulfate groups present in the nanocrystal surface, a tighter packed crystalline structure and increased intermolecular bonding within the nanocrystal structure. Namazi and Dadkhah[90] studied the thermal stability behaviour of starch, starch nanocrystals and starch nanocrystal-*g*-PCL using TGA. They observed that following the process of generating starch nanocrystals, the $T_d$ value of starch granules (290 °C) decreased after acid hydrolysis as a result of disruption of the amorphous zone.

Sun *et al.*[73] recently investigated the thermal stability behaviour of native starch and nanocrystals using TGA. The maximum degradation rate for native starch was reported to be about 304 °C with the degradation occurring over a relatively narrow temperature range between 285 and 320 °C. This temperature range was reduced to 260–293 °C when the nanocrystals were prepared *via* debranching and recrystallization with other methods resulting in a broader temperature range. It was concluded that the native starch had a higher degradation temperature but the narrowest temperature range. They suggested that since the native starch has a compact surface and interior structure comprised of amylose and amylopectin, this requires more energy to complete the $T_g$ phase transition even though smaller quantities of crystals are present in this starch form.[73] It has also been demonstrated that the method used to obtain starch nanocrystals can affect the thermal characteristics of the resulting nanocrystals. Generally, a mild acid hydrolysis can significantly increase the melting temperature and melting enthalpy, which can be attributed to the erosion of amorphous regions, resulting in a more rigid crystalline structure. For example, the melting enthalpy of waxy maize starch nanocrystals prepared by sulfuric acid hydrolysis increased with an increase in the hydrolysis temperature and time.[57]

Lamanna *et al.*[130] reported that starch nanocrystals prepared using γ-radiation from both waxy maize and cassava starch sources had a similar thermal degradation behaviour to those obtained *via* acid hydrolysis. Moreover, both irradiation and acid hydrolysis resulted in starch nanocrystals that exhibit weight loss by TGA over wide temperature ranges. This may be due to the better dispersion of molecular weights generated by γ-radiation[131] or hydrolysis treatment[31] and is in accordance with the previous suggestion that starch nanocrystals have a high number of hydroxyl groups on their surface. The observation that starch nanocrystals will not undergo any undue thermal modification is promising for the development biodegradable composites by conventional thermal processing techniques. Sun *et al.*[73] prepared starch nanocrystals using ultrasonic-assisted oxidation and, although the DSC analysis resulted in only small DSC endotherms, they suggested that a combination of ultrasonic treatment and oxidation could be an effective process for producing nanocrystals with ideal thermal characteristics. It has been established by some workers that the surface chemical modification of starch nanocrystals can noticeably improve their thermal behaviour characteristics. For example, Lin *et al.*[44] indicated that grafting starch nanocrystals to alkyl polymer chains can considerably improve the thermal properties. Thielemans *et al.*[91] also observed that the thermal behaviour of starch nanocrystals was improved by grafting to alkyl polymer chains, which they suggested may be due to the protective crystalline layer formed by the oxygen-poor stearate surface. Namazi and Dadkhah[77] obtained similar results in relation to hydrophobically modified starch nanocrystals using octanoyl, nonanoyl and decanoyl chloride in an aqueous medium under mild conditions. They evaluated the thermal properties of the starch nanocrystals using TGA and observed that the onset decomposition temperature increased for modified starch nanocrystals, revealing their higher thermal stability than the unmodified form. It was suggested that this was probably due to the smaller amount of remaining hydroxyl groups after acylation. Ma *et al.*[81] also reported improved thermal stability of citric acid-modified starch nanocrystals compared with the starch granules and unmodified nanocrystals due to the steric hindrance of hydrogen bond interactions. They concluded that the citric acid-modified starch nanocrystals were thermally stable under processing even at higher temperatures. Yu *et al.*[132] developed modified starch nanocrystals using highly active dialdehyde groups intended for medical applications.

Recently, García *et al.*[131] developed a new three-step grafting strategy for PLA reinforced with starch nanocrystals. The grafting process involved (i) protection of the PLA hydroxyl groups by benzoylation, (ii) activation of carboxyl groups using thionyl chloride and (iii) grafting of the modified PLA onto the starch nanocrystals. A different thermal behaviour was observed for the grafted composites compared with the blended mixture of PLA with starch nanocrystals. In addition, $T_g$ was found to be much lower than that of the original PLA owing to the increase in molecular mobility induced by the benzoylation step or by the increase in the free volume due to the presence of benzoyl groups. Furthermore, the grafted material had a $T_d$ slightly lower than that of PLA, although this does not adversely affect its possible use in many packaging applications.

### 3.4.5 Morphological Characteristics

Putaux *et al.*[42] conducted perhaps the first structural observation of starch nanocrystals prepared from waxy maize starch granules. Using transmission electron microscopy (TEM), they clearly observed micrometre-sized fragments and some 200–500 nm wide compact fragments after hydrolysing the starch for 2 weeks. After 6 weeks of acid hydrolysis, the TEM results revealed a longitudinal view of lamellar fragments comprised of stacks of 5–7 nm wide elongated elements and a planar view of individual platelets.[42] It was also demonstrated that the branching chains in the amorphous domains of the semicrystalline starch structure were initially disrupted and, during the hydrolysis process, the amorphous regions between the crystalline lamellae began to hydrolyse, leading starch nanocrystals with a platelet-like morphology (see Figure 3.12). Angellier *et al.*[103] showed that under some conditions, starch nanocrystals can form aggregates with a mean size of ~4.4 µm. Recently, field electron scanning electron microscopy (FE-SEM)

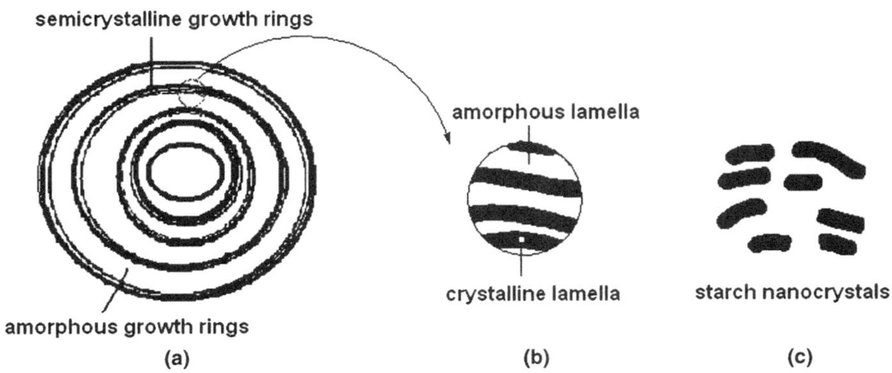


**Figure 3.12** Microstructure of a starch granule with amorphous and semicrystalline growth rings (a) and amorphous and crystalline lamellae (b) and starch nanocrystals (c).
Reproduced from Wang *et al.*[140] with permission of Springer-Verlag.

has been employed to characterize the morphology of starch nanocrystals. This allowed high-resolution images down to 0.8 nm in both the reflection (topographic) and transmission (TEM) modes to be obtained.[56,76,106] Putaux *et al.*[42] reported that some modified starch nanocrystal platelets have a thickness of 5–7 nm and an equivalent diameter between 30 and 100 nm depending on the origin of the starch. Given the nature of the platelet structure of starch nanocrystals, they can be considered an excellent candidate for reducing the permeability to grease, oxygen and water vapour when applied as a coating.[58]

In separate studies, Chen *et al.*[133] and Yu *et al.*[108] used the same method to prepare starch nanocrystals from pea starch granules, but they obtained different size ranges of 30–80 and 60–150 nm, respectively. A size of 70–100 nm[90] for waxy maize starch nanocrystals and 40–70 nm for potato starch nanocrystals[46] have been reported elsewhere. This difference may be attributed to differences in the botanical origin of the starch and to difficulties in capturing well-defined pictures of non-aggregated nanocrystals.[14] Angellier-Coussy *et al.*[31] characterized the molecular content of A-type starch nanocrystals obtained by the acid hydrolysis of waxy maize starch. The presence of different structural motifs and the degree of polymerization clearly indicated that the starch nanocrystals correspond to the crystalline lamellae present in native starch granules. Le Corre *et al.*[40] comprehensively studied the influence of the botanical origin and amylose content of five native starches (normal maize, high-amylose maize, waxy maize, potato and wheat) on the morphology and characteristics of formed starch nanocrystals. In general, variations in size were moderate but differences were more evident when the shapes and crystallinities of the nanocrystals were compared simultaneously. The different starch nanocrystals as a function of the starch source had a diameter ranging between around 50 and 120 nm and thickness from 4 to 8 nm. Whereas potato and high-amylose starch nanocrystals displayd disk-like platelets, normal maize starch nanocrystals seemed to have both round- and square-like crystals. Conversely, starch nanocrystals produced from A-type starches showed square-like particles, whereas those prepared from B-type starches rendered rounder particles. This discrepancy can be attributed to the diverse packing configurations of amylopectin chains for A- and B-type starches. It would be expected that there would be some critical differences in the rheological characteristics of starch nanocrystal suspensions prepared from different starch origins that will effectively facilitate the selection of the best source depending on their intended industrial purposes.

Le Corre *et al.*[40] investigated the morphology and particle size of native starch before and after hydrolysis and obtained the SEM images illustrated in Figure 3.13. As shown, nanocrystals from corn starch are polyhedral granules whereas those from other starches are elliptical, spherical and/or irregular shaped. The largest nanocrystal, with an average granular size of 41.3 μm, was obtained from potato starch, followed by legume starches. Interestingly, all starch nanocrystals produced *via* the acid hydrolysis

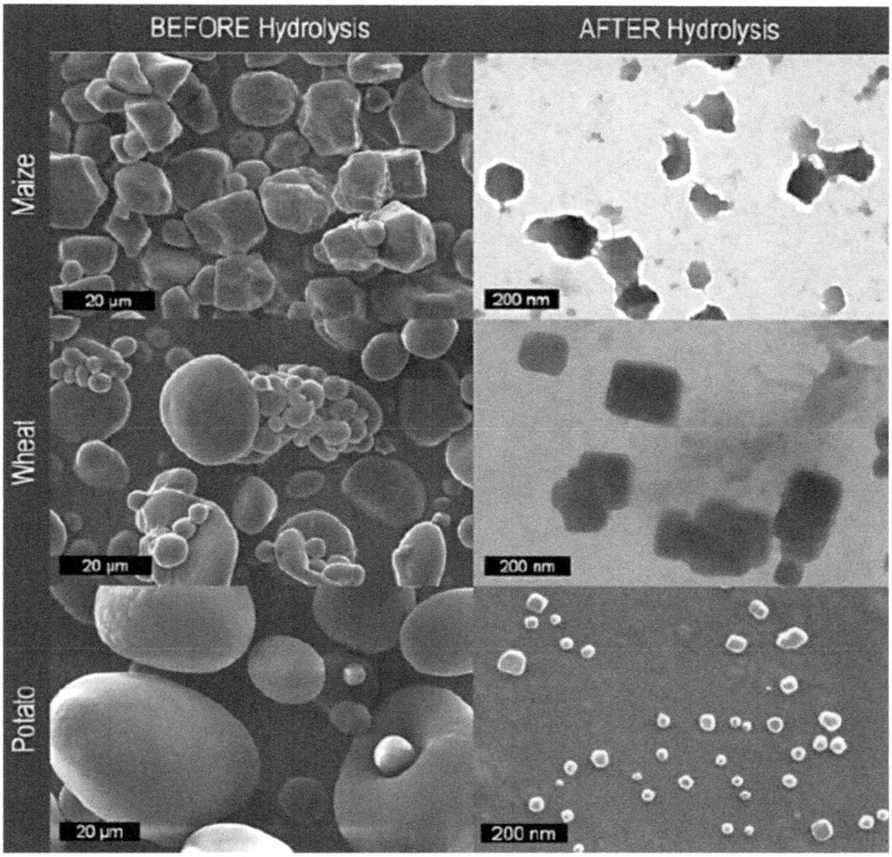


**Figure 3.13** SEM images of starches before (left) and after (right) hydrolysis for different botanic sources.
Reproduced from Le Corre *et al.*[40] with permission of Springer-Verlag.

of starch granules exhibited spherical shapes regardless of the botanical origin.

Wei *et al.*[134] recently investigated the influence of different dispersion pH values (2.07, 3.92, 5.38, 9.45, 10.25 and 10.92) on the size, morphology and aggregation behaviour of maize starch nanocrystal suspensions after 2 h (Figure 3.14). The FE-SEM results showed that when the dispersion pH was in the range 2.07–3.92, nanocrystals exhibited parallelepiped nanoplatelets with aggregates about 1.5 mm in size (Figure 3.14a and b). Monodisperse starch nanocrystals with sizes between 100 and 200 nm were obtained by increasing the dispersion pH to 9.45 (Figure 3.14c and d), which may be due to a significant increase in repulsion forces among starch nanocrystals, thereby decreasing the formation of aggregates. When the pH of the dispersion was increased above 10.25, the platelet-like nanocrystals transformed into more spherical-like shapes with sizes ranging between 50 and

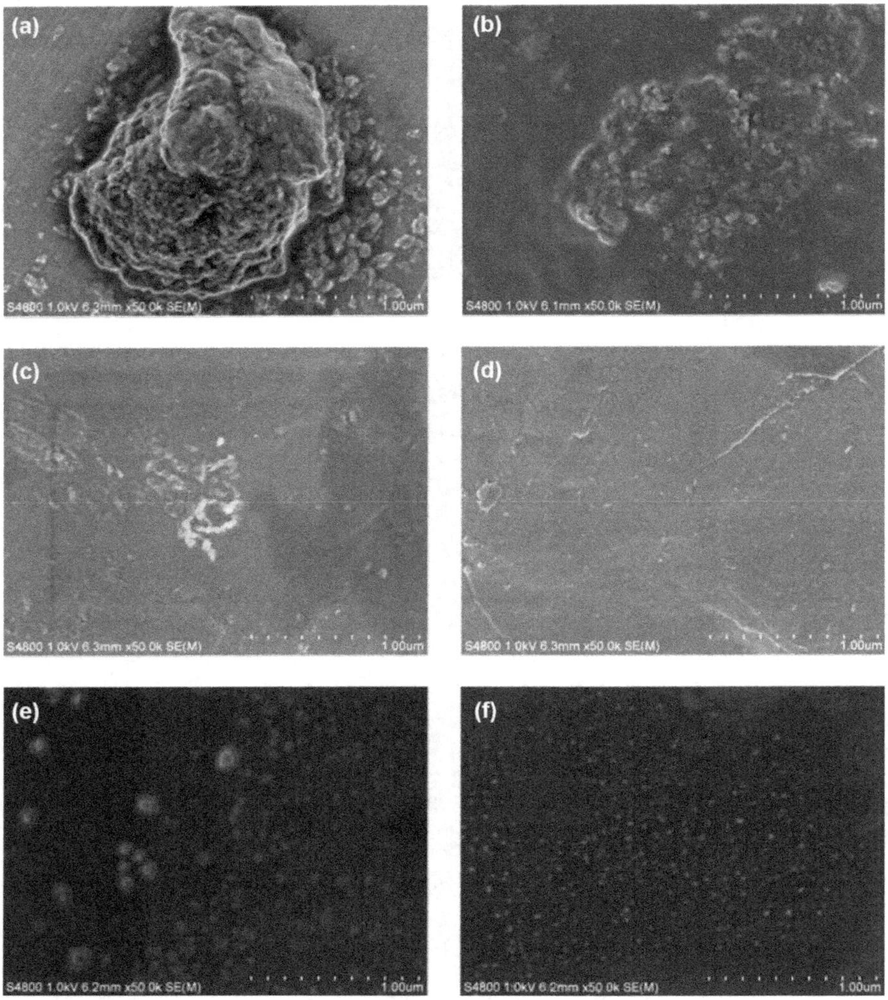


**Figure 3.14** FE-SEM images of starch nanocrystal suspensions at pH values of 2.07 (a), 3.92 (b), 5.38 (c), 9.45 (d), 10.25 (e) and 10.92 (f) after 2 h of dispersion.
Reproduced from Wei *et al.*[134] with permission of Elsevier.

100 nm (Figure 3.14e and f) as a result of the alkaline conditions promoting dissolution of the nanocrystals.[134]

Lamanna *et al.*[130] reported that γ-irradiation with a dose of 20 kGy can produce starch nanocrystals with sizes around 20–30 nm from cassava and waxy maize starch granules (Figure 3.15). They also found that nanocrystals tended to cluster together and form laminar aggregates with a relatively high specific surface area, and this observation was similar to those reported by Oostergetel and van Bruggen,[135] Capron *et al.*[136] and Waigh *et al.*[137]

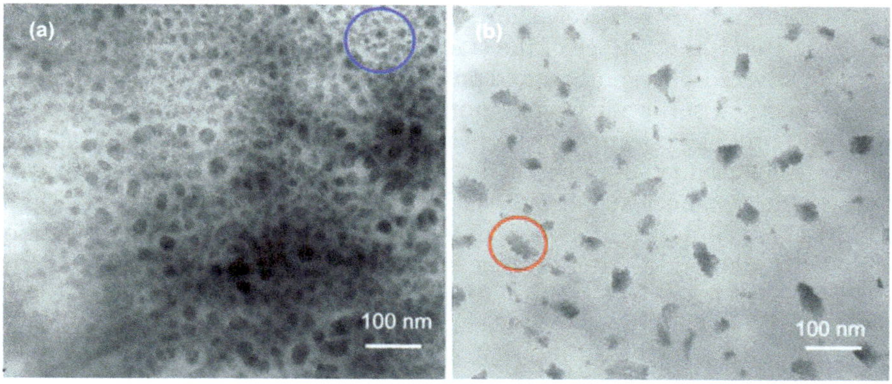


**Figure 3.15** TEM images of nanocrystals prepared from waxy maize (a) and cassava (b) starch granules using $\gamma$-irradiation.
Reproduced from Lamanna *et al.*[130] with permission of Elsevier.

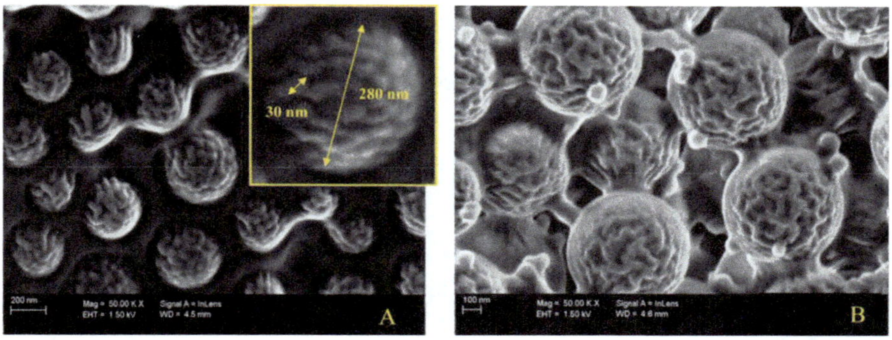


**Figure 3.16** FE-SEM images of freeze-dried nanocomposite dispersions obtained from emulsion polymerization in the presence of 8% w/w starch nanocrystals prepared through $H_2SO_4$ (A) and HCl (B) hydrolysis.
Reproduced from Bel Haaj *et al.*[138] with permission of the American Chemical Society.

Although the chemical modification of the surface of nanocrystals improves the functional characteristics and resulting morphologies, these properties can also be regulated and controlled by other methods that influence the relative molecular weights and lengths of the polymer chains.[75] For example, Bel Haaj *et al.*[138] prepared freeze-dried poly(butyl methacrylate) dispersions through emulsion polymerization in the presence of 8% w/w starch nanocrystals obtained from $H_2SO_4$ and HCl hydrolysis. FE-SEM observations showed that the polymer particles were homogeneous spheres with a rubbery-like morphology due to the strong attachments of the starch nanocrystals on the polymer particles (Figure 3.16).

Wang *et al.*[139] investigated the copolymerization of starch nanocrystal-*g*-polystyrene and reported that in the initial grafting stage, the polystyrene graft side chain was relatively short but became lengthened with increasing reaction. Hence the surface of the starch nanocrystals in the initial stages is partially covered with short grafted polystyrene chains, but after longer times the nanocrystal surface becomes completely covered with hydrophobic side chains to produce essentially spherical particles.[139] In another study,[140] the same group evaluated the structure, morphology and characteristics of benzyl starch nanocrystals. Analysis by SEM showed evidence of morphological changes during nanocrystal formation and following modification with benzyl chloride. After 3 days of acid hydrolysis, native corn starch granules with a size of 5–40 μm were fragmented into lamellae with sizes below 10 μm and a "snowflake"-like appearance (Figure 3.17b), and further treatment of the residues resulted in 50–150 nm particles (Figure 3.17c). Figure 3.17d shows that the size of the benzyl chloride-modified starch nanocrystals was decreased in comparison with the unmodified types. Poorly substituted benzyl starch nanocrystals had a homogeneous distribution owing to the weakness of hydrogen bond interactions between starch macromolecules and hydrophobic benzyl groups.[140] Zou *et al.*[141] reported that the reinforcement of WPU filled with starch nanocrystals was improved after the addition of increasing amounts of the nanocrystals. Piyada *et al.*[105] found that rice starch films reinforced with starch nanocrystals became

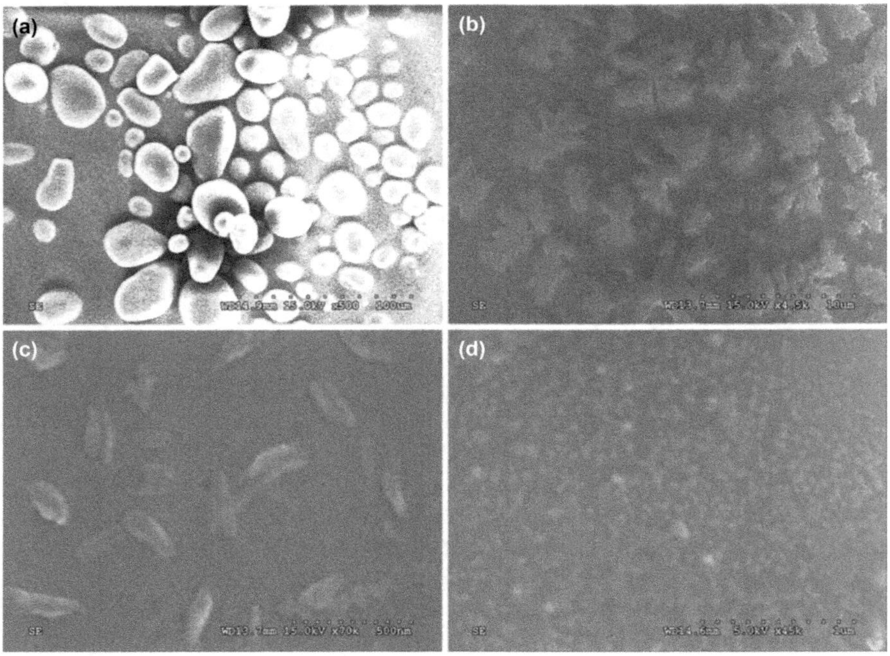


**Figure 3.17** SEM images of natural corn starch (a), corn starch hydrolysed after 3 days (b), starch nanocrystals (c) and benzyl starch nanocrystals (d). Reproduced from Wang *et al.*[140] with permission of Springer-Verlag.

rougher as the nanocrystal loading level increased, with no adverse morphological effects being observed when the concentration increased from 5 to 20% w/w. However, when the nanocrystal concentration was greater than 25% w/w, nanopores appeared in the film structure due to the aggregation of starch nanocrystals and microphase separation between the nanoparticles and the matrix.[133]

### 3.4.6 Emulsification Characteristics

In recent years, the presence of solid particles during Pickering emulsion polymerization has been utilized as an easy and quick method to develop and stabilize complex colloidal structures.[138] The solid particles used in Pickering emulsions have dimensions ranging from several nanometres to several micrometres with intermediate wettability that can adequately attach to liquid/liquid interfaces and provide long-term emulsion stability.[142] The use of these environmentally friendly particles can reduce the need for conventional emulsifying agents and hazardous surfactants in various emulsion systems.[143] The application of Pickering emulsion polymerization is therefore of great interest in the formulation of food, nutraceutical and cosmetic packaging products, since common surfactants can adversely affect the strength resistance and WVP of biodegradable films.[144,145] Moreover, the broad availability and moderately facile production of these particles, along with their biodegradability, high biocompatibility and non-toxic nature, can also warrant the use of this sustainable class of nanoparticles as stabilizers for replacing synthetic surfactants.[146] Starch nanocrystals can also be used as particle emulsifiers and they are now attracting considerable attention because these materials are inexpensive, abundant, biocompatible, biodegradable and non-toxic. They have also shown outstanding reinforcement in other organic and inorganic materials such as soybean protein, thermoplastic starch, pullulan, natural rubber, polyurethane and PVA, among others. The synthesis of starch nanocrystals with various sizes and shapes has been widely reported,[77,92] with the industrially important platelet starch nanocrystals obtained from hydrolysis of native starch granules relatively easy to obtain with thicknesses of 6–8 nm, lengths of 40–60 nm and widths of 15–30 nm.[146]

Given the favourable size of these starch nanocrystals (<100 nm), their platelet shape and the ease of surface modification to obtain different hydrophobicities, these particles have a high potential for use as emulsifiers for stabilizing Pickering emulsions. As such, emulsion-stabilized starch nanocrystals can lead to the formation of well-dispersed nanocomposites upon drying, with improved homogeneity. This enhanced homogeneity can in turn improve the reinforcing and barrier effects of the nanocomposites. The use of this functional material can provide other benefits, such as a short-term formulation cycles, the elimination of the essential dilution steps after mixing of the polymer emulsion and nanoparticle dispersion and reduced use or omission of conventional organic emulsifying agents.[138]

Improved interactions between starch nanocrystals and polymer particles under these conditions can provide nanoparticle individualization and better tracking of the nanocomposite properties on the nanoscale level and can thus form strong chemical linkages between the polymer and reinforcement *via* suitable reactive functional groups.[147] This can significantly improve the transparency of films and increase the nanoparticle reinforcing potential and barrier properties, especially in the case of platelet-like particles.[148]

The application of different types of starch particles as emulsifiers in both oil-in-water (O/W) and water-in-oil (W/O) emulsions has been reported, and the potential application of partially hydrophobic micron-sized starch particles from various sources to emulsify tetradecane in water has been studied.[149] Marku *et al.*[150] attempted to stabilize O/W emulsions based on Miglyol, paraffin and shea nut oil employing modified quinoa starch granules. Li *et al.*[146] demonstrated that liquid paraffin-in-water emulsions could be stable to both creaming and coalescence with a starch nanocrystal content of 3%. Kargar *et al.*[151] also used modified starch particles with similar origins and of the same size to produce stable sunflower O/W emulsions. Tan *et al.*[152] concluded that acetylated starch phthalic ester can be used as an emulsifier to stabilize undecanol–water emulsions and to develop monodisperse nanospheres through a precipitation method. Timgren *et al.*[153] investigated the emulsifying capacity and physical stability of various types of native starch granules with different sizes and amylose/amylopectin ratios. They suggested that modified starch can greatly stabilize Pickering emulsions against the coalescence of droplets for a minimum of 2 years. Tan *et al.*[154] investigated the stability of glyceryl trioctanoate–water emulsions using novel starch-based nanospheres by evaluating the critical structural components in terms of particle concentration, oil/water ratio and salt concentration. The results showed that starch nanospheres can be highly adsorbed at the liquid interface to form a dense film around the dispersed drops, impeding coalescence. The size of the oil droplets was decreased by increasing the concentration of starch nanoparticles up to a specific level. The O/W emulsions could be converted to W/O emulsions by increasing the volume fraction of the oil phase to >0.65. Li *et al.*[155] used starch nanocrystals prepared *via* the acid hydrolysis of amorphous regions of waxy-maize starch granules below their gelatinization temperature to prepare and stabilize Pickering emulsions. It was found that starch nanocrystals could be adequately adsorbed at the oil/water interface to retard the creaming and coalescence mechanisms due to the formation of an interparticle network in the emulsions. The droplet size of starch nanocrystal-stabilized emulsions could be tailored by changing the pH value or salt concentration. Since the starch nanocrystals were negatively charged in the aqueous dispersion due to the ionization of the sulfate groups at the surface, their electrostatic repulsion was decreased at low pH values or high salt contents. This phenomenon further caused droplet aggregation and a significant increase in the particle size of the starch nanocrystal-stabilized emulsions. In a recent study, Bel Haaj *et al.*[138] reported that starch nanocrystals obtained from HCl hydrolysis provide a better stabilization effect with a lower average particle size compared with those hydrolysed using $H_2SO_4$.

### 3.4.7 Crystallinity Characteristics

Reported data pertaining to the crystallinity of starch granules are clearly very important for the large-scale optimized production of starch nanocrystals. One of the major properties of the crystalline regions in starch granules is the polymorphism of the α-glucans. Since native starch granules have crystalline regions, their presence can be detected using XRD.[77] The appearance of XRD patterns depends on the water content of the starch granules during the analysis: greater starch hydration results in thinner diffraction ring patterns up to a certain point. Water content is therefore one of the major and more critical variables in the organization of the crystalline regions in starches. The transition from crystalline to amorphous occurs primarily between 60 and 70 °C in water by a process known as gelatinization.[10] In general, native starches comprise between 15 and 45% crystalline content and XRD patterns have permitted the classification of three crystalline types, named A, B and C.[10] To clarify the difference between the A and B types of starches, a comprehensive model was presented to describe the double-helix packing configuration.[156] The A-type structure is strongly packed with water molecules between each double-helical structure, whereas the B-type structure is more open with the surrounding water molecules within the central cavity constructed by six double helices. Jane *et al.*[157] reported the possible existence of different branching patterns in various types of starch granules. The branching points in the amylopectin of B- and A-type starches are clustered with smaller amorphous lamellae and are scattered between the amorphous and crystalline regions with the increased flexibility of closely packed double helices.[10] Other important parameters in the crystalline development of starch granules include the branching density inside each cluster and the distance between the $\alpha(1 \rightarrow 6)$ linkages.[158] Overall, it has been hypothesized that crystalline polymorphism can affect the amylopectin chain length.[159] Therefore, the A-type starch with clusters comprised of many short chains and short linkages can provide densely packed structures, whereas the B-type can be produced with both longer chains and linkages. An XRD analysis showed that C-type starch is a polymorphic mixture of B-type structures at the granule centre and A-type structures surrounding the granules.[28] Wang *et al.*[160] hydrolysed and characterized C-type starch granules and found that the amorphous regions of B-type polymorphs in the core were more hydrolysed than the crystalline regions of A-type polymorphs in the framework. Halley and Avérous[161] observed that a decrease in the crystallinity can lead to an increase in the amylose content in maize, barley, wheat and rice starches (Figure 3.18) *via* the transition in the double-helix packing arrangement from A-type to

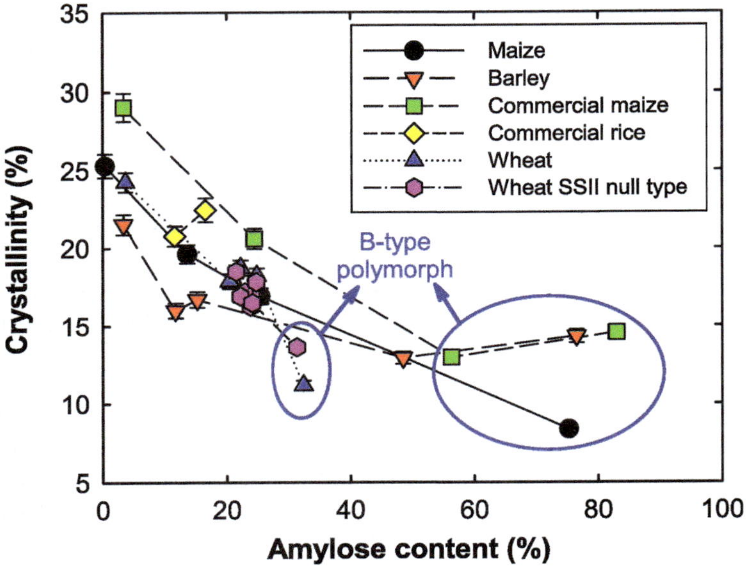


**Figure 3.18** The difference in the degree of crystallinity with amylose content for maize, barley, wheat and rice starches.
Reproduced from Halley and Avérous[161] with permission of Elsevier.

B-type. These findings confirm the predominant role of amylopectin in supporting the crystalline framework in starch granules.

Xu *et al.*[45] used XRD analysis to demonstrate that the starches derived from corn, barley and tapioca have a typical A-type crystalline structure with $2\theta$ diffraction peaks at about 15°, an unresolved doublet at 17.0 and 17.6° and 22.8° and a small peak at 19.5°. Le Corre *et al.*[58] reported that both waxy maize and wheat native starches have A-type crystallinity polymorphs, with waxy maize starches more crystalline than wheat starches. Piyada *et al.*[105] found that rice starch powders have a characteristically typical A-type crystalline pattern with strong reflections at 16.8, 18.0 and 22.7°. Xu *et al.*[45] found that the potato starch granules with a B-type pattern exhibited a strong $2\theta$ reflection peak at 17° and several medium $2\theta$ peaks at around 5.4, 14.4, 19.3, 21.8 and 23.7°. Starch nanocrystals from native starches often present similar diffraction patterns with A- or B-type diffraction patterns or even a combination of the two (C-type), depending on the packing configuration of the crystallites. They have closely packed amylopectin side chains because they are more crystalline than their native counterparts, suggesting considerable hydrolysis of amorphous regions.[40] Jivan *et al.*[60] showed that the crystallinity of starch nanocrystals generated from potato starch granules (62%) was almost double that of the native granular starch (35%). Singh *et al.*[162] and Xu *et al.*[45] also reported relative crystallinities of 50% and 32% for starch nanocrystals with the source potato starch, respectively. They also reported B-type scattering patterns for both samples with clear diffraction peaks at 17.2° (strong) and 19.9° (weak) and a double peak at 22.4–24.2°. This result is in accordance with that of Piyada *et al.*,[105] who reported a 30.36% crystallinity of starch nanocrystals obtained from 23.19% crystalline native rice starches.

The changes in diffraction patterns of starch nanocrystals after acid hydrolysis correspond to the nature of the original botanical source and to the level of crystallinity in the native starch granules.[45] Other factors, such as crystal size, orientation of the double helices within the crystalline domains and the extent of interaction between double helices, may also be responsible for the differences in the degree of crystallinities reported between the various starches.[163] Xu *et al.*[45] studied the diffraction patterns of starch nanocrystals from B-type (potato) starch, which they compared with the corresponding native starch. They confirmed that the starch nanocrystals did not lose their crystallinity after a chemical grafting treatment and this could be a reason for the improvement in thermal stability, compatibility and reinforcing effect of starch nanocrystals. Other reports have suggested that B-type starch is more sensitive to acid hydrolysis than the A- or C-type starches.[29] Some workers have also used XRD to explore the presence of starch nanocrystals in polymeric matrices.[53,76] For example, the crystallinity of rice starch films (13.3%) was considerably decreased by the addition of more than 20% w/w rice starch nanocrystals, mainly due to the destruction of crystalline starch granules during gelatinization.[164] Nonetheless, the crystallinity was still much greater than that of unfilled rice starch films and it was concluded that starch nanocrystals can increase the crystallinity of the resultant films. Chen *et al.*[46] studied the crystallinity of PVA films reinforced with pea starch nanocrystals. The nanocrystals displayed the typical C-type crystalline pattern containing A- and B-type polymorphs like other legume starches. Interestingly, it was observed that the crystalline structure of starch nanocrystals remain unchanged in the blend films owing to their preservation during the blending process under the conditions of the experiment (room temperature and evaporation process at 40 °C). However, the nanocrystals were weakened in the extrusion film-forming process, suggesting that the diffraction pattern of PVA dominated in the composite and the crystallinity of the starch nanocrystals was partly inhibited by the strong interactions between PVA and the nanocrystal molecules.[133]

## 3.5 Conclusion

Starch can be considered a suitable polymer for preparing biodegradable nanoparticles owing to its abundance in nature and low cost. Several methods for preparing starch nanocrystals have been developed and optimized over recent years, with improvements in techniques, yields and resulting properties. Throughout this chapter, it has become evident that the semicrystalline nature of starch is one of most critical characteristic properties allowing the fabrication of nanocrystals. This chapter has summarized some of the important technologies involved in producing these nanocrystals, which is typically achieved *via* sulfuric or hydrochloric acid hydrolysis of starch granules. New and improved solutions have been discussed that aim to optimize the preparation of starch nanocrystals; however, despite the recent progress in synthesizing nanocrystals, there are still many critical issues that need to be addressed. In addition to the synthesis of starch nanocrystals, a number of techniques have also been developed to understand and characterize the behaviour of various polymer matrices in which the particles have been filled. Several very recent and ongoing studies in this research area have found some industrial applications, although the expected commercial success has yet to be documented.

## References


1. D. Le Corre and A. Dufresne, Preparation of starch nanoparticles, in *Biopolymer Nanocomposites: Processing, Properties, and Applications*, ed. A. Dufresne, S. Thomas and L. A. Pothen, Wiley, Hoboken, NJ, 2013, pp. 153–180.
2. R. Tharanathan, *Trends Food Sci. Technol.*, 2003, **14**, 71–78.
3. D. Salarbashi, S. Tajik, S. Shojaee-Aliabadi, M. Ghasemlou, H. Moayyed, R. Khaksar and M. S. Noghabi, *Food Chem.*, 2014, **146**, 614–622.
4. D. Le Corre, A. Dufresne, M. Rueff, B. Khelifi and J. Bras, *J. Appl. Polym. Sci.*, 2014, **131**, DOI: 10.1002/APP.39826.
5. S. Tajik, Y. Maghsoudlou, F. Khodaiyan, S. M. Jafari, M. Ghasemlou and M. Aalami, *Carbohydr. Polym.*, 2013, **97**, 817–824.
6. A. Arora and G. Padua, *J. Food Sci.*, 2010, **75**, R43–R49.
7. S. Shojaee-Aliabadi, M. A. Mohammadifar, H. Hosseini, A. Mohammadi, M. Ghasemlou, S. M. Hosseini, M. Haghshenas and R. Khaksar, *Int. J. Biol. Macromol.*, 2014, **69**, 282–289.
8. M. Mariano, N. El Kissi and A. Dufresne, *J. Polym. Sci., Part B: Polym. Phys.*, 2014, **52**, 791–806.
9. M. A. S. Azizi Samir, F. Alloin and A. Dufresne, *Biomacromolecules*, 2005, **6**, 612–626.
10. D. Le Corre, J. Bras and A. Dufresne, *Biomacromolecules*, 2010, **11**, 1139–1153.
11. I. J. Tetlow, *Seed Sci. Res.*, 2011, **21**, 5–32.
12. S. C. Zeeman, J. Kossmann and A. M. Smith, *Annu. Rev. Plant Biol.*, 2010, **61**, 209–234.
13. S. H. Imam, D. F. Wood, M. A. Abdelwahab, B.-S. Chiou, T. G. Williams, G. M. Glenn and W. J. Orts, Starch: chemistry, microstructure, processing, and enzymatic degradation, in *Starch-Based Polymeric Materials and Nanocomposites: Chemistry, Processing and Applications*, ed. J. Ahmed, B. K. Tiwari, S. H. Imam and M. A. Rao, CRC Press, Boca Raton, 2012, pp. 5–32.
14. H. Angellier, L. Choisnard, S. Molina-Boisseau, P. Ozil and A. Dufresne, *Biomacromolecules*, 2004, **5**, 1545–1551.
15. C. G. Oates, *Trends Food Sci. Technol.*, 1997, **8**, 375–382.

16. D. J. Gallant, B. Bouchet and P. M. Baldwin, *Carbohydr. Polym.*, 1997, **32**, 177–191.
17. M. Yamaguchi, K. Kainuma and D. French, *J. Ultrastruct. Res.*, 1979, **69**, 249–261.
18. A. Buléon, P. Colonna, V. Planchot and S. Ball, *Int. J. Biol. Macromol.*, 1998, **23**, 85–112.
19. H. Angellier, S. Molina-Boisseau, L. Lebrun and A. Dufresne, *Macromolecules*, 2005, **38**, 3783–3792.
20. S. Pérez and E. Bertoft, *Starch Stärke*, 2010, **62**, 389–420.
21. R. F. Tester, J. Karkalas and X. Qi, *J. Cereal Sci.*, 2004, **39**, 151–165.
22. J. Dang and L. Copeland, *J. Cereal Sci.*, 2003, **37**, 165–170.
23. D. Le Corre, J. Bras and A. Dufresne, *Carbohydr. Polym.*, 2012, **87**, 658–666.
24. S. Hizukuri, T. Kaneko and Y. Takeda, *Biochim. Biophys. Acta, Gen. Subj.*, 1983, **760**, 188–191.
25. A. Imberty, A. Buléon, V. Tran and S. Péerez, *Starch Stärke*, 1991, **43**, 375–384.
26. S. Wang and L. Copeland, *Crit. Rev. Food Sci. Nutr.*, 2015, **55**, 1081–1097.
27. R. Sousa, V. Correlo, S. Chung, N. Neves, J. Mano and R. Reis, Processing of starch-based blends for biomedical applications, in *Natural-Based Polymers for Biomedical Applications*, ed. R. L. Reis, N. M. Neves, J. F. Mano, M. E. Gomes, A. P. Marques and H. S. Azevedo, Woodhead Publishing, Abington, 2008, pp. 85–105.
28. T. Y. Bogracheva, V. Morris, S. Ring and C. Hedley, *Biopolymers*, 1998, **45**, 323–332.
29. H. Y. Kim, J. H. Lee, J. Y. Kim, W. J. Lim and S. T. Lim, *Starch Stärke*, 2012, **64**, 367–373.
30. R. R. Koshy, L. A. Pothan and S. Thomas, Starch-based bionanocomposite: processing techniques, in *Biopolymer Nanocomposites: Processing, Properties, and Applications*, ed. A. Dufresne, S. Thomas and L. A. Pothen, Wiley, Hoboken, NJ, 2013, pp. 203–226.
31. H. Angellier-Coussy, J.-L. Putaux, S. Molina-Boisseau, A. Dufresne, E. Bertoft and S. Perez, *Carbohydr. Res.*, 2009, **344**, 1558–1566.
32. A. Vazquez, V. P. Cyras, V. A. Alvarez and J. I. Moran, Starch/clay nano-biocomposites, in *Environmental Silicate Nano-Biocomposites*, ed. L. Avérous and E.Pollet, Springer, London, 2012, pp. 287–321.
33. J. P. Maran, V. Sivakumar, R. Sridhar and K. Thirugnanasambandham, *Carbohydr. Polym.*, 2013, **92**, 1335–1347.
34. K. K. Kuorwel, M. J. Cran, K. Sonneveld, J. Miltz and S. W. Bigger, *LWT-Food Sci. Technol.*, 2013, **50**, 432–438.
35. K. K. Kuorwel, M. J. Cran, K. Sonneveld, J. Miltz and S. W. Bigger, *Packag. Technol. Sci.*, 2014, **27**, 149–159.
36. A. Jiménez, M. J. Fabra, P. Talens and A. Chiralt, *Food Hydrocolloids*, 2012, **26**, 302–310.
37. J. Viguié, S. Molina Boisseau and A. Dufresne, *Macromol. Biosci.*, 2007, **7**, 1206–1216.

38. M. Ghasemlou, N. Aliheidari, R. Fahmi, S. Shojaee-Aliabadi, B. Keshavarz, M. J. Cran and R. Khaksar, *Carbohydr. Polym.*, 2013, **98**, 1117–1126.
39. S. Song, C. Wang, Z. Pan and X. Wang, *J. Appl. Polym. Sci.*, 2008, **107**, 418–422.
40. D. Le Corre, J. Bras and A. Dufresne, *J. Nanopart. Res.*, 2011, **13**, 7193–7208.
41. A. Dufresne, *Molecules*, 2010, **15**, 4111–4128.
42. J.-L. Putaux, S. Molina-Boisseau, T. Momaur and A. Dufresne, *Biomacromolecules*, 2003, **4**, 1198–1202.
43. H. M. Palma Rodriguez, E. Agama Acevedo, G. Mendez Montealvo, R. A. Gonzalez Soto, E. J. Vernon Carter and L. A. Bello Pérez, *Starch Stärke*, 2012, **64**, 115–125.
44. N. Lin, J. Huang, P. R. Chang, D. P. Anderson and J. Yu, *J. Nanomater.*, 2011, **2011**, 20.
45. Y. Xu, E. Sismour, C. Grizzard, M. Thomas, D. Pestov, Z. Huba, T. Wang and H. L. Bhardwaj, *Cereal Chem.*, 2014, **91**, 383–388.
46. G. Chen, M. Wei, J. Chen, J. Huang, A. Dufresne and P. R. Chang, *Polymer*, 2008, **49**, 1860–1870.
47. D. Liu, Q. Wu, H. Chen and P. R. Chang, *J. Colloid Interface Sci.*, 2009, **339**, 117–124.
48. N. Lin, J. Huang and A. Dufresne, *Nanoscale*, 2012, **4**, 3274–3294.
49. P. Mélé, H. Angellier-Coussy, S. Molina-Boisseau and A. Dufresne, *Biomacromolecules*, 2011, **12**, 1487–1493.
50. H. Angellier, S. Molina-Boisseau, M. N. Belgacem and A. Dufresne, *Langmuir*, 2005, **21**, 2425–2433.
51. E. Kristo and C. G. Biliaderis, *Carbohydr. Polym.*, 2007, **68**, 146–158.
52. H. Angellier, S. Molina-Boisseau, P. Dole and A. Dufresne, *Biomacromolecules*, 2006, 7, 531–539.
53. H. Zheng, F. Ai, P. R. Chang, J. Huang and A. Dufresne, *Polym. Compos.*, 2009, **30**, 474–480.
54. E. Bouthegourd, K. Rajisha, N. Kalarical, J. M. Saiter and S. Thomas, *Mater. Lett.*, 2011, **65**, 3615–3617.
55. A. Dufresne, J.-Y. Cavaille and W. Helbert, *Macromolecules*, 1996, **29**, 7624–7626.
56. D. Le Corre, J. Bras and A. Dufresne, *Biomacromolecules*, 2011, **12**, 3039–3046.
57. H.-Y. Kim, J.-A. Han, D.-K. Kweon, J.-D. Park and S.-T. Lim, *Carbohydr. Polym.*, 2013, **93**, 582–588.
58. D. Le Corre, J. Bras and A. Dufresne, *Carbohydr. Polym.*, 2011, **86**, 1565–1572.
59. D. Le Corre, J. Bras, L. Choisnard and A. Dufresne, *Starch Stärke*, 2012, **64**, 489–496.
60. M. J. Jivan, A. Madadlou and M. Yarmand, *Food Chem.*, 2013, **141**, 1661–1666.
61. Y.-J. Wang, V.-D. Truong and L. Wang, *Carbohydr. Polym.*, 2003, **52**, 327–333.

62. W. Nägeli, *Justus Liebigs Ann. Chem.*, 1874, **173**, 218–227.
63. C. Lintner, *J. für Prakt. Chem.*, 1886, **34**, 378–394.
64. S. Wang, J. Blazek, E. Gilbert and L. Copeland, *Carbohydr. Polym.*, 2012, **87**, 1941–1949.
65. W. Li, H. Corke and T. Beta, *Carbohydr. Polym.*, 2007, **69**, 398–405.
66. Y. Nakazawa and Y.-J. Wang, *Carbohydr. Res.*, 2003, **338**, 2871–2882.
67. T. Vasanthan and R. Bhatty, *Cereal Chem.*, 1996, **73**, 199–207.
68. L. Jayakody and R. Hoover, *Food Res. Int.*, 2002, **35**, 665–680.
69. R. Hoover and H. Manuel, *J. Cereal Sci.*, 1996, **23**, 153–162.
70. D. Song, Y. S. Thio and Y. Deng, *Carbohydr. Polym.*, 2011, **85**, 208–214.
71. J.-Y. Kim, D.-J. Park and S.-T. Lim, *Cereal Chem.*, 2008, **85**, 182–187.
72. J.-Y. Kim and S.-T. Lim, *Carbohydr. Polym.*, 2009, **76**, 110–116.
73. Q. Sun, G. Li, L. Dai, N. Ji and L. Xiong, *Food Chem.*, 2014, **162**, 223–228.
74. P. R. Chang, F. Ai, Y. Chen, A. Dufresne and J. Huang, *J. Appl. Polym. Sci.*, 2009, **111**, 619–627.
75. J. Huang, Q. Huang, P. R. Chang and J. Yu, Chemical modification of starch nanoparticles, in *Biopolymer Nanocomposites: Processing, Properties, and Applications*, ed. A. Dufresne, S. Thomas and L. A. Pothen, Wiley, Hoboken, NJ, 2013, pp. 181–202.
76. N. L. García, L. Ribba, A. Dufresne, M. Aranguren and S. Goyanes, *Carbohydr. Polym.*, 2011, **84**, 203–210.
77. H. Namazi and A. Dadkhah, *Carbohydr. Polym.*, 2010, **79**, 731–737.
78. L. Cai, Y.-C. Shi, L. Rong and B. S. Hsiao, *Carbohydr. Polym.*, 2010, **81**, 385–393.
79. H. An, H. Liang, Z. Liu, H. Yang, Q. Liu and H. Wang, *J. Food Sci.*, 2011, **76**, N11–N14.
80. S. F. Chin, S. C. Pang and S. H. Tay, *Carbohydr. Polym.*, 2011, **86**, 1817–1819.
81. X. Ma, R. Jian, P. R. Chang and J. Yu, *Biomacromolecules*, 2008, **9**, 3314–3320.
82. G. Siqueira, J. Bras and A. Dufresne, *Biomacromolecules*, 2008, **10**, 425–432.
83. Y. Tan, P. Wang, K. Xu, W. Li, H. An, L. Li, C. Liu and L. Dong, *Macromol. Mater. Eng.*, 2009, **294**, 855–859.
84. J. Han, G. Borjihan, R. Bai, X. Chen and X. Jing, *J. Appl. Polym. Sci.*, 2008, **108**, 523–528.
85. J. Fang, P. Fowler, C. Sayers and P. Williams, *Carbohydr. Polym.*, 2004, **55**, 283–289.
86. J. Fang, P. Fowler, J. Tomkinson and C. Hill, *Carbohydr. Polym.*, 2002, **47**, 245–252.
87. J. Aburto, I. Alric and E. Borredon, *Starch Stärke*, 2005, **57**, 145–152.
88. L. Ren, M. Jiang, L. Wang, J. Zhou and J. Tong, *Carbohydr. Polym.*, 2012, **87**, 1874–1876.
89. Y. Habibi and A. Dufresne, *Biomacromolecules*, 2008, **9**, 1974–1980.
90. H. Namazi and A. Dadkhah, *J. Appl. Polym. Sci.*, 2008, **110**, 2405–2412.

91. W. Thielemans, M. N. Belgacem and A. Dufresne, *Langmuir*, 2006, **22**, 4804–4810.
92. Y. Xu, W. Ding, J. Liu, Y. Li, J. F. Kennedy, Q. Gu and S. Shao, *Carbohydr. Polym.*, 2010, **80**, 1078–1084.
93. L. Ren, Z. Dong, M. Jiang, J. Tong and J. Zhou, *Ind. Crops Prod.*, 2014, **59**, 115–118.
94. K. Rajisha, H. Maria, L. Pothan, Z. Ahmad and S. Thomas, *Int. J. Biol. Macromol.*, 2014, **67**, 147–153.
95. Y. Habibi, A.-L. Goffin, N. Schiltz, E. Duquesne, P. Dubois and A. Dufresne, *J. Mater. Chem.*, 2008, **18**, 5002–5010.
96. M. Labet, W. Thielemans and A. Dufresne, *Biomacromolecules*, 2007, **8**, 2916–2927.
97. S. Rooj, A. Das, V. Thakur, R. Mahaling, A. K. Bhowmick and G. Heinrich, *Materials & Design*, 2010, **31**, 2151–2156.
98. K. Gopalan Nair and A. Dufresne, *Biomacromolecules*, 2003, **4**, 666–674.
99. H. Angellier, S. Molina-Boisseau and A. Dufresne, *Macromolecules*, 2005, **38**, 9161–9170.
100. J. Bras, M. L. Hassan, C. Bruzesse, E. A. Hassan, N. A. El-Wakil and A. Dufresne, *Ind. Crops Prod.*, 2010, **32**, 627–633.
101. A. Bendahou, H. Kaddami and A. Dufresne, *Eur. Polym. J.*, 2010, **46**, 609–620.
102. X. Cao, H. Dong and C. M. Li, *Biomacromolecules*, 2007, **8**, 899–904.
103. H. Angellier, J. L. Putaux, S. Molina Boisseau, D. Dupeyre and A. Dufresne, *Macromol. Symp.*, 2005, **221**, 95–104.
104. M. Valodkar and S. Thakore, *Int. J. Polym. Anal. Charact.*, 2010, **15**, 387–395.
105. K. Piyada, S. Waranyou and W. Thawien, *Int. Food Res. J*, 2013, **20**, 439–449.
106. N. L. García, L. Ribba, A. Dufresne, M. I. Aranguren and S. Goyanes, *Macromol. Mater. Eng.*, 2009, **294**, 169–177.
107. Y. Wang, H. Tian and L. Zhang, *Carbohydr. Polym.*, 2010, **80**, 665–671.
108. J. Yu, F. Ai, A. Dufresne, S. Gao, J. Huang and P. R. Chang, *Macromol. Mater. Eng.*, 2008, **293**, 763–770.
109. K. K. Mokwena and J. Tang, *Crit. Rev. Food Sci. Nutr.*, 2012, **52**, 640–650.
110. B. Duan, P. Sun, X. Wang and C. Yang, *Starch Stärke*, 2011, **63**, 528–535.
111. P. R. Chang, R. Jian, J. Yu and X. Ma, *Carbohydr. Polym.*, 2010, **80**, 420–425.
112. D. S. Le Corre, J. Bras and A. Dufresne, *Macromol. Mater. Eng.*, 2012, **297**, 969–978.
113. D. Le Corre, C. Hohenthal, A. Dufresne and J. Bras, *J. Polym. Environ.*, 2013, **21**, 71–80.
114. A. Dufresne, *Can. J. Chem.*, 2008, **86**, 484–494.
115. A. Dufresne and J.-Y. Cavaille, *J. Polym. Sci. Part B: Polym. Phys.*, 1998, **36**, 2211–2224.
116. H. Bizot, P. Le Bail, B. Leroux, J. Davy, P. Roger and A. Buleon, *Carbohydr. Polym.*, 1997, **32**, 33–50.

117. A. Dufresne and M. R. Vignon, *Macromolecules*, 1998, **31**, 2693–2696.
118. A. Dufresne, D. Dupeyre and M. R. Vignon, *J. Appl. Polym. Sci.*, 2000, **76**, 2080–2092.
119. K. Gopalan Nair and A. Dufresne, *Biomacromolecules*, 2003, **4**, 657–665.
120. A. Celzard, E. McRae, C. Deleuze, M. Dufort, G. Furdin and J. Marêché, *Physical Review B*, 1996, **53**, 6209.
121. Y. Yin, J. Li, Y. Liu and Z. Li, *J. Appl. Polym. Sci.*, 2005, **96**, 1394–1397.
122. Y. U. Nabar, D. Draybuck and R. Narayan, *J. Appl. Polym. Sci.*, 2006, **102**, 58–68.
123. J. Yu, J. Yang, B. Liu and X. Ma, *Bioresour. Technol.*, 2009, **100**, 2832–2841.
124. A. Kaushik, M. Singh and G. Verma, *Carbohydr. Polym.*, 2010, **82**, 337–345.
125. M. Roman and W. T. Winter, *Biomacromolecules*, 2004, **5**, 1671–1677.
126. S. Srichuwong, T. C. Sunarti, T. Mishima, N. Isono and M. Hisamatsu, *Carbohydr. Polym.*, 2005, **60**, 529–538.
127. J. S. Smith, D. Bedrov and G. D. Smith, *Compos. Sci. Technol.*, 2003, **63**, 1599–1605.
128. J. Mano, D. Koniarova and R. Reis, *J. Mater. Sci.: Mater. Med.*, 2003, **14**, 127–135.
129. S. Thiebaud, J. Aburto, I. Alric, E. Borredon, D. Bikiaris, J. Prinos and C. Panayiotou, *J. Appl. Polym. Sci.*, 1997, **65**, 705–721.
130. M. Lamanna, N. J. Morales, N. L. García and S. Goyanes, *Carbohydr. Polym.*, 2013, **97**, 90–97.
131. N. García, M. Lamanna, N. D'Accorso, A. Dufresne, M. Aranguren and S. Goyanes, *Polym. Degrad. Stab.*, 2012, **97**, 2021–2026.
132. D. Yu, S. Xiao, C. Tong, L. Chen and X. Liu, *Chin. Sci. Bull.*, 2007, **52**, 2913–2918.
133. Y. Chen, X. Cao, P. R. Chang and M. A. Huneault, *Carbohydr. Polym.*, 2008, **73**, 8–17.
134. B. Wei, X. Hu, H. Li, C. Wu, X. Xu, Z. Jin and Y. Tian, *Food Hydrocolloids*, 2014, **36**, 369–373.
135. G. T. Oostergetel and E. F. van Bruggen, *Carbohydr. Polym.*, 1993, **21**, 7–12.
136. I. Capron, P. Robert, P. Colonna, M. Brogly and V. Planchot, *Carbohydr. Polym.*, 2007, **68**, 249–259.
137. T. A. Waigh, K. L. Kato, A. M. Donald, M. J. Gidley, C. J. Clarke and C. Riekel, *Starch Stärke*, 2000, **52**, 450–460.
138. S. Bel Haaj, W. Thielemans, A. Magnin and S. Boufi, *ACS Appl. Mater. Interfaces*, 2014, **6**, 8263–8273.
139. C. Wang, Z. Pan, M. Wu and P. Zhao, *J. Appl. Polym. Sci.*, 2014, **131**, DOI: 10.1002/app.40571.
140. C. Wang, Z. Pan and J. Zeng, *Arabian J. Sci. Eng.*, 2014, 1–8.
141. J. Zou, F. Zhang, J. Huang, P. R. Chang, Z. Su and J. Yu, *Carbohydr. Polym.*, 2011, **85**, 824–831.

142. P. Kralchevsky, I. Ivanov, K. Ananthapadmanabhan and A. Lips, *Langmuir*, 2005, **21**, 50–63.
143. Y. Chevalier and M.-A. Bolzinger, *Colloids Surf., A*, 2013, **439**, 23–34.
144. L. N. Butler, C. M. Fellows and R. G. Gilbert, *Prog. Org. Coat.*, 2005, **53**, 112–118.
145. J. Frelichowska, M.-A. Bolzinger, J. Pelletier, J.-P. Valour and Y. Chevalier, *Int. J. Pharm.*, 2009, **371**, 56–63.
146. C. Li, P. Sun and C. Yang, *Starch Stärke*, 2012, **64**, 497–502.
147. A. Ben Mabrouk, M. Rei Vilar, A. Magnin, M. N. Belgacem and S. Boufi, *J. Colloid Interface Sci.*, 2011, **363**, 129–136.
148. L. A. Fielding, J. Tonnar and S. P. Armes, *Langmuir*, 2011, **27**, 11129–11144.
149. A. Yusoff and B. S. Murray, *Food Hydrocolloids*, 2011, **25**, 42–55.
150. D. Marku, M. Wahlgren, M. Rayner, M. Sjöö and A. Timgren, *Int. J. Pharm.*, 2012, **428**, 1–7.
151. M. Kargar, K. Fayazmanesh, M. Alavi, F. Spyropoulos and I. T. Norton, *J. Colloid Interface Sci.*, 2012, **366**, 209–215.
152. Y. Tan, K. Xu, C. Liu, Y. Li, C. Lu and P. Wang, *Carbohydr. Polym.*, 2012, **88**, 1358–1363.
153. A. Timgren, M. Rayner, P. Dejmek, D. Marku and M. Sjöö, *Food Sci. Nutr.*, 2013, **1**, 157–171.
154. Y. Tan, K. Xu, C. Niu, C. Liu, Y. Li, P. Wang and B. P. Binks, *Food Hydrocolloids*, 2014, **36**, 70–75.
155. C. Li, Y. Li, P. Sun and C. Yang, *J. Sci. Food Agric.*, 2014, **94**, 1802–1807.
156. A. Imberty, H. Chanzy, S. Perez, A. Buleon and V. Tran, *Macromolecules*, 1987, **20**, 2634–2636.
157. J.-l. Jane, K.-s. Wong and A. E. McPherson, *Carbohydr. Res.*, 1997, **300**, 219–227.
158. C. Gérard, V. Planchot, P. Colonna and E. Bertoft, *Carbohydr. Res.*, 2000, **326**, 130–144.
159. S. Hizukuri, *Carbohydr. Res.*, 1986, **147**, 342–347.
160. S. Wang, J. Yu, F. Jin and J. Yu, *Int. J. Biol. Macromol.*, 2008, **43**, 216–220.
161. *Starch Polymers: from Genetic Engineering to Green Applications*, ed. P. Halley and L. Avérous, Elsevier, Burlington, MA, 2014.
162. J. Singh, C. Lelane, R. B. Stewart and H. Singh, *Food Chem.*, 2010, **121**, 980–989.
163. R. Hoover and W. Ratnayake, *Food Chem.*, 2002, **78**, 489–498.
164. Q.-X. Zhang, Z.-Z. Yu, X.-L. Xie, K. Naito and Y. Kagawa, *Polymer*, 2007, **48**, 7193–7200.

CHAPTER 4

# *Natural Fibre-reinforced Thermoplastic Starch Composites*

J. SAHARI,*[a] S. M. SAPUAN,[b,c] Y. A. EL-SHEKEIL,[b] M. R. ISHAK[d] AND R. AKHTAR[e]

[a] School of Science and Technology, Universiti Malaysia Sabah Jalan UMS, 88400 Kota Kinabalu, Sabah, Malaysia; [b] Department of Mechanical and Manufacturing Engineering, Universiti Putra Malaysia, 43400 UPM Serdang, Selangor, Malaysia; [c] Laboratory of Biocomposite Technology, Institute of Tropical Forestry and Forest Products, Universiti Putra Malaysia, 43400 UPM Serdang, Selangor, Malaysia; [d] Department of Aerospace Engineering, Universiti Putra Malaysia, 43400 UPM Serdang, Selangor, Malaysia; [e] School of Engineering, University of Liverpool, Liverpool L69 3GH, UK
*Email: sahari@ums.edu.my

## 4.1 Introduction

Composites are materials that consist of two or more materials; one of them serves as a matrix and one or more serve as fillers or reinforcing materials. The properties of the composite material vary from the properties of each individual component. In the 1960s, fibre-reinforced polymers were too expensive, and therefore were used only in niche markets. By the 1980s and 1990s their use had extended to broader applications.[1] Polymers are divided into petroleum-based polymers, which are popular, and bio-based polymers,

RSC Green Chemistry No. 37
Starch-based Blends, Composites and Nanocomposites
Edited by Visakh P. M. and Long Yu

Published by the Royal Society of Chemistry, www.rsc.org

which are replacing petroleum-based polymers to reduce the impact of the latter on the environment. Bio-based polymers are now gaining greater awareness for replacing petroleum-based polymers in the field of natural fibre composites. Faruk *et al.* stated that the use of biopolymers is continuously growing year by year,[2] and that the worldwide capacity of bio-based polymers of 0.36 million tonnes in 2007 is expected to expand approximately 10-fold by 2020. The main biopolymers are starch-based polymers, polylactide (PLA) and polyhydroxyalkanoates (PHA).[2] Fibres can be synthetic, such as carbon and glass fibres, and can be natural fibres, such as kenaf, jute and sisal. Owing to the huge amounts of plastics and synthetic fibre–plastic composites and their adverse effects on the environment, the shortage of landfill space and the depletion of petroleum resources – which is needed to produce plastics – the need for bio-based plastics and composites is growing. The new trend in bio-based plastics and composites is based on renewable plant and agricultural stock.[3] It will be difficult to replace petroleum-based products with 100% bio-based materials, and a viable solution is to combine petroleum and bioresources to obtain useful products.[3]

The importance of natural fibre composites has become more apparent owing to their numerous benefits, such as low cost and density. They show less abrasiveness towards equipment and they need less energy for processing compared with synthetic fibre composites. They are also renewable and biodegradable. On the other hand, there are some obstacles facing today's very promising natural fibre composites industry. The first point is the lack of fibre–polymer compatibility, which is the greatest challenge in developing o natural fibre-reinforced polymer composites. This is due to the hydrophobic nature of petroleum-based polymers and the hydrophilic behaviour of natural fibre, which makes it difficult to combine them. An important solution is to use polymers that are derived from thermoplastic starch, which have the same hydrophilic properties as natural fibres.

## 4.2 Natural Fibres

### 4.2.1 Introduction

Natural fibres are considered to be composites of hollow cellulose fibrils with lignin and hemicelluloses as the matrix.[4] Natural fibres are divided into three main types based on their origins: plants, animals and minerals. The term "natural fibres" in this chapter is directed to plant (lignocellulosic) fibres. Plant fibres include straw, bast, leaf, seed and grass fibres.

The main component of plant fibres is cellulose, followed by hemicelluloses and lignin. Cellulose is a linear macromolecule consisting of $C_6H_{11}O_5$ repeating units. Cellulose has an average molecular weight between 130 000 and 190 000 with an average degree of polymerization of approximately 800–1200. Hemicelluloses consist of polysaccharides of comparatively low molecular weight built up from hexoses, pentoses and uronic acid residues.[5] Lignin is a complex chemical compound – it is thought to be a three-dimensional copolymer of aliphatic and aromatic constituents with very high molecular weight. It gives rigidity to plants.[3,5]

Cellulose is the predominant component of natural fibres. The main component of cellulose is anhydro-D-glucose, which contains three hydroxyl groups. These hydroxyl groups make natural fibres hydrophilic in nature.[6,7] Natural fibres are increasingly being used owing to their light weight, non-abrasiveness and combustibility. They are also non-toxic, less expensive than synthetic fibres, with biodegradable properties.

### 4.2.2 Natural Fibre Composites

Natural fibre-reinforced polymers are attracting increasing attention owing to benefits such as renewability and reductions in weight and cost. Further advantages of natural fibres are:

- As plants they contribute to $CO_2$ consumption.
- At the end of natural fibre's life, if burned or landfilled, the amount of $CO_2$ is neutral.
- The abrasive nature of natural fibre is low, which make it easy to process and more recyclable.[8]

Because of worldwide environmental concern and the very rapid consumption of petroleum resources, since plant resources can be renewed $10^5$ times faster than petroleum resources, these factors are forcing the whole world to "go green."[9] Joshi *et al.* studied the life cycle of natural and glass fibre composites and found that natural fibres are environmentally superior to glass fibres in most cases.[10] The following reasons justify this conclusion: (1) processing natural fibres has less impact on the the environment; (2) any application requires a higher percentage of natural fibre than glass fibre to achieve the same performance, which reduces the percentage of polymer needed, and contributes to lowering the cost and reducing the pollution caused by the polymers; (3) the lower density of natural fibre composites results in higher efficiency and less emission in the usage stage, *e.g.* in automotive applications; and (4) burning natural fibres at the end of their life results in energy and carbon credits. However, fertilizer use in natural fibre cultivation results in higher nitrate and phosphate emissions, which can lead to a negative effect on local water resources.[11] The amount of energy consumed to produce a glass fibre is much higher than the energy of producing a natural fibre, *e.g.* the energy needed to produce a fibre such as flax is approximately 17% of that needed to produce the same amount of glass fibre.[12]

Research by Mohanty *et al.* demonstrated that natural fibre composites show mechanical properties comparable to or even better than those of glass fibre-reinforced plastics.[13] However, utilizing natural fibres and replacing glass fibres are still challenging, and much work remains to be done in order to overcome the shortcomings of natural fibres.

Some of the common drawbacks of utilizing natural fibres with polymer composites are high moisture sorption, which affects the properties of the final product, and the need for low processing temperatures owing to the degradation of natural fibres at high temperatures. The high moisture sorption limits the applications of the composites and the low processing temperatures limit the matrix selection to low-melting materials. Another problem facing natural fibre-reinforced polymers is the poor adhesion (*i.e.* incompatibility) between the fibres and polymers due to the hydrophilic nature of cellulose and the hydrophobic nature of polymers. There are three main solutions to tackle the poor adhesion: pretreating natural fibres to enhance their properties, using coupling agents or compatiblizers to modify the polymers and selecting suitable processing methods for producing the composites.[13]

The uniformity and shape of natural fibres raise a further challenge. Along the length of different fibres there are always different cross-sections, which makes the prediction of the mechanical properties of the composites difficul. Natural fibre composites are not considered totally green unless the growth, separation and processing of the fibres can be well controlled. Durability of the composite product is another issue as it should be comparable to that of corresponding glass fibre composites.[11] There are also variations in the properties of natural fibres depending on climate, soil quality, timing of harvesting, location of fields and the location of the fibres on the plant.[14] These variations in the properties of natural fibres consequently affect the quality of the composites produced.

### 4.2.3 Preparation of Natural Fibres

The preparation of natural fibres starts at the farm, with breeding, growing, harvesting and storage of the raw material. After that, the fibres are processed. Fibre processing starts with extraction of the fibres from the plant (*i.e.* retting). Fibres consist of cellulosic cemented with non-cellulosic materials such as lignin, hemicellulose and pectin. Hence retting is the process of separating fibres from woody tissue by the removal of the non-cellulosic materials. Retting can be carried out by field, mechanical, enzymatic or chemical processes, and the selection of the retting method is vital owing to its effect on the final quality of the fibres.[15,16]

Field retting is the most common method, where the stems are kept in the field for up to 6 weeks. During this period, microorganisms degrade pectin – the cementing agent for the cellulose – resulting in the separation of the fibres. Farmers should observe the stems during retting in order to prevent the retting process from affecting the cellulosic material. The quality of fibres retted by this method is dependent on the weather conditions. The mechanical retting method is also called green retting.[15] In this process, the stems are dried for 2–10 days, then the fibres are separated by a hammermill or a decorticator, which is convenient for short fibre production.[16] The quality of the fibres produced by this method is poor. Enzymatic retting is based on the idea of enzymes produced by bacteria in field retting, but here the enzymes are commercially produced. This process takes a shorter time as the amounts of enzymes are controlled. The quality of the fibres depends on the types of enzymes used and the time of processing. In the chemical retting method, retting is performed by using chemicals that dissolve the cementing agents in the stems. Common chemicals used in this process are sodium hydroxide and sodium benzoate.[17] It was reported by Paridah *et al.* that these chemicals produced a lower quality fibres compared to retting with water alone.[17]

## 4.2.4 Characterization of Natural Fibres

### *4.2.4.1 Properties of Natural Fibres*

According to Reddy and Yang, the term "fibres" refers to a bundle of individual cells with adequate strength, length and fineness required for various fibrous industrial applications. The structural model of natural fibre shows the nature of the fibrillar textures, which can be divided into three categories based on size: (1) macrofibrils, (2) fibrils and (3) microfibrils.[18]

The assumption made for the purpose of analysis, that natural fibre is homogeneous, appears to be valid. Although the presence of cells makes the material inhomogeneous on a microscopic scale, it has been shown theoretically that for the units that are large compared with the dimensions of the cells or with the width of the annual rings, the gross structure may be treated mathematically as homogeneous.[19] From a structural point of view, natural fibres are multicellular in nature, consisting of a number of continuous, mostly cylindrical, honeycomb cells which have different sizes, shapes and arrangements for different types of fibres,[20] thus providing different properties for different types of fibres.

Concerning the physical structure of natural fibres, Mohanty *et al.* reported that natural fibres are bundles of elongated, thick-walled, dead natural cells where a single natural fibre is a single cell typically with a length from 1 to 50 mm and a diameter of around 10–50 μm.[21] Bismarck *et al.* reported that natural fibres are like microscopic tubes and cell walls surrounding the central lumen,[22] and the lumen acts as a medium for water uptake through the natural fibres.[23]

With regard to the microstructure of natural fibres, Brett and Waldron reported that natural fibres comprise different hierarchical microstructures. The cell wall in a fibre is not a homogeneous membrane.[24] It is built up of several layers, and the primary cell wall is the first layer deposited during cell growth. Stamboulis *et al.* reported that the primary cell wall consists of pectin, which can be stiffened and dehydrated by the addition of lignin. It then becomes part of the adhesive system holding cells together.[25]

The next layer is the secondary wall, which represents the most important cell wall. According to Brett and Waldron, the secondary cell wall (S) consists of three layers, S1, S2 and S3. In this wall, the cellulose molecules are synthesized by enzymes, each containing about 40 molecules.[24] These primary fibrils are assembled into microfibrils about 20–40 nm in diameter, which have hemicelluloses decorating the outside. These hemicelluloses act as the connection between the microfibrils, giving the primary structural network.[25] The structural constitution of the natural fibre cell has been elucidated by Rong *et al.*[26]

For the orientation of microfibrils in general, long cellulose microfibrils lying in the plane of the cell surface form the wall's scaffolding. Microfibrils are typically 4–10 nm thick and spaced roughly 30 nm apart. The orientation of the cellulose microfibrils with respect to the fibre axis determines the stiffness of the fibres. Natural fibres are more ductile if the microfibrils have a spiral angle orientation to the fibre axis. Fibres are inflexible and rigid and have a high tensile strength if the microfibrils are oriented parallel to the fibre axis.[27,28]

Many natural cells are approximately cylindrical, with the neighbouring microfibrils positioned parallel to each other so as to form "hoops" around the cylinder. This arrangement dictates the direction in which the cell can elongate.

This model was developed by Fry to show the arrangement of microfibrils in the side walls of an approximately cylindrical plant cell. Microfibrils are inextensible, so this hoop arrangement dictates that the cylindrical cell will expand predominantly in one dimension. All the other cells in an elongating stem or root conform to this model.[29] This cylindrical arrangement had been proved by Gañán *et al.* for vascular bundles of leaf sheath, pseudostem and rachis fibres.[30] Scanning electron microscope (SEM) observations provided evidence that fibre bundles are formed from aligned elementary fibres and conducting tissue with a spiral arrangement attached to them through the sclerenchyma inside the vascular bundle.

### *4.2.4.2 Chemical Composition of Natural Fibres*

One of the important properties that determines the main properties of natural fibre is chemical composition. It consists mainly of cellulose, hemicelluloses and lignin, free sugars, starch, proteins and other organic compounds.[31–33] There are three main components that have a strong influence on the mechanical properties of the fibre, the cellulose content having the greatest influence, followed by hemicelluloses and lignin.[34] This is due to the natural structural component of cellulose which provides strength and stability to the plant cell walls and the fibres. Cellulose is the stiffest and strongest organic constituent in a natural fibre.[35]

According to Rowell *et al.*, the relative amounts of the various compounds found in natural fibres vary from plant to plant and even within different parts of the same plant.[32] As shown in Table 4.1, the chemical composition, moisture content and microfibrillar angle of some natural fibres vary considerably.[22]

**Table 4.1** Chemical composition, moisture content and mirofibrillar angle of plant fibres.[22]

| Fibre | Cellulose (wt%) | Hemicellulose (wt%) | Lignin (wt%) | Pectin (wt%) | Moisture content (wt%) | Waxes (wt%) | Microfibrillar angle (°) |
|---|---|---|---|---|---|---|---|
| Flax | 71 | 18.6–20.6 | 2.2 | 2.3 | 8–12 | 1.7 | 5–10 |
| Hemp | 70–74 | 17.9–22.4 | 3.7–5.7 | 0.9 | 6.2–12 | 0.8 | 2–6.2 |
| Jute | 61–71.5 | 13.6–20.4 | 12–13 | 0.2 | 12.5–13.7 | 0.5 | 8 |
| Kenaf | 45–57 | 21.5 | 8–13 | 3–5 | – | – | – |
| Ramie | 68.6–76.2 | 13.1–16.7 | 0.6–0.7 | 1.9 | 7.5–17 | 0.3 | 7.5 |
| Nettle | 86 | – | – | – | 11–17 | – | – |
| Sisal | 66–78 | 10–14 | 10–14 | 10 | 10–22 | 2 | 10–22 |
| Henequen | 77.6 | 4–8 | 13.1 | – | – | – | – |
| PALF | 70–82 | – | 5–12.7 | – | 11.8 | – | 14 |
| Banana | 63–64 | 10 | 5 | – | 10–12 | – | – |
| Abaca | 56–63 | – | 12–13 | 1 | 5–10 | – | – |
| Oil palm EFB[a] | 65 | – | 19 | – | – | – | – |
| Oil palm mesocarp | 60 | – | 11 | – | – | – | 46 |
| Cotton | 85–90 | 5.7 | – | 0–1 | 7.85–8.5 | 0.6 | – |
| Coir | 32–43 | 0.15–0.25 | 40–45 | 3–4 | 8 | – | 30–49 |
| Cereal straw | 38–45 | 15–31 | 12–20 | 8 | – | – | – |

[a]EFB = empty fruit bunch.

### 4.2.4.3 *Relationship Between Cellulose Content and Microfibrillar Angle*

Mechanical properties are mainly determined by the cellulose content, degree of polymerization and microfibrillar angle. The reinforcing efficiency of natural fibre is related to the nature of cellulose and its crystallinity.[27] A high cellulose content and low microfibril angle are desirable properties of a fibre to be used as reinforcement in polymer composites.[36] Semiemperical relationships to correlate the fibre elongation and microfibrillar angle and also the tensile strength and microfibrillar angle with the cellulose content were developed by Satyanarayana *et al.* as follows:[37]

$$\varepsilon = 2.78 + 7.28\times10^{-2}\theta + 7.7\times10^{-3}\theta^{2}\ldots \quad (4.1)$$

$$\sigma = 334.005 - 2.8300\theta + 12.22W \quad (4.2)$$

where $\varepsilon$ = fibre elongation, $\theta$ = microfibrillar angle, $\sigma$ = tensile strength and $W$ = cellulose content.

The structure, microfibrillar angle, cell dimensions and defects and the chemical composition of the natural fibres are the most important variables that determine the overall properties of the fibres.[38,39] In general, the tensile strength and Young's modulus of natural fibres increase with increasing cellulose content of the fibres. Flemming *et al.* studied the relationship between the cellulose content and microfibrillar angle and the average fibre tensile strength.[28]

### 4.2.5 Treatment of Natural Fibres

Methods for the treatment of fibre can be classified into (1) chemical (2) biological and (3) physical treatment.

#### 4.2.5.1 *Chemical Treatment*

Generally, chemical reagents cause changes in the surface tension of fibres and improve their dispersion in the polymer matrix, hence improving the interfacial bonding between fibre and matrix. Mercerization or alkali treatment, pretreatment with chemical reagents and grafting have been studied by numerous workers to improve the interfacial bonding between fibre and matrix.

Alkali treatment was studied by Gomes *et al.*,[40] Mwaikambo and Ansell,[41] Biagiotti *et al.*[42] and Rout *et al.*[43] to improve the wettability of natural fibre. Alkali treatment caused the partial removal of lignin and hemicellulose and allowed the separation of grass fibres into finer microfibres.[44] Sgriccia *et al.* reported that alkali treatment removed lignin and hemicellulose from the natural fibre surface and caused a reduction in fibre size.[45] Both groups reported better dispersion of grass fibre in the matrix. The dispersion improved with increased treatment time and concentration of alkali solution.

Several groups have studied pretreatment methods to improve the interfacial bonding of natural fibre composites. Pretreatment of soft wood fibreboard with 2,3-dihydroxybenzoic acid (DHBA) increased the internal bonding, modulus of rupture and modulus of elasticity of wood fibre composites.[46] Manrich and Marcondes reported a method of impregnating monomer fibres, followed by polymerization *in situ* using a catalyst, radiation or heat.[47] Herrera-Franco and Aguilar-Vega reported that preimpregnation of henequen cellulosic fibre in low-density polyethylene (LDPE)–xylene solution improved the interfacial adhesion between the fibres and the matrix.[48]

#### 4.2.5.2 *Biological Treatment*

Biological methods have been developed as a novel approach for treating natural fibre in recent years, and the substances and processes used for treatment are inexpensive. These methods are used for modifying natural fibres in a cost-effective and environmentally safe manner.[49]

One of the early methods used in the treatment of natural fibre is water retting. This process is usually used for bast fibres in which the fibres must be separated from the rest of the stalk. Retting is a biological process in which microbes act as agents that break the chemical bonds that hold the stem together and allow separation of the bast fibres from the woody core.[49]

Another method for fibre surface modification is fungal treatment. Gulati and Sain demonstrated that surface modification using the fungus *Ophiostoma ulmi* enhanced the adhesion characteristics and moisture resistance of hemp fibres/unsaturated polyester.[50] It gave the lowest level of moisture absorption while untreated fibres absorbed the highest level of moisture.

A more recently developed method is sea water treatment, which was developed by Leman and co-workers for improving the interfacial bonding of sugar palm fibre and the matrix.[51–53] The treatment with sea water changed the surface tension and improved the dispersion in the polymer matrix, which significantly affected the bonding strength between the fibre and the matrix. The process involves soaking sugar palm fibre in sea water for 30 days and the results showed that the tensile and flexural properties of sugar palm fibre-reinforced epoxy composites increased after this treatment.

#### 4.2.5.3 Physical Treatment

Common physical methods such as stretching, calendering, surface fabrilation, thermal treatment, corona treatment and cold plasma treatment change the structural and surface properties of fibres.[54,55] Physical treatments led to the removal of the waxy cuticle layer, changes in surface roughness, changes in the physical appearance of fibres and even changes in the density of fibres, which influenced the mechanical bonding between the fibres and the matrix.[56] In addition to the above methods, coating jute with poly(vinyl alcohol) and polypropylene also affects the fibre–matrix interfacial adhesion and interaction.[57]

The latest fibre treatment method is resin impregnation, which was developed by Ishak and co-workers.[58–61] Physical modification of the fibres occurs when they are impregnated with thermosetting polymer *via* a vacuum resin impregnation process. This enhances the properties of the fibres before they are used for composite reinforcement. The principle of this process is to impregnate the fibre cell lumen and cell wall and to cover the fibre surface with hydrophobic thermosetting polymer in order to prevent the absorption of water molecules on the hydrophilic fibre and to improve its mechanical properties. Studies on fibre modification *via* vacuum resin impregnation with various impregnation pressures and times and with different impregnation resins showed that there were significant improvements in the physical and mechanical properties of sugar palm fibre after impregnation. Similar findings were also obtained with its composites: after impregnating the fibre with thermosetting polymer, the moisture and water uptakes of the composites were significantly lower than those of control samples, and the composites showed better moisture repellence with increase in the impregnation loading.

### 4.2.6 Advances in Natural Fibres: Nanotechnology

The abundance of cellulose fibre material makes it viable for use in long fibre composites, microscale composites and nanoscale composites. Like other fillers, natural fibre nanofillers enhance the mechanical properties of the composites. Nanoscale cellulose composites have attracted attention not only because of the contribution to enhancing the mechanical properties but also because of the low weight, biodegradability and renewability. The preparation of nanosized cellulose is the most challenging aspect of utilizing them in polymer composite materials. Nanoscale cellulosic materials face the same challenge as microscale fibres and long fibres when combined with hydrophobic matrices, owing to the hydrophilicity of cellulosic materials. To overcome this problem, hydrophilic matrices are used or chemical pretreatments are applied to the fibres. This section briefly summarizes the sources and preparation of nanocellulose; more details of nanocellulose technology can be found in reviews.[62,63]

### 4.2.7 Sources of Nanocellulose

As reported by Siró and Plackett, sources of nanocellulose are wood, agricultural crops and by-products and bacterial cellulose.[62] Extracting wood nanofibres started in the 1980s and wood is still considered the most important source of cellulosic fibres. Agricultural crops are mainly renewable plants such as flax, kenaf, hemp and sisal. Agricultural by-products are from plants such as rice, coconut, pineapple, bananas and others. Bacterial cellulose is produced by bacteria through cellulose biosynthesis and building up of bundles of microfibrils. Bacterial cellulose has higher crystallinity and strength than even macroscale natural fibre.[62,64] Fu *et al.* reported that bacterial cellulose can be used in skin tissue repair.[64]

### 4.2.8 Preparation of Nanocellulose

The preparation of nanocellulose requires intensive mechanical treatment. Siró and Plackett reported three main methods for nanocellulose preparation:, namely mechanical, chemical and enzymatic processes.[62] Chemical and enzymatic processes are considered as pretreatments that are used to reduce the amount of energy by around 1.5–5% of the energy consumed by the mechanical procedures.[62]

Mechanical procedures are high-pressure homogenization, cryocrushing and grinding,[65–67] which consume large amounts of energy and need several repetitions to produce fine particles.[67] Hmogenization simply involves passing fibres and water through a mechanical homogenizer. In this method, a drop in pressure causes the cellulose to break down into smaller parts. Nanocellulose is also produced by cryocrushing, in which fibres are frozen and then ruptured using high shear.[62] In the grinding process, fibres are placed in a special grinding machine to produce nanocellulose.

## 4.3 Starch

### 4.3.1 Introduction

Biopolymers are polymers produced by renewable resources and living organisms, and include naturally occurring polymers such as cellulose, starch,

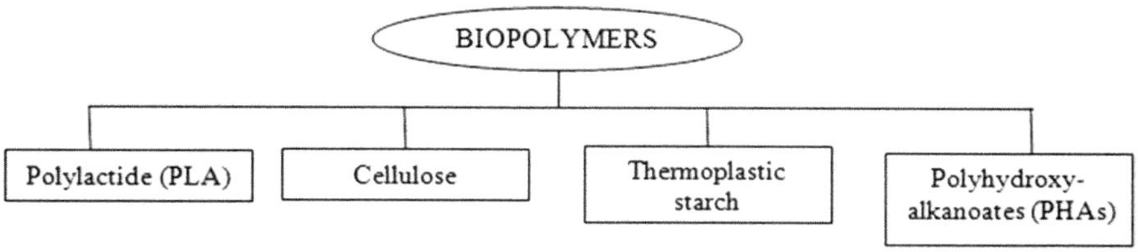


**Figure 4.1** Classifications of biopolymers.

chitin, proteins and peptides. Biopolymers are biodegradable and are broken down into $CO_2$ and water by microorganisms. Figure 4.1 shows the classification of biopolymers. As illustrated, biopolymers may be divided into three main categories based on their origin and production. Category 1 includes polymers directly extracted/removed from biomass. Examples are polysaccharides such as starch and cellulose. Category 2 are polymers produced by classical chemical synthesis using renewable bio-based monomers. A good example is polylactide (PLA), a biopolyester produced by polymerization of monomeric lactic acid. Category 3 includes polymers produced by microorganisms or genetically modified bacteria. To date, this group of biobased polymers consists mainly of the polyhydroxyalkanoates (PHAs).

Starch is a natural biopolymer material, widely found in seeds, roots and tubers and also in stems, leaves and fruits. It is the major form of stored carbohydrate in plants such as corn, wheat, rice and potatoes. Among all biopolymers, starch is considered to be one of the most promising because of its ready availability, biodegradability, low cost and renewability.[68] Starch is composed of a mixture of two polymers of α-glucose, *i.e.* linear amylose and highly branched amylopectin (Figure 4.2). Amylose molecules consist of 200–20 000 glucose units, which form a helix as a result of the bond angles between the glucose units. Amylopectin is a highly branched polymer containing short side chains of 30 glucose units attached to every 20–30 glucose units along the chain. Amylopectin molecules may contain up to two million glucose units.[69]

Starches from various sources are chemically similar and their granules are heterogeneous with respect to their size, shape and molecular constituents. The proportion of the polysaccharides amylose and amylopectin is the most critical criterion that determines starch behaviour.[70,71] Most amylose molecules (molecular weight $\sim 10^5$–$10^6$ Da) consist of $(1 \rightarrow 4)$-linked α-D-glucopyranosyl units in a linear chain. However, a few molecules are branched to some extent with $(1 \rightarrow 6)$ α-linkages.[72,73]Amylose molecules can vary in their molecular weight distribution and in their degree of polymerization (DP), which affects their solution viscosity during processing and their retrogradation/recrystallization behaviour, which is important for product performance. Amylopectin is the highly branched polysaccharide component of starch that consists of hundreds of short chains formed of α-D-glucopyranosyl residues with $(1 \rightarrow 4)$ linkages. These are interlinked by $(1 \rightarrow 6)$ α-linkages, 5–6% of which occur at the branch points. As a result,

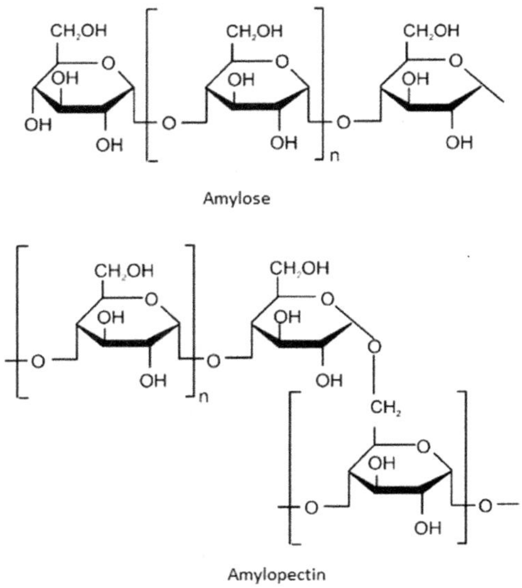


**Figure 4.2** Unit structures of amylose and amylopectin.

amylopectin has a high molecular weight ($10^7$–$10^9$ Da) and its intrinsic viscosity is very low (120–190 mL $g^{-1}$) because of its extensively branched molecular structure.[72,73]

The physicochemical properties of starches are dependent on the ratio of amylose to amylopectin, hence this ratio in starch affects the physico-mechanical properties of materials produced from it. It is observed that the elongation at break of materials decreases with decrease in amylopectine content.[74] This type of changing pattern is governed by the increase in flexibility of the polymer network caused by the highly branched structure of amylopectin. More recent investigations have been focused on the development and characterization of biopolymers derived from starch since conventional synthetic plastic materials are resistant to microbial attack and biodegradation.[75,76]

### 4.3.2 Preparation of Starch

The common industrial starches are typically derived from cereals (corn, wheat, rice, sorghum), tubers (potato, sweet potato), roots (cassava) and legumes (mung bean, green pea). Sago palm and sugar palm starch are examples of commercial starch derived from another source, the palm stem. The extraction of sugar palm starch (SPS) was described by Sahari *et al.*[77] They obtained sugar palm log sections from two heights, *i.e.* the base (1 m from the ground) and mid-height (5 m from the ground) of the sugar palm trunk. The bark (approximately 2 cm thick) was removed manually using a machete and the remaining pith was chopped into 1–2 cm cubes and blended in distilled water. Extraction of starch was carried out by squeezing the blended pith mixture using muslin cloth. The filtrate was allowed to stand for the starch to settle, after which the supernatant was decanted. The sedimented starch was washed with water and filtered through a 250 mm sieve. The filtered starch was washed twice, then dried in a convection oven at 50 °C for 2 days and then at 105 °C until constant weight. Figure 4.3 shows the extraction method for obtaining SPS.

In 2009, sorghum starch was extracted by Singh *et al.*[78] Sorghum grain (100 g) was steeped in 200 mL of NaOH solution (0.25% w/v) at 5 °C for 24 h. The steeped grains were washed and ground with an equal volume of water using a blender for 3 min. The slurry was filtered through a 200-mesh screen and the residue on the sieve was rinsed with water. Grinding and filtration were repeated three times on this material. After rinsing, the residue was discarded and the filtrate was allowed to stand for 1 h. The filtrate was centrifuged at 6000 rpm for 10 min and the top grey-colored, protein-rich layer was removed using a spatula. Excess water was added to resuspend the sample and centrifugation was repeated for 5 min. Washing and centrifugation were repeated several times until the top starch layer was white. The final starch product was dried for 24 h at 40 °C.

Flour extraction was conducted by Alves *et al.* in 2002.[79] The tubers were peeled, washed, cut into 1–2 cm cubes and sliced into thick chips (5 mm). These chips were then soaked in sodium metabisulfite solution (0.075%) for 5 min and oven dried at 30 °C for 40 h until they reached 13% moisture. Subsequently, the dried chips were milled into flour and sifted through a 300 μm sieve. The flour was then packed into a closed container and stored under dry conditions at room temperature until used for further applications.

**Figure 4.3** Extraction of sugar palm starch.

### 4.3.3 Characterization of Starch

#### 4.3.3.1 Chemical Properties

Table 4.2 shows the chemical properties of SPS and other commercial starches. SPS and other commercial native starches contain 10–20% moisture under normal atmospheric conditions. As starch is very sensitive to water, the amount of water absorbed by starch granules is dependent on the relative humidity (RH) and the temperature of the atmosphere in which they have been stored. The starches isolated from SPS contain comparable amounts of amylose (37.60%), higher than those in tapioca (17%), sago (24–27%), potato (20–25%), wheat (26–27%) and maize (26–28%) starches. All kinds of starch contain two types of glucose polymers, *i.e.* the essentially linear molecule amylose and the highly branched polymer amylopectin (see Figure 4.2 and its description).

SPS has a 0.2 % ash content, the same as those of tapioca, sago and wheat starches. The ash content can be determined as the residue after ignition of the starch at 600 °C. As shown in Table 4.2, potato starch has a relatively high ash content because of the presence of phosphate groups. With regard to metals, the ash of native starches contains mainly calcium, potassium, magnesium and sodium.[80]

#### 4.3.3.2 Thermal Properties

Thermogravimetric analysis (TGA) can be used to measure the mass change, thermal decomposition and thermal stability of materials. In previous studies, we found that the molecular structure of SPS is degraded on heating.[81] The mass loss below 100 °C may be attributed to evaporation of absorbed moisture.[88] The substantial degradation at 310 °C appears to involve further elimination of polyhydroxyl groups, accompanied by depolymerization and decomposition.[89] At higher temperatures, depolymerization of the macromolecules takes place with the formation of β-(1,6)-anhydro-D-glucopyranose (levoglucosan), 2-furaldehyde (furfural) and a range of lower molecular-weight volatile and gaseous fragmentation products. A carbonaceous (substance rich in carbon) residue remains after all the volatile products have been driven off.[90] From the differential scanning

**Table 4.2** Chemical properties of SPS and other commercial starches.

| Starch | Density | Water content (%) | Amylose (%)$^a$ | Ash (%) | Ref. |
|---|---|---|---|---|---|
| Sugar palm | 1.54 | 15 | 37.60 | 0.2 | 80 |
| Tapioca | 1.446–1.461 | 13 | 17 | 0.2 | 80, 82 |
| Sago | – | 10–20 | 24–27 | 0.2 | 83, 84 |
| Potato | 1.54–1.55 | 18–19 | 20–25 | 0.4 | 80, 85, 86 |
| Wheat | 1.44 | 13 | 26–27 | 0.2 | 80, 85, 87 |
| Maize | 1.5 | 12–13 | 26–28 | 0.1 | 80, 85, 87 |

$^a$Amylopectin is calculated as 100% – amylose content.

calorimetric (DSC) curve, we found that the glass transition temperature ($T_g$) and melting temperature ($T_m$) of SPS were 242.14 and 279.84 °C, respectively, which were higher than those of Indica rice starch (237 and 276 °C, respectively).[91] Luiz *et al.* reported the $T_m$ of cassava starch as 169.2 °C.[92] Thermal characteristic studies using TGA and DSC showed that SPS was more thermally stable than other starches.

### *4.3.3.3 Particle Size*

Starches from various sources are chemically similar and their granules are heterogeneous with respect to their size, shape and molecular constituents. Figure 4.4 (top) shows the particle size analysis of SPS. SPS has the largest mean diameter (36.308 μm) with a particle size distribution of 0.020–1000 μm. The results observed for the SPS peak are similar to those of tapioca starch, as shown in Figure 4.4 (bottom). The particle size of starch is one of the most important characteristics, and may influence other physicochemical properties such as swelling power and water-binding capacity.[78]

The starch granules of SPS when viewed by SEM were round and oval-shaped granules with smooth surfaces. From the micrographs (Figure 4.5), it

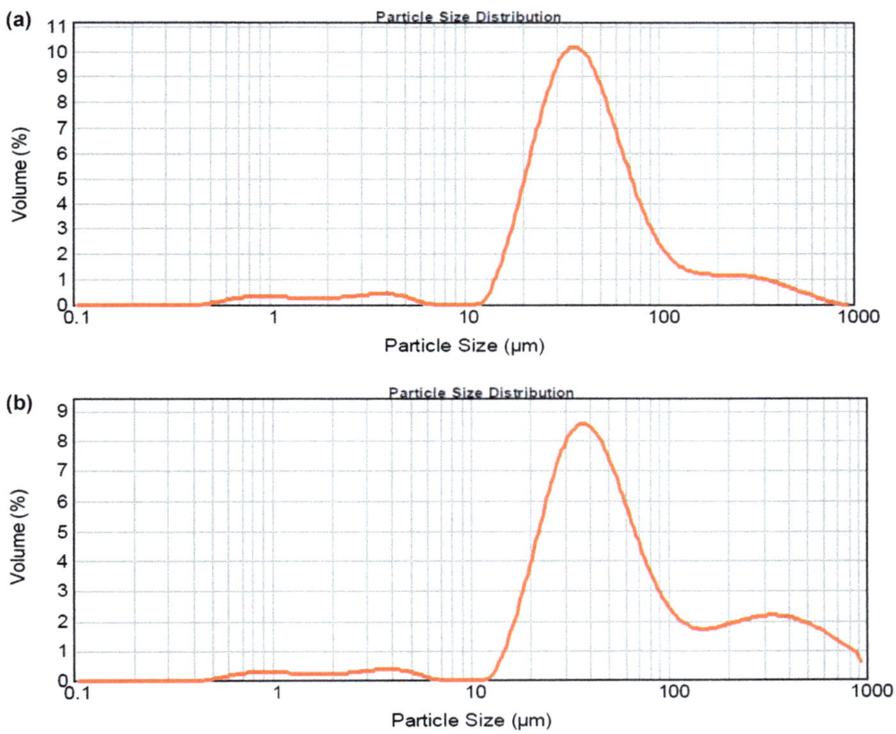


**Figure 4.4** Particle size distribution of (top) SPS and (bottom) tapioca starch.

**Figure 4.5** Scanning electron micrograph of SPS.

is estimated that the granule size ranges from approximately 10 to 100 μm, which is in agreement with the study on particle size.

# 4.4 Thermoplastic Starch (TPS)

## 4.4.1 Introduction

Starch is attractive because it is a cheap material and shows very fast biodegradability. Under high temperature and shear, starch can be processed into a mouldable thermoplastic known as thermoplastic starch (TPS). TPS is plasticized starch that has been processed (typically using heat and pressure) to destroy completely the crystalline structure to form an amorphous thermoplastic material as shown in Figure 4.6. During the gelatinization process, water contained in the starch and the added plasticizers plays an indispensable role because the plasticizers can form hydrogen bonds with the starch, replacing the strong interactions between the hydroxyl groups of the starch molecules, thus making the starch thermoplastic. In most studies on thermoplastic starch, polyols were usually used as plasticizers, such as glycerol, sorbitol and urea.[93]

Plasticizers are the most important material for increasing the flexibility and processibility of TPS and many studies have been carried out on the plasticization of TPS using glycerol, sorbitol, urea, formamide, dimethyl sulfoxide and low molecular weight sugars.[94–97] The properties of TPS also depend considerably on moisture. As water has a plasticizing effect, the behaviour of amaterial changes according to the relative humidity of the air through a sorption–desorption mechanism.[98]

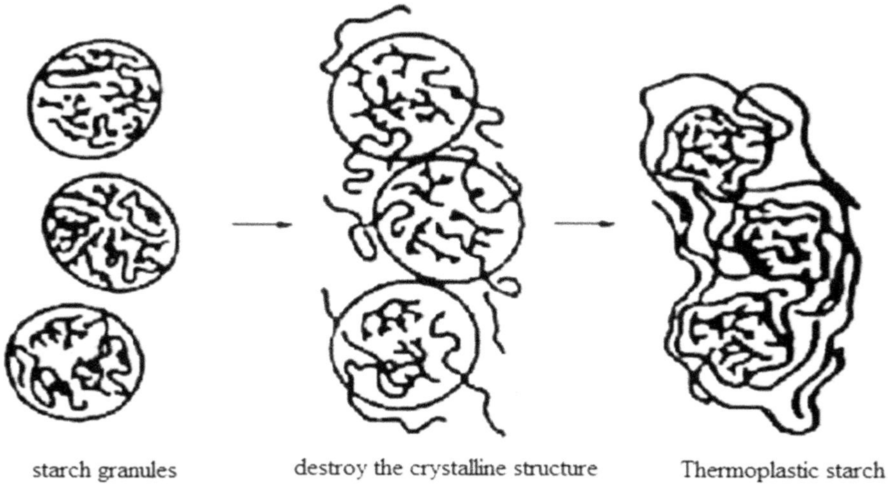


**Figure 4.6** Starch gelatinization process.

The various properties of TPS products such as mechanical strength, water solubility and water absorption can be adjusted by changing the plasticizer content.[99] Glycerol is the most commonly used plasticizer, and its effectiveness in producing biodegradable blend films from TPS is most likely due to its small size, which allows it to be more readily inserted between the polymer chains and consequently exert more influence on the mechanical properties. McHugh *et al.* suggested that, owing to its small size, glycerol was more effective than sorbitol in plasticizing starch.[100] Many studies have been performed on the preparation of TPS using glycerol as plasticizer. Park *et al.* developed biodegradable thermoplastic potato starch by using 30% of glycerol as plasticizer.[101] Bourtoom studied the properties of biodegradable blend film derived from glycerol and rice starch.[102] Avérous and Boquillon prepared thermoplastic wheat starch by using glycerol as plasticizer.[103] They concluded that the optimum combination of starch and glycerol was 70:30 wt%. At 35% glycerol content, the tensile strength of thermoplastic starch starts to decrease.

## 4.4.2 Characterization of TPS

### *4.4.2.1 Thermal Properties*

The glass transition temperature ($T_g$) is a very important parameter for determining the mechanical properties. Table 4.3 shows $T_g$ values of plasticized SPS determined by DSC. For dry SPS, $T_g$ reached 242.14 °C and decreased with addition of glycerol. This value was higher than that of Indica rice starch, with a $T_g$ value of 237 °C.[91] Myllärinen *et al.* reported that for dry starch, $T_g$ reaches 227 °C.[104] The sample with high glycerol concentrations showed lower $T_g$ values and the $T_g$ values of starch without plasticizer were higher than those of samples with glycerol. This behaviour was also observed

**Table 4.3** Glass transition temperatures ($T_g$) of plasticized SPS.

| | $T_g$ (°C) | |
|---|---|---|
| Sample[a] | Onset | Mid-point |
| Native SPS | 237.91 | 242.14 |
| SPS/G15 | 225.68 | 229.26 |
| SPS/G20 | 206.44 | 217.90 |
| SPS/G30 | 189.57 | 187.65 |
| SPS/G40 | 176.71 | 177.03 |

[a]G values represent the glycerol concentration (wt%).

by Mali *et al.*[105] for yam starch and by Forssell *et al.*[94] for films based on barley starch, in both cases with glycerol as plasticizer. According to Guilbert and Gontard, plasticization decreases the intermolecular forces between polymer chains, consequently changing the overall cohesion and leading to a decrease in $T_g$.[106] Mitrus suggested that a plasticizer decreases $T_g$ because it facilitates chain mobility.[107] Brittleness is one of the major problems connected with starchy material owing to its high $T_g$.[108] In the absence of plasticizers, starches are brittle. The addition of plasticizers improves the flexibility and extensibility of the polymers.

Degradation of the molecular structure occurs when plasticized SPS is heated. The extent of the changes induced depends on the temperature and, under extreme conditions, may result in a complete loss of carbohydrate character. Because the boiling point of the plasticizer is more than 100 °C, the mass loss below 100 °C was mainly ascribed to water loss.[109] A mass loss below 290 °C indicated the vaporization of glycerol. Pure glycerol has a boiling point of 290 °C but in this case, starch possesses a hydrophilic nature with a helical structure of all hydroxyl groups that are directed outside the ring.[110,111] This structure promotes direct interactions of starch and glycerol by forming hydrogen bonds and subsequently lowers the volatility of glycerol in plasticized SPS. The substantial degradation at 310 °C appears to involve further elimination of the polyhyroxyl groups, accompanied by depolymerization and decomposition.[89] It has been documented earlier that at higher temperatures, depolymerization of the macromolecules takes place with the formation of β-(1,6)-anhydro-D-glucopyranose (levoglucosan), 2-furaldehyde (furfural) and a range of lower molecular weight volatile and gaseous fragmentation products. A carbonaceous residue remains after all the volatile products have been driven off.[90]

### *4.4.2.2 Mechanical Properties*

Figure 4.7 shows the tensile strength of plasticized SPS. It can be seen that the tensile strength of SPS/G30 showed the highest value of 2.42 MPa compared with the other concentrations of the plasticizer. The higher the concentration of the plasticizer, the higher is the tensile strength of plasticized SPS, and optimum glycerol concentration was 30 wt%. The tensile strength decreased to 0.5 MPa when the concentration of plasticizer was

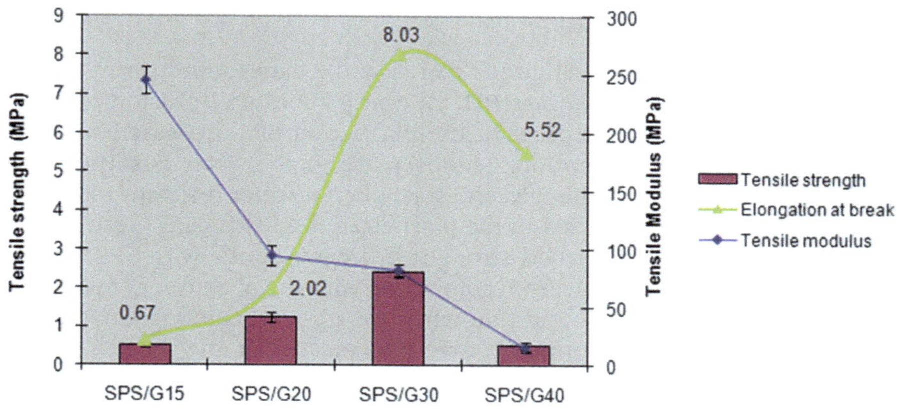


**Figure 4.7** Tensile strength of plasticized SPS.

40 wt%. As the plasticizer content increased to 40 wt%, there was insufficient SPS to be well bonded with glycerol and therefore poor adhesion occurred, which reduced the mechanical properties of SPS/G40. This result well agreed with the finding of Laohakunjit and Noomhorm, who reported that films outside this range are either too brittle (<20 wt%) or too tacky (>45 wt%).[112]

The tensile modulus or elastic modulus of plasticized SPS is also shown in Figure 4.7, which clearly reveals that the tensile modulus decreases from 245.24 to 15 MPa with increase in plasticizer concentration from 15 to 40 wt%. Elongation at break analysis indicated that the relative elongation increased significantly (from 0.67 to 8.03%) with increase in the content of plasticizer from 15 to 30 wt%, as shown in the Figure 4.7. It is interesting that 30 wt% of glycerol led to very high elongation, but the elongation decreased to 5.52% when the glycerol concentration was 40 wt%. A similar finding was reported by Laohakunjit and Noomhorm, who found that the relative elongation increased significantly with increase in the conentration of plasticizer for both plasticizers (sorbitol and glycerol) and the elongation decreased when the glycerol concentration was 35 wt%.[112] The increase in the elongation of plasticized SPS can be ascribed to the nature of the glycerol molecule, which is relatively hydrophilic owing to the large number of hydroxyl groups which can establish hydrogen bonding. Generally, as the plasticizer concentration increases, the tensile strength and elongation of plasticized SPS increase, whereas the tensile modulus decreases. This indicates that plasticized SPS is more flexible when subjected to tension or mechanical stress. The results from this study confirm the finding of Beerler and Finney, who reported that plasticizers such as glycerol will interfere withthe arrangement of the polymer chains and the hydrogen bonding.[113] It also probably affects the crystallinity of starch by decreasing the polymer interaction and cohesiveness, thus making the plasticized SPS become more flexible with increase in glycerol concentration.

#### 4.4.2.3 Water Absorption

The moisture content of plasticized SPS reached a plateau, indicating that it became equilibrated with storage RH, following a Fickian diffusion process. In general, the water absorption of the plasticized SPS decreased with increase in glycerol concentration. The decrease in the water absorption in plasticized SPS indicated that glycerol was a better water-resistant material, less water gain being observed in the plasticized SPS. This can be attributed to the fact that glycerol formed stronger hydrogen bonds with SPS, which restricted the water molecule from combining with the plasticizer or with the SPS.[114] The results also showed that when the glycerol content was higher, stronger hydrogen bonds formed between the plasticizer and the SPS. It was more difficult for water molecules to penetrate into the plasticized SPS and, as a result, plasticizer–water and starch–water interactions were very weak, so that the plasticized SPS showed lower water absorption at high glycerol concentrations.

#### 4.4.2.4 Scanning Electron Microscopy (SEM)

Fracture surface analysis of plasticized SPS was carried out by SEM. The SEM images shown in Figure 4.8 clearly reveal the smooth surface area increase in plasticized SPS caused by the addition of glycerol. At high concentrations of glycerol (30 and 40 wt%), the specimens were very soft, whereas at lower glycerol concentrations, they were very brittle. The plasticizer effectively reduced internal hydrogen bonding while increasing intermolecular spacing, thereby decreasing brittleness.[114]

## 4.5 Natural Fibre-reinforced Thermoplastic Starch Composites (NFTPSs)

### 4.5.1 Introduction

So far, biocomposites yet to become widely available. The depletion of petroleum resources coupled with the awareness of global environmental problems provides a need for new green materials with constituents that are compatible with the environment and independent of petroleum-based resources. The term "green composites" is intended to encourage researchers to find the opportunities to develop new materials based on environmentally friendly criteria. Figure 4.9 shows the classifications of biocomposites.

Green composites are biocomposites where both the matrix and reinforcement derive from renewable resources. In 2003, the first green composite appeared in a commercial product, namely the spare tyre cover of a Toyota car, which was developed from PLA-reinforced kenaf fibres. Automotive applications represent the best opportunity for biocomposites, because their low density allows the production of light-weight materials with the advantage of fuel and cost savings. Potential applications in this field are

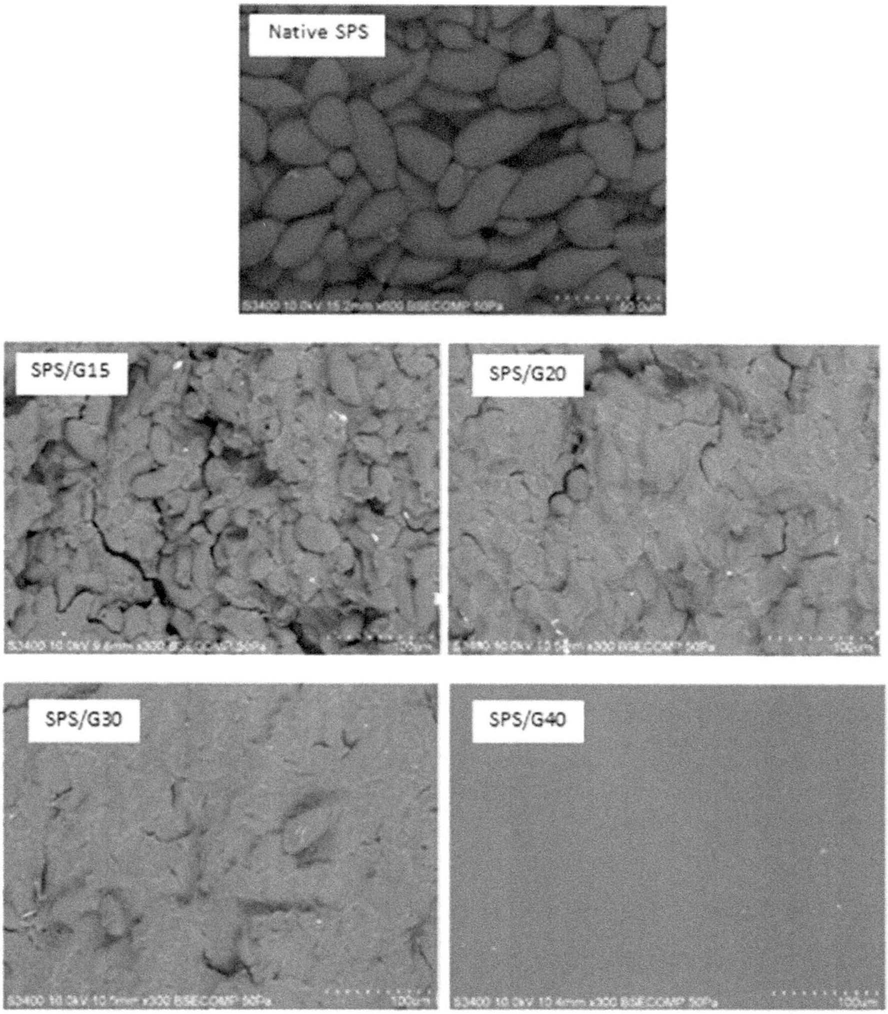


**Figure 4.8** SEM images of plasticized SPS.

interior structures such as seatback linings, package shelves, seat bottoms, seatback cushions, head restraints, under-floor body panels, door panel inserts and arm rests. Nowadays, car manufacturers are beginning to use biocomposite materials in car interiors, where traditional reinforcing fibres such as glass fibre are replaced with natural fibres. Along this line, the first prototype of a "green car" was developed at the University of Warwick with the "world's first Formula 3 racing car" in 2009. Green composites can also find applications in a range of widely different fields such as toys, packaging, musical instruments and cases for electronic devices. Building is also an important and expanding application field for green composites, especially

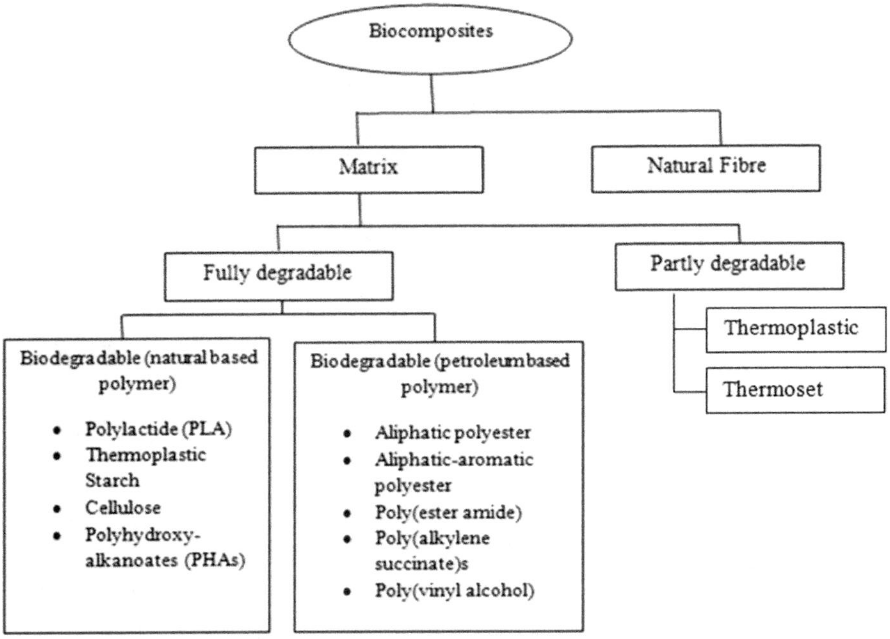


**Figure 4.9** Classifications of biocomposites.

those containing wood as reinforcement (WPCs): decking, window and door frames, panels for both indoor and outdoor uses and a number of structural components in buildings.[115]

## 4.5.2 Characterization of NFTPSs

### *4.5.2.1 Mechanical Properties*

Biocomposites with starch used as a matrix are one of the most popular biodegradable biocomposites and have been widely investigated.[116,117] Biodegradable matrices were reinforced with natural fibres to improve the properties of the composites and these composites provide positive environmental advantages, have good mechanical properties and are light weight.[118,119]

In recent years, studies have been performed on the properties of biodegradable composites made from plasticized starch reinforced with natural fibres. Vilaseca *et al.* developed composite materials from biodegradable starch and jute strands fabricated using an injection moulding process.[120] The mechanical properties of starch-based polymer and its composites with different percentages of untreated jute strands and alkali-treated jute strands were determined. The tensile strength of untreated jute-reinforced starch-based biopolymer composites increased with increasing fibre contents of 10, 20 and 30 wt%. Increases in the strength and stiffness were also observed for alkali-treated jute strand–starch composites.

Avérous and Boquillon studied the thermal and mechanical behaviour of composites made from TPS reinforced with agro-materials (cellulose and lignocellulose fibres).[103] The modulus of the TPS composites displayed a regular behaviour: the reinforcement effect increased with the fibre length from short-length fibre (SF) to medium length fibre (MF) and fibre content whereas the elongation at break decreased with increase in fibre contents and length. However, the result for elongation at break clearly conflicts with the results obtained by Soykeabkaew *et al.*, where the elongation at break increased with increase in fibre content.[121]

Vallejos *et al.* studied the potential use of fibrous materials obtained from ethanol–water fractionation of bagasse as reinforcements of TPS in order to improvethe mechanical properties of TPS and it was also reported that there were significant improvements due to fibre treatment and fibre loading on the mechanical properties of the composites.[122] In general, it has been shown that there is high compatibility between TPS and natural fibres in composites.[123]

Figure 4.10 demonstrates the effect of fibre loading on the tensile properties of sugar palm fibre (SPF)-reinforced SPS biocomposites. The tensile strength and modulus of SPF–SPS biocomposites increased with increase in SPF loading, while the addition of the SPF made the elongation decrease from 8.03 to 3.32%. A considerable increase in tensile strength with increase in SPF concentration indicated that the natural cellulose fibre has a great impact on the formation of good bonding between SPS and SPF. This was due to the remarkable intrinsic adhesion of the fibre/matrix interface caused by the chemical similarity of starch and the cellulose fibre.[109] SPF is believed to work as a carrier of load in the SPS matrix. The stronger bonding between the matrix and the fibres leads to enhanced interfacial adhesion between

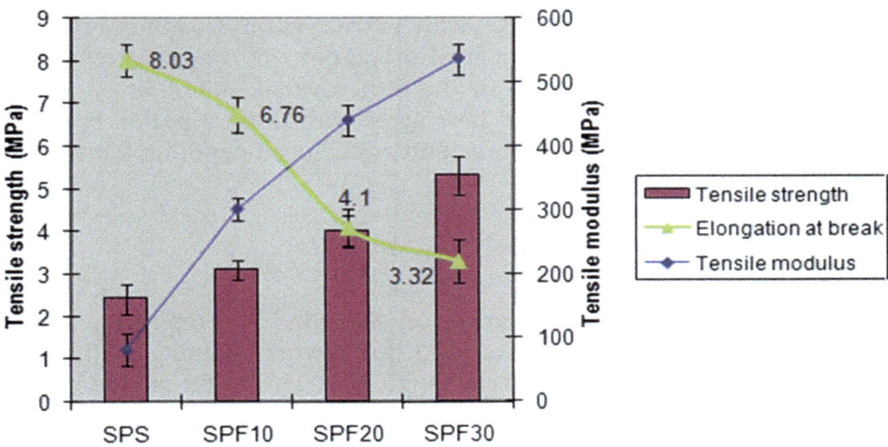


**Figure 4.10** Effect of fibre loading on tensile properties of SPF–SPS biocomposites.

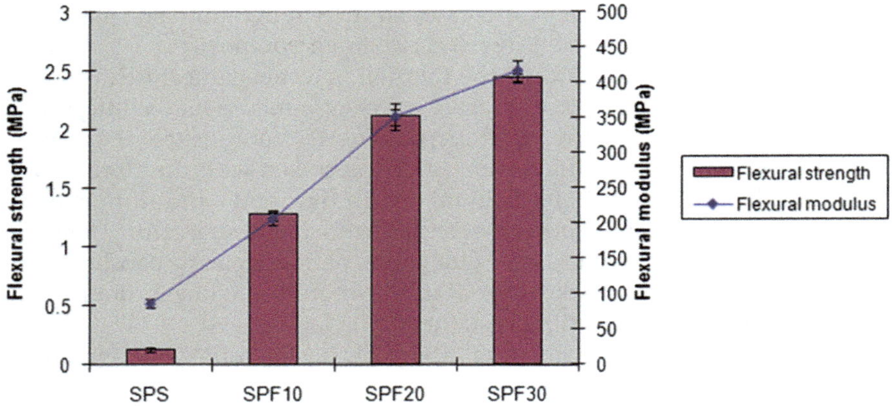


**Figure 4.11** Effect of fibre loading on flexural properties of SPF–SPS biocomposites.

them and therefore a greater transfer of stress from the matrix to the fibres during tensile testing.[124] This is supported by SEM studies of the fractured tensile samples. Increasing the fibre loading resulted in an increase of tensile modulus. The SPF–SPS biocomposite consists of a low-stiffness matrix and high-stiffness fibres, thus increasing the weight percentage of fibres increases the stiffness of the biocomposites. Further, elongation showed a decreasing trend with increase in fibre loading. This is also a normal consequence of the increase in fibre weight percentage, with low strain compared with the rubbery starch material SPS.[125]

Figure 4.11 shows the effect of fibre loading on the flexural properties of SPF–SPS biocomposites. The flexural strength and modulus increased with increase in fibre loading. It is normal to observe an increase in flexural strength and modulus on introducing fibres into an SPS matrix. A similar trend was also found for the impact strength of SPF–SPS biocomposites, where the impact strength increases with increase in fibre loading, as shown in Figure 4.12. Like other mechanical properties, the impact strength of biocomposites is highly affected by the fibre. Overall, the observed mechanical property results presented here agree with those reported by other researchers for lignocellulosic fibres and/or cellulose nanofibres incorporated in thermoplastic starches.[126–129]

### *4.5.2.2 Thermal Properties*

Figure 4.13 shows thermal behaviour of SPF–SPS biocomposites. It was found that the fibre loading increased the thermal stability of the biocomposite. The first mass loss, in the range 31–100 °C, is due to evaporation of moisture. The subsequent mass losses between 150 and 380 °C for SPF–SPS biocomposites are due to the decomposition of the three major constituents of the natural fibres, hemicellulose, cellulose and lignin.[130]

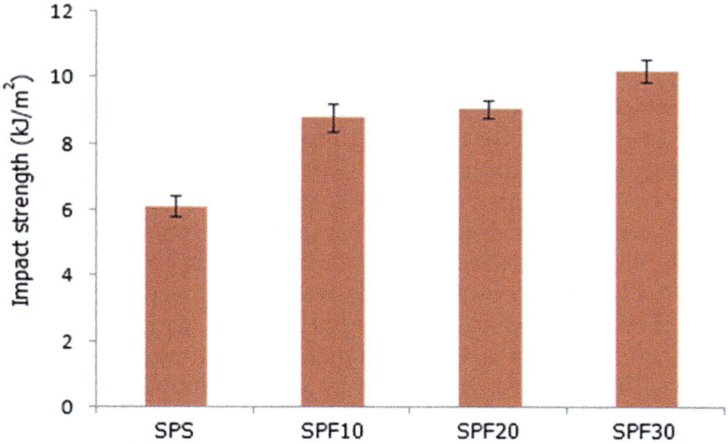


**Figure 4.12** Effect of fibre loading on impact properties of SPF–SPS biocomposites.

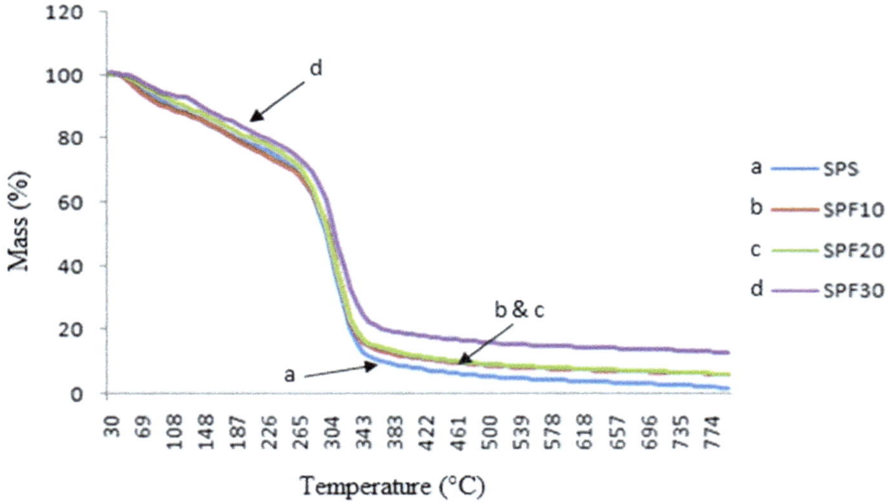


**Figure 4.13** Effect of fibre loading on the thermal properties of SPF–SPS biocomposites.

In general, the thermal decomposition of these fibres consists of four phases. The first phase is decomposition of hemicelluloses, followed by cellulose, lignin and lastly their ash.[131] Yang *et al.* reported that hemicelluloses start to decompose at 220 °C and decomposition is substantially completed at 315 °C.[132] As soon as the hemicelluloses have completely decomposed, the decomposition of cellulose will take place as the second phase of decomposition. Because of the highly crystalline nature of their cellulose chain compared with the amorphous form, the cellulose is relatively thermally stable and it does not start to decompose until the hemicelluloses have completely decomposed and it normally starts at temperatures higher than about 315 °C.[132] This is supported by the results of Kim *et al.*, who they reported that the critical temperature of decomposition of crystalline cellulose is 320 °C.[133] The third phase is the decomposition of lignin, which is more difficult to decompose than hemicelluloses and cellulose. Although the decomposition of lignin starts as early as 160 °C, it is slow and extends to temperatures as high as 900 °C for complete decomposition.[132] This is ascribed to lignin being a very tough component that gives rigidity to plant materials. Finally, when the lignin has completely decomposed, the component that is left is inorganic material in the fibres, which can be assumed to be the ash content. This is due to the presence of inorganic materials such as silica ($SiO_2$) in the fibre, which can only be decomposed at a very high temperature of 1723 °C. Figure 4.13 clearly shows that an increase in fibres increased the lignin and ash content compared with pure plasticized SPS. It was found that the lignin and ash contents of SPF were 31.5 and 4%, respectively.[134] The substantial degradation at 310 °C that occurs with plasticized SPS is due to the elimination of the polyhydroxyl groups, accompanied by depolymerization and decomposition of the starch.[135]

### 4.5.2.3 Water Absorption

Figure 4.14 shows the effect of fibre loading on water content of SPF–SPS biocomposites. The water uptake or absorption of the SPF–SPS biocomposite decreases with increase in fibre loading. Similar trends were also found for moisture content. This could be attributed to the hydrophobic behaviour of

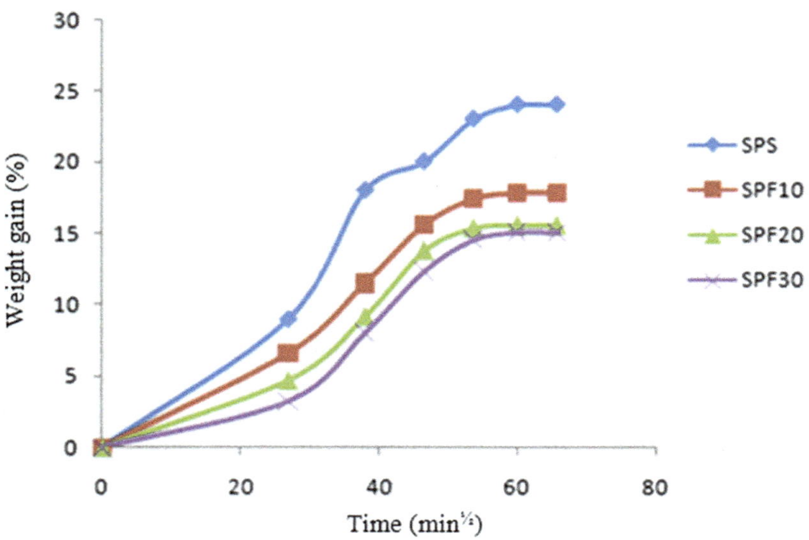


**Figure 4.14** Water absorption behaviour of SPS and SPF–SPS biocomposites.

SPF compared with SPS. Ma *et al.* noted that starch is a multi-hydroxyl polymer with three hydroxyl groups per monomer.[109] The remarkable decrease in the water content of SPF–SPS biocomposites compared with SPS indicated that SPF is a better water-resistant material than plasticized SPS. From the present results it generally be said that the greater the fibre content, the lower is the water absorption in the biocomposite. This phenomenon can also be attributed to better interfacial bonding between the matrix and fibres and the hindrance to absorption caused by the fibres.

Moisture absorption by the composites was also studied by leaving untreated jute–starch composite samples in the open air the for 72 days. The amount of water absorbed was very low and it was concluded that an increase in the percentage of reinforcement did not substantially affect the moisture absorption.[120]

### 4.5.2.4 Scanning Electron Microscopy (SEM)

Figure 4.15 shows a series of SEM images of the fractured surface of SPF–SPS biocomposites containing different amounts of SPF. It is indicative of the relatively homogeneous nature of these biocomposites. One can observed smooth fracture surfaces with a homogeneous matrix and also good adhesion of the SPF to the thermoplastic SPS matrix, which play an important

**Figure 4.15** SEM of SPF–SPS biocomposites.

role in improving the mechanical performance of biocomposites, as explained earlier.

## 4.6 Conclusion

Natural fibre-reinforced thermoplastic starch composites were successfully prepared with different amounts of sugar palm fibre by using glycerol as plasticizer for the starch. From this investigation, it was found that the mechanical properties of plasticized SPS improved with the incorporation of fibres. This was due to the remarkable intrinsic adhesion of the fibre/matrix interface caused by the chemical similarity of starch and the cellulose fibres. An increase in fibre loading also increased the thermal stability of the biocomposites in this investigation. This is contributed by lignin and inorganic materials such as silica, which is a very tough component that can only be decomposed at a very high temperatures. Water uptake and the moisture content of SPF–SPS biocomposites decreased with the incorporation of fibres, which is due to better interfacial bonding between the matrix and fibres and the hindrance to absorption caused by the fibres. SEM fractographic studies showed a homogeneous distribution of fibres and matrix and good adhesion, which play an important role in improving the mechanical properties of biocomposites.

## References

1. C. E. Bakis, E. Cosenza, J. J. Lesko and A. Machida, *J. Compos. Constr.*, 2002, **6**, 73.
2. O. Faruk, A. K. Bledzki, H.-P. Fink and M. Sain, *Prog. Polym. Sci.*, 2012, **37**, 1552.
3. S. M. Sapuan and M. A. Maleque, *Mater. Des.*, 2005, **26**, 65.
4. M. Z. Rong, M. Q. Zhang, Y. Liu, G. C. Yang and H. M. Zeng, *Compos. Sci. Technol.*, 2001, **61**, 1437.
5. R. M. Rowell and H. P. Stout, Jute and kenaf, in *A Handbook of Fiber Chemistry*, ed. M. Lewin and E. M. Pearce, Marcel Dekker, New York, 2nd edn, 1998, pp. 465–504.
6. Y. Habibi, W. K. El-Zawawy, M. M. Ibrahim and A. Dufresne, *Compos Sci. Technol.*, 2008, **68**, 1877.
7. I. S. Aji, Mechanical and thermal characterization of hybridized short kenaf/pineapple fibre reinforced high density polyethylene composites, PhD thesis, Universiti Putra Malaysia, 2011.
8. A. K. Bledzki and J. Gassan, *Prog. Polym. Sci.*, 1999, **24**, 221.
9. A. N. Netravali and S. Chabba, *Mater Today*, 2003, **6**, 22.
10. S. V. Joshi, L. T. Drzal, A. K. Mohanty and S. Arora, *Appl. Sci. Manufact.*, 2004, **35**, 371.
11. J. Summerscales, N. Dissanayake, A. Virk and W. Hall, *Appl. Sci. Manufact.*, 2010, **41**, 1336.
12. J. Holbery and D. Houston, *J. Min. Me.t Mat. S.*, 2006, **58**, 80.

13. A. K. Mohanty, M. Misra and L. T. Drzal, *J. Polym. Environ.*, 2002, **10**, 19.
14. V. J. R. R. Pita, E. E. M. Sampaio and E. E. C. Monteiro, *Polym. Test.*, 2002, **21**, 545.
15. H. Y. Sastra, J. P. Siregar, S. M. Sapuan, Z. Leman and M. H. M. Ahmad, *Am. J. Appl. Sci.*, 2005, 21.
16. M. T. Paridah, B. A. Amel, O. A. S. Syeed and Z. Ahmed, *BioRes*, 2011, **6**, 5260.
17. H. Djidjelli, A. Boukerrou, R. Founas, A. Rabouhi, M. Kaci and J. Farenc, *J. Appl. Polym. Sci.*, 2007, **103**, 3630.
18. N. Reddy and Y. Yang, *Trends Biotechnol.*, 2005, **11**, 22.
19. J. M. Dinwoodie, *Wood: Nature's Cellular, Polymer Fibre-Composite*, Institute of Metals, London, 1989, p. 56.
20. K. G. Satyanarama, B. C. Pai, K. Sukumaran and S. G. K. Pillai, Fabrication and properties of lignocellulosic fibre-incorporated polyester composites, in *Handbook of Ceramics and Composites*, ed. N. P. Cheremisioff, Marcel Dekker, New York, 1990, vol. 1, ch. 12, p. 341.
21. A. K. Mohanty, M. Misra, L. T. Drzal, S. E. Selke, B. R. Harte and G. Hinrichsen, Natural fibres, biopolymers, and biocomposites: an introduction, in *Natural Fibers, Biopolymers and Their Biocomposites*, ed. A.K. Mohanty, M. Misra and L.T. Drzal, CRC Press, Boca Raton, FL, 2005, pp. 1–36.
22. A. Bismarck, S. Mishra and T. Lampke, Plant fibres as reinforcement for green composites, in *Natural Fibers, Biopolymers and Their Biocomposites*, ed. A. K. Mohanty, M. Misra and L. T. Drzal, CRC Press, Boca Raton, FL, 2005, pp. 37–108.
23. J. G. Cook, *Handbook of Textile Fibres*, Woodhead Publishing, Abington, 5th edn, 1984, pp. 35–75.
24. C. Brett and K. Waldron, *Physiology and Biochemistry of Plant Cell Walls*, 2nd edn, Chapman and Hall, London, 1996.
25. A. Stamboulis, C. Baillie and Z. Schulz, Interfacial characterization of flax fibre-thermoplastic polymer composites by the pull-out test, Paper presented at the 2nd International Wood And Natural Fibre Composites Symposium, Kassel, June 1999.
26. Z. Rong, M. Q. Zhang, Y. Liu, G. C. Yang and H. M. Zeng, *Compos. Sci. Technol.*, 2001, **61**, 1437.
27. M. J. John and S. Thomas, *Carbohydr. Polym.*, 2008, **71**, 343.
28. M. Flemming, G. Ziegman and S. Roth, Faserverbundbauweisen, *Fasern und Matrices*, Springer, Berlin, 1995.
29. S. C. Fry, Plant cell walls, in *Handbook of Plant Sciences*, ed. K. Roberts, Wiley, Chichester, 2007, pp. 266–276.
30. P. Gañán, S. Garbizu, R. Llano-Ponte and I. Mondragon, *Polym. Compos.*, 2005, **26**, 121.
31. R. M. Rowell, Property enhanced natural fibre composite material based on chemical modification, in *Science and Technology of Polymers and Advanced Materials*, ed. P. N. Prasad, J. E. Mark, S. H. Kandil and Z. H. Kafafi, Plenum Press, New York, 1998, pp. 717–732.

32. R. M. Rowell, J. S. Han and J. S. Rowell, Characterization and factors effecting fibre, in *Natural Polymers and Agrofibres Composite*, ed. E. Frollini, A.L. Leao and L.H.C. Mattoso, Embrapa Instrumentacao Agropecuaria, San Carlos, 2000, pp. 115–134.
33. R. H. Kirby *Vegetable Fibres*, Interscience, New York, 1963.
34. Y. Habibi, W. El-Zawawy, M. M. Ibrahim and A. Dufresne, *Compos. Sci. Technol.*, 2008, **68**, 1877.
35. D. Ray and J. Rout, Thermoset biocomposites, in *Natural Fibers, Biopolymers and Biocomposites*, ed. A. K. Mohanty, M. Misra and L. T. Drzal, CRC Press, Boca Raton, FL, 2005, pp. 291–345.
36. S. Ramarad, Preparation and properties of kenaf bast fibre filled (plasticized) poly(lactic acid) composites, MS thesis, Universiti Sains Malaysia, 2008.
37. K. G. Satyanarayana, C. K. S. Pillai, K. Sukumaran, S. G. K. Pillai, P. K. Rohatgi and K. Vijayan, *Mater Sci.*, 1982, **17**, 2453.
38. K. G. Satyanarayana, K. K. Ravikumar, K. Sukumaran, P. S. Mukherjee, S. G. K. Pillai and A. K. Kulkarni, *Mater Sci.*, 1986, **21**, 57.
39. P. S. Mukherjee and K. G. Satyanarayana, *Mater. Sci.*, 1986, **21**, 51.
40. A. Gomes, K. Goda and J. Ohgi, *JSME*, 2004, **47**, 541.
41. L. Y. Mwaikambo and M. P. Ansell, *J. Appl. Polym. Sci.*, 2002, **84**, 2222.
42. J. Biagiotti, D. Puglia, L. Torre, J. M. Kenny, A. Arbelaiz, G. Cantero, C. Marieta, R. Llano-Ponte and I Mondragon, *Polym. Compos.*, 2004, **25**, 470.
43. J. Rout, S. S. Tripathy, M. Misra, A. K. Mohanty and S. K. Nayak, *Polym. Compos.*, 2004, **22**, 468.
44. W. Liu, A. K. Mohanty, L. T. Drzal, P. Askel and M. Misra, *Mater. Sci.*, 2004, **39**, 1051–1054.
45. N. Sgriccia, M. C. Hawley and M. Misra, *J. Compos.:* Part A, 2008, **39**, 1632.
46. D. Yelle, B. Goodell, D. J. Gardner, A. Amirbahman, P. Winistorfer and S. Shaler, *Forest Prod. J.*, 2004, **54**, 73.
47. S. Manrich and J. A. Marcondes, *J. Appl. Polym. Sci.*, 1989, **37**, 1777.
48. P. J. Herrera-Franco and M. J. Aguilar-Vega, *J. Appl. Polym. Sci.*, 1997, **65**, 197.
49. Z. Leman, S. M. Sapuan and S. Suppiah, *Key Eng. Mater.*, 2010, **471**, 1034.
50. D. Gulati and M. Sain, *Polym. Environ.*, 2006, **14**, 347.
51. Z. Leman, S. M. Sapuan, A. M. Saifol, M. A. Maleque and M. M. H. M. Ahmad, *Mater. Des.*, 2008, **29**, 1666.
52. Z. Leman, H. Y. Sastra, S. M. Sapuan, M. M. H. M. Hamdan and M. A. Maleque, *J. Appl .Technol.*, 2005, **3**, 14–19.
53. Z. Leman, S. M. Sapuan, M. Azwan, M. M. H. M. Ahmad and M. A. Maleque, *Polym. Plast. Tech. Eng.*, 2008, **47**, 606.
54. J. O. Karlsson, J. F. Blachot, A. Peguy and P. Gatenholm, *J. Polym. Compos.*, 1996, **17**, 300.

55. S. Dong, S. Sapieha and H. P. Schreiber, *J. Polym. Eng. Sci.*, 1992, **32**, 1734.
56. M. S. Sreekala, M. G. Kumaran, S. Joseph, M. Jacob and S. Thomas, *J. Appl. Compos. Mater.*, 2000, **7**, 295.
57. O. A. Khondker, U. S. Ishiaku, A. Nakai and H. Hamada, *J. Polym. Environ.*, 2005, **13**, 115.
58. M. R. Ishak, Enhancements of properties of sugar palm (Arenga pinnata Merr.) reinforced unsaturated polyester composites via vacuum resin impregnation, PhD thesis, Universiti Putra Malaysia, 2013.
59. M. R. Ishak, Z. Leman, S. M. Sapuan, M. Z. A. Rahman and U. M. K. Anwar, *J. Therm. Anal. Calorim.*, 2013, **111**, 1375.
60. M. R. Ishak, S. M. Sapuan, Z. Leman, M. Z. A. Rahman and U. M. K. Anwar, *Carbo. Polym.*, 2013, **91**, 699.
61. M. R. Ishak, Z. Leman, S. M. Sapuan, M. Z. A. Rahman and U. M. K. Anwar, *Fibers Polym.*, 2013, **14**, 250–257.
62. I. Siró and D. Plackett, *Cellulose.*, 2010, **17**, 459.
63. S. Alila, I. Besbes, M. R. Vilar, P. Mutjé and S. Boufi, *Ind. Crops Prod.*, 2013, **41**, 250.
64. L. Fu, J. Zhang and G. Yang, *Carbohydr. Polym.*, 2013, **92**, 1432.
65. J. Zhao, W. Zhang, X. Zhang, X. Zhang, C. Lu and Y. Deng, *Carbohydr. Polym.*, 2013, **97**, 695.
66. A. Chakraborty, M. Sain and M. Kortschot, *Holzforschung*, 2005, **59**, 102.
67. M. Jonoobi, A. P. Mathew and K. Oksman, *Ind. Crops Prod.*, 2012, **40**, 232–238.
68. Y. Xu, K. Kim, M. Hanna and D. Nag, *Ind. Crops Prod.*, 2005, **21**, 185.
69. S. S. Ray and M. Bousmina, *Prog. Mater. Sci.*, 2005, **50**, 962.
70. R. P. Ellis, *J. Sci. Food Agric.*, 1998, **77**, 289.
71. H. F. Zobel, *Starch*, 1998, **40**, 44.
72. A. Buleon, *Int. J. Biol. Macromol.*, 1998, **23**, 85.
73. R. L. Whistler and J. R. Daniel, Molecular structure of starch, in *Starch: Chemistry and Technology*, ed. Whistler, R. L., BeMiller, J. N., Paschall, E. F., Academic Press, Orlando, FL, 2nd edn, 1984, p. 153–182.
74. S. T. Ganjali, F. Motiee, E. Shakeri and A. Abbasian, *J. Appl. Chem. Res.*, 2010, **4**, 53.
75. I. Arvanitoyannis, *J. Macromol. Sci.*, 1999, **39**, 205.
76. J. Fang, P. Fawler, C. Eserig, R. González, J. Costa and L. Chamudis, *Carbohydr. Polym.*, 2005, **60**, 39.
77. J. Sahari, S. M. Sapuan, E. S. Zainudin and M. A. Maleque, *Procedia Chem.*, 2012, **4**, 254.
78. N. Singh, J. Singh, L. Kaur, N. S. Sodhi and B. S. Gill, *Food Chem.*, 2003, **81**, 219.
79. R. M. Alves, M. V. Grossmann, C. Ferrero, N. E. Zaritzky, M. N. Martino and M. R. Sierakoski, *Starch*, 2002, **54**, 476.
80. J. J. M. Swinkels, *Starch*, 1985, **37**, 1.

81. J. Sahari, Characterization and development of biopolymer derived from arenga pinnata and their biocomposites, PhD thesis, Universiti Putra Malaysia, 2013.
82. G. Alebiowu and A. K. Osinoiki, *FARMACIA*, 2010, **58**, 341.
83. P. T. Adrina, A. K. Alias, M. Dos and A. Manan, *Starch*, 2008, **60**, 408.
84. F. B. Ahmad, P. A. Williams, J. L. Doublier, S. Durand and A. Buléon, *Carbohydr. Polym.*, 1999, **38**, 361.
85. D. Bachtiar, S. M. Sapuan and M. M. Hamdan, *Mater. Des.*, 2008, **29**, 1285.
86. D. R. Isleib and J. Americ, *Potato Res.*, 1958, **35**, 428.
87. O. J. Olayemi, A. R. Oyi and T. S. Allagh, *J. Pharm. Sci.*, 2008, **7**, 131.
88. X. Ma, J. Yu and F. Jin, *Polym. Int.*, 2004, **53**, 1780.
89. A. Poonam and D. David, *Thermochim. Acta*, 1998, **319**, 17.
90. R. H. Marchessault, M. G. Taylor and C. A. Fyfe, *Carbohydr. Res.*, 1985, **144**, C1.
91. S. Vasudeva, O. Hiroshi, T. Hidechika, I. Seiichiro and O. Ken'ichi, *J. Agric. Food Chem.*, 2000, **48**, 2639.
92. G. L. Luiz, R. A. Rafael, M. D. Ivo, A. S. C. F. Marco, C. V. Eliane, L. W. Adenise, B. Gilbert, S. Egon and R. S. Carlos, Braz. Arch. Biol. *Technol.*, 2009, **52**, 143.
93. X. Ma and J. Yu, *J. Appl. Polym. Sci.*, 2004, **93**, 1769.
94. P. M. Forssell, J. M. Mikkia, G. K. Moates and R. Parker, *Carbohydr. Polym.*, 1997, **34**, 275.
95. S. Gaudin, D. Lourdin, D. Le Botlan, J. L. Ilari and P. Colonna, *J. Cereal Sci.*, 1999, **29**, 273.
96. S. Nakamura and A. V. Tobolsky, *J. Appl. Polym. Sci.*, 1967, **11**, 1371.
97. M. T. Kalichevsky, E. M. Jaroszkiewicz and J. M. V. Blanshard, *Polymer*, 1993, **34**, 346.
98. O. Vilpoux and L. Avérous, Starch-based plastics, in *Technology, Use and Potentialities of Latin American Starchy Tubers*, ed. M. P. Cereda and O. Vilpoux, NGO Raízes and Cargill Foundation, São Paolo, 2004, pp. 521–553.
99. A. K. Mohanty, M. Misra and G. Hinrichsen, *Macromol. Mater. Eng.*, 2000, **276**, 1.
100. T. H. McHugh, J. F. Aujard and J. M. Krochta, *J. Food Sci.*, 1994, **59**, 416.
101. H. M. Park, X. Li, C. Z. Jin, C. Y. Park, W. J. Cho and C. S. Ha, *Macromol Mater. Eng.*, 2002, **287**, 553.
102. T. Bourtoom, *Songklanakarin J. Sci. Technol.*, 2008, **30**, 149.
103. L. Avérous and N. Boquillon, *Carbohydr. Polym.*, 2004, **56**, 111.
104. P. Myllärinen, R. Partanen, J. Seppälä and P. Forsella, *Carbohydr. Polym.*, 2002, **50**, 355.
105. S. Mali, M. V. E. Grossmann, M. A. Garcia, M. N. Martino and N. E. Zaritzky, *Carbohydr. Polym.*, 2002, **50**, 379.
106. S. Guilbert and N. Gontard, (1995), Technology and applications of ebible protective films, in VII, *Biotechnology and Food Research, New Shelf-life Technologies and Safety Assessments*, VTT Symposium 148,

Helsinki, ed. R. Ahvenainen, T. Ohlsson and T. Mattila-Sandholm, VTT, Espoo, pp. 49–60.
107. M. Mitrus, *Int. Agrophys.*, 2005, **19**, 237.
108. R. A. De Graaf, A. P. Karman and L. P. B. M. Janssen, *Starch*, 2003, **55**, 80.
109. X. Ma, J. Yu and F. K. John, *Carbohydr. Polym.*, 2005, **62**, 19.
110. Acros Organics, *Reference Handbook of Fine Chemicals*, Acros Organics, Geel, 2006.
111. S. H. Imam, S. H. Gordon and R. V. Greene, (1996), Starch biodegradation (in starch–plastic blends), in *Polymeric Materials Encyclopedia*, ed. J. C. Salamone, C. R. C. Press, Boca Raton, F. L., vol. 10, pp. 7892–7901.
112. N. Laohakunjit and A. Noomhorm, *Starch*, 2004, **56**, 348.
113. A. D. Beerler and D. C. Finney, Plasticizers, in *Modern Plastics Encyclopedia*, McGraw-Hill, New York, 1983.
114. E. R. Lieberman and S. G. Gilbert, *J. Polym. Sci.*, 1973, **41**, 33.
115. E. Zini and M. Scandola, *Polym. Compos.*, 2011, **12**, 1905.
116. A. S. Hermann, J. Nickel and U. Riedel, *Polym. Degrad. Stab.*, 1998, **59**, 251.
117. C. Bastioli, *Macromol. Symp.*, 1998, **130**, 379.
118. R. Narayan, *ACS Symp. Ser.*, 1992, **476**, 1.
119. A. K. Rana and K. Jayachandran, *Mol. Cryst. Liq. Cryst.*, 2000, **353**, 35.
120. F. Vilaseca, J. A. Mendez, A. Pèlach, M. Llop, N. Cañigueral, J. Gironès, X. Turon and P. Mutjé, *Process Biochem.*, 2007, **42**, 329.
121. N. Soykeabkaew, P. Supaphol and R. Rujiravanit, *Carbohydr. Polym.*, 2004, **58**, 53.
122. M. E. Vallejos, A. A. S. Curvelo, E. M. Teixeira, F. M. Mendes, A. J. F. Carvalho, F. E. Felissia and M. C. Area, *Ind. Crops Prod.*, 2011, **33**, 739.
123. L. Avérous and P. J. Halley, *Biofuel, Bioprod. Biorefin.*, 2009, **3**, 329.
124. P. J. Herrera Franco, A. Valdez-González, Fiber–matrix adhesion in natural fibre composites, in *Natural Fibers, Biopolymers and Their Biocomposites*, ed. A. K. Mohanty, M. Misra and L. T. Drzal, CRC Press, Boca Raton, FL, 2005, pp. 177–230.
125. M. G. Lomelí Ramírez, K. G. Satyanarayana, S. Iwakiri, G. Bolzon de Muniz, V. Tanobe and T. S. Flores-Sahagun, *Carbohydr. Polym.*, 2011, **86**, 1712.
126. A. A. S. Curvelo, A. J. F. Carvalho and J. A. M. Agnelli, *Carbohydr. Polym.*, 2001, **45**, 183.
127. M. Gaspar, Z. Benko, G. Dogossy, K. Reczey and T. Czigany, *Polym. Degrad. Stab.*, 2005, **90**, 563.
128. J. L. Guimaraes, F. Wypych, C. K. Saul, L. P. Ramos and K. G. Satyanarayana, *Carbohydr. Polym.*, 2010, **80**, 130.
129. C. M. O. Muller, J. B. Laurindo and F. Yamashita, *Carbohydr. Polym.*, 2009, 77, 293.
130. S. Mohanty, S. K. Verma and S. K. Nayak, *Compos. Sci. Technol.*, 2006, **66**, 538.

131. M. R. Ishak, S. M. Sapuan, Z. Leman, M. Z. A. Rahman and U. M. K. Anwar, *J. Therm. Anal. Calorim.*, 2012, **109**, 981.
132. H. Yang, R. Yan, H. Chen, D. H. Lee and C. Zheng, *Fuel*, 2007, **86**, 1781.
133. D. Y. Kim, Y. Nishiyama, M. Wada, S. Kuga and T. Okano, *Holzforschung*, 2001, **55**, 521.
134. J. Sahari, S. M. Sapuan, Z. N. Ismarrubie and M. Z. A. Rahman, *Fibres Text. East. Eur.*, 2012, **20**, 23.
135. P. Aggarwal and D. Dollimore, *Thermo Acta*, 1998, **319**, 17.

CHAPTER 5

# Applications of Starch Nanocrystal-based Blends, Composites and Nanocomposites

VERÔNICA M. A. CALADO* AND ANDRESA RAMOS

Centro de Tecnologia, Escola de Química, Universidade Federal do Rio de Janeiro, Cidade Universitária, Av. Athos da Silveira Ramos 149, Ilha do Fundão, 21941-909 Rio de Janeiro, RJ, Brazil
*Email: calado@eq.ufrj.br

## 5.1 Introduction

There are increasing worldwide concerns about the environment, global warming and water scarcity, and one of the major environmental problems is the disposal of non-biodegradable plastics. The plastics industry has been continuously growing for more than 50 years, and in 2012, 288 million tonnes of plastics were produced worldwide, which represents an increase of 2.8% compared with 2011.[1]

In 2012, of all the plastics produced, 26.3% was recycled, 35.6% was used for energy recovery and 38.1% was disposed of in landfills. The total recovery (recycling and energy recovery) increased by 4% and this growth shows a continuously strong trend. At the same time, there was a reduction of 5.5% in landfilled plastics, which also shows a general positive development. The waste generated from plastics comes from the packaging industries,

RSC Green Chemistry No. 37
Starch-based Blends, Composites and Nanocomposites
Edited by Visakh P. M. and Long Yu

Published by the Royal Society of Chemistry, www.rsc.org

representing 62.2% of the total. Building and construction, electrical and electronic products and agriculture represent 5–6% each.[1]

Many researchers are searching for biodegradable plastics and the answer they are reaching is biopolymers, which are a class of material that comes from natural sources, such as polysaccharides, carbohydrates or synthesis by microrganisms [polyhydroxybutyrate (PHB), polyhydroxyalkanoates (PHAs), polylactide (PLA), produced by the fermentation of sugars extracted from plants]. It is worth noting, however, that the energy costs for producing PHB, PHA and PLA are high.[2]

The global production capacity for bioplastics in general was more than 1.5 million tonnes in 2012 and the projection for 2017 is over than 6 million tonnes per year.[3] We need to differentiate between degradable and non-degradable bioplastics – the former class is the one that researchers are looking for. The latter class, such as "green polyethylene", represents bioplastics that come from a natural source (*e.g.*, ethanol from sugar cane) but are still non-degradable.

Considering these issues, biopolymers from polysaccharides appear to be a promising alternative that needs to be carefully analyzed. Starch, a natural, renewable and biodegradable polymer, is the second most abundant biomass material in Nature and a promising material because of its versatility, low price, wide availability and total compostability without toxic residues. It exists in granular forms inherently composed of nanoscale crystalline blocklets in a range of diameters from 20 to 500 nm, depending on their botanical origin and the location in the granules.[4]

Starch has been investigated widely for the potential manufacture of products such as water-soluble pouches for detergents and insecticides, flushable liners and bags and medical delivery systems and devices.[5]

The main sources are plant roots, stalks, crop seeds and staple crops such as rice, corn (82%), wheat (8%), cassava (5%) and potato (5%).[6–8]

Starch, like many other polymers, can be produced by extrusion, molding, wet grinding, sieving, drying and thermoforming.[9] This polymer has attracted considerable attention as a biodegradable thermoplastic polymer, although it is not truly thermoplastic; it can be converted into a continuous polymeric entangled phase by mixing with sufficient water or plasticizer (generally a polyol, such as glycerol).[6]

Starch is a source of stored energy in vegetable plastids formed by D-glucopyranose polymers linked together by α-1,4 and α-1,6 glycosidic bonds. Carbon C1 on a D-glucopyranose molecule reacts with C4 or C6 from an adjacent D-glucopyranose molecule (Figure 5.1). Because the aldehyde group on one end of a starch polymer is always free, starch polymers always have one reducing end.[10]

Generally, starch consists primarily of highly branched and linear chains of glucose molecules, namely amylopectin and amylose, respectively.[11] A preponderance of amylose in starches gives stronger films.[12] The branched structure of amylopectin generally leads to films with different mechanical properties, such as decreased tensile stress.[13]

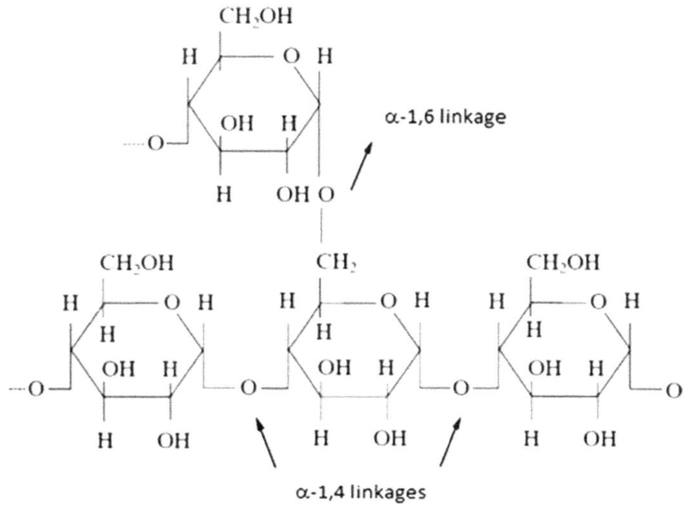


**Figure 5.1** The α-1,4 and α-1,6 glycosidic bonds of starch. Adapted from Thomas and Atwell.[12]

Amylopectin is a very large polysaccharide with extensive branching formed by a large number of shorter chains that are bound together at their reducing end side by a 1,6-linkage. Amylose consists only of either a single or a few long chains of 1,4-linked-D-glucosyl residues, thus making the molecule linear or slightly branched.[14]

Starch granules have a semicrystalline structure: approximately 70% as an amorphous part and about 30% as a crystalline part. The amorphous regions contain the main part of the amylose but also a considerable part of the amylopectin. The crystalline region consists primarily of amylopectin.[15] The amorphous regions are highly susceptible to hydrolysis and, under controlled conditions, may be removed, leaving the crystalline regions intact.

It is believed that the crystalline region is created by the intertwining of amylopectin side chains.[16] Amylose molecules are thought to occur in the granules as individual molecules, randomly interspersed among amylopectin molecules. Depending on the botanical origin of starch, amylose is preferably found in the amorphous region (*e.g.*, wheat starch),[17] interspersed among amylopectin clusters in both the amorphous and crystalline regions (*e.g.*, normal maize starch),[18] in bundles among amylopectin clusters or co-crystallized with amylopectin chains (*e.g.*, potato starch).[17] The amylose content should therefore considerably influence the crystalline organization. Depending on their X-ray diffraction pattern, reflecting long-range ordering in the granules, starches are categorized in three crystalline types (polymorphs) called A, B and C. However, a property common to most starches is gelatinization, which refers to the collapse (disruption) of molecular order (breaking of H-bonds) within the granule, along with all concomitant and irreversible changes in properties such as water uptake, granular swelling, crystallite melting, birefringence loss, starch solubilization and viscosity development. It is now widely accepted that it can be viewed as a melting process, which can be evidenced as an endothermic thermal transition in differential scanning calorimetric (DSC) measurements. Depending on the application, this property is sought for or not.[17]

Table 5.1 presents the amylase and amylopectin contents of some starch sources. Yam and pea are the starch sources with the highest content of amylopectin and amylase, respectively.

Starch-based materials have some drawbacks, such as poor processability and properties (*e.g.*, weak mechanical properties, poor long-term stability and high water sensitivity).[9,10] The hydrophilic nature of thermoplastic starches makes them susceptible to moisture attack, resulting in variability in dimensional stability and mechanical properties. Further, retrogradation and crystallization of the mobile starch chains lead to undesired variations of its thermomechanical properties.[7] Other natural polymers that could be useful as green materials nowadays are not so useful because they also have poor properties.

In order to improve the performance and the mechanical properties (such as strength, modulus and dimensional stability), thermal stability, barrier properties, surface appearance, physical properties (such as permeability, optical clarity and fire retardancy), to add new functionalities to polymeric materials and to decrease the gas and water permeability of the resultant composites, some nanosized fillers (at least one dimension in the nanometer range, *i.e.*, 1–100 nm) can be added, thus obtaining the well-known polymer nanobiocomposites.[7,29] Polysaccharide nanofillers represent the second popular group of nanofillers used for these materials. Cellulose, starch, chitin and chitosan can be used as nanofillers (nanowhiskers,

**Table 5.1** Amylose and amylopectin contents of various materials.

| Starch source | Amylose (%) | Amylopectin (%) | Ref. |
|---|---|---|---|
| Waxy maize starch | ~0 | ~100 | García *et al.*[19] |
| Pea (granule size 30 μm) | 35.0 | 65.0 | Yu *et al.*[20]; Zhang *et al.*[21] |
| Rice | 30.4 | 69.6 | Piyada *et al.*[10] |
| Cassava | 28.0 | 72.0 | García *et al.*[19] |
| Wheat | 23.0 | 77.0 | Chivrac *et al.*[22] |
| Jackfruit seed | 32.14 | 67.86 | Rengsutthi and Charoenrein[23] |
| Corn | 24.30 | 75.70 | |
| Potato | 28.08 | 71.92 | |
| Taro | 19.2 | 80.8 | Tattiyakul *et al.*[24] |
| Sago | 21.9 | 78.1 | Srichuwong *et al.*[25] |
| Yam | 14.2 | 85.8 | |
| Barley | 15.0 | 85.0 | Tan *et al.*[26] |
| Chickpea | 29.2 | 69.1 | Polesi and Sarmento[27] |
| Pinhão (*Araucaria angustifolia* seeds) | 26.3 | 83.7 | Muccillo[28] |

nanoparticles and nanocrystals) in different forms. Cellulose nanowhiskers (CNWs) and starch nanoparticles (SNPs) have already been widely used, whereas the use of chitin nanoparticles has been less investigated.[26] These nanoparticles display different shapes depending on the polysaccharide source; these are rigid rod-like particles for cellulose and platelets for starch.[30] The crystalline nanoplatelets are about 6–8 nm thick, with a length of 20–40 nm and a width of 15–30 nm.[10] Other nanofillers include phyllosilicates, carbonaceous nanofillers, natural sodium montmorillonite and organomodification of the clay platelets.[31] They have different geometries (size and shape) and surface chemistry. Three distinct shape types of nanofillers can be observed: nanoparticles, nanotubes and nanolayers.[32] Although "nanoparticles" is a term frequently used in general, they are particulate materials having low aspect ratio. Nanotubes have a higher aspect ratio, which gives better mechanical properties. Nanolayers, such as nanoclays and nanocrystals, are layered polymer nanocomposites, with different degrees of dispersion, *i.e.*, intercalated nanocomposites (intercalated polymer chains between layered nanocomposites), exfoliated nanocomposites (separation of individual layers) and flocculated or phase-separated nanocomposites, having poorer physical properties.[8] Nevertheless, the term "nanoparticles" is also frequently used in a broad sense to describe a nanofiller regardless of its shape.

The main advantages of these nanofillers are their renewable nature, availability, high specific strength, non-abrasive nature that allows easier processing even at high filling levels, biodegradability and a relatively reactive surface.[33] They come from cheap, abundant, renewable and biodegradable biopolymers. They are versatile, light-weight and have a high aspect ratio and high specific strength and modulus.[29] Starch nanocrystals (SNCs) are non-toxic, biodegradable, biocompatible, of high reactivity and have easy processability owing to their non-abrasive nature. These properties, added to their capacity for exhibiting a reinforcing function similar to that of inorganic nanofillers, make them of interest for incorporation into polymeric matrices, in order to improve the mechanical properties.[35]

Starch nanoparticles and nanocrystals have many potential applications, such as plastic fillers, food additives, drug carriers, implant materials, vehicles for carrying bioactive substances and nutraceuticals, fillers in biodegradable composites, coating binders, adhesives and a source of energy at the end of their life cycle.[6] The SNCs can also be used in biomedical, biochemical and technological applications, and as vehicles for carrying bioactive substances and nutraceuticals. However, they tend to aggregate and settle down in aqueous solutions, which is a limits their application in most biological and food systems.[34]

We can ask whether starch from any source could result in nanocrystals. Le Corre *et al.*[17] answered this question and they also examined whether the final properties of SNCs depend on the amylose content and/or botanical origin. They concluded that botanical origin of starch has only a moderate influence on the final properties of SNCs for the same amylose content and crystalline type. However, for the same botanical origin (maize), there was a strong influence of the amylopectin content and molecular structure because of differences in the nanocrystal size. Thus, differences in size were rather moderate whereas differences were more pronounced when comparing shapes and crystallinity.

Figure 5.2 shows micrographs of native starch and SNCs with platelet-like morphology (thickness 5–8 mm) from different sources and Table 5.2 presents their characteristics. As indicated above, Le Corre *et al.*[17] concluded that the amylase and amylopectin content is more important than the starch source, which has only a moderate influence on the final SNC properties.

According to Le Corre *et al.*,[8] starch crystallites, starch nanocrystals, microcrystalline starch and hydrolyzed starch all refer to the crystalline part of starch obtained by hydrolysis but to different extents (from the most to the least). It should be distinguished from starch nanoparticles, which can be amorphous.

SNCs are crystalline platelets obtained by the acid hydrolysis of amorphous parts of starch.[34] They are believed to correspond to the starch granule's crystallites.[17] They are layered particles and when applied in polymeric matrices, they can show different degrees of dispersion.[8] Layered particles not well dispersed in the polymer matrix because of self-aggregation are shown in Figure 5.3a. An ordinary microcomposite is formed, whose properties are not affected by the high surface energy of nanoparticles. In this case, X-ray diffraction shows that polymer composites display refraction angles the same as or higher than those of layered particles. When the polymer chains penetrate or intercalate between the layers forming a multilayer structure ordered with a uniform distance of a few nanometers, without however completely exfoliating, a nanocomposite is formed, as shown in Figure 5.3b.[36,37] X-ray diffraction shows that intercalated nanocomposites exhibit reflection below the characteristic peak of the layered particles angles. The intercalation can be helpful and sometimes a precursor of exfoliation.[38] Exfoliated nanocomposites are characterized by delamination of the layered particle and a random dispersion in the polymer matrix, which makes it impossible to identify a peak on the X-ray diffraction pattern (Figure 5.3c).[37,39] Nanocomposites formed by exfoliation of layered particles show a greater improvement in physical properties[8].

Table 5.3 presents a summary of some characteristics of starch-based nanoparticles.

Dufresne[40] stated that the dispersion of polysaccharide nanocrystals in non-aqueous media is possible using surfactants or chemical grafting as they possess a reactive surface covered with hydroxyl groups, providing the possibility of extensive chemical modification. The reinforcing phenomenon of polysaccharide nanocrystals results from the formation of a percolating network because of hydrogen bonding forces. Grafting of long chains instead of small molecules can be used to preserve the mechanical properties of the material.

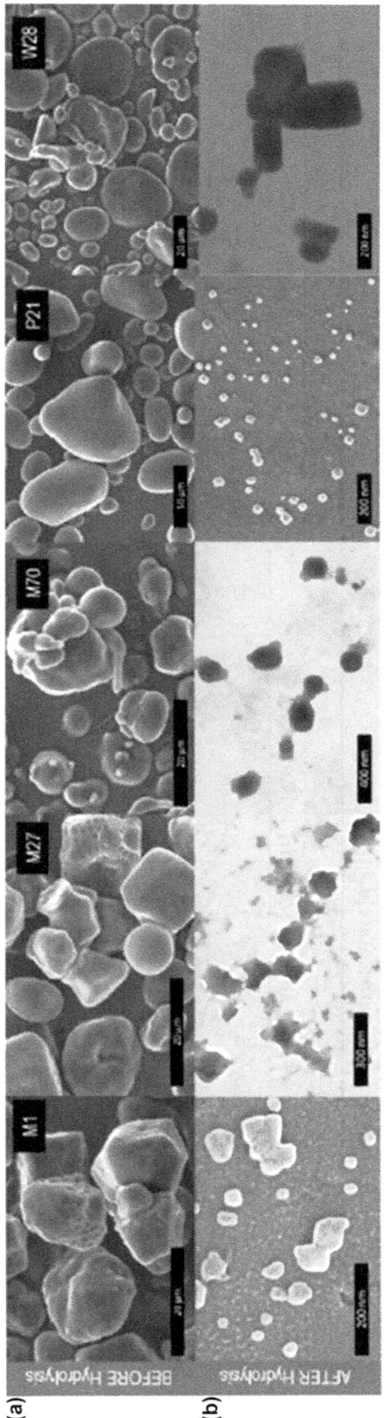


**Figure 5.2** Micrographs of native starches and corresponding SNCs: (a) before hydrolysis and (b) after hydrolysis. M1, waxy maize; M27, normal maize; M70, high amylase; P21, potato; W28, wheat. From Le Corre *et al.*[17]

**Table 5.2** Some characteristics of some native starches and their SNCs. Source: Le Corre *et al.*[17]

| Botanical origin | Referred to as | Granular size (μm) | Amylose (%) | Crystalline type | SNC diameter | | SNC thickness | |
|---|---|---|---|---|---|---|---|---|
| | | | | | nm | Counts | nm | Counts |
| High-amylose maize | M70 | 5–20 | 65–75 | B | $118 \pm 53$ | 190 | $5 \pm 1.6$ | 76 |
| Normal maize | M27 | 5–20 | 27 | A | $58 \pm 36$ | 576 | $8.3 \pm 3.1$ | 112 |
| Waxy maize | M1 | 5–20 | 1 | A | $47 \pm 42$ | 71 | $6.1 \pm 1.9$ | 14 |
| Wheat | W28 | 2–30 | 28 | A | $100 \pm 50$ | 71 | $3.7 \pm 0.6$ | 30 |
| Potato | P21 | 5–80 | 21 | B | $52 \pm 4$ | 951 | $7.6 \pm 1.6$ | 31 |

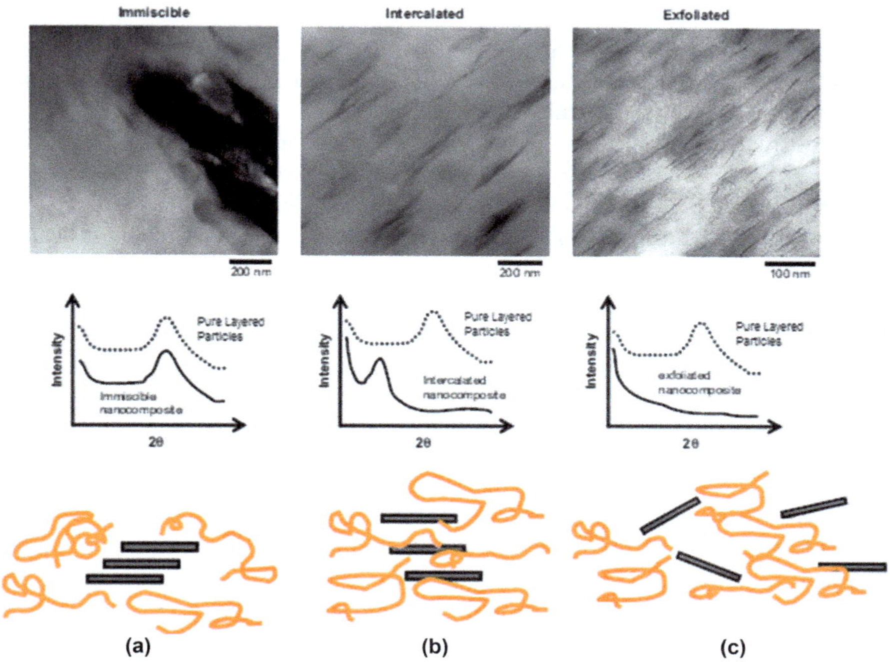


**Figure 5.3** Different states of dispersion of layered nanoparticles in polymers. From Paul and Robeson.[38]

**Table 5.3** Summary of characteristics of starch-based nanoparticles.

| Source | SNC diameter (nm) | SNC thickness (nm) | Ref. |
|---|---|---|---|
| Waxy maize | 50 | 6.0–8.0 | Xie *et al.*[7] |
| | 5–89 | 4.2–8.0 | Le Corre *et al.*[17] |
| Regular maize | 50–100 | — | Xie *et al.*[7] |
| | 22–94 | 5.2–11.4 | Le Corre *et al.*[17] |
| High-amylose maize | 65–171 | 3.4–6.6 | |
| Wheat | 50–150 | 3.1–4.3 | |
| Potato | 48–56 | 6.0–9.2 | |

For many types of fillers, bad compatibility occurs between the hydrophilic polysaccharide and the hydrophobic polymer matrix. To improve the compatibility between the two phases, it is necessary to perform physical or chemical modifications of the nanoparticle surfaces.[41–43] One of the advantages of using polysaccharide nanomaterials is the similar chemical structure between these nanomaterials and the polysaccharide matrices, improving nanofiller–matrix interactions.[7]

Wei *et al.*[44] stated that SNCs are crystalline platelets originating from breakdown of the semicrystalline structure of starch granules by acid hydrolysis of amorphous parts. Because of their unique properties, SNCs have been widely used as particle emulsifiers[45] to prepare Pickering emulsions[36] and as reinforcement to prepare nanocomposites, such as biodegradable films and natural polymers.[46–48]

In contrast to the rod-like whiskers of cellulose and chitin, SNCs show a distinct platelet-like structure similar to that of exfoliated layered silicate. The stability of an SNC suspension is, to a large extent, dependent on the surface charge of the SNCs. Generally, the stability of a disperse system depends on the balance of electrostatic repulsive energy and the van der Waals attraction energy.[49] However, because of their natural hydrophilicity, SNCs are easily self-aggregated and settled in water, forming agglomerates on the micrometer scale, inhibiting their function in composites.[35] The aggregation behavior of SNCs greatly limits their applications, since a procedure involving mixing SNC aqueous suspensions with matrix solution is necessary in the casting fabrication of nanocomposites.[50] Consequently, a homogeneous dispersion of SNCs is required for high mechanical performance of nanocomposites.

Fortunately, SNCs have reactive surfaces suitable for chemical derivatization and grafting reactions. Such modification would facilitate the dispersion of SNCs, manipulate the hydrophobicity of surface and improve the miscibility between the SNCs and the polymer matrix. In particular, grafting produces long tails on the surface of SNCs which may penetrate into the polymer matrix and exert a stronger interfacial interaction.[35] Zheng *et al.*[47] reported an increase in strength and Young's modulus, together with a decrease in elongation at break, at low pea starch nanocrystal (PSN) loading levels (lower than 2 wt%) in a soy protein isolate (SPI) matrix, which was attributed to uniform dispersion of the nanocrystals. Chen *et al.*[51] reported a low reinforcing effect and a slight improvement of properties when the PSN content was lower than 10 wt% and a decrease in both the elongation at break and strength when the nanocrystal content was higher than 10 wt%, using poly(vinyl alcohol) (PVA) as matrix. Indeed, Wang and Zhang[52] reported the preparation of a high-strength water-borne polyurethane (WPU)-based elastomer reinforced with 1–5% waxy maize SNCs. Another interesting phenomenon reported in some composites is the observation of a constant elongation at break with increasing filler content. Low filler content caused a better dispersion of the SNCs in the WPU matrix, resulting in stronger interactions.[53]

In order to reduce the aggregation of SNCs in aqueous solutions, Jivan *et al.*[34] suggested crosslinking SNCs with sodium hexametaphosphate by intra- and inter-ester linkages, as the hydroxyl groups at the reactive surface of nanocrystals become oxygen anions under alkaline conditions, resulting in water-dispersive SNCs. The chemically crosslinked nanocrystals form a much stronger lattice than the unmodified nanocrystals, resisting the phase transitions induced by heating, as commented on later in Section 5.3.2.2.1.

## 5.2 Preparation of Starch Nanocrystals

Different methodologies for preparing SNCs have been reported in the literature (Table 5.4), but they all agree that starch from different sources should be subjected to an acid hydrolysis under various conditions.

SNCs can be obtained by acid hydrolysis (normally with sulfuric acid) of native starch granules by strictly controlling the temperature, acid and starch concentrations, hydrolysis duration and stirring speed.[7] The use of sulfuric acid to prepare nanocrystals leads to more stable aqueous suspensions than those prepared using hydrochloric acid.[33,44] The $H_2SO_4$-prepared nanoparticles present a negatively charged surface whereas the HCl-prepared nanoparticles are not charged.

Angellier *et al.*[48] found that different acid concentrations and hydrolysis times can affect the size of SNCs during the hydrolysis of starch granules. For 2.2 M HCl, smaller crystals were obtained after 50 days (89 nm) whereas when using 3.7 M HCl this occurred after 24 days (48 nm). Jayakody and Hoover[58] proposed a two-stage hydrolysis: an initial fast step, presumably on account of the hydrolysis of amorphous regions of granules, followed by a second slower step, probably because of the hydrolysis of crystalline regions. An extended time of acid hydrolysis is mostly utilized in the preparation of SNCs, but the conditions of hydrolysis during the extraction process (such as the type of acid, acid concentration, temperature and duration) are different.[37]

In order to decide how many days would be sufficient for nanocrystal formation during hydrolysis, Le Corre *et al.*[57] studied the kinetics and found that SNCs were already seen from the first day on (Figure 5.4). The size of such nanoparticles obtained after 1 day was 25–70 nm, with an average length of 50 nm and an average width of 35 nm. These values correspond to the dimensions reported in the literature for SNCs obtained after 5 days of hydrolysis. It was concluded that separation processes, such as centrifugation, could increase the final yield.

To improve the properties of SNCs, some workers have proposed some modifications of SNCs. García *et al.*[54] used a strategic synthetic pathway consisting of three reaction steps to graft polylactate (PLA) onto the surface of SNCs to make them miscible. They first performed benzoylation of PLA, followed by activation of the PLABz carboxyl groups and grafting of the modified chains onto the SNCs. It was concluded that the presence of the benzoyl group in the PLA modifies its ability for crystallization and this process is strongly affected by the addition of nanocrystals only when they are chemically bound to PLA.

Thielemans *et al.*[59] chemically modified starch-derived nanocrystals with poly(ethylene glycol) methyl ether (PEGME) and stearoyl chloride, without changing the crystalline structure. There was extensive crystallization of the stearate moieties grafted at the starch nanoparticle surface, forming a crystalline hydrophobic shell around the hydrophilic SNCs. These modifications had a large effect on the individualization of the nanocrystals

**Table 5.4** Summary of techniques for the preparation of starch nanocrystals.

*General process*
36.725 g of starch powder were mixed with 250 mL of 3.16 M $H_2SO_4$, at 40 °C under constant stirring at 100 rpm for 5 days. The ensuing insoluble residue was washed in distilled water and separated by successive centrifugations at 10 000 rpm and 5 °C up to pH 7.0. The aqueous suspensions of starch nanoparticles were stored at 4 °C with several drops of chloroform. Finally, the nanocrystals were lyophilized and stored until further use.

| Study | Starch source | Variations |
|---|---|---|
| García *et al.*[6,19,54] | Waxy maize | None |
| Zheng *et al.*[47] Zou *et al.*[53] | Pea | None |
| Lin *et al.*[29] | Pea | After the classical procedure, the nanocrystals were subjected to an overnight dialysis against distilled water. Finally, the loose SNC powder was obtained by freeze-drying |
| Chang *et al.*[35] | Pea | 14.69 g of pea starch granules. After centrifugation, a small amount of ammonia was added followed by dialysis overnight with distilled water. The nanocrystals in powder form were then obtained by lyophilization |
| Chen *et al.*[51] | Pea | Starch concentration of 15%. After centrifugation, the SNCs were dialyzed for 3 days and then diluted with distilled water to obtain a PSN dispersion with a concentration of SNCs of around 2 wt% |
| Chen *et al.*[55] | Potato | Isolated SNCs had a width of 10–20 nm and a length of 40–70 nm. After neutralization, the SNC precipitate was ultrasonically homogenized in an aqueous solution for 3 min using an ultrasonic instrument |
| Jivan *et al.*[34] | Potato | Starch was dispersed (5% w/w) in either 2.2 or 3.7 M HCl and incubated at 35 °C for up to 50 days while being shaken at 120 rpm. At various time intervals, the solution was allowed to rest for 15 min and the supernatant was centrifuged at 15 600 *g* for 6 min. The sediment was washed with doubly distilled water several times and reprecipitated by centrifugation at 18 000 *g* for 5 min. The neutralized SNCs were lyophilized to give a powder for further processing |
| Bouthegourd *et al.*[43] | Potato | 36 g of native starch in 25 mL of 3.16 N $H_2SO_4$. After centrifugation, the dispersion was further ultrasonicated for 10 min. The resulting aqueous suspension was constituted of starch fragments with a homogeneous distribution. The solid fraction of this aqueous suspension had a concentration of about 3.4 wt%. |
| Piyada *et al.*[10] | Rice | 36 g of native starch. The resulting suspension was freeze-dried using freeze-dryer. The resulting nanocrystals were generally observed in the form of aggregates having an average size of around 3.97 μm, as measured with a laser particle size analyzer |
| Wei *et al.*[44] Angellier *et al.*[56] | Waxy maize | 50 g of starch powder were mixed with 500 mL of 3.16 M $H_2SO_4$ under constant stirring at 200 rpm for 7 days. The resulting samples from centrifugation were dispersed again using an Ultra-Turrax for 3 min at 13 500 rpm to avoid aggregates and stored at 4 °C with several drops of chloroform |
| Le Corre *et al.*[57] | Waxy maize | 147 g of waxy maize in 1 L of 3 M $H_2SO_4$; agitation of 400 rpm. The resulting suspension was redispersed using an Ultra-Turrax for 5 min at 13 000 rpm to avoid aggregates. The suspensions obtained were filtered on a filter tissue (40 μm). Sodium azide was added to the suspensions before storage at 4 °C to avoid microbial growth. The experiments were carried out during 1, 3 and 5 days. |

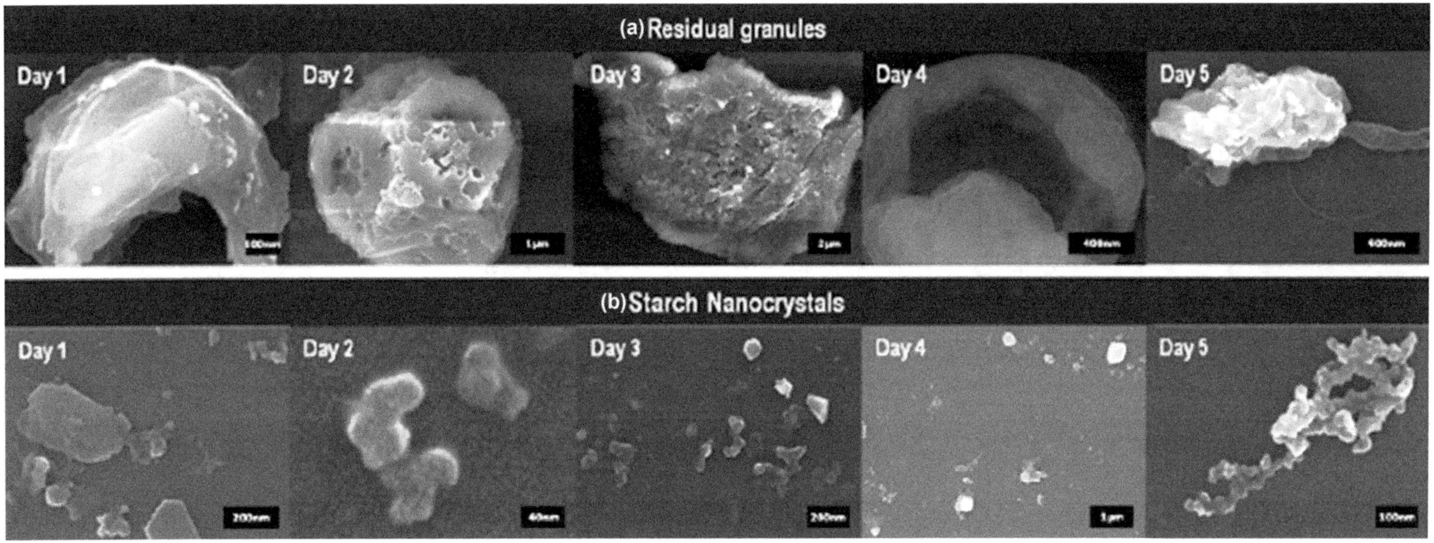


**Figure 5.4** Field emission gun (FEG)–SEM images of residual granules and SNCs in different stages of hydrolysis (no separation process). Source: Le Corre *et al.*[57]

because of the reduced hydrogen bonding and polar interactions between the individual particles.[40]

## 5.3 Composites

The literature is abundant in papers about composites and nanocomposites manufactured with polymers and nanoparticles/nanocrystals. Natural nanofillers present a relatively active surface, which makes it easier to perform chemical derivatization and grafting or to form strong physical interactions and to achieve comparatively easy processability because of their non-abrasive nature and other resultant functional properties. SNCs have successfully reinforced many natural and synthetic materials, such as starch,[6] PLA,[20] natural rubber latex,[43] soy protein,[47] poly(vinyl chloride),[51] water-borne polyurethane[53,60] and pullulan.[61] Especially SNCs hydrolyzed from starch granules show an excellent reinforcing function as a nanofiller for starch plastics.[55]

Some common matrices used in biocomposites include natural rubber (NR), an apolar matrix such as polyethylene (PE) and polypropylene (PP), PLA and poly(ε-caprolactone). A larger latex particle size results in better mechanical properties as the polymeric particles hamper the entry of polysaccharide nanoparticles during the film formation because of their high viscosity. It is difficult to achieve good dispersion of nanoparticles in PE or PP matrices, in general, leading to poor mechanical performance.[40]

### 5.3.1 Preparation of Composites

Several papers have been published on composites involving SNCs and some thermoplastic polymers, and modifications of SNCs in order to improve their mechanical and/or thermal properties have been proposed (Table 5.5).

Dufresne[40] noted that very few studies have been reported concerning the processing of polysaccharide nanocrystal-reinforced nanocomposites by extrusion methods. His paper reviews different processing techniques for polysaccharide (mainly cellulose) nanocrystal-reinforced polymer nanocomposites.

There are some well-known techniques for preparing composites. Casting allows polymeric films to be formed by evaporating the solvent.[62] In recent years, some workers have opted for this method at 40 °C for 24 h,[17] while others[19] have reported casting/evaporation at 50 °C after constant stirring of SNCs in thermoplastic starch at 90 °C for 30 min. Most studies used a solution casting method to produce starch-based nanobiocomposites. According to Xie *et al.*,[7] this is because (1) nanowhiskers/nanoparticles tend to aggregate because of association by strong hydrogen bonding (especially cellulose nanowhiskers and starch nanoparticles); (2) the nanofiller structure may be destroyed at high temperatures, which affects the reinforcing ability; and (3) the nanofiller is prepared in aqueous condition with the

**Table 5.5** Modifications and the associated composite systems.

| Study | System | Composite preparation method |
|---|---|---|
| García *et al.*[6] | **Waxy maize:** thermoplastic starch was processed by casting a mixture of 10 g of waxy maize starch granules and 5 g of glycerol in 185 g of distilled water<br>The mixture was heated from room temperature at a rate of 1.59 °C $min^{-1}$ under mechanical stirring for 28 min up to gelatinization (~70 °C). The gel was degassed for 30 min under vacuum mechanical pumping | **Nanocomposite films:** prepared by adding an aqueous suspension of waxy maize SNCs in the desired quantity (2.5 wt.% relative to the dry total mass, starch + plasticizer + nanocrystals). Then, the mixture was stirred for 10 min at 250 rpm and degassed for a further 1 h. The ensuing mixture was cast in a plastic mold and evaporated in a ventilated oven at 50 °C for 48 h. The solid films obtained had a thickness between 200 and 300 μm. Matrix and composite films were stored at 43% relative humidity (RH) ($K_2CO_3$ saturated solution) for 2 weeks before characterization and testing |
| Piyada *et al.*[10] | **Starch nanocrystals (SNCs):** rice starch<br>**Matrix:** rice starch.<br>**Composites:** rice starch with rice SNC contents of 5, 10, 15, 20, 25 and 30 wt% | Films were prepared by mixing rice starch solution, with a concentration of 3% w/v, sorbitol (40%) and 5, 10, 15, 20, 25, and 30% of an SNC dispersion. The mixture was stirred for 2 min and cast in flat, leveled, non-stick trays. The trays were held overnight at 55 °C for 10 h undisturbed, and then cooled to ambient temperature before peeling the films off the plates. The film samples were stored in plastic bags and held in desiccators at 60% RH for further testing. All treatments were performed in triplicate |
| García *et al.*[19] | **System:** waxy maize starch nanocrystals, cassava starch (72% amylopectin and 28% amylase) and glycerol as a plasticizer.<br>15 g of a mixture of starch and glycerol (2 : 1 w/w) were dispersed in 185 g of distilled water. The mixture was heated from room temperature at a rate of 1.59 °C $min^{-1}$ under mechanical stirring for 28 min up to gelatinization, which occurred at ~70 °C. After gelatinization, the gel was degassed for 30 min with a vacuum mechanical pump | The composite films were prepared by adding a suspension of waxy maize SNCs in the desired amounts (2.5 wt% relative to the total mass, starch, plasticizer and nanocrystals). Subsequently, the mixture was stirred for 10 min at 250 rpm and degassed for a further 1 h. Then, the mixture was cast in a plastic mold and evaporated in a ventilated oven at 50 °C for 24 h. Solid films having a thickness between 300 and 400 μm were obtained. Matrix and composite films were stored at 43% RH ($K_2CO_3$ saturated solution) for 2 weeks before characterization and testing |

**Table 5.5** (*Continued*)

| Study | System | Composite preparation method |
|---|---|---|
| Chang *et al.*[35] | The SNC powder (from pea: 35% amylase and 65% amylopectin, with an average particle size of about 30 μm) and ε-caprolactone (CL) monomer, in a 1 : 10 weight ratio of SNC to CL, were placed in an ampoule. $Sn(Oct)_2$ catalyst was added at a level of 5%, based on the CL weight. The mixture was homogenized using a compact shaker and then vacuum degassed for 30 min. The ampoule containing the reactant was conditioned under microwave irradiation of 255 W power for 5 min. Finally, the crude product was dissolved in dichloromethane followed by precipitation with methanol.<br>Most of the SNC-*g*-PCL nanoparticles (SNCs grafted with polycaprolactone) were platelet-like in structure and less than 100 nm in size | Films of SNC-*g*-PCL and WPU were prepared by a casting process. PCL2000 and TDI were placed into a three-necked round-bottomed flask equipped with a mechanical stirrer and a reflux condenser. The contents were heated to ∼65 °C, and stirred for 2 h under dry nitrogen. DMPA, a chain extender, was dispersed in acetone and added to the flask to give an NCO : OH molar ratio of 1.8. The temperature was elevated to 75 °C for a further 1 h. During this period, acetone was occasionally added to lower the viscosity of the reactant. The resulting product was cooled to below 40 °C before TEA was added to neutralize the –COOH of DMPA in the polyurethane chains. The designated amount of SNC-*g*-PCL was dispersed in acetone and added to the mixture under vigorous mechanical stirring to produce a blend. Iced water was then added to form an emulsion with a solids content of 20 wt%. The WPU emulsion containing SNC-*g*-PCL was cast into a Teflon mold after vacuum degassing, and solidified as a film (0.5 mm thick) at 50 °C by evaporation |
| Bouthegourd *et al.*[43] | Potato starch nanocrystals and natural rubber latex | Suspensions of different compositions of natural rubber latex and SNCs were mixed at room temperature using Ultra-thorax. They were then mixed for 15 min to obtain a uniform mixture, degassed and cast on a glass plate. The films were then dried in an air oven for 1 week at 60 °C. The prepared films had a thickness of 1.5 mm |

| | | |
|---|---|---|
| Lin *et al.*[46] | Pea starch (35% amylase and 65% amylopectin), with an average granule size of about 30 μm.<br>Poly(butylene succinate) – PBS | 2.0 g of PBS pellets were dissolved in 30 mL of trichloromethane with stirring at ambient temperature to obtain an emulsion. At the same time, the desired amount of SNCs (5, 8, 10, 15 and 20%) was dispersed in 10 mL of trichloromethane with ultrasonic dispersion for 30 min. The SNC suspension was then added to the PBS emulsion with vigorous stirring for 2 h to facilitate the dispersion of nanocrystals in the polymer matrix. The resulting white viscous liquid was vaporized at 40 °C to remove the solvent, and the solid residue was collected and vacuum dried for 24 h to obtain the nanocomposite mixture. The PBS–SNC solid mixtures were compression molded as sheets on a hot-press (70×70×0.30 mm) at 120 °C under 20 MPa pressure for 5 min, and then air cooled to about 25 °C before the pressure was released for demolding |
| Zheng *et al.*[47] | **SNCs:** pea starch composed of 35% amylose and 65% amylopectin, with an average granule size of about 30 μm<br>**Matrix**: soy protein isolate (SPI)<br>**Composites:** SSNC-*x*, where $x = 1, 2, 3, 4, 8, 12$ and 16 wt% of SNCs | An SNC aqueous suspension was diluted to 100 mL with distilled water and then homogenized in an ultrasonic bath for 20 min. Subsequently, SPI powder was slowly added to the SNC suspension under mechanical stirring for 2 h to achieve a homogeneous distribution of SNC and SPI components in suspension. The resulting well-mixed suspensions were freeze-dried for 48 h to obtain a series of nanocomposite powders. All the nanocomposite powders were mixed with the plasticizer glycerol in an intensive mixer. The weight ratio of each solid powder to glycerol was maintained at 70 : 30 throughout the process. Subsequently, the glycerol-plasticized powders were compression molded at 120 °C and 20 MPa for 5 min, then air-cooled to 50 °C for 30 min before releasing the pressure for demolding. All the nanocomposite sheets were conditioned in a desiccator with silica gel desiccant for 1 week at room temperature before characterization |

**Table 5.5** (*Continued*)

| Study | System | Composite preparation method |
|---|---|---|
| Chen *et al.*[51] | **NPS:** native pea starch (35% amylose and 65% amylopectin)<br>**PSN:** pea starch nanocrystals<br>**PVA:** poly(vinyl alcohol), with $M_w = 1.15 \times 10^5$ and a minimum degree of hydrolysis of 87%<br>**PVA/NPS-*n* films:** $n = 5$, 10, 15, 20, 25, 30 and 40 wt% of NPS<br>**PVA/PSN-*n* films:** $n = 5$, 10, 15, 20, 25, 30 and 40 wt% of PSN | **Manufacturing method of PVA/NPS and PVA/PSN films:** solution casting and evaporation process. A solution containing 8.0 wt% of PVA and 2.0 wt% of glycerol was prepared by dissolving PVA and glycerol in distilled water. NPS granules and glycerol were mixed and dispersed in distilled water to obtain a suspension containing 8.0 wt% NPS and 2.0 wt% glycerol. The suspension of NPS was then charged into a round-bottomed flask equipped with a stirrer and heated at 100 °C for 30 min until the mixture gelatinized. Then, the desired weight of PVA solution was added and the mixture was stirred for another 30 min at 100 °C. The mixture was then filtered quickly using a cellulose sieve, cast in a polystyrene Petri dish and dried at 40 °C overnight. Films of PVA/NPS-*n* ($n = 5$, 10, 15, 20, 25, 30 and 40 wt%) were prepared with a thickness of about 0.2 mm<br>**PVA/PSN films:** A designated amount of PVA solution was mixed with PSN dispersion and glycerol at room temperature for 30 min. The mixture was filtered and cast in a polystyrene Petri dish and then dried at 40 °C overnight. Films of PVA/PSN-*n* ($n = 5$, 10, 15, 20, 25, 30 and 40 wt%) were prepared<br>All films were kept at room temperature in a conditioning desiccator of 43% RH (conditioned with $K_2CO_3$ saturated solution) for 7 days before being tested |
| Zou *et al.*[53] | A homogeneous SNC slurry (from PSN powder) was obtained by adding a certain amount of SNCs to a calculated amount of water and stirring for 30 min. Thereafter, the mixture was dispersed ultrasonically for 1 h at 25 °C to obtain a uniform emulsion<br>Water-borne polyurethane (WPU), that was prepared by dissolving 4.98 g of PCL2000 and 2.23 g of MDI in 10 g of butanone, was placed in a three-necked round-bottomed | Uniform SNC emulsion was added to WPU latex and stirred for 30 min. The WPU latex was cast in a Teflon mold after vacuum degassing, and allowed to solidify as a film at 60 °C *via* water evaporation. The thickness of the resulting films was about 200 μm. By varying the SNC content to 0, 5, 10, 20, and 30 wt%, a series of nanocomposite films were obtained, which were stored at room temperature in amylase containing $P_2O_5$ with 0% RH for 5 days |

| | | |
|---|---|---|
| | flask equipped with a mechanical stirrer, a reflux condenser and a dry nitrogen inlet. The solution was then heated to ~70 °C and kept under a nitrogen atmosphere for 2 h with mechanical stirring. Thereafter, 0.46 g of DMPA, a chain extender, was dispersed in butanone and added to give an NCO : OH molar ratio of 1.6. The reaction temperature was elevated to 85 °C and held for 1 h, during which time butanone was occasionally added in order to decrease the viscidity of the solution. Subsequently, the product was cooled to below 40 °C, and the –COOH of DMPA in the polyurethane chains was neutralized by adding 0.34 g of TEA. Emulsification was then carried out by adding iced water under severe shear to produce WPU latex with a solids content of 20 wt% | |
| García *et al.*[54] | Waxy maize SNCs with PLA pellets (90% L-LA, 10% D-LA; $M_n = 49\,860$ g mol$^{-1}$):<br>1. PLA with benzoyl chloride (PLABz).<br>2. Activation of PLABz carboxyl groups (PLAM).<br>3. Grafting of PLAM onto the starch nanoparticles (PLASTARCH).<br>4. Physical blends were prepared by mixing PLA powder with 3.2 wt% SNCs to obtain a visually homogeneous material | 0.21 g (5.2 wt%) of SNCs (previously sonicated in 10 mL for 10 min) was added to PLAM to perform the esterification reaction. After 3 days, distilled water was used to stop the esterification reaction. The precipitate was washed with ethanol several times, then this solid was centrifuged with methyl ethyl ketone four times (3000 rpm, 15 min) until SNCs were not observed in the supernatant. The solid obtained (PLASTARCH) was left at room temperature for 24 h, dried in a vacuum oven and stored under vacuum. The content of SNCs in the PLASTARCH was estimated as 3.2 wt% |
| Chen *et al.*[55] | Potato starch nanocrystals and WPU | The WPU latex was cast in a Teflon mold after vacuum degassing and then solidified as a film at 60 °C *via* evaporating water<br>Addition of SNCs to the WPU latex occurred in three stages of synthesizing WPU:<br>1. Method I: post-emulsification<br>2. Method II: during emulsification<br>3. Method III: chain extension of polyurethane prepolymer |

resulting dispersion easily incorporated into a starch solution. For the dispersion of nanowhiskers/nanoparticles in a starch solution, some additional treatments such as ultrasonication and homogenization might help.

Very few studies have been reported concerning the processing of polysaccharide nanocrystal-reinforced nanocomposites by extrusion methods. The hydrophilic nature of polysaccharides causes irreversible agglomeration during drying and aggregation in non-polar matrices because of the formation of additional hydrogen bonds between amorphous parts of the nanoparticles.

Thermoplastic extrusion and hot-pressing are classified as dry process methods of film developing and are based on the thermoplastic properties of biopolymers when plasticized and heated above their glass transition temperature ($T_g$) under low water content conditions.[63] The drawback of the hot-pressing method is the potential melting of SNCs, and one might query whether starch nanoparticles are still crystalline after this treatment. A similar issue should be considered if classical polymer processing methods, *e.g.*, extrusion or injection molding, are used.

Most recently, compression molding was used to prepare glycerol-plasticized SNC composites at 120 °C from a freeze-dried powder.[47]

Angellier *et al.*[42] prepared films incorporating SNC suspensions treated using an Ultra-Turrax with latex of a copolymer of styrene (34 wt%) and butyl acrylate (64 wt%) in fractions varying from 0 to 50 wt%. The mixtures were freeze-dried, heated at 90 °C for 10 min and hot-pressed for 1 min at 2 MPa in order to obtain films 1 mm thick.

Yu *et al.*[20] grafted poly(ε-caprolactone) (PCL) to the surface of SNCs *via* microwave-assisted ring-opening polymerization and then filled the resulting nanoparticles into PLA as a matrix. SNCs play a reinforcing role, improving the mechanical performance of PLA/PCL-based materials. Films were made by a casting/evaporation method using dichloromethane or a water-borne polyurethane matrix. Both the elongation at break and strength of PLA were enhanced on adding 5 wt% of nanoparticles. The grafted PCL chains were found to improve the association of SNCs with the PLA matrix and to form an interfacial layer able to transfer stress. The rubbery PCL phase provided sufficient flexibility to improve the ductility of the material but decreased the Young's modulus.[40]

High loading levels (30%) of SNCs have been incorporated as the reinforcing filler in natural rubber[41] and thermoplastic starch,[42] resulting in significant enhancements of strength and modulus. Zou *et al.*[53] strongly believe that inhibition of aggregation in aqueous solution and control of the SNC dispersion in the matrix can be performed in order to utilize this bioresource fully. They inhibited the sedimentation and self-aggregation of SNCs by incorporating high levels of SNCs in WPU to prepare WPU–SNC nanocomposites.

Poly(butylene succinate) (PBS), one of most significant aliphatic polyesters, is a potential material for producing biodegradable plastics. In addition to biodegradability, PBS has many outstanding features, including melt processability and thermal and chemical resistance; other attributes of PBS, such as softness, gas barrier properties and melt viscosity, are often too weak for practical applications. Moreover, the modulus of PBS plastic is relatively low, typically 300–500 MPa, and it is much more expensive than traditional thermoplastics. For that reason, many attempts have been focused on creating PBS-based composites with improved mechanical or thermal properties and higher gas barrier properties at lower cost by blending with natural polymers,[29] such as cellulose acetate,[64] cellulose,[65] corn starch,[66] chitosan,[67] silk[68] and plant and red algae fibers.[69,70]

Dufresne and co-workers[71,72] used the copolymer latex obtained by co-polymerization of styrene and butyl acrylate [poly(S-*co*-BuA)] and potato SNCs. Solid nanocomposite films could be obtained by mixing and casting two aqueous suspensions followed by water evaporation performed above the glass transition temperature of the polymer.

Working with PCL-grafted waxy maize SNCs, by using the "grafting onto" approach with an isocyanate-mediated reaction, Habibi and Dufresne[73] found that grafted polymeric chains formed a crystalline structure at the surface of the nanoparticles. Nanocomposite films were processed from both unmodified and PCL-grafted nanoparticles and PCL as matrix using a casting–evaporation technique. It was shown that the mechanical properties of the resulting films were notably different. Compared with unmodified nanoparticles, the grafting of PCL chains on the surface resulted in lower modulus values but significantly higher strain at break. This unusual behavior clearly reflects the originality of the reinforcing phenomenon of polysaccharide nanocrystals resulting from the formation of a percolating network thanks to chain entanglements and co-crystallization.[40]

The dispersion of nanoparticles in the nanocomposite film depends strongly on the processing technique and conditions adopted. The matrix–filler and filler–filler interactions affect the mechanical behavior of polysaccharide nanocrystal-reinforced nanocomposites. Dufresne[40] observed that in polysaccharide nanocrystal-based composite materials, the higher the affinity between the polysaccharide filler and the host matrix the lower were the mechanical performances. This unusual fact can be explained by hydrogen bonding forces, responsible for the formation of a percolating network in the reinforcing phenomenon of polysaccharide nanocrystals.

Angellier *et al.*[48] studied the reinforcing properties of waxy maize SNCs [tensile tests, dynamic mechanical analysis (DMA)] with a filler content between 0 and 15 wt% in films with a thermoplastic waxy maize starch matrix. In this case, the reinforcing effect of SNCs could be attributed to strong filler–filler and filler–matrix interactions because of the establishment of hydrogen bonding. The presence of SNCs was responsible for slowing of the recrystallization of the matrix during aging in a humid atmosphere. In a later study, García *et al.*,[19] using another kind of matrix (cassava starch), showed that the addition of nanocrystals not only improved the mechanical properties but also reduced the water vapor permeability by 40%.

As soy protein has been considered a promising ingredient for bioplastics, by virtue of its thermoplasticity, biodegradability, non-toxicity and ready availability,[81–83] Zheng *et al.*[47] prepared soy protein isolate (SPI) films by compression with SNCs with a platelet-like structure. This structure is different from the rod-like whiskers of cellulose and chitin and the oblate nanophase of lignin. Initially, they dispersed SNCs in an aqueous suspension in 100 mL of distilled water followed by homogenization in an ultrasonic bath for 20 min. Subsequently, the SPI powder was slowly added to the SNC suspension under mechanical stirring for 2 h. The resulting well-mixed suspensions were freeze-dried for 48 h to obtain a series of nanocomposite powders. All the nanocomposite powders were mixed with glycerol plasticizer in an intensive mixer. Subsequently, the glycerol-plasticized powders were compression-molded as sheets at 120 °C under a pressure of 20 MPa for 5 min and then air-cooled to 50 °C for 0.5 h before releasing the pressure for demolding.

Chen *et al.*[51] studied the effect of pea starch in PVA films (with a thickness of about 0.2 mm containing 8.0 wt% of PVA and 2.0 wt% of glycerol, dissolving both in distilled water) using two different methods. One method utilized the direct blending of native pea starch with PVA to prepare low-cost biodegradable composites. The other involved the preparation of SNCs from native pea starch by an acid hydrolysis process, followed by blending of the obtained pea SNCs with PVA to provide property-enhanced composites. The different effects of native pea starch and pea SNCs on the structure and properties of the two series of PVA-based composites were compared. The results revealed that pea SNCs, compared with native pea starch, had much smaller sizes and dispersed more homogeneously in the PVA matrix, resulting in stronger interactions with the matrix.

Chen *et al.*[55] investigated potato SNCs and water-borne polyurethane. The addition of the nanocrystals to the polymer was carried out using three different approaches. *Method I:* post-emulsification; in this case, 1, 2, 4, 5 and 8 wt% of SNCs (US-1, US-2, US-4, US-5, US-8) were added to form the film. *Methods II and III:* during emulsification and chain extension, respectively, of polyurethane prepolymer, only 5 wt% of SNCs (US-5-II and US-5-III) were added to the films. All results are presented later in Tables 5.6 and 5.11. It was concluded that the addition of SNCs increased the strength, elongation and Young's modulus. The film with 2 and 5 wt% of SNCs presented the maximum elongation (1406.6%) and the highest tensile strength (51.5 MPa), respectively. Many researchers have used high nanofiller contents to improve the mechanical properties, but Chen *et al.*[51] proposed a low loading level (less than 8 wt%). They stated that the simultaneous reinforcing and toughening were caused by the stress of rigid SNCs and the strong interaction between the SNC nanofiller and WPU matrix. Chemical grafting did not enhance the strength and elongation because there is inhibition of the formation of physical interactions on the SNC surface and an increase in the network density of nanocomposites. Conclusions were drawn based on only one experiment, as no replicates were reported. There was no trend on $T_g$ as the percentage of SNCs was increased. This may suggest that the amount of SNCs does not matter, but no statistical test was carried out. However, it is clear that the addition of SNCs increased $T_g$ and decreased $\tan\delta$.

Several groups have studied blends of PVA with native starch, such as corn,[74] potato,[75] sago[76] and wheat.[77] However, in general, the addition of starch to PVA does not improve the physical properties because of the poor compatibility between PVA and starch.[78] In order to overcome this drawback and also to improve mechanical properties, transparency and water resistivity or swelling properties, while retaining biodegradability,[51] some approaches have been proposed: (a) PVA or starch can be chemically modified;[79] (b) PVA–starch composites can be chemically or physically modified during or after the blending process, such as with crosslinking reactions or grafting;[80] and (c) surface modification can be applied.

Kristo and Biliaderis[61] produced a film-forming solution by mixing the appropriate amount of an aqueous suspension of SNCs (0–40 wt% of pullulan) with an aqueous solution of pullulan and addition of sorbitol as plasticizer; films were formed by casting in polystyrene plates and evaporation at 35 °C in an oven until the formation of films.

Wang and Zhang,[52] preparing WPU–SNC films, first dispersed the SNCs in water for 15 min to avoid aggregation before incorporating them on the film. Then, the suspensions of nanocrystals and WPU were mixed, followed by stirring for 5 h. Subsequently, the resulting mixtures were degassed for 3 h under vacuum. Nanocomposite films, with a mean thickness of 0.2 mm, were prepared by drying the mixtures on hydrophobic glass plates at room temperature for 2 weeks.

García *et al.*[6] processed thermoplastic starch by casting a mixture of waxy maize starch granules, glycerol and distilled water. The mixture was heated from room temperature at a rate of 1.59 °C $min^{-1}$ under mechanical stirring for 28 min until gelatinization, which occurred at ~70 °C. After gelatinization, the gel was degassed for 30 min under vacuum mechanical pumping. Nanocomposite films were prepared by adding the aqueous suspension of waxy maize SNCs in the desired quantity (2.5 wt% relative to the dry total mass, starch + plasticizer + nanocrystals). The mixture was then stirred for 10 min at 250 rpm and degassed for a further 1 h. The ensuing mixture was cast in a plastic mold and evaporated in a ventilated oven at 50 °C for 48 h. Solid films having a thickness between 200 and 300 μm were obtained. Matrix and composite films were stored at 43% relative humidity (RH) (saturated $K_2CO_3$ solution) for 2 weeks before characterization and testing.

Bouthegourd *et al.*[43] mixed suspensions of different compositions of natural rubber latex and SNCs for 15 min and cast them on glass plates. Films 1.5 mm thick were obtained by drying the suspensions for 1 week at 60 °C.

The effects of SNCs on the properties of rice starch films were analyzed by Piyada *et al.*[10] The films were prepared by casting of a starch solution, with a concentration of 3% (w/v), in distilled water and sorbitol. The experimental conditions are given in Table 5.5. The film samples were stored in plastic bags and held in desiccators at 60% RH for all the characterizations. In general, it was found that the addition of SNCs increased the mechanical properties, the crystalline peak structure, the thermal properties and the water barrier properties of the films.

### 5.3.2 Techniques for Characterizing Composites

Numerous techniques for characterizing SNCs and resin–SNC composites have been described and are summarized here, including those based on microscopy [scanning electron microscopy (SEM), transmission electron microscopy (TEM), infrared (IR), X-ray diffraction (XRD)], thermal analyses [differential scanning calorimetry (DSC), thermogravimetric analysis (TGA), dynamic mechanical analysis (DMA)), water vapor permeability and mechanical properties. A summary of experimental results obtained from many different papers is presented in Tables 5.6, 5.8–5.10, 5.13 and 5.14.

#### *5.3.2.1 Techniques Using Images*

Here we discuss papers related to diffraction, Fourier transform infrared (FTIR), SEM, atomic force microscopy (AFM) and TEM techniques and a summary is presented in Table 5.6.

**5.3.2.1.1 X-ray Diffraction.** XRD is used for investigating the crystallinity of materials.[84] Angellier *et al.*,[48] studying films incorporated with waxy SNCs (crystallinity type A), observed that on adding SNCs to a thermoplastic waxy starch matrix, typical peaks of the A-type crystals appeared and the magnitude of the peaks increased with increase in filler content, revealing that the crystalline structure of the SNCs is preserved. However, at 5 wt%, the lowest filler concentration used, the peaks associated with the type A pattern were not very prominent.

The following figures are related to studies discussed in Table 5.6, where all details of the systems can be found. Figure 5.5 shows XRD patterns for native pea starch (NPS), pea starch nanocrystals (PSNs) and their composites with poly(vinyl alcohol) (PVA), presented by Chen *et al.*[51] Chang *et al.*[35] and Zou *et al.*[53] worked with WPU and SNCs from peas. They used different nanocrystal contents and obtained the XRD patterns shown in Figures 5.6 and 5.7, respectively.

**5.3.2.1.2 Fourier Transform Infrared Spectroscopy.** FTIR spectroscopy is a physical method for characterizing composite materials. It allows the identification of each component in the blend by the specific movement of each atom, which changes according to structure, composition and vibration mode of the sample.[85,86]

When starch granules are hydrolyzed, most of the changes in the FTIR spectra are related to the hydrogen bonding in starch molecules, indicating

**Table 5.6** Summary of results from image analyses.

| Study | Systems | TEM | SEM | FTIR | XRD |
|---|---|---|---|---|---|
| García *et al.*[6] | **Film of thermoplastic starch:** processed by casting a mixture of waxy maize starch granules (10 g), glycerol (5 g) and distilled water (185 g)<br>**Nanocomposite films:** 2.5 wt% (relative to the dry total mass, starch + plasticizer + nanocrystals) of aqueous suspension of waxy maize starch nanocrystals added to the other components (starch granules, glycerol and water) | **Average size of the primary nanoparticles:** <50 nm.<br>**Aggregates:** ~1 μm.<br>**Reason for aggregation:** hydrogen bonding forces among the nanocrystals | The unplasticized sample displays a rather smooth homogeneous surface associated with a brittle fracture<br>For the plasticized composite specimen, a nanometric fibrillar structure (nanothreads) is observed. These nanothreads may be formed by nanocrystals, glycerol and transcrystallized amylopectin | No significant difference was observed between the two specimens (glycerol-plasticized waxy maize starch sample and related composite reinforced with 2.5 wt% SNCs) in the two FTIR spectral regions<br>Nanocrystal aggregates contribute to the local order only in the presence of glycerol | **Matrix and composite films:** diffraction peaks of B-type crystalline structure, with peaks at 17 and 20° $2\theta$, corresponding to *d* spacings of 5.2 and 4.4 Å, respectively. The peaks at 5.6, 22 and 24° (*d* spacings of 15.8, 4 and 3.7 Å) are hardly noticeable<br>**Nanocomposite film:** additional diffraction peaks characteristic of the A-type allomorph. Peaks around 15, 17–18 and 23° are observed |
| Piyada *et al.*[10] | Films of native rice starch, with amylose and moisture contents of about 30.40 and 12.19%, and rice SNCs | The authors did not perform this analysis | **Rice starch films reinforced with SNCs:** became rougher with increased SNC content<br>No visible nanocrystal agglomerates or microphase separation up to 20 wt% of nanocrystals<br>Presence of nanopores at 25 and 30 wt% of nanocrystals | The authors did not performed this analysis | **Crystallinity values:**<br>**Rice starch:** 23.19%<br>**Rice starch films:** 13.3%<br>**Rice starch films reinforced with 20% of SNCs:** 17.35%<br>**SNCs:** 30.36%<br>**Observation (OBS):** for SNC contents ≥20 wt%, there was a decrease in crystallinity because of agglomerate formation |

**Table 5.6** (*Continued*)

| Study | Systems | TEM | SEM | FTIR | XRD |
|---|---|---|---|---|---|
| García *et al.*[19] | Composite of cassava starch (72% amylopectin and 28% amylose) and 2.5 wt% of waxy maize SNCs | Nanocrystals have an average size below 50 nm and form aggregates of 1–5 μm. See Figure 5.15<br>**Reason for aggregation:** high density of OH groups, resulting in association because of hydrogen bonding forces | **Cryofractured surfaces:**<br>**Neat film:** rough (glycerol- and starch-rich phases)<br>**Composite film:** smooth surface with a vein pattern, characteristic of a composite deformation controlled by the amorphous homogeneous matrix | FTIR spectra of the composite and matrix showed only slight differences. At high wavenumbers (3260 $cm^{-1}$) there was a very slight broadening with incorporation of SNCs. The band at 1015 $cm^{-1}$, associated with the C–OH bond show considerable changes in the relative intensity of the doublet | **Composite film:** reduced the B–V-type peaks and weakened the appearance of A-type peaks<br>Composite is more amorphous than the neat starch, maybe because of the higher water equilibrium content in the composites |
| Yu *et al.*[20] | Poly(ε-caprolactone) (PCL) was grafted to the surface of SNCs. The resulting nanoparticles were then incorporated into PLA<br>**Matrix films:** these were coded as PLA/SNC-*g*-PCL(5), PLA/SNC-*g*-PCL(10), PLA/SNC-*g*-PCL(15), PLA/SNC-*g*-PCL(20), PLA/SNC-*g*-PCL(25) and PLA/SNC-*g*-PCL(30), where the numbers represent the weight percentage of SNC-g-PCL in the nanocomposites | The authors did not perform this analysis | For the fractured surfaces of PLA/SNC-*g*-PCL(5) and PLA/SNC-*g*-PCL(15), many small microparticles, with diameters of ~1000 nm, were dispersed on a porous plane, which resulted from the removal of microparticles. With increasing SNC-*g*-PCL content, the size of dispersed particles increased, even forming large agglomerates for PLA/SNC-*g*-PCL(30) | An increase in SNC-*g*-PCL content resulted in an increased extent of peak intensities located at 1720.9 $cm^{-1}$, resulting from more severe self-aggregation at higher loading-levels of SNC-*g*-PCL. The absorption assigned to the SNC-*g*-PCL, such as the peaks located at 2945.5, 1720.9, 1417.2, 1292.2 and 1238.8 $cm^{-1}$, gradually became stronger | PLA/SNC-*g*-PCL nanocomposites had a distinct diffraction located at 16.6° $2\theta$ in similar diffuse patterns to neat PLA-F, which might result from the combination of PCL and PLA. In addition, when the agglomerates of SNC-*g*-PCL formed in the nanocomposite with higher loading-levels of SNC-*g*-PCL, the crystalline character of PCL can be observed, namely a weak diffraction occurred at 21.2° |

| | | | | | |
|---|---|---|---|---|---|
| Chang *et al.*[35] | **SNC-*g*-PCL:** PSNs grafted with PCL diol ($M_n = 2000$), PCL2000<br><br>**Water-borne polyurethane film:** WPU-F<br>**Composite film:** WPU + SNC-*g*-PCL($x$), where $x$ represents 5, 10, 15, 20, 25 or 30 wt% of SNC-*g*-PCL | The authors did not perform this analysis | **Fractured surfaces:**<br>**WPU-F:** showed a fluctuating fracture surface with many pleats<br>For higher contents of SNC-*g*-PCL, a faultage-like structure occurred and the density of this structure increased, because of the crystalline domains based on self-aggregated SNC-*g*-PCL. This is a brittle-fracture nature<br>See Figure 5.10 | Hydrogen bonding in the WPU matrix was partly destroyed after introducing SNC-*g*-PCL<br>As the SNC-*g*-PCL content increased, the fraction of –C=O absorption in the crystalline domain increased, but in the amorphous region it decreased | The PCL chains grafted onto SNC-*g*-PCL showed crystalline diffraction similar to PCL, but stronger diffraction intensities<br>For WPU–SNC-*g*-PCL(5), the diffraction pattern was similar to that of WPU-F. As the SNC-*g*-PCL content increased, the intensities of diffraction peaks also increased, which may be attributed to self-aggregation of SNC-*g*-PCL as greater crystalline domain<br>See Figure 5.6 |
| Bouthegourd *et al.*[43] | Composites of natural rubber and 0, 5, 10, 15 and 20 wt% of potato SNCs | **Average diameter of the nanocrystals:** 3.45 nm<br>No aggregate was observed | The authors did not perform this analysis | The peak of C–O stretching at 1085–1050 $cm^{-1}$, characteristic of potato starch, was found for the composite | The authors did not perform this analysis |
| Lin *et al.*[46] | **Nanocrystals:** pea starch (35% amylase and 65% amylopectin), with an average granule size of about 30 µm<br>**Matrix:** poly(butylene succinate) (PBS)<br>**Composite:** compression molding of PBS and SNCs (2, 5, 10, 15, and 20 wt%) | SNCs exhibited platelet morphology (see Figure 5.14). Some SNC aggregates were observed, but distinct SNCs could also be observed with a length of 40–60 nm and width of 15–30 nm | **Fractured surfaces:**<br>**TPBS-SNC-5:** nanoparticles well dispersed and presented a fractured surface similar to that of PBS<br>For 8 and 10 wt% of SNCs, there was formation of a rigid | **Composite**: an intense peak of an ester group at about 1,724 $cm^{-1}$<br>**Composites with SNCs >5 wt%:** the interaction between nanofillers caused the formation of a rigid three-dimensional | The addition of SNCs did not change the structure of PBS. There are two distinct peaks at 19 and 23° $2\theta$ |

**Table 5.6** (*Continued*)

| Study | Systems | TEM | SEM | FTIR | XRD |
|---|---|---|---|---|---|
| | mixtures at 120 °C and 20 MPa for 5 min, and air-cooled to about 25 °C before the pressure was released for demolding | | three-dimensional continuous network that caused rougher fracture surfaces<br>For higher SNC contents, there was a severe self-aggregation of nanoparticles | continuous network that destroyed the ordered structure of the PBS crystalline component and induced the disordered motion of polymers in the amorphous region | |
| Zheng *et al.*[47] | **SNCs:** pea starch composed of 35% amylase and 65% amylopectin, and with average granule size of about 30 μm<br>**Matrix:** soy protein isolate (SPI)<br>**Composites:** SSNC-*x*, where *x* represents 1, 2, 3, 4, 8, 12 and 16 wt% of SNCs | SNCs were visualized in the form of aggregates and isolated, with a length of 60–150 nm and a width of 15–30 nm | **Fractured surfaces:**<br>**SSNC-1:** lamellar structure (good dispersion of SNCs), similar to the neat SPI sheet<br>**SSNC-2:** lamellar structure with some thin and dense imprints (interaction between SNCs and SPI)<br>**SNCs higher than 3 wt%:** gradually smoother fractured surfaces because of the greater destruction of the SNC domains in the structure of the SPI matrix<br>**SNCs higher than 3 wt%:** gradually smoother fractured<br>See Figure 5.11 | The authors did not perform this analysis | **SNCs:** weak double peaks at 10.3 and 11.5°; a strong peak at 15.3°, strong double peaks at 17.2 and 18.2°, and a strong peak at 23.2°<br>**SPI:** two diffuse semicrystalline peaks located at 8.8 and 19.0°<br>**Composites**: *lower than 3 wt% of SNCs*: practically no changes<br>*SSNC-8:* shift of the peak at 8.8° to 10.2°<br>*Higher contents of SNCs:* self-aggregation of SNCs<br>*SSNC-16:* appearance of a peak at 17.2° from SNCs and a weak peak at 9.2° from SPI |

| | | | | | |
|---|---|---|---|---|---|
| Zou *et al.*[53] | **SNCs:** Pea SNCs<br>**Film of water-borne polyurethane:** WPU-F<br>**Composite film:** WPU + SNC ($x$), where $x$ represents 5, 10, 20 or 30 wt% of SNCs | Although the SNCs tended to self-aggregate, isolated SNCs were observed to have a width of 15–20 nm and a length of 40–70 nm | **Fractured surfaces of the WPU–SNC nanocomposites and neat WPU:** completely different fracture morphology<br>All the WPU–SNC nano-composites showed similar fracture morphology<br>Increasing the SNC loading level caused the size of the nanoparticles to become larger and gradually to increase in number<br>The dispersion of SNCs in the WPU matrix was very good as the loading level increased | New hydrogen bonds between the –C=O of the WPU component and the –OH on the SNC surface were formed. With an increase in the SNC loading level, improved hydrogen bonding occurred mainly in the amorphous region. Hydrogen bonding in the ordered domain was usually constructed by hard-segments in the WPU matrix, suggesting that adding SNC destroyed the original ordered alignment of hard-segments | **WPU-F:** a diffuse peak located at 20.1° $2\theta$, in spite of the fact that –C=O-based hydrogen bonds constructed a small-scale ordered domain in the WPU matrix<br>**Composite films:**<br>(a) Diffraction peak of SNCs (SNCs are semicrystalline), located at 23.6° $2\theta$, appeared in all nanocomposites<br>(b) Diffraction peaks of SNCs, located at 15.0 and 17.9° $2\theta$, were shielded by the diffuse diffraction of the WPU matrix<br>(c) On addition of SNCs to WPU, a new strong peak, located at 21.4° $2\theta$, appeared because of SNCs that changed and improved the crystalline character of WPU<br>See Figure 5.7 |
| Garcia *et al.*[54] | Waxy maize SNCs with PLA pellets:<br>1. PLA with benzoyl chloride (PLABz)<br>2. Activation of PLABz carboxyl groups (PLAM) | The authors did not perform this analysis | The authors did not perform this analysis | **Peak at 1750 cm$^{-1}$:** C=O of the ester groups from all samples<br>**PLA:** small and broad peak at 3400 cm$^{-1}$: | The authors did not perform this analysis |

**Table 5.6** (*Continued*)

| Study | Systems | TEM | SEM | FTIR | XRD |
|---|---|---|---|---|---|
| | 3. Grafting of PLAM onto the starch nanoparticles (PLASTARCH)<br>4. Physical blends were prepared by mixing PLA powder with 3.2 wt% SNCs to obtain a visually homogeneous material | | | –OH of the carboxyl function<br>Small peaks above 3500 $cm^{-1}$: stretching of –OH<br>**PLABz:** no signals of –OH stretching, because of the benzoylation reaction<br>**PLASTARCH:** No broad peaks around 3500 $cm^{-1}$, due to the ester formation between of PLABz and SNC | |
| Chen *et al.*[55] | Potato SNCs and WPU. The WPU latex was cast on a Teflon mold after vacuum degassing and then solidified as a film at 60 °C *via* evaporating water<br>Addition of SNCs to the WPU latex occurred at three stages of synthesizing WPU:<br>1. Method I: post-emulsification<br>2. Method II: during emulsification | Although the SNCs tended to self-aggregate, distinguished isolated SNCs were observed having a width of 10–20 nm and a length of 40–70 nm | **Nanocomposites with SNC ≥ 4 wt%:** fractured surface gradually became smooth with densely fluctuating thin stripes<br>**Nanocomposites with SNC ≥ 4 wt%:** fractured surface gradually became smooth with densely fluctuating thin stripes | **WPU and nanocomposites**: two peaks in the 3300–3500 μm range (–NH) and three peaks in the 1690–1735 μm range (–C=O) | **Nanocomposites with SNCs < 5 wt%:** absence of two diffraction peaks (17.06° and 22.3° 2$\theta$) and relatively uniform dispersion<br>**Nanocomposites with SNCs ≥ 8 wt%:** a shoulder peak appeared at 17.6° 2$\theta$<br>**Nanocomposites with SNCs ≥ 8 wt%:** a shoulder peak appeared at 17.6° 2$\theta$ |

| | | | | | |
|---|---|---|---|---|---|
| | 3. Method III: chain-extension of polyurethane prepolymer | | With increase in chemical grafting (methods II and III), the fractured surface also gradually became smoother, showing a relatively brittle character | | |
| Duan *et al.*[111] | **Matrix:** carboxymethyl chitosan, glycerol as plasticizer<br>**Composites:** 3, 6, 10, 15, 20, 30, 40 wt% of maize starch | The insoluble hydrolysis residues of native starch granules presented polygonal structures with a size of 40–80 nm, but a small number of aggregates larger than 100 nm could be found<br>Films were not analyzed using this technique | SNCs could not be observed when their contents were below 10 wt%. When the SNC contents were between 10 and 30 wt%, they appeared like white dots, which could be found dispersing uniformly in the carboxymethyl chitosan matrix. However, when the SNC contents were above 30 wt%, they formed large aggregates and phase separation between the SNC aggregates and carboxymethyl chitosan matrix was observed | The absorption peaks of –OH and $-NH_2$ of nanocomposites shifted to a lower wavenumber of 3358 $cm^{-1}$ and the peaks at 1600.1 and 1637.4 $cm^{-1}$ corresponding to the C–O of carboxymethyl chitosan and the –OH of SNCs were superposed to become the peak at 1611 $cm^{-1}$ for nanocomposites<br>This result indicates that SNCs formed a strong hydrogen bonding interaction with the carboxymethyl chitosan | The crystallinity of insoluble hydrolysis residues increased with time of hydrolysis, but after 10 days of hydrolysis only 10 wt% of insoluble SNCs remained, whereas with a 6 days of hydrolysis, 24 wt% of insoluble SNCs with a size of 40–80 nm and a relative crystallinity of 63% could be obtained |

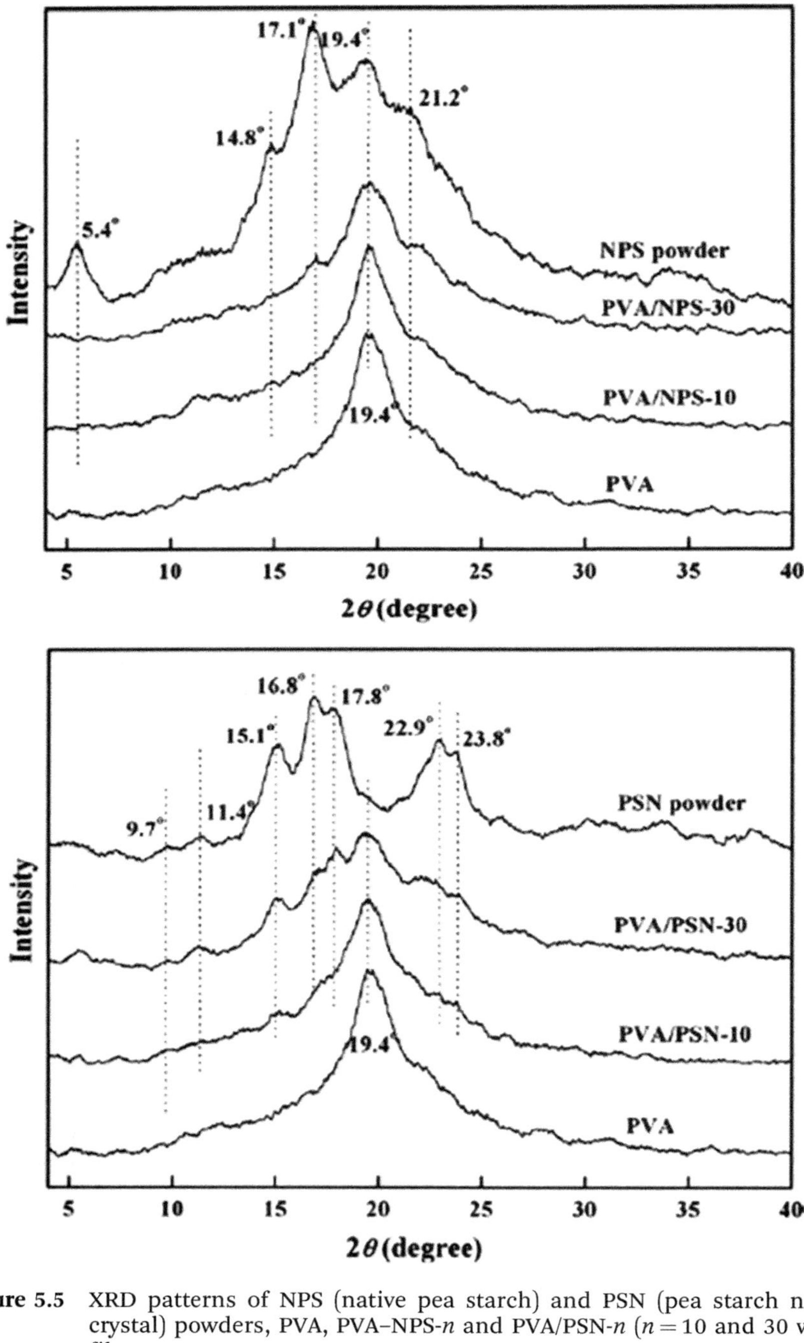


**Figure 5.5** XRD patterns of NPS (native pea starch) and PSN (pea starch nanocrystal) powders, PVA, PVA–NPS-*n* and PVA/PSN-*n* ($n = 10$ and 30 wt%) films.

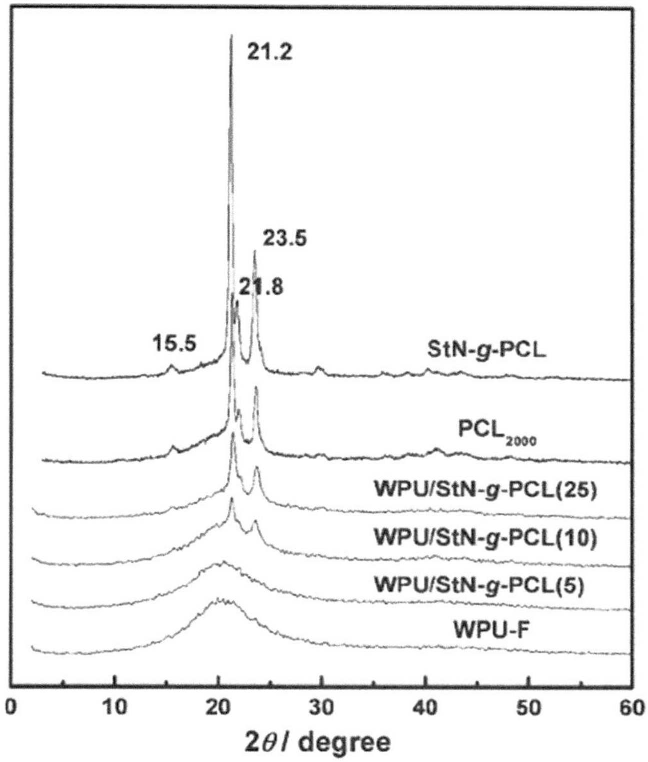


**Figure 5.6** XRD patterns of WPU–SNC-*g*-PCL nanocomposite films containing various loading levels of SNC-*g*-PCL and SNC-*g*-PCL, WPU-F (neat resin) and PCL2000 as references. On the traces, StN denotes SNC.
Source: Chang *et al.*[35]

that the hydrolysis treatment significantly changed the inter- and intramolecular hydrogen bonding of starch molecules.[51]

On comparing the spectra of native starch granules and SNCs from potato starch, Jivan *et al.*[34] found that the intensity of the peak at 2928 $cm^{-1}$, characteristic of C–H stretching associated with the ring methane hydrogen atoms, was substantially higher in SNCs than in native starch. Also, a broad band appeared at 3445 $cm^{-1}$ in the starch spectrum because of hydrogen-bonded hydroxyl groups. See Figure 5.8.

Table 5.7 and Figure 5.9, from Chen *et al.*,[51] present the peaks found in native pea starch (NPS), in pea starch nanocrystals (PSNs) and in films of PVA–NPS and PVA–PSN. It can be seen that there was no difference among the films. The authors also observed that the shape and location of the main peaks in PVA–NPS and PVA–PSN films were closer to those in the PVA spectrum, which means that the interactions of PVA–PVA molecules dominated and were stronger than those of PVA–starch molecules and starch–starch molecules in the blending system.

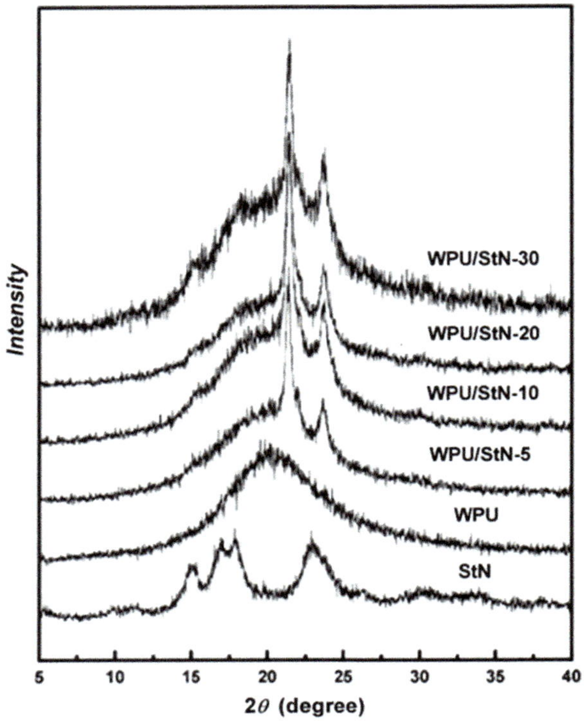


**Figure 5.7** XRD patterns of WPU–SNC nanocomposites, for various SNC contents, WPU-F and freeze-dried SNC powder. On the traces, StN denotes SNC. Source: Zou *et al.*[53]

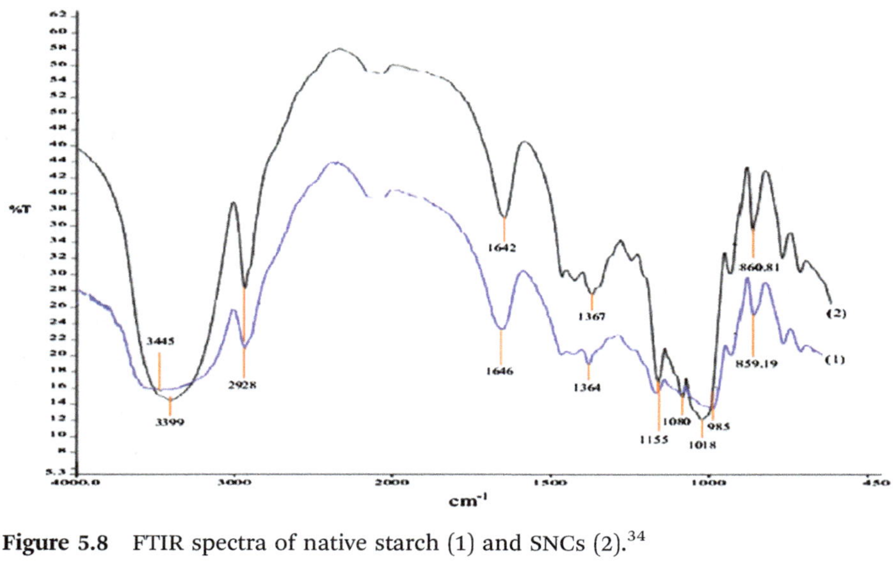


**Figure 5.8** FTIR spectra of native starch (1) and SNCs (2).[34]

**Table 5.7** Wavenumbers ($cm^{-1}$) for native pea starch (NPS) and pea starch nanocrystal (PSN) powders and PVA–NPS and PVA–PSN films.[51]

| Peak assignment[a] | NPS powder | PSN powder | PVA–NPS-10 | PVA–NPS-30 | PVA–PSN-10 | PVA–PSN-30 | PVA |
|---|---|---|---|---|---|---|---|
| –OH (st) | 3276 | 3368 | 3287 | 3281 | 3292 | 3280 | 3292 |
| C=O (st) (characteristic | — | — | 1733 | 1733 | 1733 | 1733 | 1733 |
| carbonyl vibration in the residue acetate in PVA) | — | — | 1713 | 1713 | 1713 | 1713 | 1713 |
| –OH (b) | 1650 | 1644 | 1656 | 1643 | 1656 | 1652 | 1654 |
| C–O (st) in C–O–H | 1150 | 1155 | 1143 | 1143 | 1143 | 1143 | — |
| | 1077 | 1078 | 1090 | 1078 | 1087 | 1074 | 1091 |
| C–O (st) in C–O–C | 990 | 1016 | 1026 | 1026 | 1022 | 1013 | 1035 |
| C–O (b) (characteristic C–O–C ring vibration in starch) | 760 | 768 | 760 | 760 | 765 | 773 | — |

[a]st, stretching; b, bending.

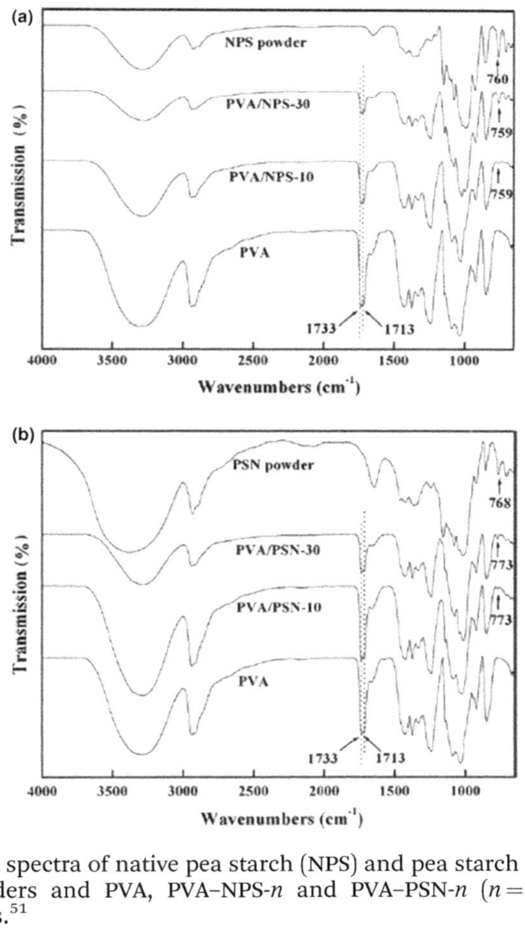


**Figure 5.9** FTIR spectra of native pea starch (NPS) and pea starch nanocrystal (PSN) powders and PVA, PVA–NPS-$n$ and PVA–PSN-$n$ ($n = 10$ and 30 wt%) films.[51]

**5.3.2.1.3 Scanning Electron Microscopy.** SEM is a technique that allows the shape and surface features of starch granules and films to be viewed in three dimensions. When a scanning electron microscope is used, the sample is coated with a thin layer of a reflective metal and then irradiated with a beam of electrons. The electrons are reflected back to a sensor, allowing surface features of starch granules to be seen in great detail.[87] This technique has a resolution power at least 10 times greater than that of light microscopy.[88]

Figure 5.10 presents micrographs of fractured surfaces of water-borne polyurethane film (WPU-F) and composite films of WPU, SNCs and PCL[35] and Figure 5.11 those of fractured surfaces of pea starch and soy protein isolate (SPI).[47] The system details are given in Table 5.6.

Chen *et al.*[51] present excellent SEM and field emission gun (FEG)–SEM images obtained for native pea starches (NPS) and their corresponding pea starch nanocrystals (PSNs) (Figure 5.12) and for the fractured surfaces of films made of PVA–NPS and PVA–PSN (Figure 5.13). All PSNs exhibit a platelet-like morphology with thickness 5–8 nm. Therefore, although platelets with a square-like shape should be preferred for barrier properties, all PSNs can be considered as potential fillers in nanocomposites. It was observed that the PVA film had smooth cross-sections, PVA–NPS-*n* and PVA–PSN-*n* films presented rougher surfaces for higher NPS/PSN contents, PVA–NPS-*n* had a much rougher surface than PVA–PSN-*n*, PVA–PSN-*n* film, with

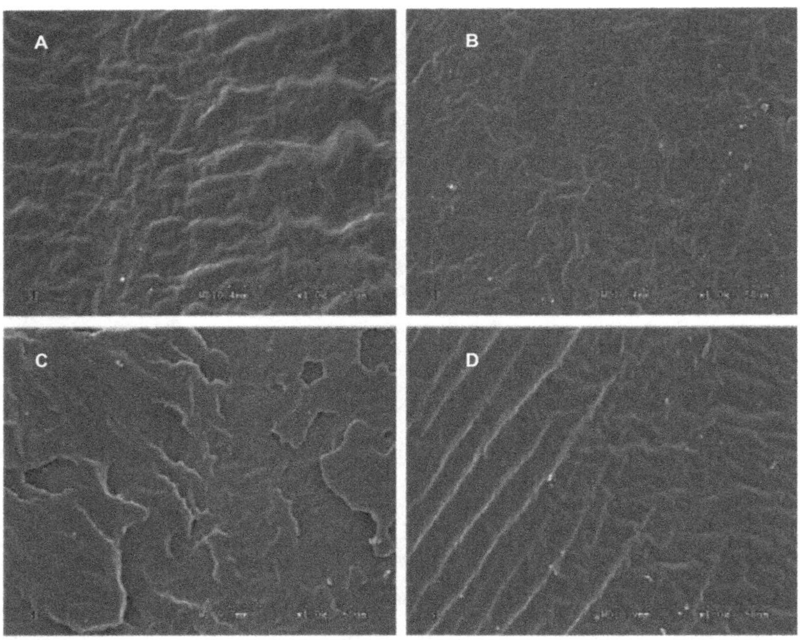


**Figure 5.10** SEM images of fractured surfaces of WPU-F (A), WPU/SNC-*g*-PCL(5) (B), WPU/SNC-*g*-PCL(10) (C) and WPU/SNC-*g*-PCL(25) (D).
Results from Chang *et al.*[35]

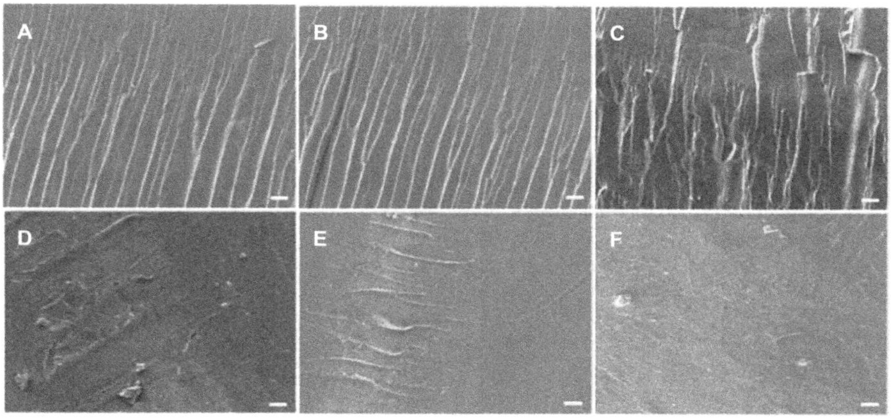


**Figure 5.11** SEM images of fractured surfaces of the SPI sheet (A) and SSNC-1 (B), SSNC-2 (C), SSNC-3 (D), SSNC-8 (E) and SSNC-16 (F). The scale bar is 10 μm. Results from Zheng *et al.*[47]

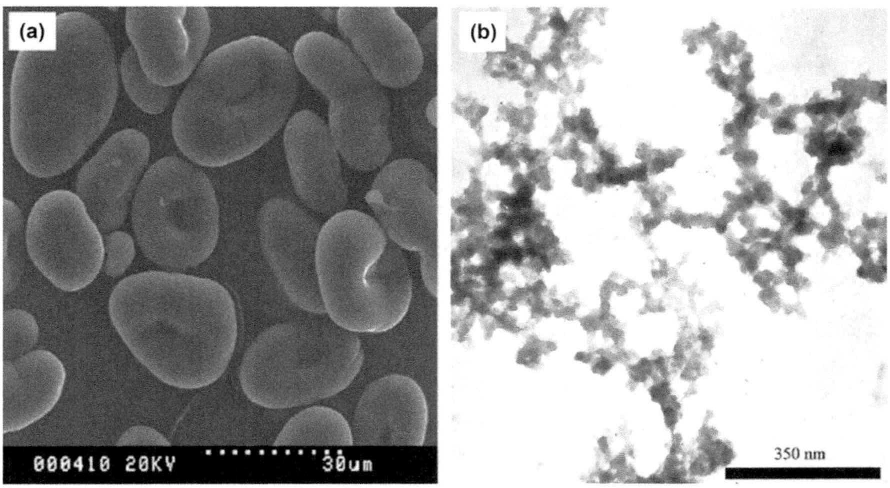


**Figure 5.12** (a) Native pea starch granules (SEM) and (b) pea starch nanocrystals (TEM). Results from Chen *et al.*[51]

$n < 30$ wt%, showed no visible aggregations of nanocrystals and microphase separation and PVA/NPS-40 film had starch aggregations and microphase separation between PVA and starch.

**5.3.2.1.4 Atomic Force Microscopy.** AFM involves scanning a conductive or non-conductive surface of a sample with a probe to obtain topographic images with atomic resolution and to map certain mechanical and physicochemical properties of the materials.[89]

Jivan *et al.*[34] analyzed the topographic surface of SNCs by AFM. They observed that the starch structure is composed of larger blocklets with a size

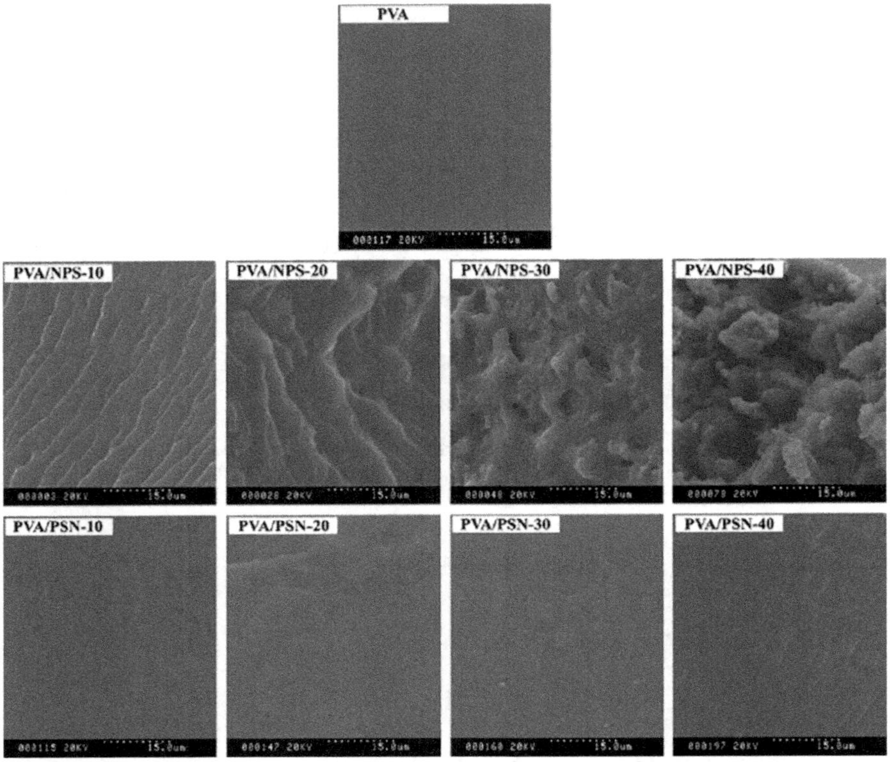


**Figure 5.13** SEM images of the cross-sections of PVA, PVA–NPS-*n* and PVA–PSN-*n* ($n = 10$, 20, 30 and 40 wt%) films.
Results from Chen *et al.*[51]

of 100–200 nm separated by gaps; each of these blocklets is composed of smaller blocklets 20–30 nm in size separated by shallower depressions. Baldwin *et al.*[90] reported that these small blocklets are amylopectin superhelical clusters 18 nm in diameter. Nanosized structures, with diameters of 17–50 nm, are observed, reminiscent of small amylopectin blocklets.

**5.3.2.1.5 Transmission Electron Microscopy.** TEM is characterized by a resolving power 1000 times greater than that of light microscopy.[88] It is commonly used together with XRD to determine if an intercalated or exfoliated nanocomposite was formed or if the nanofiller and polymeric matrix were miscible.[82]

In 2003, the morphology of SNCs extracted from waxy maize was revealed. TEM observations showed (a) a longitudinal view of lamellar fragments consisting of stack of elongated elements, with a thickness of 5–7 nm, and (b) a planar view of individualized platelets after hydrolysis. Shapes and lateral dimensions were derived from observations of individual platelets in planar view, and marked 60–65° acute angles for parallelepiped blocks with a length of 20–40 nm and a width of 15–30 nm were reported. However, more

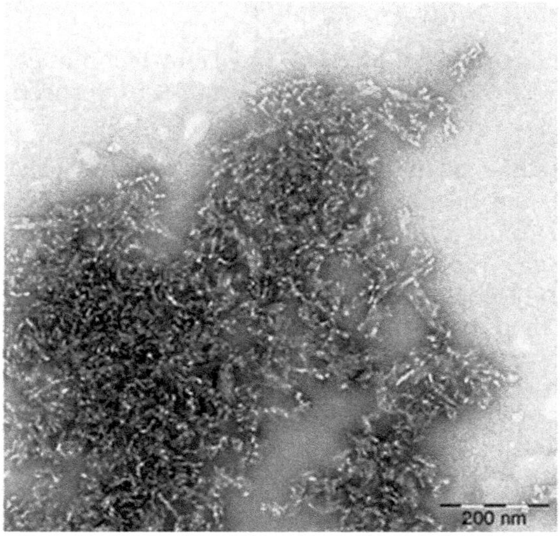


**Figure 5.14** TEM image of negatively stained PSNs.[46]

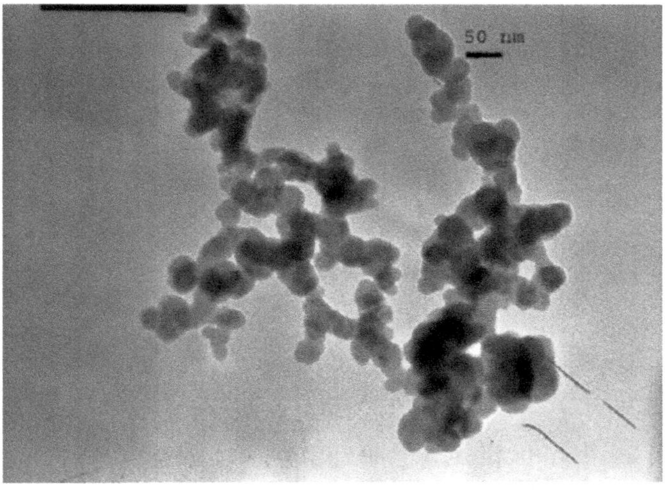

**Figure 5.15** TEM image of waxy maize-SNCs.
From García *et al.*[19]

recent publications reported larger SNCs, with round edges and found as grape-like aggregates of 1–5 μm. The heterogeneity in particle size could be explained by the differences in the starch botanical origin and also by the difficulty of obtaining well-defined pictures of nanoaggregated nanocrystals.[57]

Figures 5.12, 5.14 and 5.15 present TEM images of SNCs from potato,[51] pea[46] and waxy maize,[19] respectively, where it is possible to see different forms of nanocrystals. The system details are given in Table 5.6.

### 5.3.2.2 Thermal Analysis Methods

Here, investigations using DSC, TGA and DMA techniques are discussed. A summary is presented in Tables 5.8, 5.9 and 5.10, respectively.

**5.3.2.2.1 Differential Scanning Calorimetry.** DSC is used to determine the temperature and heat of different transitions that occur when a substance is heated under constant temperature (isothermal mode) or with a variable heating rate (non-isothermal mode).

Jivan *et al.*[34] and Hoover *et al.*[91] found a higher glass transition temperature for nanocrystals than for starch and the former group also observed two separate endotherms at 26 and 125 °C for SNCs. They observed that the first endotherm of SNCs, arranged in a regular lattice, was easily destroyed by heating to ambient temperature. The second endotherm is due to the well-ordered SNC structure requiring more heat. The crosslinking of nanocrystals with sodium hexametaphosphate preserved the lattice of nanocrystals during heating, hence no phase transition occurred for this sample from −6 to 150 °C.

Le Corre *et al.*[17] studied the thermal behavior of SNCs from five native starches (waxy maize starch, normal maize starch, high-amylose maize, potato starch and wheat starch) in excess of water (70%). Table 5.11 summarizes the characteristics of native starch and the respective SNCs in excess of water and in dry conditions (conditioned at 50% RH).

Native starches do not have a second endothermic peak with either 0 or 70% of water. Le Corre *et al.*[17] and many other workers[92–96] proved that the thermal properties of starch are very dependent on the water content. As it increases, the peak temperature shifts to lower values and a second peak appears at lower temperatures, related to starch gelatinization. Le Corre *et al.*[17] presented an excellent discussion of the reasons for these two peaks.

Table 5.12 shows the characteristic transition peaks for the samples studied by Le Corre *et al.*[17] From Figures 5.16–5.18, it is possible to conclude that the effect of water content on the onset and maximum temperatures for TGA experiments is much greater for nanocrystals than for native starch. Also, these temperatures are higher for dry samples. Maize and pea nanocrystals have the highest onset and maximum temperatures for dry samples. Native samples did not present a second peak.

The authors did not observe any influence of starch source, amylase content or crystalline type, but SNCs showed higher temperatures values. For SNCs, Angellier[97] stated that the first peak is related to the nanocrystal gelatinization and the shift up in temperature to its crystalline nature. There are many explanations to justify the higher temperatures for SNCs, but Tester and Morrison[98] suggested that amylose acts as a plasticizer (dilutant) in the starch granules and, after hydrolysis, most of the amorphous starch has been hydrolyzed, limiting its effect as a plasticizer. There is an increase in amylopectin content on hydrolyzing starch into SNCs and consequently a decrease in amylose content. When there is an excess of water, it can penetrate the SNCs and lower their thermal

**Table 5.8** Summary of results using DSC.

<table>
<tr><th>Study</th><th>System</th><th>DSC</th><th>Comments</th></tr>
<tr><td>Piyada et al.[10]</td><td>Films of native rice starch, with amylase and moisture contents of about 30.40 and 12.19%, respectively, and rice SNCs.Film samples were cast by transferring the mixtures to plastic bags after peeling and placed in desiccators</td><td>Endothermic peak starts to appear as SNCs are added</td><td>The authors did not give any value for the glass transition temperature ($T_g$)</td></tr>
<tr><td>Yu et al.[20]</td><td>PCL was grafted to the surface of SNCs. The resulting nanoparticles were then incorporated into a PLA matrix. Films were coded as PLA/SNC-g-PCL(x), with $x = 5, 10, 15, 20, 25$ or 30, representing the weight percentage of SNC-g-PCL in the nanocomposites</td><td>
<table>
<tr><th>Sample</th><th>$T_g$ (°C)</th><th>$T_{m1}$ (J g$^{-1}$ K$^{-1}$)</th><th>$T_{m2}$ (J g$^{-1}$ K$^{-1}$)</th></tr>
<tr><td>PLA</td><td>57.7</td><td>–</td><td>149.1</td></tr>
<tr><td>PLA/SNC-g-PCL(5),</td><td>47.8</td><td>54.4</td><td>145.2/150.8</td></tr>
<tr><td>PLA/SNC-g-PCL(15)</td><td>–</td><td>50.9/55.1</td><td>145.9/151.0</td></tr>
<tr><td>PLA/SNC-g-PCL(30)</td><td>–</td><td>50.7/55.8</td><td>145.5/150.8</td></tr>
<tr><td>SNC-g-PCL</td><td>–</td><td>55.0</td><td>–</td></tr>
</table>
</td><td>After adding SNC-g-PCL, the $T_g$ assigned to PLA shifted down to 47.8 °C for PLA/SNC-g-PCL(5), and was even absent under higher loading-levels of SNC-g-PCL (i.e., higher than 15 wt.%), which is characteristics of an increasing on crystallinity of films.<br>The double-peaks of the PCL component were located at ~51 and ~55 °C. The 55 °C value is close to that of neat SNC-g-PCL, and represented isolated crystalline domains based on grafted PCL chains.<br>The melting transition peak at 51 °C resulted from the PCL component in the interfacial layer composed of PCL and PLA. Similarly, the double peaks of the PLA matrix were assigned to the interfacial layer based on the PCL of SNC-g-PCL and PLA (low $T_{m2}$), and the PLA matrix (high $T_{m2}$ and close to that of neat PLA)</td></tr>
</table>

**Table 5.8** (*Continued*)

| Study | System | DSC | | | Comments |
|---|---|---|---|---|---|
| Chang *et al.*[35] | **SNC-*g*-PCL:** PSNs grafted with polycaprolactone diol ($M_n = 2000$), PCL2000.<br>**WPU-F**<br>**Composite film:** WPU + SNC-*g*-PCL($x$), where $x$ represents 5, 10, 15, 20, 25 or 30 wt% of SNC-*g*-PCL | Sample | $T_g$ (°C) | $T_m$ (°C) | The glass transition of SNC-*g*-PCL was inhibited by the high crystallinity of nanofillers.<br>The $T_g$ values of WPU/SNC-*g*-PCLs were lower than that of neat WPU-F, which was attributed to the cleavage of the hydrogen bonds between the soft and hard segments in the WPU matrix. With an increase in the SNC-*g*-PCL content, the melting transition assigned to the PCL component occurred for WPU/SNC-*g*-PCL(25) |
| | | WPU-F | −53.79 | – | |
| | | WPU/SNC-*g*-PCL(5) | −54.95 | – | |
| | | WPU/SNC-*g*-PCL(10) | −56.94 | – | |
| | | WPU/SNC-*g*-PCL(25) | −59.35 | 42.17 | |
| | | SNC-*g*-PCL | – | 55.02 | |
| | | PCL | −66.83 | 42.49/48.85 | |
| Lin *et al.*[46] | **Nanocrystals:** pea starch (35% amylase and 65% amylopectin), with an average granule size of about 30 μm.<br>**Matrix:** poly(amylase succinate) (PBS).<br>**Composite:** compression molding of PBS and SNCs (2, 5, 10, 15 and 20 wt%) mixtures at 120 °C and 20 MPa for 5 min, and air-cooled to about 25 °C before the pressure was released for demolding | Sample | $T_m$ (°C) | $\chi_c$ (%) | For 5 wt% of SNCs, the percolating network did not form and the interactions between SNC nanoparticles and PBS were stronger than the interactions between the nanofillers themselves, preserving the intact and ordered structure of the PBS. Nucleation of rigid nanocrystals caused crystalline properties in PBS based nanocomposite. The destruction caused by the percolating network dominated the decrease in crystallinity |
| | | PBS | 113.3 | 74.1 | |
| | | PBS/SNC-5 | 117.2 | 80.4 | |
| | | PBS/SNC-8 | 115.1 | 75.6 | |
| | | PBS/SNC-10 | 111.6 | 70.4 | |
| | | PBS/SNC-15 | 114.6 | 80.0 | |
| | | PBS/SNC-20 | 115.5 | 83.1 | |

| | | Sample | $T_g$ (°C) | $\Delta C_p$ (J g$^{-1}$ K$^{-1}$) | |
|---|---|---|---|---|---|
| Zheng *et al.*[50] | **SNCs:** Pea starch composed of 35% amylase and 65% amylopectin, and with an average granule size of about 30 μm.<br>**Matrix:** soy protein isolate (SPI).<br>**Composites:** SSNC-$x$, where $x = 1$, 2, 3, 4, 8, 12 and 16 wt% of SNCs | SPI | − 64.7 | 0.642 | After the addition of SNCs, there was a decrease in $T_g$, suggesting that the entangling of SPI segments was cleaved by the SNCs and its aggregates to gain higher motion freedom.<br>An increase in SNC content resulted in an initial increase in $T_g$ up to SSNC-8, because of the steric hindrance of SNC domains that restricted the motion of SPI segments. For SNC-16, the expansion of the SNC domains destroyed the ordered structure in the SPI matrix and resulted in the lowest $T_g$ |
| | | SSNC-1 | − 69.0 | 0.423 | |
| | | SSNC-2 | − 69.1 | 0.410 | |
| | | SSNC-3 | − 68.2 | 0.398 | |
| | | SSNC-8 | − 65.5 | 0.407 | |
| | | SSNC-16 | − 72.1 | 0.412 | |
| | | Sample | $T_g$ (°C) | $\Delta C_p$ (J g$^{-1}$ K$^{-1}$) | |
| Chen *et al.*[51] | **Potato starch nanocrystals and WPU films:**<br>**Method I:** US-$n$, where $n$ represents the SNC content.<br>**Method II:** US-$n$-II, where $n$ represents the SNC content.<br>**Method III:** US-$n$-III, where $n$ represents the SNC content | WPU | − 41.3 | 0.412 | The $T_g$ of all the nanocomposites were higher than that of WPU, suggesting that the restriction of SNCs to the mobility of soft segments was dominant in spite of the inevitable cleavage of the original interaction between hard and soft segments in the WPU matrix. For SNC contents less than 2 wt%, the rigid SNCs could dissolve homogeneously into the WPU matrix, and then inhibit the motion of soft segment mediated with newly formed interactions with the WPU component.<br>The authors did not run replicates and the difference among the samples could be on account of experimental errors. No clear trend was found |
| | | US-1 | − 41.0 | 0.384 | |
| | | US-2 | − 40.1 | 0.361 | |
| | | US-4 | − 40.6 | 0.391 | |
| | | US-5 | − 41.0 | 0.355 | |
| | | US-8 | − 40.5 | 0.362 | |
| | | US-5-II | − 39.8 | 0.338 | |
| | | US-5-II | − 40.3 | 0.381 | |

**Table 5.8** (*Continued*)

| Study | System | DSC | | | Comments |
|---|---|---|---|---|---|
| Zou *et al.*[53] | **SNCs:** PSNs.<br>**WPU film.**<br>**Composite film:** WPU + SNC ($x$), where $x$ represents 5, 10, 20 or 30 wt% of SNC amylopectin | Sample | $T_g$ (°C) | $T_m$ (°C) | The addition of SNCs increased the $T_g$ for all contents, because of a motion reduction of the soft segment caused by the steric hindrance of the rigid SNC nanophase and by hydrogen bonding on the active SNC surface |
| | | WPU | −46.7 | 45.2 | |
| | | WPU/SNC-5 | −46.4 | 45.2 | |
| | | WPU/SNC-10 | −46.0 | 46.5 | |
| | | WPU/SNC-20 | −45.9 | 47.3 | |
| | | WPU/SNC-30 | −45.4 | 47.4 | |
| García *et al.*[54] | Waxy maize SNCs with PLA pellets (90% L-LA, 10% D-LA; $M_n = 49\,860$ g mol$^{-1}$)<br>1. PLA with benzoyl chloride (PLABz).<br>2. Activation of PLABz carboxyl groups (PLAM).<br>3. Grafting of PLAM onto the starch nanoparticles (PLASTARCH).<br>4. Physical blends were prepared by mixing PLA powder with 3.2 wt% SNCs to obtain a visually homogeneous material | PLA and PLA-SNC blend samples show only one endothermic peak, associated with the melting transition.<br>PLABz shows a broader melting peak that is the result of two superimposed endothermic peaks. This suggests that the incorporation of benzoyl groups leads to an increase in the local free volume, which increases the molecular mobility of the PLA chains allowing the recrystallization of the material (clearly seen in the second heating scan).<br>See Figure 5.19.<br>$T_g$ of PLA-SNC: ∼62 °C;<br>$T_g$ of PLABz: ∼40 °C;<br>$T_g$ of PLASTARCH: ∼59 °C | | | Different thermal profiles between first and second heatings.<br>A physical mixture (PLA–SNC blend) offers a different thermal profile compared with the grafted starch (PLASTARCH), which presents two endothermic peaks associated with the melting of two different crystalline arrangements. These results suggest that the influence of SNCs depends on whether they are chemically attached to PLA or not.<br>PLASTARCH chars at 500 °C higher than the chars obtained in the degradation of PLA and also higher than that of the physical mixture (PLA–NC blend), which is an indication of covalent functionalization |

**Table 5.9** Summary of results using TGA.

| Study | System | TGA | Comments |
|---|---|---|---|
| García *et al.*[6] | **Film of thermoplastic starch:** processed by casting a mixture of waxy maize starch granules (10 g), glycerol (5 g) and distilled water (185 g).<br>**Starch nanocrystals:** 2.5 wt% (relative to the dry total mass, starch + plasticizer + nanocrystals) of an aqueous suspension of waxy maize starch nanocrystals.<br>**Nanocomposite films:** 2.5 wt.% (relative to the dry total mass, starch + plasticizer + nanocrystals) of an aqueous suspension of waxy maize starch nanocrystals added to the other components (starch granules, glycerol and water) | Three main steps:<br>1. First stage: corresponds to the loss of water and low-molecular-weight compounds.<br>2. Second stage: decomposition of the glycerol-rich phase and nanocrystals.<br>3. Third stage: oxidation of the partially decomposed starch.<br>**Plasticized composite film:** significant mass loss (around 36%), within 80–280 °C, because of nanocrystal and glycerol decomposition (expected degradation temperature: 180 °C) | Composite showed lower thermal stability, maybe because of a closer association of glycerol with starch nanoparticles.<br>There is practically no effect of the inclusion of nanocrystals in the unplasticized matrix compared with the glycerol-plasticized material |
| Piyada *et al.*[10] | Films of native rice starch, with amylase and moisture contents of about 30.40% and 12.19%, respectively, and rice starch nanocrystals.<br>Film samples were cast by transferring the mixtures to plastic bags after peeling and placed in desiccators | All film samples: 250 °C (dehydration of the saccharide rings and depolymerization).<br>Maximum rate of decomposition: at about 340 °C<br>Char yield:<br>(a) starch nanocrystals: ~19.23%;<br>(b) rice starch film: <12.7%;<br>(c) rice starch films reinforced with starch nanocrystals: 13.4–14.5% | The results indicated that addition of SNCs increased the thermal stability of rice starch films |
| García *et al.*[19] | Composite of cassava starch (72% amylopectin and 28% amylase) and 2.5 wt% of waxy maize-starch nanocrystals | **First stage:** loss of water:<br>*Matrix:* 17.14% and $T_{max} \approx 113$ °C<br>*Composite:* 21.39% and no $T_{max}$. | First stage: the films with SNCs have a higher water content with than the neat starch matrix, because of the increased availability of OH groups (probably |

**Table 5.9** (*Continued*)

| Study | System | TGA | Comments |
|---|---|---|---|
| | | **Second stage:** decomposition of the glycerol-rich phase:<br>*Matrix:* 54.32%<br>*Composite:* 51.14%<br>*For both:* $T_{max} = 240$ °C (decomposition of glycerol-rich phase) and $T_{max} = 290$ °C (degradation of starch-rich phase).<br>**Third stage:** oxidation of the partially decomposed starch:<br>*Matrix:* 10.63%<br>*Composite:* 9.16% | arising from the starch matrix), forming hydrogen bonds with the absorbed water.<br>The mass loss in the other stages is similar for both systems when it is considered on a dry basis |
| García *et al.*[54] | Waxy maize starch nanocrystals with PLA pellets (90% L-LA, 10% D-LA; $M_n = 49\,860$ g mol$^{-1}$):<br>1. PLA with benzoyl chloride (PLABz);<br>2. Activation of PLABz carboxyl groups (PLAM);<br>3. Grafting of PLAM onto the starch nanoparticles (PLASTARCH);<br>4. Physical blends were prepared by mixing PLA powder with 3.2 wt% SNCs to obtain a visually homogeneous material | **PLA:** single degradation stage at 307–377 °C and a maximum degradation rate at 358 °C.<br>**SNCs:** multiple degradation steps: one below 100 °C (water loss) and two more with maximum loss rates at 233 °C (initial decomposition of nanocrystals) and 350 °C (degradation of the partially decomposed starch).<br>**PLA-SNC blend:** the degradation is at a lower temperature than for PLA, but the residual char is higher.<br>**PLABz:** first degradation step: 100–280 °C (maximum at 137 °C), on account of the scission of free end carboxylic groups. There is another degradation step between 280 and 370 °C (maximum at 355 °C), which corresponds to the degradation of PLA (ester scission).<br>**PLASTARCH:** degradation step: 278 – 366 °C, with a maximum rate at 353 °C | The presence of SNCs reduces the onset of thermal degradation as a consequence of the higher content of OH groups, which can be hydrogen bonded to PLA.<br>PLASTARCH: has a degradation temperature slightly lower than that of PLA, but still high |

**Table 5.10** Summary of results using DMA.

| Study | System | DMA | Comments |
|---|---|---|---|
| García *et al.*[6] | **Film of thermoplastic starch:** processed by casting a mixture of waxy maize starch granules (10 g), glycerol (5 g) and distilled water (185 g).<br>**Starch nanocrystals:** 2.5 wt% (relative to the dry total mass, starch + plasticizer + nanocrystals) of an aqueous suspension of waxy maize starch nanocrystals.<br>**Nanocomposite films:** 2.5 wt% (relative to the dry total mass, starch + plasticizer + nanocrystals) of an aqueous suspension of waxy maize starch nanocrystals added to the other components (starch granules, glycerol and water) | **Low-temperature relaxation:** associated with the glass transition of glycerol-rich domains.<br>**High-temperature relaxation:** corresponds to starch-rich domains.<br>**Starch nanocrystals:** induces a shift towards higher temperatures.<br>At 50 °C, $E'$ of the nanocomposite film with 2.5 wt% starch nanocrystals is roughly 470% higher | The type of starch used as the matrix has a remarkable effect on the morphology and corresponding phase relaxations.<br>SNCs are much more compatible with the glycerol-rich phase than with the waxy maize starch-rich phase |
| Piyada *et al.*[10] | Films of native rice starch, with amylase and moisture contents of about 30.40% and 12.19%, respectively, and rice starch nanocrystals.<br>Film samples were cast by transferring the mixtures to plastic bags after peeling and placed in desiccators | **Rice starch film:**<br>$E' = 2.0 \times 10^8$ MPa<br>Tan $\delta$ peak = 170 °C<br>**Rice starch film reinforced with 30% of rice nanocrystals:**<br>$E' = 1.7 \times 10^9$ MPa<br>Tan $\delta$ peak = 202 °C | All films showed an increase in storage modulus and $T_g$ on increasing the SNC content. For temperatures below $T_g$, the composite film is in the glassy state and the storage modulus decreases slightly with increase in temperature |
| García *et al.*[19] | Composite of cassava starch (72% amylopectin and 28% amylose) and 2.5 wt% of waxy maize-starch nanocrystals | **Rubbery storage modulus at 50 °C:**<br>Cassava matrix: $3.80 \times 10^7$ Pa<br>Composite: $1.47 \times 10^8$ Pa<br>Peaks:<br>1. Around −60 °C (glycerol)<br>2. Between −20 and 60 °C (starch) | The presence of nanocrystals contributes to the establishment of hydrogen bonding between glycerol and starch nanocrystals. The peak related to the relaxation of the glycerol-rich phase is slightly shifted to higher temperature and its intensity is decreased as the nanocrystals are added |

**Table 5.10** (*Continued*)

| Study | System | DMA | | | | | Comments |
|---|---|---|---|---|---|---|---|
| Lin *et al.*[29] | **Nanocrystals**: Pea starch (35% amylase and 65% amylopectin), with an average granule size of about 30 μm.<br>**Matrix**: poly(amylase succinate) (PBS)<br>**Composite**: compression molding of PBS and SNCs (2, 5, 10, 15 and 20 wt%) mixtures at 120 °C and 20 MPa for 5 min, and air-cooled to about 25 °C before the pressure was released for demolding | Sample | $T_{onset}$ (°C) | Log $E'$ | $T_{max}$ (°C) | Tan $\delta$ | There was practically no significant difference among the composites with respect to the storage modulus, the maximum temperature and tan $\delta$ values |
| | | PBS | − 46.8 | 3.64 | − 22.9 | 0.089 | |
| | | PBS/SNC-5 | − 45.5 | 3.63 | − 23.9 | 0.094 | |
| | | PBS/SNC-8 | − 45.8 | 3.64 | − 23.8 | 0.087 | |
| | | PBS/SNC-10 | − 44.9 | 3.61 | − 23.9 | 0.089 | |
| | | PBS/SNC-15 | − 45.7 | 3.64 | − 24.1 | 0.088 | |
| | | PBS/SNC-20 | − 44.3 | 3.63 | − 23.9 | 0.086 | |
| Chang *et al.*[35] | Pea starch composed of 35% amylase and 65% amylopectin, with an average particle size of about 30 μm.<br>Polycaprolactone diol, PCL2000 ($M_n = 2000$) and dimethylolpropionic acid (DMPA) were dried under vacuum at 80 °C for 12 h | Sample | $T_g$ (°C) | Log $E'$ | | | With an increase in SNC-*g*-PCL loading-level (> 10 wt%), the magnitude of loss-peaks for the WPU/SNC-*g*-PCL nanocomposites decreased. This means that the interactions between the soft segments in the WPU matrix and the SNC-*g*-PCL nanofillers decreased the mobility of soft segments |
| | | WPU-F | − 25.22 | 3.27 | | | |
| | | WPU/SNC-*g*-PCL(5) | − 38.68 | 3.29 | | | |
| | | WPU/SNC-*g*-PCL(10) | − 31.80 | 3.31 | | | |
| | | WPU/SNC-*g*-PCL(25) | − 29.52 | 3.10 | | | |
| Zheng *et al.*[47] | **Starch nanocrystals:** Pea starch composed of 35% amylase and 65% amylopectin, and with an average granule size of about 30 μm.<br>**Matrix**: soy protein isolate (SPI)<br>**Composites:** SSNC-*x*, where $x = 1, 2, 3, 4, 8, 12$ and 16 wt% of SNCs | Sample | $T_g$ (°C) | Tan $\delta$ | | | **Initial decrease of $T_g$:** resulted from untangling of SPI segments.<br>**Increase of $T_g$:** sufficient active area of SNC domains.<br>**Final decrease of $T_g$:** decrease of effective active area and decrease of the original structure order of the SPI matrix.<br>The authors did not run any replicates and a statistical analysis of significant differences was not performed |
| | | SPI-S | − 61.8 | 0.272 | | | |
| | | SSNC-1 | − 63.6 | 0.274 | | | |
| | | SSNC-2 | − 65.5 | 0.290 | | | |
| | | SSNC-3 | − 60.9 | 0.249 | | | |
| | | SSNC-8 | − 58.9 | 0.262 | | | |
| | | SSNC-16 | − 62.2 | 0.251 | | | |

<table>
<tr>
<td>Chen et al.[55]</td>
<td>Potato starch nanocrystals and WPU.<br>The WPU latex was cast on a Teflon mold after vacuum degassing and then solidified as a film at 60 °C via evaporation of water.<br>Addition of SNCs to the WPU latex occurred at three stages of synthesizing WPU:<br>1. Method I: post-emulsification.<br>2. Method II: during emulsification.<br>3. Method III: chain extension of polyurethane prepolymer</td>
<td><table>
<tr><th>Sample</th><th>$T_g$ (°C)</th><th>Tan $\delta$</th></tr>
<tr><td>WPU</td><td>−37.1</td><td>0.263</td></tr>
<tr><td>US-1</td><td>−36.5</td><td>0.223</td></tr>
<tr><td>US-2</td><td>−35.9</td><td>0.244</td></tr>
<tr><td>US-4</td><td>−36.9</td><td>0.262</td></tr>
<tr><td>US-5</td><td>−35.4</td><td>0.239</td></tr>
<tr><td>US-8</td><td>−36.5</td><td>0.232</td></tr>
<tr><td>US-5-II</td><td>−35.1</td><td>0.254</td></tr>
<tr><td>US-5-II</td><td>−34.1</td><td>0.267</td></tr>
</table></td>
<td>The energy dissipation process is slowed by the SNCs, which means that the interactions between the soft segments in the WPU matrix and the SNCs are stronger and therefore the mobility of soft segments decreased<br>For methods II and III to produce covalent bonding between SNCs and WPU components, chemical grafting distinctly resulted in an increase in $T_g$</td>
</tr>
<tr>
<td>Zou et al.[53]</td>
<td>SNCs: PSNs<br>Film of water-borne polyurethane: WPU-F<br>Composite film: WPU + SNC ($x$), where $x$ represents 5, 10, 20 or 30 wt% of SNC amylopectin</td>
<td><table>
<tr><th>Sample</th><th>$T_g$ (°C)</th><th>Log $E'$</th><th>$T_g$ (°C)</th><th>Tan $\delta$</th></tr>
<tr><td>WPU</td><td>−50.6</td><td>3.29</td><td>−33.1</td><td>0.22</td></tr>
<tr><td>WPU/SNC-5</td><td>−49.7</td><td>3.31</td><td>−32.2</td><td>0.23</td></tr>
<tr><td>WPU/SNC-10</td><td>−49.6</td><td>3.38</td><td>−32.0</td><td>0.19</td></tr>
<tr><td>WPU/SNC-20</td><td>−49.0</td><td>3.38</td><td>−30.6</td><td>0.22</td></tr>
<tr><td>WPU/SNC-30</td><td>−48.5</td><td>3.41</td><td>−30.5</td><td>0.25</td></tr>
</table></td>
<td>The increase in $T_g$ was attributed to the formation of interactions between the polar groups on the SNC surface and the WPU matrix</td>
</tr>
</table>

**Table 5.11** Characteristics of native starch and the respective starch nanocrystals.[17]

| Botanical origin | Granular size (μm) | Amylose (%) | Crystalline type | SNC diameter (nm) | SNC thickness (nm) |
|---|---|---|---|---|---|
| High-amylose maize (M70) | 5–20 | 65-75 | B | $118 \pm 53$ | $5 \pm 1.6$ |
| Normal maize (M27) | 5–20 | 27 | A | $58 \pm 36$ | $8.3 \pm 3.1$ |
| Waxy maize (M1) | 5–20 | 1 | A | $47 \pm 42$ | $6.1 \pm 1.9$ |
| Wheat (W28) | 2–30 | 28 | A | $100 \pm 50$ | $3.7 \pm 0.6$ |
| Potato (P21) | 5–80 | 21 | B | $52 \pm 4$ | $7.6 \pm 1.6$ |

transitions compared with dry SNCs; the first peak is attributed to the first stage of crystallite melting (unpacking of the double helices) and the second transition to the second stage of crystallite melting (unwinding of the helices). In the dry state, the peaks are attributed to crystallite melting, with a direct transition from packed helices to unwound helices, and the presence of two peaks is attributed to heterogeneity in the crystallite quality.

With an excess of water, Le Corre *et al.*[17] did not find any correlation between amylase content and peak temperatures, in contrast to dry starches. In the latter case, they found a slightly higher maximum peak temperature for 25% of amylase. Further, for maize starches, the two peaks were further apart than for wheat and potato starches, indicating that the crystallites were less homogeneous. It was demonstrated that SNCs in excess water should not be processed at temperatures higher than 80 °C for wheat and waxy maize starches, 85 °C for normal maize starch and 90 °C for potato and high-amylase starches.

García *et al.*[54] reported a very interesting study in which they compared physical and chemical mixtures of SNCs and PLA. PLA is a biodegradable thermoplastic polymer, available from annually renewable sources and composites made of starch, and is a promising component for starch-based packaging. There have been about a few studies of starch–PLA physical blends.[99,100] However, as PLA and starch are thermodynamically immiscible, some modifications need to carried out. García *et al.*[54] proposed chemical grafting as a new and better alternative to overcome the incompatibility problems between starch and PLA. In comparison with the physical blends, this new system (PLASTARCH) has a very different thermal response. They studied three consecutive reactions and one mixture: (1) PLA with benzoyl chloride (PLABz); (2) activation of PLABz carboxyl groups (PLAM); (3) grafting of PLAM onto the starch nanoparticles (PLASTARCH); and (4) physical mixture of PLABz and SNCs (PLABz–SNC). Physical blends were prepared by mixing PLA powder with 3.2 wt% SNCs to obtain a visually homogeneous material. Thus, the authors could evaluate the differences between a physical mixture and PLASTARCH obtained from chemical modification. Figure 5.19 presents the DSC results, showing the first heating (Figure 5.19a) and the second heating (Figure 5.19b). It is possible to see a crystallization

**Table 5.12** Characteristic transition temperatures of starches and starch nanocrystals (30%) in excess water (70%). Adapted from Le Corre *et al.*[17]

| | Native starch[a] | | | | | Nanocrystals[a] | | | Comparison: |
|---|---|---|---|---|---|---|---|---|---|
| Starch type | WC (%) | Peak | $T_{onset}$ (°C) | $T_{max}$ (°C) | $T_f$ (°C) | $T_{onset}$ (°C) | $T_{max}$ (°C) | $T_f$ (°C) | shift of peak (°C) |
| M70 (high-amylose maize) | 0 | 1 | 94.6 | 122.6 | 231.1 | 116.0 ± 2.2 | 148.3 ± 2.8 | 178.2 ± 2.5 | 25.7 |
| | | 2 | – | – | – | 193.3 ± 3.3 | 205.4 ± 3.8 | 246.0 ± 0.5 | |
| | 70 | 1 | 69.1 | 72.5 | 78.3 | 84.5 ± 12.5 | 93.0 ± 2.1 | 99.9 ± 1.0 | 20.5 |
| | | 2 | – | – | – | 106.7 ± 6.3 | 103.1 ± 3.5 | 112.9 ± 5.2 | |
| M27 (normal maize) | 0 | 1 | 109.6 | 148.7 | 255.0 | 105.0 ± 7.8 | 126.1 ± 6.3 | 161.0 ± 0.2 | − 22.6 |
| | | 2 | – | – | – | 247.2 ± 1.4 | 253.5 ± 1.0 | 256.9 ± 0.0 | |
| | 70 | 1 | 63.9 | 69.4 | 81.9 | 67.6 ± 2.2 | 85.5 ± 1.3 | 99.7 ± 0.5 | 16.0 |
| | | 2 | – | – | – | 100.2 ± 0.1 | 106.6 ± 2.1 | 112.9 ± 3.2 | |
| M1 (waxy maize) | 0 | 1 | 107.6 | 124.7 | 223.5 | 123.4 ± 11.2 | 154.0 ± 2.2 | 180.3 ± 10.2 | 29.3 |
| | | 2 | – | – | – | 188.0 ± 9.1 | 200.0 ± 10.2 | 234.0 ± 7.4 | |
| | 70 | 1 | 65.0 | 71.4 | 83.4 | 54.5 ± 14.9 | 79.7 ± 11.6 | 99.2 ± 0.3 | 8.3 |
| | | 2 | – | – | – | 99.9 ± 0.5 | 104.0 ± 1.6 | 108.6 ± 0.8 | |
| W28 (wheat) | 0 | 1 | 104.8 | 130.7 | 241.4 | 186.6 ± 2.5 | 199.0 ± 9.3 | 212.8 ± 12.0 | 68.3 |
| | | 2 | – | – | – | 215.9 ± 13.2 | 218.9 ± 12.6 | 248.3 ± 6.4 | |
| | 70 | 1 | 53.3 | 59.1 | 76.4 | 59.0 ± 9.1 | 77.7 ± 11.8 | 100.7 ± 1.7 | 18.6 |
| | | 2 | – | – | – | 101.7 ± 1.1 | 105.5 ± 0.3 | 111.3 ± 2.2 | |
| P21 (potato) | 0 | 1 | 111.6 | 147.6 | 204.8 | 178.9 ± 3.0 | 191.4 ± 9.2 | 205.4 ± 23.1 | 43.8 |
| | | 2 | – | – | – | 207.0 ± 21.7 | 215.3 ± 9.7 | 245.5 ± 5.6 | |
| | 70 | 1 | 59.6 | 64.0 | 82.5 | 82.1 ± 10.2 | 91.3 ± 4.3 | 100.3 ± 0.5 | 27.3 |
| | | 2 | – | – | – | 101.6 ± 0.4 | 105.5 ± 0.1 | 110.0 ± 1.6 | |

[a]WC = water content; $T_{onset}$ = onset temperature; $T_{max}$ = peak temperature; $T_f$ = final temperature.

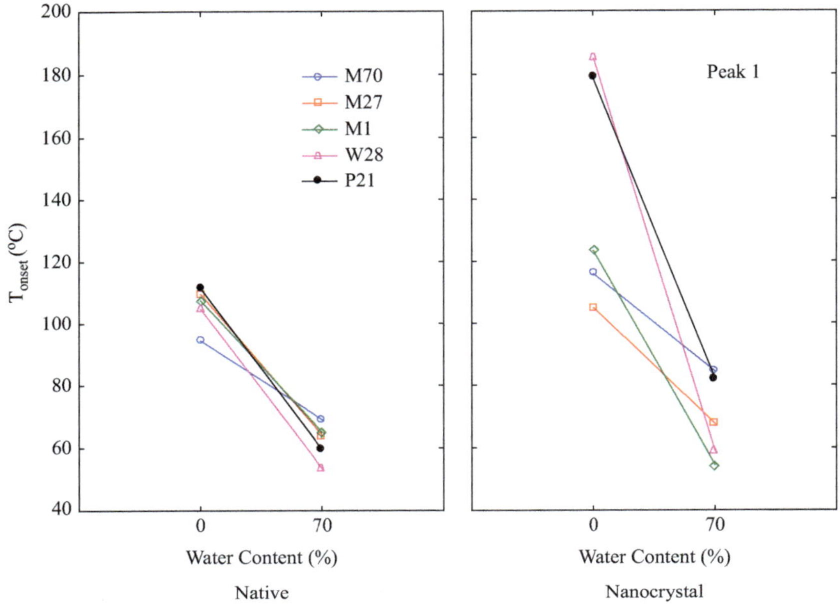


**Figure 5.16** Means plot of onset temperature, $T_{onset}$, for peak 1 related to TGA experiments. M70, high-amylose maize; M27, normal maize; M1, waxy maize; W28, wheat; P21, potato.[17]

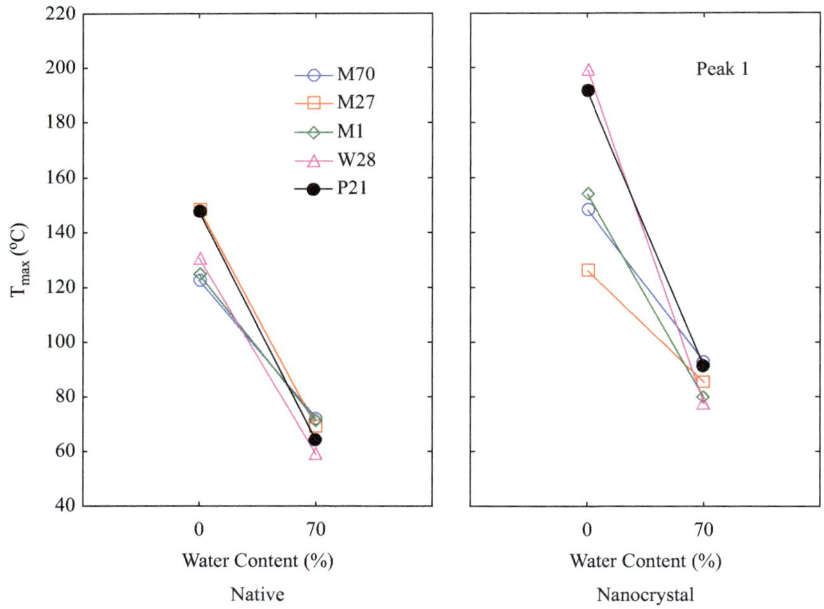


**Figure 5.17** Means plot of maximum temperature, $T_{max}$, for peak 1 related to TGA experiments. M70, high-amylose maize; M27, normal maize; M1, waxy maize; W28, wheat; P21, potato.[17]

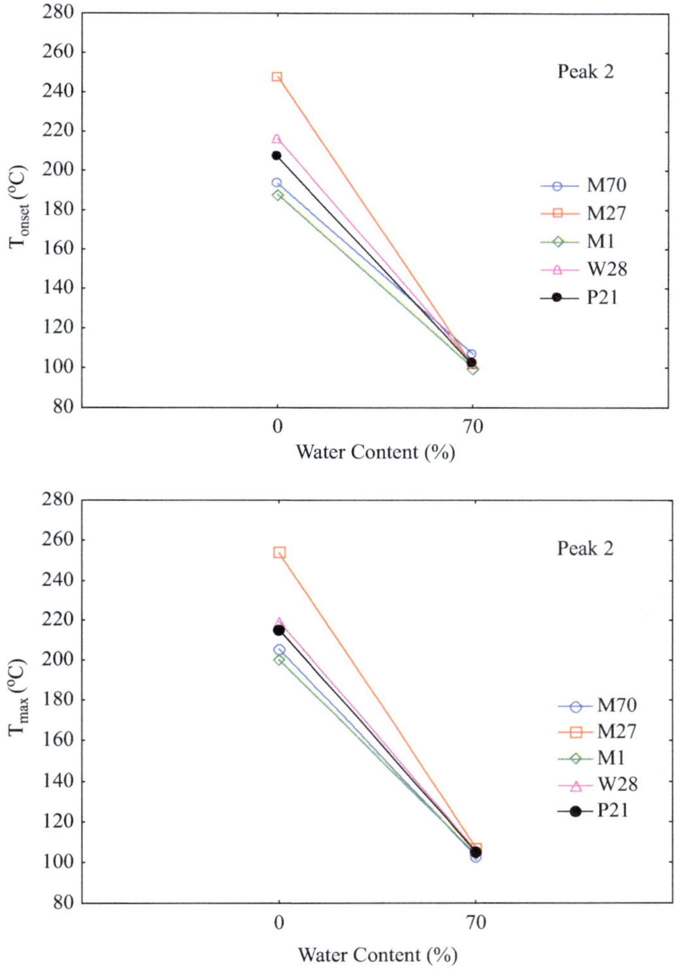


**Figure 5.18** Means plot of (left) onset temperature, $T_{onset}$, and (right) maximum temperature, $T_{max}$, for peak 2 related to TGA experiments. M70, high-amylose maize; M27, normal maize; M1, waxy maize; W28, wheat; P21, potato.[17]

(exothermic peak) of the grafted material. The last peak is broad and appears to have an overlapped small second endothermic peak. The chemical bonding of SNCs to PLA affects the amorphous component (because of changes in $T_g$) in addition to recrystallization. The results also show that the widths of the two endothermic peaks of PLASTARCH are much narrower than in the case of PLABz or PLABz–SNC blend, which suggests the formation of crystals with more uniform sizes in PLASTARCH compared with those formed in the other two cases. It can be concluded that PLASTARCH has a very different thermal response to that of the physical mixture. In

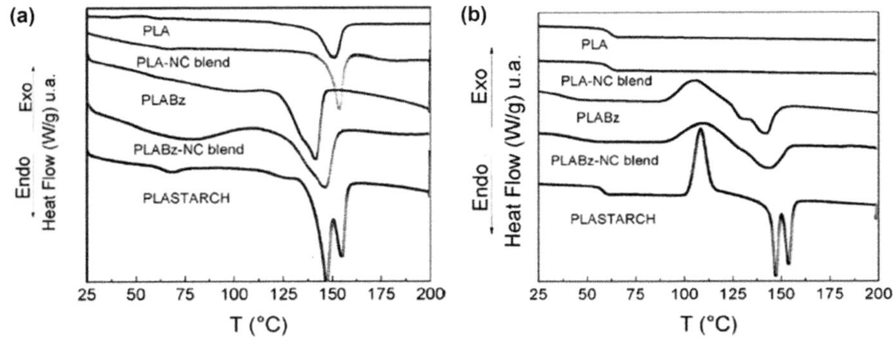


**Figure 5.19** DSC thermograms: (a) first heating scan; (b) second heating scan. PLA, neat polylactate; PLABz, PLA with benzoyl chloride; PLABz–NC, physical mixture of PLABz and SNCs; PLASTARCH, grafting of PLAM (PLABz carboxyl groups activated) onto the starch nanoparticles.[54]

particular, it was demonstrated that the presence of the benzoyl group in the PLA modifies its ability for crystallization and this process is strongly affected by the addition of SNCs only when these are chemically bound to PLA. In summary, the results suggest that the SNCs plays a very different role depending on whether they are chemically attached to PLA or only physically to amylose. More details can be found in Table 5.8.

Zou *et al.*[53] investigated water-borne polyurethane (WPU) and different percentages of pea SNC powder. The DSC results are given in Figure 5.20 and Table 5.8. It can be seen that the addition of nanocrystals increased $T_g$ for all contents and shifted the peak of α-relaxation, which was concluded to be due to the soft segment motion being suppressed by the steric hindrance of the rigid SNC nanophase and to hydrogen bonding on the active SNC surface.

Figure 5.21 presents the results of Lin *et al.*[29] The addition of only 5 wt% SNC filler did not form a network and the presence of rigid particles and the increase in crystallinity of the nanocomposites inhibited the free motion of PBS polymer chains, which resulted in the demand for more energy for the thermal transformation, leading to a higher melting temperature, $T_m$, of PBS/SNC-5 nanocomposites in comparison with the neat PBS-F. Table 5.8 gives values of the glass transition and melting temperatures.

**5.3.2.2.2 Thermogravimetric Analysis.** TGA is s used to determine degradation temperatures, maximum temperatures of the degradation rate and the residual masses that occur when a substance is heated under constant temperature (isothermal mode) or with a variable heating rate (non-isothermal mode). The degradation kinetics can also be determined in the non-isothermal mode.

Le Corre *et al.*[17] used TGA measurements to elucidate the thermal transitions in DSC. The degradation temperature is around 300 °C for native starch and SNCs for dry samples, independent of the starch type. They concluded that all thermal events in DSC occurred before the degradation

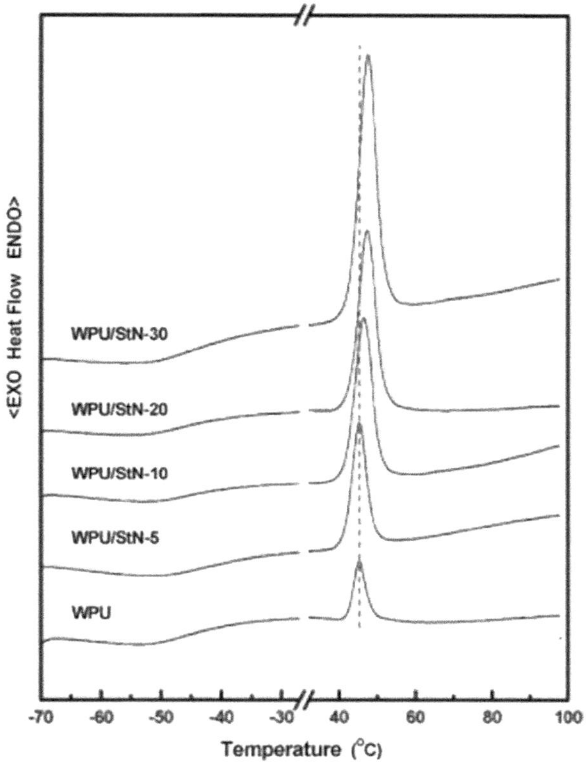


**Figure 5.20** DSC thermograms of WPU–SNC nanocomposites with WPU film as reference.
From Zou *et al.*[52]

event. They observed that the degradation temperature was higher for native starch than for SNCs as the SNC depolymerization starts earlier. Angellier[97] attributed this fact to the presence of sulfate groups at the surface of SNCs which catalyze the reaction. Around 300 °C, the depolymerization rate decreases, causing the thermal decomposition of sulfonic acid groups. There is no difference between native starch and M70 SNC, because of its larger particle size giving a smaller specific surface area; hence fewer sulfate groups can catalyze the reaction. They also showed that the degradation temperatures were the same for different native starch types. With respect to SNCs from different starch sources, practically the same degradation temperatures were found. It was concluded that the SNCs could be used in dry processes at temperatures lower than 180 °C, such as "cold" extrusion.

Piyada *et al.*[10] studied films made of rice starch and rice nanocrystals. These films were characterized by mechanical [tensile strength (TS) and elongation at break (ε)] and thermal (TGA, DSC and DMA) tests, by water vapor permeability (WVP), by XRD and by SEM. Their results are summarized in Tables 5.8–5.10. Figure 5.22 presents TGA curves, showing that all

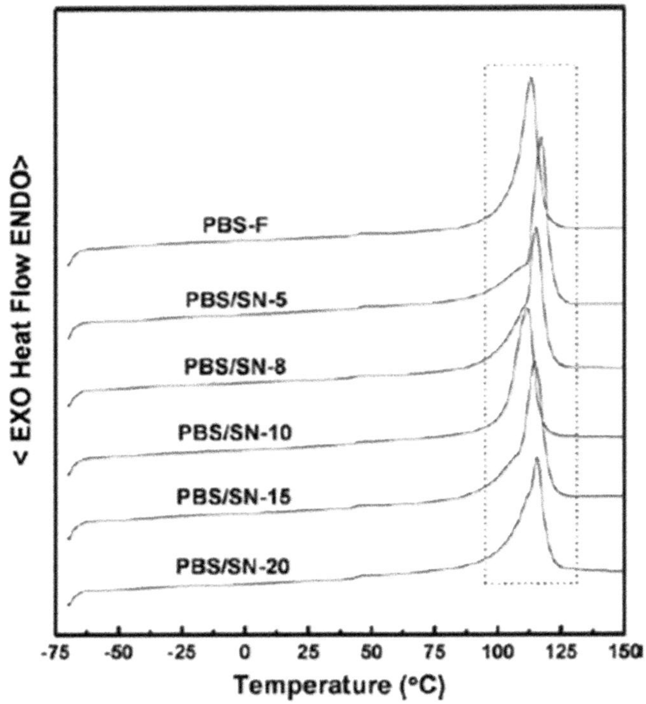


**Figure 5.21** DSC thermograms of PBS–SNC nanocomposites with various SNC contents and neat PBS-F as reference.
From Lin *et al.*[29]

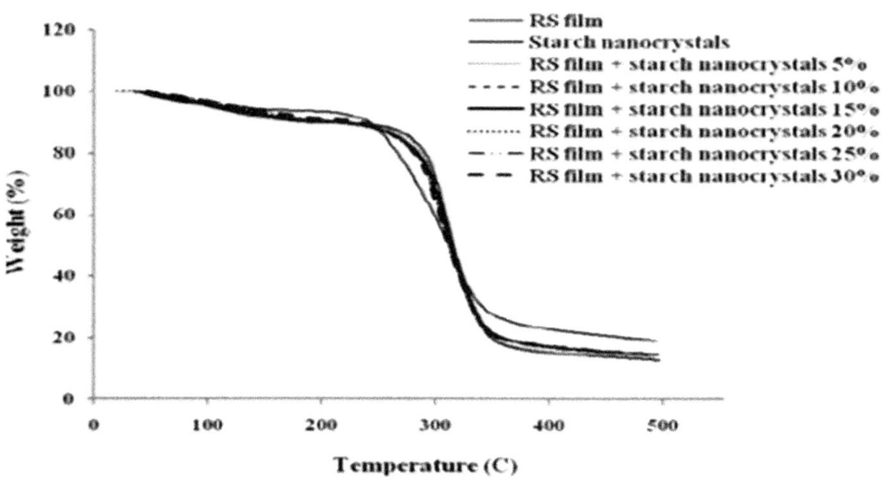


**Figure 5.22** TGA results for rice SNCs, neat rice starch films and rice starch films reinforced with 5–30 wt% of SNCs.

rice starch films reinforced with SNCs are stable up to 250 °C with a maximum rate of decomposition occurring at about 340 °C. The char yield of rice starch film (12.7%) was less than that of rice starch films reinforced with SNCs (13.4–8.5%). The char yield of rice starch film was enhanced as the SNC filler content increased from 0 to 30%. These results indicated that addition of SNCs increased the thermal stability of rice starch films.

5.3.2.2.3 **Dynamic Mechanical Analysis.** DMA is used to determine many properties, such as storage and loss moduli, tan $\delta$, $T_g$, tensile stress and deformation. Three different modes – flexure, tension and compression – can be used, depending on which property is desired. The flexure mode is usually used as it gives storage and loss moduli, tan $\delta$ and $T_g$. From the peak of the tan $\delta$ curve or the onset temperature on the storage modulus curve, it is possible to obtain the $T_g$ of the material, and this technique is more precise than DSC.

Table 5.10 presents DMA results for different systems involving SNCs and their composites. It can be concluded that when SNCs are added as a reinforcement, there is a shift to higher $T_g$, because of the strong interactions due to hydrogen bonding between the amorphous polymeric matrices and these particles, reducing the mobility of amylopectin chains.[10] These bonds cause strong absorption of polymer chains on the surface of nanoparticles. The same trend was found for many systems, *e.g.*, (a) glycerol-plasticized potato starch film and chitin nanoparticles;[101] (b) thermoplastic waxy maize starch film and SNCs;[48] and (c) rice starch.[10] Piyada *et al.*[10] also observed a double-peak of tan $\delta$ for rice starch films containing 20–30% SNCs, which suggested aggregation of SNCs. Aggregation of fillers would result in heterogeneous collections of polymer matrix and filler.[101]

Figure 5.23 presents graphs of storage modulus and tan $\delta$ for SPI and different percentages of PSNs. There was an initial decrease in $T_g$ with an increase in SNC content, resulting from untangling of SPI segments. Subsequently, there was an increase in $T_g$, caused by a sufficient active area of SNC domains. The final decrease in $T_g$ occurred because of a decrease in the effective active area and of a decrease in the original structure order of the SPI matrix. More details are given in Table 5.10.

### 5.3.2.3 *Water Absorption*

Only a few papers on water absorption for composite materials with SNCs have been published. García *et al.*[19] presented a figure showing the water uptake of previously dried films and after conditioning them at 98% RH and 25 °C. It was observed that all samples absorbed water during the experiment and, as usual, at the beginning (in this case, less than 20 h) the absorption is fast. For longer times, an equilibrium was obtained, represented by a plateau. The system was left for 30 days to confirm the equilibrium. The composite system (with 2.5 wt% of SNCs) absorbed more water than the unfilled

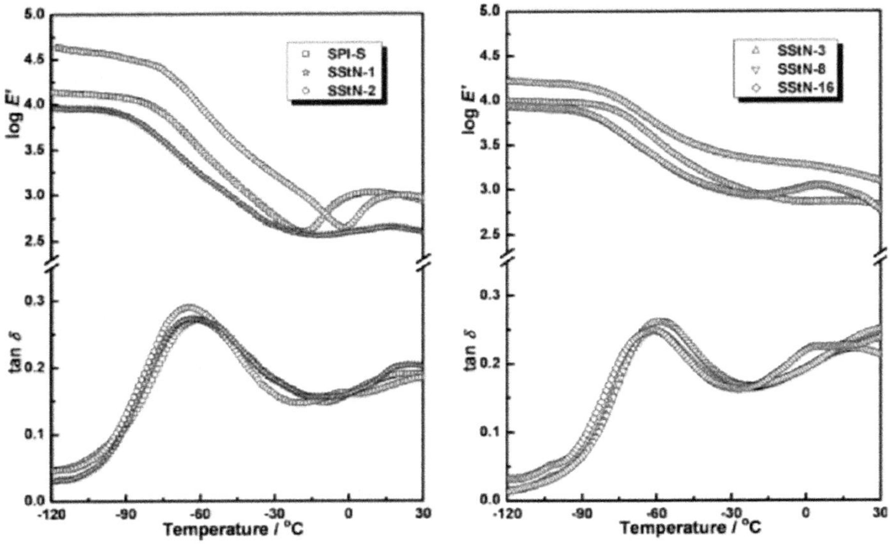


**Figure 5.23** Storage modulus and tan δ for composites of soy protein isolate (SPI) and different percentages of PSNs (SStN).[47]

starch. The equilibrium water uptake was high: ~50% for the composite and ~35% for the unfilled starch film.

Chen *et al.*[51] studied the water uptake in PVA–SNC (PSN) films. They observed that the moisture uptake (MU) of these films at 7 days under 98% was lower than that of the PVA film. In general, the MU decreased with increase in SNC content. Compared with the neat matrix (PVA), the decrease was about 20.5% for films with 40 wt% of SNCs.

Zheng *et al.*[47] studied nanocomposites of soy protein (SPI) and PNCs (system described in Table 5.5) and presented a graph (Figure 5.24) with two distinct zones, zone I ($t<50$ h) and zone II ($t>50$ h), for high and low water absorption rates, respectively. All the nanocomposites (29–32% water uptake) showed a similar behavior to SPI (29% water uptake) because of the strong interfacial interaction between the nanocrystals and the SPI matrix.

### 5.3.2.4 Water Vapor Permeability

Gas permeability is a very important property of food packaging films as one should avoid, or at least reduce, water vapor, oxygen and carbon dioxide transfer between food and the surroundings.

Water vapor permeability is constant and considered to be independent of the water vapor pressure gradient applied across the films.[10] However, hydrophilic materials such as protein films deviate from this ideal behavior because of the interactions of permeating water molecules with polar groups in the film structure.[102]

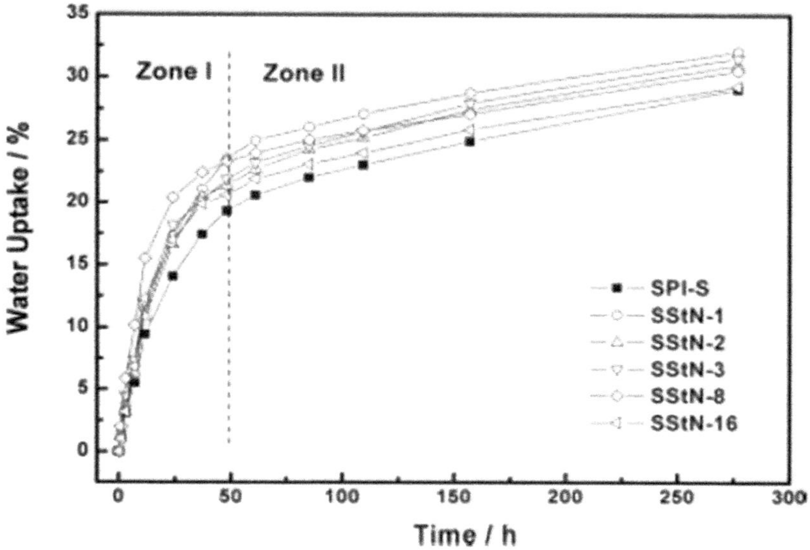


**Figure 5.24** Water uptake under an RH of 98% for nanocomposites (SStN-*x*) with different SNC contents (*x*) and SPI sheet (SPI-S).
From Zheng *et al.*[47]

The idea when adding nanoparticles in general is to create a tortuous path or compact network structure for water molecules to pass through the film. The longer diffusive path that the penetrating molecules must travel, the greater the reduction in permeability.[103]

Table 5.13 summarizes some studies in this area.

### 5.3.2.5 Mechanical Properties

Packaging films generally require high stress with deformation. In general, food packaging must be an undeformable material to provide structural integrity or to reinforce the food structure, or a deformable film for other applications.[66] TS is the maximum tensile stress sustained by the sample during the tension test. If maximum tensile stress occurs at either the yield point or the breaking point, it is designated the TS at yield or at break, respectively. Owing to the very high specific surface area and higher surface energy provided by nanoparticles, when they are added to a matrix, there is an effective transfer of stress by a shear mechanism from the matrix to the nanoparticles. This enhances the composite strength.[103–109]

Another very important property for packaging films is the elongation at break (ε), as this gives an indication of the film flexibility and stretchability (extensibility). This is determined at the point when the film breaks under tensile testing and is expressed as the percentage of stretching from the original length of the film. When nanoparticles are added, they are expected to lead to a decrease in elongation values because of differences in the rigidity

**Table 5.13** Summary of the results of studies on water vapor permeability.

| Study | System | Water vapor permeability (WVP) | Comments |
|---|---|---|---|
| García *et al.*[6] | **Film of thermoplastic starch:** processed by casting a mixture of waxy maize starch granules (10 g), glycerol (5 g) and distilled water (185 g).<br>**Nanocomposite films:** 2.5 wt.% (relative to the dry total mass, starch + plasticizer + nanocrystals) of an aqueous suspension of waxy maize starch nanocrystals added to the other components (starch granules, glycerol and water) | Unfilled plasticized film: $(3.8 \pm 0.3) \times 10^{-10}$ g $m^{-1}$ $s^{-1}$ $Pa^{-1}$.<br>Plasticized film reinforced with 2.5% of SNCs: $(6.8 \pm 0.1) \times 10^{-10}$ g $m^{-1}$ $s^{-1}$ $Pa^{-1}$.<br>It can be seen that this result is opposite to that obtained using a cassava starch matrix[16] and in natural rubber films,[38] where the nanocrystals were well distributed | The permeability to water vapor increases by 79% when only 2.5 wt% of waxy maize starch nanocrystals are added |
| Piyada *et al.*[10] | Films of native rice starch, with amylase and moisture contents of about 30.40 and 12.19%, respectively, and rice starch nanocrystals | **Starch films:** $\mathbf{7.23 \times 10^{-11}}$ **g** $\mathbf{m^{-1}}$ $\mathbf{s^{-1}}$ $\mathbf{Pa^{-1}}$<br>**Starch film with 20% of starch nanocrystals:** $2.96 \times 10^{-11}$ g $m^{-1}$ $s^{-1}$ $Pa^{-1}$<br>Starch films with more than 20% were brittle | WVP of rice starch films decreased significantly with the addition of SNCs and decreased slowly with an increase in SNC content |

| | | | |
|---|---|---|---|
| García *et al.*[19] | Composite of cassava starch (72% amylopectin and 28% amylase) and 2.5 wt% of waxy maize-starch nanocrystals.<br>System with 0% of water at 58% RH | **Cassava starch:** $(4.5 \pm 0.6) \times 10^{-10}$ g $m^{-1}$ $s^{-1}$ $Pa^{-1}$.<br>**Composite with 2.5% of nanocrystals:** $(2.7 \pm 0.7) \times 10^{-10}$ g $m^{-1}$ $s^{-1}$ $Pa^{-1}$ | Decrease of 40% in the water permeability of cassava starch filled with only 2.5 wt% of SNCs. The presence of SNCs introduced a tortuous route for the molecule of water to diffuse through the films |
| Kristo and Biliaderis[61] | **Matrix:** aqueous solution of pullulan + sorbitol as plasticizer + SNCs in aqueous suspensions. The content of SNCs ranged from 0 to 40 wt% [starch nanoparticles/(pullulan + sorbitol)], namely 0, 3, 6, 10, 15, 20, 30 and 40 wt% | **SNC content up to 20%:** $(12.5 \pm 0.3) \times 10^{-10}$ g $m^{-1}$ $s^{-1}$ $Pa^{-1}$.<br>**Films reinforced with 30 and 40% of SNCs:** $(9.5 \pm 0.2) \times 10^{-10}$ g $m^{-1}$ $s^{-1}$ $Pa^{-1}$ | No significant differences ($p > 0.05$) were observed in WVP of unfilled samples and those containing up to 20 wt% SNCs. Adding 30 and 40 wt% of SNCs resulted in a significant decrease in WVP |
| Duan *et al.*[111] | **Matrix:** carboxymethyl chitosan, glycerol as plasticizer.<br>**Composites:** 3, 6, 10, 15, 20, 30 and 40 wt% of maize starch nanocrystals | **Carboxymethyl chitosan films:** $13 \times 10^{-13}$ g $m^{-1}$ $s^{-1}$ $Pa^{-1}$.<br>**Composite with 20% of nanocrystals:** $3 \times 10^{-13}$ g $m^{-1}$ $s^{-1}$ $Pa^{-1}$ | WVP of nanocomposites decreased radically with increase in SNC content up to 20 wt% and then remained almost unchanged with further increases. Higher contents of SNCs would tend to form aggregates that could not contribute to reduce the WVP |

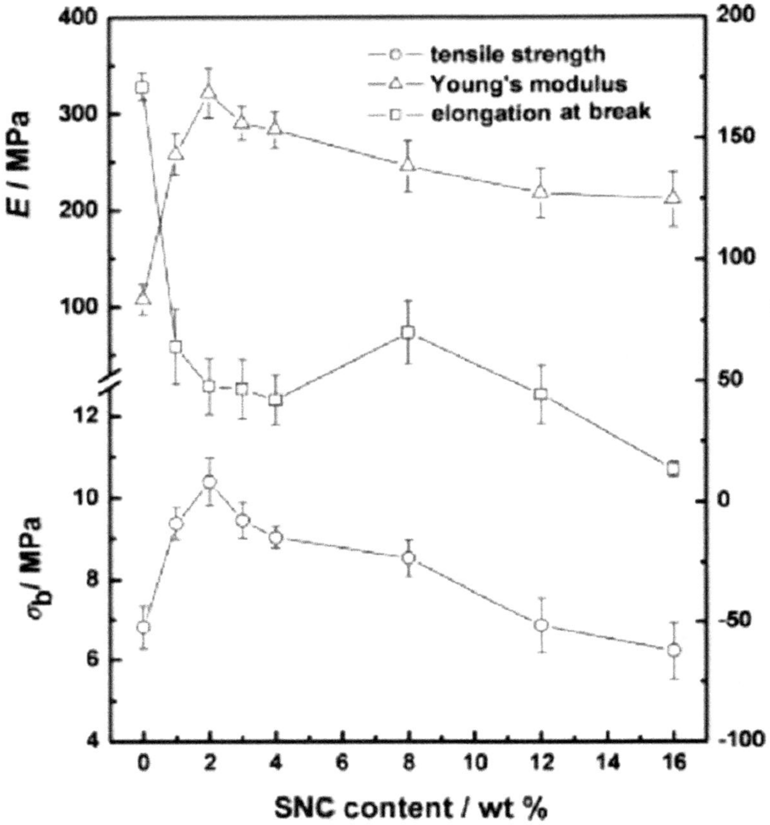


**Figure 5.25** Effects of the SNC content on mechanical properties of SPI and pea starch nanocrystals (PSNs).
From Zheng *et al.*[47]

between the matrix and fillers. Because of the rigid nature of the fillers, most of the system deformation under high strain comes from the polymer.[10]

Figure 5.25 presents the results of Zheng *et al.*[47] related to composites of SPI and different amounts of PSNs. The higher value of $\sigma_B$ was probably caused by the uniform distribution of SNCs. For higher SNC contents, self-aggregation occurred, decreasing the effective active SNC surface area for interaction with the SPI matrix. This destroyed the ordered structure in the SPI matrix. More details are given in Table 5.14 for this system and for others.

### *5.3.2.6 Fractured Morphologies of Nanocomposites*

In order to analyze the dispersion of nanoparticles in matrices after the tension experiments, the fractured composites were studied using SEM.

From Figures 5.26 and 5.27, it can be seen that good dispersion occurs when low contents of nanocrystals are introduced into the matrices.[29,55]

**Table 5.14** Summary of results for mechanical properties.

| Study | System | Tensile strength, $\sigma_B$ (MPa) | Young's modulus, $E$ (MPa) | Elongation at break, $\varepsilon_B$ (%) | Comments |
|---|---|---|---|---|---|
| Chang *et al.*[35] | **SNC-*g*-PCL:** PSNs grafted with polycaprolactone diol ($M_n = 2000$), PCL2000. **Film of water-borne polyurethane:** WPU-F. **Composite film:** WPU + SNC-*g*-PCL($x$), where $x$ represents 5, 10, 15, 20, 25 or 30 wt% of SNC-*g*-PCL | **WPU-F:** 31.12. **WPU–SNC-*g*-PCL(5):** maximum value 40.19 | **WPU-F:** ~1340 **WPU–SNC-*g*-PCL(5):** maximum value ~1410 | **WPU-F:** ~10. **WPU–SNC-*g*-PCL(20):** maximum value: 220 | The addition of SNC-*g*-PCL enhanced the rigidity of the nanocomposite materials. As the SNC-*g*-PCL content increased, the tensile strength gradually decreased, reaching values lower than for the neat film when more than 20 wt% of SNC-g-PCL was used |
| Zou *et al.*[53] | **SNCs:** PSNs. **Film of water-borne polyurethane:** WPU-F. **Composite film:** WPU + SNC ($x$), where $x$ represents 5, 10, 20 or 30 wt% of SNCs | **WPU-F:** 11.3. **WPU/SNC(10):** 31.1 (maximum value) | **WPU-F:** 3.0 **WPU/SNC(30):** 204.6 (the highest value) | **WPU-F:** 830. **WPU/SNC($x$):** decreased for all $x$ values | For more than 10 wt% of SNCs, $\sigma_B$ decreases. For all percentages of SNCs, $E$ increases and $\varepsilon_B$ decreases. For 10 wt% of SNCs, $\sigma_B = 30$ MPa, $E = 110.1$ MPa and $\varepsilon_B = 620.6\%$ |
| Piyada *et al.*[10] | Films of native rice starch, with amylase and moisture contents of about 30.40 and 12.19%, respectively, and rice starch nanocrystals. Film samples were cast by transferring the mixtures to plastic bags after peeling and placed in desiccators | **Starch film:** 7.12 MPa. **Starch film with 20%:** 16.43 MPa (maximum value) | The authors did not perform this analysis | **Starch film:** 53.46%. **Starch film with 30%:** 2.48% (lowest value) | When the SNCs were added to rice starch films up to 20 wt%, the tensile strength (TS) of the resulting films increased with the SNC content. However, when the SNC content exceeded 20 wt%, the TS of the films tended to decrease, but it was still much higher than that of the control film |

**Table 5.14** (*Continued*)

| Study | System | Tensile strength, $\sigma_B$ (MPa) | Young's modulus, $E$ (MPa) | Elongation at break, $\varepsilon_B$ (%) | Comments |
|---|---|---|---|---|---|
| Yu *et al.*[20] | Poly(ε-caprolactone) (PCL) was grafted to the surface of SNCs. The resulting nanoparticles were then incorporated into PLA. **Matrix:** films were coded as PLA/SNC-*g*-PCL($x$), where $x$ represents 0, 5, 10, 15, 20, 25 and 30 wt% SNC-*g*-PCL | Maximum value for films with 5 wt% SNC-*g*-PCL. Minimum for films with 30 wt% SNC-*g*-PCL | Maximum for films with 0 wt% SNC-*g*-PCL. Minimum for films with 30 wt% SNC-*g*-PCL | Maximum value for films with 5 wt% SNC-*g*-PCL. Minimum for films with 30% SNC-*g*-PCL | The TS of the films increased on adding 5 wt% SNC-*g*-PCL, but decreased for higher contents of nanofiller. The same effect was observed for elongation at break. The optimal loading-level was 5 wt% of SNC-*g*-PCL, which gave simultaneous enhancement of elongation at break and TS |
| Lin *et al.*[29] | Poly(butylene succinate) (PBS)/SNC nanocomposites with various SNC contents were prepared and coded as PBS/SNC-5, PBS/SNC-8, PBS/SNC-10, PBS/SNC-15, and PBS/SNC-20, where the numerals represent the theoretical SNC content (wt%) of the nanocomposites | **PBS:** 26. **Maximum value:** 28 for PBS/SNC-5. **Minimum value:** 15 for PBS/SNC-20 | **PBS:** 583. **Maximum value:** 706 for PBS/SNC-20. **Minimum value:** 583 for PBS/SNC-0 | **PBS:** 6 **Maximum value:** 8 for PBS/SNC-5 **Minimum value:** 3 for PBS/SNC-20 | The addition of 5 wt% SNCs resulted in simultaneous enhancement of the strength, toughness and rigidity of the nanocomposites. In comparison with the mechanical performance of neat PBS, the $\sigma_B$, $\varepsilon_B$ and $E$ values of PBS/SNC-5 nanocomposite all increased, by about 8, 28, and 3%, respectively. When the loading level exceeded 5 wt%, $\sigma_B$ and $\varepsilon_B$ gradually decreased, whereas the $E$ value of the nanocomposites increased sharply |

| | | | | | |
|---|---|---|---|---|---|
| Chen *et al.*[55] | Potato starch nanocrystals and WPU | **WPU**: 11.0.<br>**Maximum value**: 51.5 MPa for 5 wt% of SNCs | **WPU**: 2.7.<br>**Maximum value**: 5.2 MPa for 5 wt% of SNCs | **WPU**: 814.4.<br>**Maximum value:** 1406.6% for 2 wt% of SNCs | For SNC content ≥5 wt%, all the mechanical properties decreased because of the formation of SNC agglomerates |
| Zheng *et al.*[47] | **SNCs:** pea starch composed of 35% amylase and 65% amylopectin, and with an average granule size of about 30 μm.<br>**Matrix:** SPI.<br>**SNCs:** SSNC-*x*, where *x* = 1, 2, 3, 4, 8, 12 and 16 wt% of SNCs | **Neat SPI:** 6.82.<br>**Composite:** SSNC-1, 9.37; SSNC-2, 10.39.<br>For the other SNCs, see Figure 5.26.<br>All the other SNCs had $\sigma_B$ smaller than the neat SPI. | Behavior similar to $\sigma_B$, but the values were higher than the neat SPI for all SNCs.<br>See Figure 5.26 | **Neat SPI:** 171.5.<br>**Composite:** SSNC-1: 64.4<br>Slight increase up to SSNC-8, followed by a second decrease with increase in SNC content.<br>See Figure 5.26 | The higher value of $\sigma_B$ was probably caused by the uniform distribution of SNCs. For higher SNC contents, self-aggregation occurred, decreasing the effective active SNC surface area for interacting with the SPI matrix. This destroyed the ordered structure in the SPI matrix |
| Kristo and Biliaderis[61] | **Matrix:** aqueous solution of pullulan + sorbitol as plasticizer + SNCs in aqueous suspensions. The content of starch nanocrystals was 0, 6 and 40 wt%.<br>Films were conditioned at 11, 43, 53 and 75% RH for 10 days before the analysis | Maximum for films with 0 and 6 wt% of SNCs and 11% RH.<br>Minimum for films kept at 75% RH | Maximum for films with 6 and 20 wt% of SNCs and 11% RH.<br>Minimum for films with 0 and 6 wt% of SNCs and 75% RH | Maximum for films with 0 wt% of SNCs at 53 and 75% RH.<br>Minimum for films with 40 wt% of SNCs at 11% RH | A drastic increase in the Young's modulus and the maximum tensile strength and a significant decrease in the elongation at break with increase in nanofiller content were observed after sample conditioning in diverse RH environments (from 43 to 75%). These findings supported the crucial role of strong interfacial bonding between the starch nanofiller and the pullulan matrix. The relative reinforcing effect of SNCs was more significant at high hydration levels |

**Table 5.14** (*Continued*)

| Study | System | Tensile strength, $\sigma_B$ (MPa) | Young's modulus, $E$ (MPa) | Elongation at break, $\varepsilon_B$ (%) | Comments |
|---|---|---|---|---|---|
| Le Corre *et al.*[110] | **Natural rubber (NR):** | | | | Waxy starch nanocrystals with 5 wt% SNCs presented higher tensile strength than NR + 5 wt% SNCs. When glycerol was the plasticizer, waxy starch/5% waxy SNC films presented higher $E$ and lower $\sigma_B$ than other nanoSNCs formed with 5 wt% SNCs |
| | NR + 5% high-amylose SNCs | 1.50 | 1.35 | 0.82 | |
| | NR + 5% normal maize SNCs | 2.20 | 1.53 | 1.23 | |
| | NR + 5% potato SNCs | 2.11 | 1.15 | 1.28 | |
| | NR + 5% wheat SNCs | 1.43 | 1.15 | 1.17 | |
| | NR + 5% waxy SNCs | 1.95 | 2.46 | 0.91 | |
| | NR + 30% waxy SNCs | 2.41 | 119.69 | 0.47 | |
| | Waxy starch/sorbitol 25% + 5% waxy SNCs | 2.61 | 2.1 | 0.90 | |
| | Waxy starch/glycerol 25% + 5% waxy SNCs | 3.60 | 7.3 | 0.30 | |
| | Waxy starch/glycerol 25% + 25% waxy SNCs | 9.80 | 22.0 | 0.07 | |
| Duan *et al.*[111] | **Matrix:** carboxymethyl chitosan, glycerol as plasticizer. **SNCs:** 3, 6, 10, 15, 20, 30, 40 wt%. of maize starch nanocrystals | **0 wt% SNCs:** 16 MPa. **Maximum value**: 28 MPa for 30 wt% SNCs | The authors did not perform this analysis | **0 wt% SNCs:** 180%. **Minimum value:** 60% for 40 wt% SNCs | The tensile strength of the films increased with increasing in the SNC content up to 30 wt% and then decreased with further increase in the SNC content. The elongation at break (EB) of the SNCs decreased with increase in SNC content. When the SNC content was as high as 30 wt%, the EB was still about 90%, indicating that the nanocomposite film had high ductility |

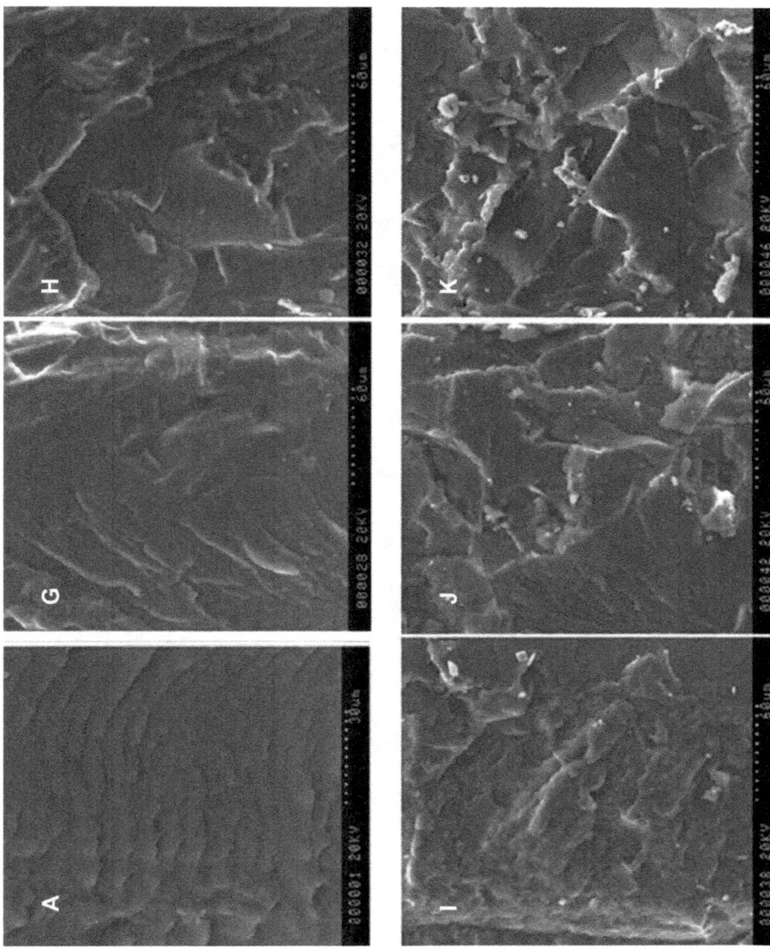


**Figure 5.26** SEM images of fractured surfaces of the neat reference, PBS-F (A), and PBS/SNC-5 (G), PBS/SNC-8 (H), PBS/SNC-10 (I), PBS/SNC-15 (J), PBS/SNC-20 (K) sheets.
From Lin *et al.*[29]

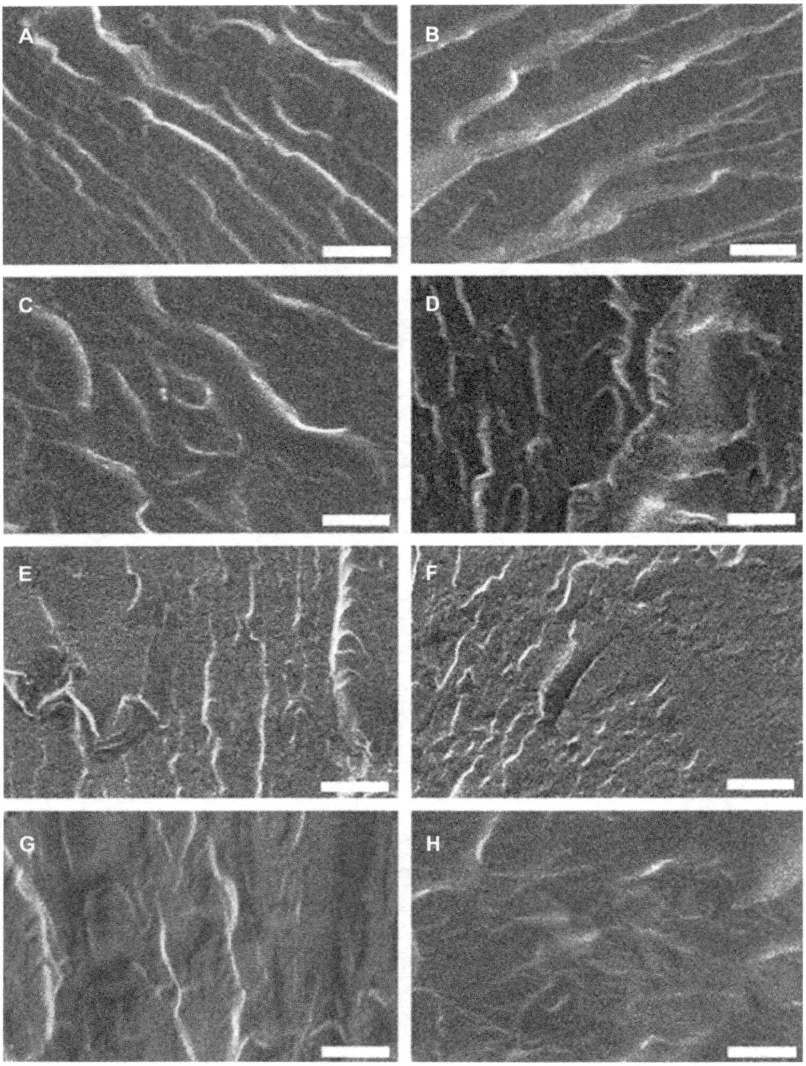


**Figure 5.27** SEM images of fractured surface of WPU nanocomposites with various SNC contents prepared by different methods. (A) Neat WPU. Method I: (B) US-1; (C) US-2; (D) US-4; (E) US-5; (F) US-8. Method II: (G) US-5-II. Method III: (H) US-5-III. US-*n* represent composites of WPU with *n* wt% of SNCs. All scale bars are 10 mm.
From Chen *et al.*[55]

Figures 5.26[29] shows fracture surfaces of neat resin (A) and PBS–SNC nanocomposites. With only 5 wt% SNCs (G), the nanoparticles were well dispersed in the matrix and exhibited a similar fractured surface to the neat PBS-F sheet. For higher loadings of SNCs, such as 8 and 10 wt%, the formation of a rigid three-dimensional continuous network was responsible for the coarser fracture surfaces of PBS/SNC-8 (H) and PBS/SNC-10 (I). For higher SNC contents, there was severe self-aggregation of the nanoparticles (J, K).

Chen *et al.*[55] presented SEM images of the fractured surface of nanocomposites of water-borne polyurethane (WPS) with various SNC contents prepared by different methods, as explained in Table 5.5. The fractured surface of composites with SNC content lower than 2 wt%, cases B and C in Figure 5.27, was similar to that of the neat WPU, suggesting a homogeneous dispersion of the nanocrystals. For SNC content higher than 4 wt%, the fractured surface became smooth with some thin stripes attributed to the breakage of the original structure in the WPU matrix after adding higher loadings of SNCs. For samples with chemical grafting, the fractured surface of US-5-II and US-5-III also gradually became smoother, showing a relatively brittle character. This was in agreement with the decrease in strength and elongation at break shown in Table 5.14.

## 5.4 Conclusion

The main advantages of SNCs are their renewable nature, low cost, high barrier properties, availability, compatibilization with biopolymers, high specific strength, non-abrasive and non toxic nature that allow easier processing even at high filling levels, biodegradability and a relatively reactive surface. They are edible, versatile and light-weight and have a high aspect ratio, high specific strength and high modulus.

Many workers have demonstrated that the thermal properties of starch are very dependent on the water content. The degradation temperature is around 300 °C for native starch and SNCs for dry samples, independent of the starch type. It has been shown that when SNCs are added as a reinforcement, there is a shift to higher glass transition temperatures, because of the strong interactions due to hydrogen bonding between the amorphous polymeric matrices and these particles, reducing the mobility of amylopectin chains.

The moisture uptake decreases with increase in SNC content.

Owing to the very high specific surface area and higher surface energy provided by nanoparticles, when they are added to a matrix, there is an effective transfer of stress by a shear mechanism from the matrix to the nanoparticles, which enhances the composite strength. When nanoparticles are added, there is expected to be a decrease in elongation values because of differences in the rigidity between the matrix and fillers. Because of the rigid nature of the fillers, most of the system deformation under high strain comes from the polymer.

The presence of SNCs leads to a substantial increase in the stiffness of the material, a significant improvement of its barrier properties and a decrease in the water vapor permeability.

SNCs have many applications, but the industrial use of SNCs is still a difficult issue because of some drawbacks, such as the tendency to aggregate and settle down in aqueous solutions, which is a limitation to application in most biological and food systems.

## References

1. PlasticsEurope, *Plastics – the Facts, 2013. An Analysis of European Latest Plastics Production, Demand and Waste Data*, PlasticsEurope, Brussels, 2013, p. 38.
2. *Recycling Plastic*. University of Cambridge, Cambridge, 2005, vol. 3, p. 27. http://www-g.eng.cam.ac.uk/impee/topics/RecyclePlastics/files/Recycling%20Plastic%20v3%20PDF%20WITH%20NOTES.pdf.
3. European Bioplastics, *Production Capacity*, http://en.european-bioplastics.org/market/market-development/production-capacity/, Accessed 28 February 2008.
4. D. J. Gallant, B. Bouchet and P. M. Baldwin, *Carbohydr. Polym.*, 1997, **32**, 177–191.
5. M. L. Fishman, D. R. Coffin, R. P. Konstance and C. I. Onwulata, *Carbohydr. Polym.*, 2000, **41**, 317–325.
6. N. L. García, L. Ribba, A. Dufresne, M. Aranguren and S. Goyanes, *Carbohydr. Polym.*, 2011, **84**, 203–210.
7. F. Xie, E. Pollet, P. J. Halley and L. Avérous, *Prog. Polym. Sci.*, 2013, **38**, 1590–1628.
8. D. Le Corre, J. Bras and A. Dufresne, *Biomacromolecules*, 2010, **11**, 1139–1153.
9. H. Liu, F. Xie, L. Yu, L. Chen and L. Li, *Prog. Polym. Sci.*, 2009, **34**, 1348–1368.
10. K. Piyada, S. Waranyou and W. Thawien, *Int. Food Res. J.*, 2013, **20**, 439–449.
11. R. N. Tharanathan, *Trends Food Sci. Technol.*, 2003, **14**, 71–78.
12. D. J. Thomas and W. A. Atwell, in *Starches*, ed. D. J. Thomas and W. A. Atwell, American Association of Cereals Chemists, St Paul, MN, 1999, pp. 1–11.
13. *Handbook of Hydrocolloids*, ed. G. O. Phillips and P. A. Williams, Woodhead Publishing, Abington, 2000.
14. E. Bertoft, in *Starch in Food: Structure, Function and Applications*, ed. A. C. Eliasson, C. R. C. Press, Boca Raton, F. L., 2004, p. 605.
15. M. G. Sajilata, R. S. Singhal and P. R. Kulkarni, *Compr. Rev. Food. Sci. Food Saf.*, 2006, **5**, 1–17.
16. C. G. Oates, *Trends Food Sci. Technol.*, 1997, **8**, 375–382.
17. D. Le Corre, J. Bras and A. Dufresne, *Carbohydr. Polym.*, 2012, **87**, 658–666.
18. T. Kasemsuwan and J. Jane, *Cereal Chem.*, 1994, **71**, 282–287.
19. N. L. García, L. Ribba, A. Dufresne, M. I. Aranguren and S. Goyanes, *Macromol. Mater. Eng.*, 2009, **294**, 169–177.
20. J. H. Yu, F. J. Ai, A. Dufresne, S. J. Gao, J. Huang and P. R. Chang, *Macromol. Mater. Eng.*, 2008, **293**, 763–770.
21. X. L. Zhang, J. Huang, P. R. Chang, J. L. Li, Y. M. Chen, D. X. Wang, J. H. Yu and J. H. Chen, *Polymer*, 2010, **51**, 4398–4407.

22. F. Chivrac, E. Pollet, P. Dole and L. Avérous, *Carbohydr. Polym.*, 2010, **79**, 941–947.
23. K. Rengsutthi and S. Charoenrein, *LWT – Food Sci. Technol.*, 2011, **44**, 1309–1313.
24. J. Tattiyakul, P. Pradipasena and S. Asavasaksakul, *Starch Stärke*, 2007, **59**, 342–347.
25. S. Srichuwong, T. C. Sunarti, T. Mishima, N. Isono and M. Hisamatsu, *Carbohydr. Polym.*, 2005, **60**, 529–538.
26. I. Tan, P. J. Torley and P. J. Halley, *Carbohydr. Polym.*, 2008, **72**, 272–286.
27. L. F. Polesi and S. B. S. Sarmento, *Starch Stärke*, 2011, **63**, 226–235.
28. R. C. S. T. Muccillo, Universidade Federal do Rio Grande do Sul, 2009.
29. N. Lin, J. H. Yu, P. R. Chang, J. L. Li and J. Huang, *Polym. Compos.*, 2011, **32**, 472–482.
30. M. A. S. Azizi Samir, F. Alloin and A. Dufresne, *Biomacromolecules*, 2005, **6**, 612–626.
31. F. Chivrac, E. Pollet, M. Schmutz and L. Avérous, *Biomacromolecules*, 2008, **9**, 896–900.
32. A. P. Kumar, D. Depan, N. Singh Tomer and R. P. Singh, *Prog. Polym. Sci.*, 2009, **34**, 479–515.
33. M. Neus Anglès, J. Salvadó and A. Dufresne, *J. Appl. Polym. Sci.*, 1999, **74**, 1962–1977.
34. M. J. Jivan, A. Madadlou and M. Yarmand, *Food Chem.*, 2013, **141**, 1661–1666.
35. P. R. Chang, F. J. Ai, Y. Chen, A. Dufresne and J. Huang, *J. Appl. Polym. Sci.*, 2009, **111**, 619–627.
36. V. P. Cyras, L. B. Manfredi, M.-T. Ton-That and A. Vázquez, *Carbohydr. Polym.*, 2008, **73**, 55–63.
37. H. M. C. Azeredo, *Food Res. Int.*, 2009, **42**, 1240–1253.
38. D. R. Paul and L. M. Robeson, *Polymer*, 2008, **49**, 3187–3204.
39. S. Sinha Ray and M. Bousmina, *Prog. Mater. Sci.*, 2005, **50**, 962–1079.
40. A. Dufresne, *Molecules*, 2010, **15**, 4111–4128.
41. H. Angellier, S. Molina-Boisseau and A. Dufresne, *Macromolecules*, 2005, **38**, 9161–9170.
42. H. Angellier, J. L. Putaux, S. Molina-Boisseau, D. Dupeyre and A. Dufresne, *Macromol. Symp.*, 2005, **221**, 95–104.
43. E. Bouthegourd, K. R. Rajisha, N. Kalarical, J. M. Saiter and S. Thomas, *Mater. Lett.*, 2011, **65**, 3615–3617.
44. B. Wei, X. Hu, H. Li, C. Wu, X. Xu, Z. Jin and Y. Tian, *Food Hydrocolloids*, 2014, **36**, 369–373.
45. C. Li, P. D. Sun and C. Yang, *Starch Starke*, 2012, **64**, 497–502.
46. N. Lin, J. Huang and A. Dufresne, *Nanoscale*, 2012, **4**, 3274–3294.
47. H. Zheng, F. Ai, P. R. Chang, J. Huang and A. Dufresne, *Polym. Compos.*, 2009, **30**, 474–480.
48. H. Angellier, S. Molina-Boisseau, P. Dole and A. Dufresne, *Biomacromolecules*, 2006, 7, 531–539.
49. Z. Adamczyk and P. Weroński, *Adv. Colloid Interface Sci.*, 1999, **83**, 137–226.

50. L. Ren, M. Jiang, L. Wang, J. Zhou and J. Tong, *Carbohydr. Polym.*, 2012, **87**, 1874–1876.
51. Y. Chen, X. Cao, P. R. Chang and M. A. Huneault, *Carbohydr. Polym.*, 2008, **73**, 8–17.
52. Y. Wang and L. Zhang, *J. Nanosci. Nanotechnol.*, 2008, **8**, 5831–5838.
53. J. W. Zou, F. Zhang, J. Huang, P. R. Chang, Z. M. Su and J. H. Yu, *Carbohydr. Polym.*, 2011, **85**, 824–831.
54. N. L. García, M. Lamanna, N. D'Accorso, A. Dufresne, M. Aranguren and S. Goyanes, *Polym. Degrad. Stab.*, 2012, **97**, 2021–2026.
55. G. Chen, M. Wei, J. Chen, J. Huang, A. Dufresne and P. R. Chang, *Polymer*, 2008, **49**, 1860–1870.
56. H. Angellier, L. Choisnard, S. Molina-Boisseau, P. Ozil and A. Dufresne, *Biomacromolecules*, 2004, **5**, 1545–1551.
57. D. Le Corre, J. Bras and A. Dufresne, *Biomacromolecules*, 2011, **12**, 3039–3046.
58. L. Jayakody and R. Hoover, *Food Res. Int.*, 2002, **35**, 665.
59. W. Thielemans, M. N. Belgacem and A. Dufresne, *Langmuir*, 2006, **22**, 4804–4810.
60. Y. Wang, H. Tian and L. Zhang, *Carbohydr. Polym.*, 2010, **80**, 665–671.
61. E. Kristo and C. G. Biliaderis, *Carbohydr. Polym.*, 2007, **68**, 146–158.
62. N. Gontard, S. Guilbert and J.-L. Cuq, *J. Food Sci.*, 1992, **57**, 190–195.
63. S. Guilbert and N. Gontard, in *Innovations in Food Packaging*, ed. H. H. Jung, Academic Press, London, 2005, pp. 263–276.
64. T. Uesaka, K. Nakane, S. Maeda, T. Ogihara and N. Ogata, *Polymer*, 2000, **41**, 8449–8454.
65. C.-F. Kuan, C.-C. M. Ma, H.-C. Kuan, H.-L. Wu and Y.-M. Liao, *Compos. Sci. Technol.*, 2006, **66**, 2231–2241.
66. T. Ohkita and S.-H. Lee, *J. Appl. Polym. Sci.*, 2005, **97**, 1107–1118.
67. V. M. Correlo, L. F. Boesel, E. Pinho, A. R. Costa-Pinto, M. L. Alves da Silva, M. Bhattacharya, J. F. Mano, N. M. Neves and R. L. Reis, *J. Biomed. Mater. Res., Part A*, 2009, **91A**, 489–504.
68. S. M. Lee, D. Cho, W. H. Park, S. G. Lee, S. O. Han and L. T. Drzal, *Compos. Sci. Technol.*, 2005, **65**, 647–657.
69. Z. Liang, P. Pan, B. Zhu, T. Dong and Y. Inoue, *J. Appl. Polym. Sci.*, 2010, **115**, 3559–3567.
70. M. W. Lee, S. O. Han and Y. B. Seo, *Compos. Sci. Technol.*, 2008, **68**, 1266–1272.
71. A. Dufresne, J.-Y. Cavaillé and W. Helbert, *Macromolecules*, 1996, **29**, 7624–7626.
72. A. Dufresne and J.-Y. Cavaillé, *J. Polym. Sci. Part B: Polym. Phys.*, 1998, **36**, 2211–2224.
73. Y. Habibi and A. Dufresne, *Biomacromolecules*, 2008, **9**, 1974–1980.
74. S. H. Imam, P. Cinelli, S. H. Gordon and E. Chiellini, *J. Polym. Environ.*, 2005, **13**, 47–55.
75. P. Cinelli, E. Chiellini, J. W. Lawton and S. H. Imam, *Polym. Degrad. Stab.*, 2006, **91**, 1147–1155.

76. M. A. Khan, S. K. Bhattacharia, M. A. Kader and K. Bahari, *Carbohydr. Polym.*, 2006, **63**, 500–506.
77. N. Follain, C. Joly, P. Dole and C. Bliard, *J. Appl. Polym. Sci.*, 2005, **97**, 1783–1794.
78. Siddaramaiah, B. Raj and R. Somashekar, *J. Appl. Polym. Sci.*, 2004, **91**, 630–635.
79. Y. U. Nabar, D. Draybuck and R. Narayan, *J. Appl. Polym. Sci.*, 2006, **102**, 58–68.
80. M. K. Beliakova, A. A. Aly and F. A. Abdel-Mohdy, *Starch Stärke*, 2004, **56**, 407–412.
81. R. Kumar, V. Choudhary, S. Mishra, I. K. Varma and B. Mattiason, *Ind. Crops Prod.*, 2002, **16**, 155–172.
82. P. Kumar, K. P. Sandeep, S. Alavi, V. D. Truong and R. E. Gorga, *J. Food Eng.*, 2010, **100**, 480–489.
83. J. Zhang, P. Mungara and J. Jane, *Polymer*, 2001, **42**, 2569–2578.
84. W. H. Ferreira, R. R. Khalili, M. J. M. Figueira Junior and C. T. Andrade, *Ind. Crops Prod.*, 2014, **52**, 38–45.
85. T. Jawhari, L. Quintanilla and J. M. Pastor, *J. Appl. Polym. Sci.*, 1994, **51**, 463–471.
86. M. C. Evora, O. L. Gonçalez, R. C. L. Dutra, M. F. Diniz, H. Wiebeck and L. G. de Andrade e Silva, *Polímeros*, 2002, **12**, 60–68.
87. D. J. Thomas and W. A. Atwell, *Starches*, American Association of Cereal Chemists, St paul, MN, 1999, pp. 13–24.
88. R. A. Simão and B. R. Cordenunsi, in *Starches: Characterization, Properties and Applications*, ed. A. C. Bertolini, CRC Press, Boca Raton, FL, 2010, pp. 21–32.
89. A. A. P. Ferreira and H. Yamanaka, *Quím. Nova*, 2006, **29**, 137–142.
90. P. M. Baldwin, J. Adler, M. C. Davies and C. D. Melia, *J. Cereal Sci.*, **27**, 255–265.
91. R. Hoover, G. Swamidas and T. Vasanthan, *Carbohydr. Res.*, 1993, **246**, 185–203.
92. C. G. Biliaderis, C. M. Page, T. J. Maurice and B. O. Juliano, *J. Agric. Food Chem.*, 1986, **34**, 6–8.
93. P. Colonna and A. Buleon, in *Starches: Characterization, Properties and Applications*, ed. A. C. Bertolini, CRC Press, Boca Raton, FL, 2010, pp. 71–102.
94. S. L. Randzio, I. Flis-Kabulska and J.-P. E. Grolier, *Macromolecules*, 2002, **35**, 8852–8859.
95. Z. Zhong and X. S. Sun, *J. Food Eng.*, 2005, **69**, 453–459.
96. T. A. Waigh, M. J. Gidley, B. U. Komanshek and A. M. Donald, *Carbohydr. Res.*, 2000, **328**, 165–176.
97. H. Angellier, Université Joseph Fourier, Grenoble, 2005.
98. R. F. Tester and W. R. Morrison, *Cereal Chem.*, 1992, **69**, 654–658.
99. E. D. Teixeira, A. A. S. Curvelo, A. C. Correa, J. M. Marconcini, G. M. Glenn and L. H. C. Mattoso, *Ind. Crops Prod.*, 2012, **37**, 61–68.
100. H. T. Liao and C. S. Wu, *Mater. Sci. Eng., A*, 2009, **515**, 207–228.

101. P. R. Chang, R. Jian, J. Yu and X. Ma, *Carbohydr. Polym.*, 2010, **80**, 420–425.
102. R. D. Hagenmaier and P. E. Shaw, *J. Agric. Food Chem.*, 1990, **38**, 1799–1803.
103. S. Sinha Ray and M. Okamoto, *Prog. Polym. Sci.*, 2003, **28**, 1539–1641.
104. B. Wetzel, F. Haupert and M. Q. Zhang, *Compos. Sci. Technol.*, 2003, **63**, 2055–2067.
105. G. M. Brown and F. Ellyin, *J. Appl. Polym. Sci.*, 2005, **98**, 869–879.
106. M. Abdelmouleh, S. Boufi, M. N. Belgacem, A. Dufresne and A. Gandini, *J. Appl. Polym. Sci.*, 2005, **98**, 974–984.
107. H. Angellier, S. Molina-Boisseau, L. Lebrun and A. Dufresne, *Macromolecules*, 2005, **38**, 3783–3792.
108. L. Avérous, C. Fringant and L. Moro, *Polymer*, 2001, **42**, 6565–6572.
109. D. Liu, T. Zhong, P. R. Chang, K. Li and Q. Wu, *Bioresour. Technol.*, 2010, **101**, 2529–2536.
110. D. B. Le Corre, J. Bras and A. Dufresne, presented at the FlexPakRenew Workshop, Lyon, 2011.
111. B. Duan, P. D. Sun, X. L. Wang and C. Yang, *Starch Starke*, 2011, **63**, 528–535.

CHAPTER 6

# *Chemical Modification of Thermoplastic Starch*

ANTONIO JOSÉ FELIX CARVALHO

Department of Materials Engineering, Engineering School of São Carlos, University of São Paulo, Av. Trabalhador São-Carlense 400, 13566-590 São Carlos, SP, Brazil
Email: toni@sc.usp.br

## 6.1 Introduction

Starch is a natural polymer that can be considered a homopolymer of anhydroglucose (α-D-glucopyranosyl residues) (molecular weight = 162 g $mol^{-1}$), formed by enzymatic condensation reactions. However, it is constituted by two main macromolecular components, the mostly linear amylose and branched amylopectin[1–3] in which the monomer residues are linked by α-(1→4) glycoside bonds in the linear portion and by α-(1→6) bonds in the branch points.[1] Starch is the main carbohydrate reserve in plants; in its native state, it has a granular structure, composed partially of crystalline regions, and it can be isolated from seeds, roots, tubers, leaves and fruits. The granules are insoluble in cold water and adopt the form of spheres, polygons or platelets, from 0.5 to 175 μm in size, depending on the source.[4]

Starch has found two main uses: in food as the most important source of carbohydrate (wheat, maize, rice, barley, oats, potato, cassava and others) and in non-food applications including materials such as glues and the production of liquid fuels, such as ethanol, by fermentation.[1,5,6] It is estimated that approximately 20–50% of the starch produced is used for these non-food applications, depending on the region in the world.

RSC Green Chemistry No. 37
Starch-based Blends, Composites and Nanocomposites
Edited by Visakh P. M. and Long Yu

Published by the Royal Society of Chemistry, www.rsc.org

Starch is an important strategic material as a alternative raw-material to satisfy the growing need for non-oil-derived materials, since it can be produced from several sources, in all kinds climates throughout the world, and it is one of the most efficient products of carbon fixation and energy converters in Nature. Another advantage is that starch can easily be separated in a very pure form from the plant source, owing to its granular structure.

Amylose, predominantly in the form of α-(1→4)-D-glucopyranosyl linear macromolecules, exhibits a degree of polymerization (DP) of 1000–6000, whereas amylopectin, the highly branched macromolecule, composed mainly of α-(1→4)-D-glucopyranose units with α-(1→6) linkages at intervals of approximately 20 units, has a very high molecular weight, with DP in the range 300 000–3 000 000.[7,8] Figure 6.1 shows the most frequently adopted schematic architecture of amylopectin.[9]

Native starch molecules form granules, which are semicrystalline, insoluble in cold water and resemble spherulites with alternating amorphous and crystalline (or semicrystalline) lamellae.[10] The granule shape and size

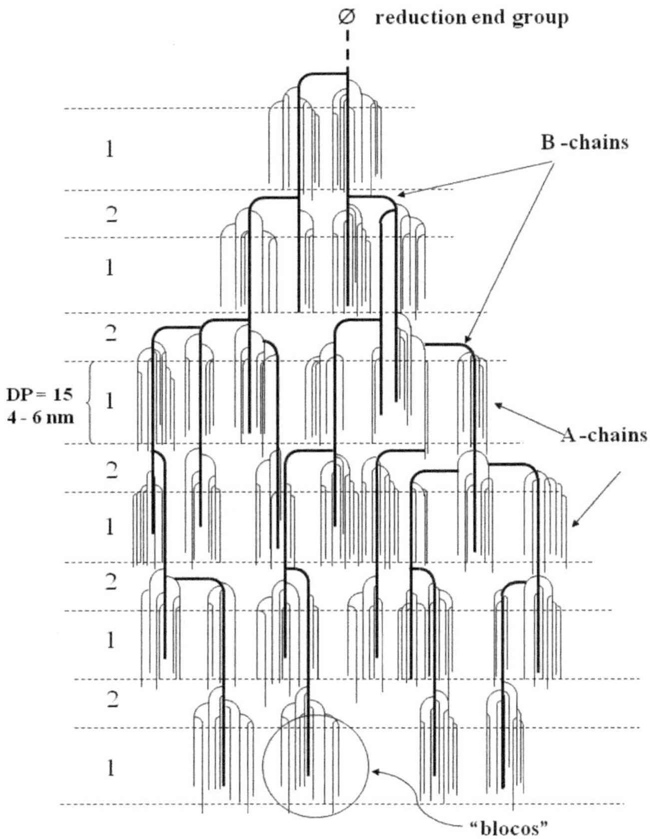


**Figure 6.1** The architecture of amylopectin according to Robin and Mercier.[9]

depend on the starch source and vary considerably.[1,5] Starch granules are not composed of pure amylose and amylopectin, but contain minor components such as lipids, proteins and inorganic compounds.

An interesting observation is that amylopectin, which, in general, comprises around 75% of the granule, is responsible for most of the crystallinity.

Distinctive crystalline structures that depend on the source and can easily be identified by X-ray diffraction are observed in different plants. Three polymorphisms have been described, denoted A, B and C,[3,11] the most commonly observed structures being A and B; A is associated mainly with cereal starches whereas B generally dominates in tuber starches, but also occurs in maize starches with more than 30–40% amylose.[3] An intermediate structure between A and B, denoted C, is usually associated with bean and root starches.[11,12]

Amylopectin is a very large molecule which can be modeled as a hyperbranched structure,[2,3,13,14] as proposed by Robin and Mercier[9] and illustrated in Figure 6.1, being considered a clear-cut example of a natural dendrimer[13] with a high degree of branching and a spherical shape. The A-chains, which are short segments of 15 D-glucopyranosyl residues, are the portion responsible for the crystalline structure of the compact areas of amylopectin; made up of A-chains (DP $\sim$ 15) that are the origin of starch crystallites.

## 6.2 Starch as an Industrial Raw Material

Starch has been used extensively in several non-food applications, in its native state or modified by physical and/or chemical processes. The main industrial applications are in adhesives,[5] paper,[5,6] production of chemicals such as organic acids and amino acids,[15] ethanol as a fuel,[15] drilling fluids for petroleum extraction, textile sizing,[6] as a binder in medicines and in hot melt adhesive compositions.[14,16–18] One relatively new area of application of starch is in polymer and rubber compositions[19] and foams[20] and as a raw material for the production of chemicals for the synthesis of other polymers. The productions of chemicals from starch has been reviewed by Robertson *et al.*,[21] Koutinas *et al.*,[22] Kennedy *et al.*[23] and Otey and Doane.[24]

A summary of the main starch uses is given below:

- Conversion by a fermentative process into ethanol for fuels.
- Production of glues and sizing agent for textiles.
- Conversion into chemicals such as ethanol, acetone, organic acids and ethane, used in the production of chemicals and synthetic polymers.
- Filler in other natural and synthetic polymers.
- Production of biopolymers by fermentative processes.
- Hydrolysis and glycolysis to give compounds of low molar weight and a high number of hydroxyls, used, for example, to produce polyurethanes.

- Destructuration and plasticizing to give a thermoplastic-like material named thermoplastic starch (TPS).
- Chemical modification or grafting with a variety of reagents to produce new polymers, used as such or as fillers for other polymers.

It is interesting that starch is more readily broken down into small saccharide molecules than, for example, cellulose, making it a better option for the production of monomers or as a raw material in a wide range of applications. Another advantage over cellulose and other natural polymers such as chitin and chitosan is the possibility of converting starch to thermoplastic materials by physical processing in the presence of process plasticizers, such as water, or "real" plasticizers, such as glycerol and urea.[25,26]

Regarding the use of starch in polymeric materials, three main approaches are followed:

1. Starch is converted into monomers used in the synthesis of polymers, such as poly(lactic acid) (PLA) from lactic acid and polyethylene from ethylene prepared by ethanol dehydration or it is directly converted into new polymers, such as polyhydroxyalkanoates (PHAs), by fermentative processes.
2. As a raw material for the production low molecular weight hydroxylated compounds, such as dextrins and glycolized products, both used in, for example, polyurethane compositions to give sources of glycols or heavily modified by grafting to give new polymers.
3. As native starch, used as a filler in other plastics or as thermoplastic starch, thus retaining its macromolecular nature unmodified.

In the first approach (1), the starch structure is completely changed, leading to polymers that are completely different from the starting material, starch being a raw material from which monomers are produced. The advantage of this approach is the possibility of producing a wide range of structures, but it has the disadvantage of high cost. This kind of process, which can be considered biorefining, is more viable for low-cost materials such as agricultural waste, for example, sugar cane bagasse.

The second approach (2) involves radical modification of the starch, while retaining its basic monomeric structure. The yield is almost 100% but the costs are still high, limiting the range of applications of the material produced.

The third approach (3) is the most economical, since starch is used as granules or modified only by continuous physical processes such as extrusion. Recently, new materials have been derived from starch by this kind of processing, since it is not only much more economical but also simple, needing only the same equipment as used in the thermoplastic transformation industry.

A new approach emerged recently that can be considered intermediate between those in which starch is used as a raw material for the preparation of new monomers without any chemical modification, the most relevant being the reactive extrusion (REX) of TPS. This process combines the advantages of approaches (1) and (2), since it can lead to significant chemical modification of starch, while using a continuous process with high yield, with low energy consumption and with no waste or by-products that need further treatment.[26]

## 6.3 Destructuration of Starch by Extrusion: Thermoplastic Starch

The starch gelatinization and destructuration process can be divided into two stages: diffusion of "solvent" (moisture or a plasticizer) into the granules, followed by melting of the starch crystallites. The second step, gelatinization, involves loss of molecular order and loss of birefringence and crystallinity, which can be observed by X-ray diffraction. Water is the commonest solvent used for starch gelatinization, but other liquids may be used, such as liquid ammonia, dimethyl sulfoxide, organic acids, sugars and alcohols, particularly polyhydric alcohols such as glycerol and glycols. These liquids are able to break the intermolecular interactions, mainly due to hydrogen bonding, between the starch molecules,[4] leading to a loss of crystallinity. From the thermodynamic point of view, the process of gelatinization can be described as the melting of the crystalline regions of starch granules. The gelation occurs above a characteristic temperature called the gelatinization temperature; however, the process of gelatinization or melting occurs in different ways, depending on the amount of solvent present during heating.[3]

The order–disorder transition that occurs during gelatinization or destructuration has been studied by several techniques, including X-ray diffraction, light scattering, optical microscopy (crossed polarizers using birefringence), thermomechanical analysis, nuclear magnetic resonance (NMR) spectroscopy and, subsequently, small-angle X-ray scattering (SAXS), wide-angle X-ray scattering (WAXS) and small-angle neutron scattering (SANS).[11,27,28] From the results obtained by such techniques, Jenkins and Donald[27] concluded that total loss of crystallinity occurs as gelation nears completion, thus establishing a direct relationship between the phenomena of loss of order and gelatinization.[27]

Thermal analysis methods, such as differential thermal analysis (DTA) and differential scanning calorimetry (DSC),[10,27] have been used to investigate the gelatinization of starch and one of the pioneering studies of this kind was reported by Donovan.[10] In this study, DSC was used to analyze aqueous suspensions of potato starch while varying the volume fraction of water from $v = 0.81$ to 0.28. It was demonstrated that two endothermic transitions occur, at lower and higher temperatures. At volume fractions of water below $\sim 0.38$, only one peak was observed. As the water content ($v > 0.7$), the single peak observed at low water contents split into two endothermic transitions.[10] Whereas the low-temperature transition occurred at a fixed temperature, irrespective of water content, the temperature of the higher-temperature peak depended on water concentration. The low-temperature peak is attributed to starch gelatinization (66 °C for potatoes).[28] The enthalpy associated with this transition ranges between 3.4 and 5.3 cal $g^{-1}$. At volume fractions of water $\nu > 0.7$, a shoulder appears at higher temperatures and, when $\nu$ is small, this shoulder grows and transforms into peak at higher temperatures, while the gelatinization peak at 66 °C shrinks, disappearing at $\nu \approx 0.38$.

This result shows that two different processes occur during the gelatinization of a starch granule. The endothermic peak at higher temperatures can be modeled in thermodynamic terms, since the temperature at the peak depends on the volume fraction of the solvent and is independent of the heating rate, as observed by Donovan.[10] This behavior was treated quantitatively by employing the Flory relationship between the melting point of the crystalline phase and the amount of added water.[10,29] This relationship is given by eqn (6.1), which is rearranged into eqn (6.2):

$$\frac{1}{T_m} - \frac{1}{T_m^0} = \left(\frac{R}{\Delta H_u}\right)\left(\frac{V_u}{V_1}\right)(\nu_1 - \chi_1 \nu_1^2) \tag{6.1}$$

$$\frac{\frac{1}{T_m} - \frac{1}{T_m^0}}{\nu_1} = \left(\frac{R}{\Delta H_u}\right)\left(\frac{V_u}{V_1}\right)(1 - \chi_1 \nu_1) \tag{6.2}$$

where $R$ is the gas constant, $\Delta H_u$ is the enthalpy of fusion per repeat unit (molar mass = 162 g $mol^{-1}$), $V_u/V_1$ is the ratio between the molar volume of the repeat unit and that of diluent, $\nu_1$ is the volume fraction of the solvent, $\chi$ is the Flory–Huggins parameter of the polymer–solvent interaction and $T_m^0$ is the melting point in the absence of diluent. $T_m^0$ is taken as the melting point of the most perfect crystallites. $T_m^0$ and $\Delta H_u$ were determined graphically from graphs of $1/T_m$ *versus* $\nu_1$, where $\chi = 0$ [eqn (6.1)] and against $\nu_1$ [eqn (6.2)], respectively.[10] The values of $\Delta H_u$ and $T_m^0$ determined for various starches were approximately 14 kcal $mol^{-1}$ per anhydroglucose residue and 168–197 °C, respectively.

Since the process of destructuration of starch for TPS production is conducted in the presence of limited quantities of plasticizer, it is governed by a high-temperature process like that for synthetic semicrystalline polymers, this being a melting process as proposed by Liu *et al.*[30] The melt mass obtained can recrystallize; however, the crystals are distinct from those of the starting native starch.

Perry and Donald concluded that, regardless of the type of plasticizer/solvent used in the process of gelatinization or destructuration of the starch, the final material shows the same structure, only the gelatinization/melting temperature varying with the solvent.[31]

In an extrusion process, where shear forces and high pressures are involved, the overall process is certainly much more complex. However, in either process where the starch is destructured or gelatinized in excess or limited quantities of water, the final product is an amorphous mass of amylose and amylopectin macromolecules that can partly recrystallize under storage conditions depending on its glass transition temperature.

## 6.4 Chemical Modification of Starch by Reactive Extrusion

When starch is processed in the molten state (TPS), its chemical modification has been achieved by reactive extrusion and there is an increasing number of publications in this field. The reactive extrusion of starch has been used for TPS compatibilization in polymer blending or for starch modification in order to produce new materials with new properties.

In reactive extrusion (REX), chemical reactions occur during the processing of the melt. A reagent or other desired material can be added at various stages of the process and simultaneously it is possible to remove reaction by-products at several points along the screw, where zones of fusion, transport, mixing and low pressure or degassing are found. Reactive extrusion is mainly used for polymer modification, but complete polymerization processes during extrusion, starting directly from monomers or from pre-polymerized low molar mass polymers or oligomers, are also used. The extruder operates as a continuous-flow reactor, which gives it a great advantage in that it allows particularly high-viscosity processing, without the use of solvents.[32,33] In summary, reactive extrusion can be used to conduct various types of reaction, such as polymerization of monomers or oligomers, free radical polymerization, polycondensation and coordination polymerization, degradation or depolymerization, controlled crosslinking, functionalization of conventional polymers and modification of polymers by grafting and other methods. Figure 6.2 shows in outline a basic system for reactive extrusion.

Although the process of extrusion of starch is often considered the precursor of the polymer extrusion technique and has been widely used since

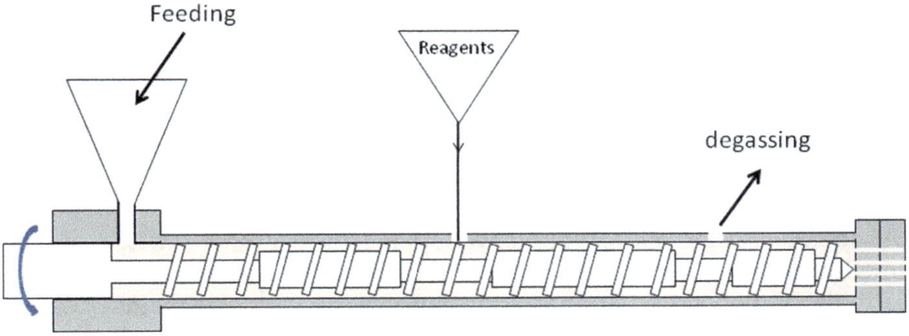


**Figure 6.2** Reactive extrusion system (REX).

the 1970s for the modification of native starch, the chemical modification of TPS by REX appeared only more recently and has been reviewed over the last decade by Xie *et al.*,[34] Kalambur and Rizvi,[35] Raquez *et al.*,[36] Liu *et al.*[30] and, more recently, Moad.[37]

Reactive extrusion of TPS has been used for compatibilization of TPS blends with other synthetic polymers, such as PLA, polyethylene, PHAs and aliphatic polyesters such as poly(butylene succinate) (PBS). The compatibilization of blends by reactive extrusion can be performed in a single step, in which polymers and coupling agents are processed together, or may be performed in two extrusion steps. In the latter case, one or both polymers to be mixed are modified by extraction separately and then mixed together in a second extrusion step. This process can also be carried out simultaneously through two connected extruders, where one of the polymers is modified and receives the second one. In following sections, the main reactions involved in the reactive compatibilization of TPS are described.

### 6.4.1 Grafting with Maleic Anhydride (Starch-*g*-MA)

This is one of the most traditional and relevant techniques used to compatibilize non-polar polymers, such as polyolefins [low- and high-density polyethylene (LDPE, HDPE), polypropylene (PP)], and polar material, usually containing hydroxyl reactive groups, such as starch. Maleic anhydride (MA) can be introduced into the structure of the non-polar polymer by copolymerization or by grafting *via* reactive extrusion. The reactions involved in the process are outlined in simplified form in Figure 6.3 for polyethylene, as an example.

Other non-polar polymers, such as poly(ε-caprolactone) (PCL), PBS, poly(butylene adipate-*co*-terephthalate) (PBAT) and PLA,[37] can also be modified by this process. Huneault and Li described the compatibilization of TPS–PLA blends by grafting MA onto PLA successfully.[38]

Wang *et al.*[39] described the compatibilization of linear low-density polyethylene (LLDPE) and TPS plasticized with 30% glycerol, where all components (native starch, glycerol, polyethylene, MA and peroxides) were processed on a single screw extruder. It was demonstrated that the grafting of polyethylene and compatibility with starch occurred. A similar process was also undertaken for TPS and PLA by Wang *et al.*, employing dicumyl peroxide (DCP)–MA, also in a single-screw extruder.[40]

### 6.4.2 Diisocyanates

Diisocyanates, such as 4,4′-diphenylmethane diisocyanate (MDI) and hexamethylene diisocyanate (HDI), have been employed for the reactive compatibilization of starch with polyesters such as PLA.[41–45] The main drawback of this process is the reactivity of isocyanates with the water present in starch[37,46] or with the plasticizer,[45] decreasing its efficiency. For polyesters

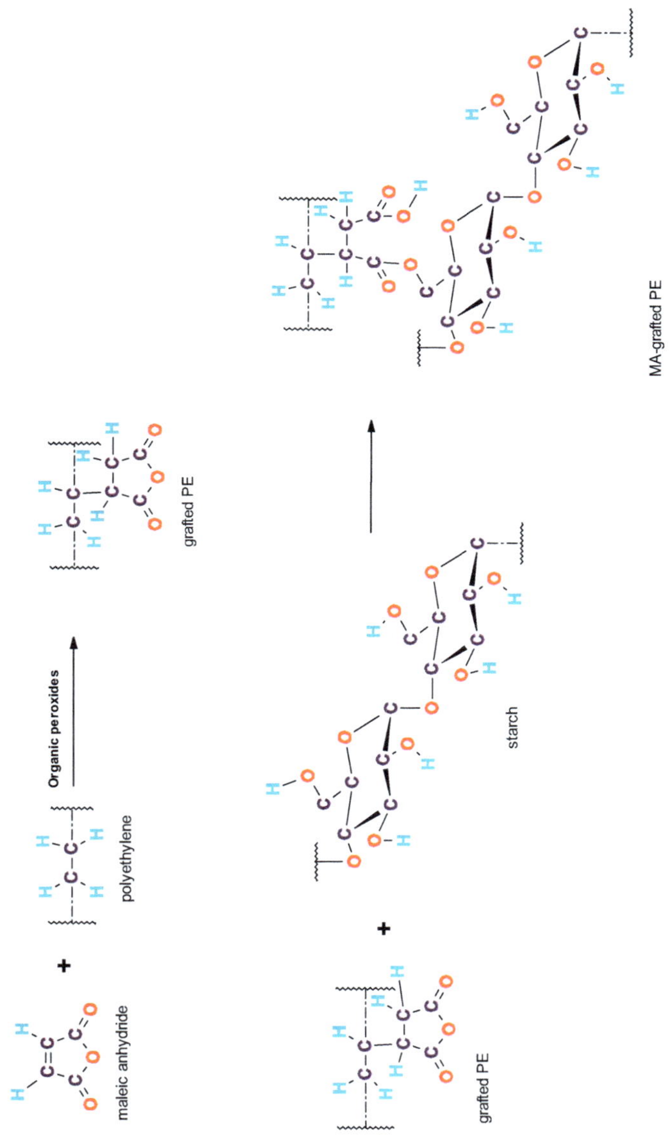


**Figure 6.3** Grafting of maleic anhydride (MA) onto polyethylene (PE) and coupling with starch.

such as PLA, isocyanate reacts with terminal hydroxyl or acid groups, giving rise to urethane or amide links, the former being the most probable.

### 6.4.3 Transesterification

Processing TPS in the presence of polyesters, such as PLA, PBAT[36] and PCL, with organic acids or peroxides and MA can lead to covalent bridges between the starch chains and the reactive polyester, thus promoting the compatibilization of the blend.[37]

### 6.4.4 Esterification

Starch can be esterified by processing TPS with acid anhydrides and catalysts. The main anhydrides used are acetic, propionic, maleic and succinic anhydride.[47] Other reagents, such as formic acid,[48,49] acid halides and vinyl acetate, have been reported.[37] Other reactions may occur during the esterification process, such as hydrolysis and glycosylation with glycerol.[36]

### 6.4.5 Epoxidation

Hydroxyalkylation reactions with epoxides under basic conditions (NaOH) have been used successfully for the epoxidation of starch. Propylene oxide is the most commonly used reagent in the preparation of hydroxypropyl starch.[50,51] Reaction with 3-chloro-2-hydroxypropyltrimethylamine chloride using REX, in the presence of 15 wt% glycerol, has been used to prepare cationic starch.[52]

### 6.4.6 Reaction with Vinyl Monomers

Reactions with vinyl acetate,[53] styrene[54] and acrylamide[55,56] have been described in processes for the grafting of starch in the molten state.

### 6.4.7 Hydrolysis and Glycolysis

Reduction of the molar mass of starch by glycolysis, catalyzed by inorganic or organic acids, has been used to prepare modified TPS with lower melt viscosity.[57] Citric acid was used to improve the compatibility of TPS and other polymers, including LDPE.[58–60] The compatibilization observed with citric acid has been attributed to a reduction of the interfacial tension between TPS and PE and also to the esterification of starch, but recent investigations by our group showed that esterification did not occur under the conditions in which TPS was processed, at least not to a significant extent. In view of the importance that the depolymerization of starch during TPS preparation or processing has gained recently, it is covered in more detail in the following section.

## 6.5 Depolymerization of Starch by Reactive Extrusion of TPS

Depolymerization of starch normally occurs during TPS preparation, owing to the mechanical work, and is influenced by the concentration of plasticizers and the presence of other materials such as fillers, which influence the melt viscosity. The reduction in the molecular weight (MW) determined by gel permeation chromatography by Carvalho *et al.*[19] was more marked in the fractions of higher MW, corresponding to amylopectin. Despite the fact that in general a fall in the MW of a polymer during its processing is undesirable, in the case of amylopectin, which displays extremely high values of MW, between $10^7$ and $10^9$ g $mol^{-1}$,[61] and a globular shape, a reduction in MW may be advantageous. Such a change can produce materials with better mechanical properties and better processability and contributes to a reduction in the melt viscosity.

When the molar mass of starch has to be reduced for industrial purposes, for example, in the production of dextrins, starch can be treated with enzymes[62] or starch powder can be treated by heating it with the addition of small amounts of inorganic acids[63] and treated with acids in aqueous suspension.[64]

van Soest *et al.*[65] described a process to decrease the MW of potato starch by hydrolysis of a suspension in HCl. They studied the effect of acid concentration, time and temperature on the final molar mass of the starch. The modified starch was isolated by washing and drying at low temperatures to avoid gelatinization. Although the process used was laborious, the authors obtained interesting results and demonstrated that the mechanical properties of TPS can be improved when the MW is reduced.

Citric acid has been used to modify starch by preparing starch esters (citrates), with the purpose of, for example, changing its non-exchange properties and, more recently, producing resistant starch as a food ingredient.[66] Starch citrates are formed in the absence of water at temperatures that cause the dehydration of citric acid to anhydrides, which react with starch. The subsequent esterification reactions can lead to the cross-linking of starch.[67] The reactions are outlined in Figure 6.4.

## 6.6 Starch Modification with Citric Acid in the Melt State: Reactive Extrusion of TPS

The effect of organic acids on starch in the melt state was investigated by Carvalho *et al.*[57] TPS was prepared in the presence of 30% glycerol and several concentrations of citric acid and ascorbic acid with various processing times, in an internal mixer. Yu *et al.*[68] studied the mechanical and rheological properties of a similar system, also using citric acid. The main effect of using organic acids during TPS processing was a sharp fall in molar mass, which was estimated by Carvalho *et al.*[69] by high-performance

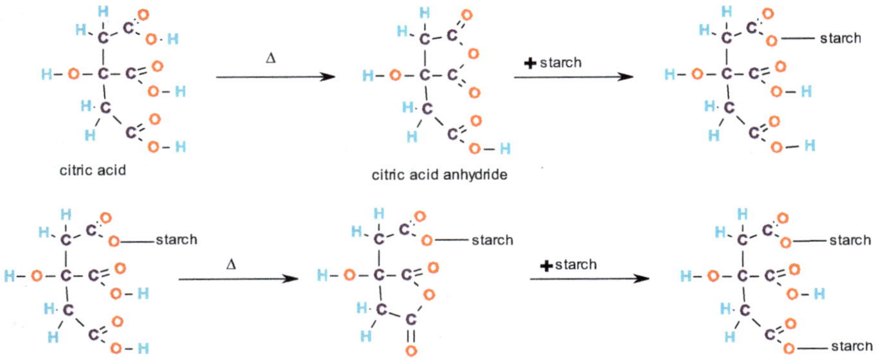


**Figure 6.4** Esterification of starch. Adapted from Wing.[67]

size-exclusion chromatography (HPSEC). Citric acid proved more effective than ascorbic acid, with the additional advantage of its low cost. Citric acid showed high efficiency in the starch chain scission because it is stronger than ascorbic acid. In that study, no signs of starch esterification were observed, leading to the conclusion that citric acid acted only as a hydrolytic scission agent.[57]

A system composed of TPS and citric acid was also studied by Yu *et al.*,[68] who concluded that the addition of citric acid caused a reduction in the extent of recrystallization of starch and a lower melt viscosity. It also induced a decrease in tensile strength and an increase in elongation at break. On the other hand, van Soest *et al.* observed an increase in tensile strength and elastic modulus with molar mass reduction.[65] An excess of citric acid also acts act as a plasticizer, which could explain this divergence.

Other studies in which citric acid was used for TPS modification include solution blending of starch and poly(vinyl alcohol) (PVA) and glutaraldehyde,[70] extrusion blending[71,72] and as a compatibilizer in blends of TPS with biodegradable polymers such as PLA[73] and PBAT.

Blends with LLDPE have been described by Wang *et al.*[60] In these studies, it was shown that citric acid can indeed improve the compatibility of the system. A recent study by our group also investigated blends of starch modified with citric acid and polyethylene in a two-stage process;[58] it was observed that the citric acid, in concentrations of about 1–3%, can significantly reduce the interfacial tension between phases and improve the compatibility between the polyethylene and the TPS. There was also a significant reduction in the absorption of water by the blends: the moisture content decreased from 9.5 to 3.3% of the TPS in the blends.

Citric acid has also been used in nanocomposites of TPS with clays such as montmorillonite (MMT), for the purpose of modifying MMT to increase its rate of exfoliation[74–76] and to alter the properties of TPS so has increase the wettability of the clay. Blends have been prepared both by solution blending ("casting")[77] and by melt processing.[78]

Shi *et al.*[79] studied a system with concentrations of citric acid higher than those used previously for TPS, in the order of 10–40% by mass with respect to the TPS. They reported that esterification of starch took place, and also a reduction in molar mass; however, the reduction in MW observed was significantly less than in our study,[57] which was attributed by the authors to the lower temperature processing of 130 °C compared with 160 °C.

Recently, Da Róz *et al.*[69] studied the effect of varying the molar mass of starch on the thermal and mechanical properties by DSC.[69] This study revealed that the melting temperature varies depending on the molar mass with a linear dependence of $1/T$ (K) with $1/M_n$ or $1/M_w$ according to the Flory equation:[80]

$$\frac{1}{T_m} - \frac{1}{T_m^0} = \frac{R}{\Delta H_f}\frac{2M_0}{M_n} \tag{6.3}$$

where $T_m$ is the average melting temperature, $T_m^0$ is the melting temperature of the polymer in its pure state equivalent to a polymer sample of infinite molar mass, $\Delta H_f$ is the heat of fusion per repeat unit, $R$ is the gas constant and $2M_0/M_n$ is the fraction of the end chains, $M_0$ being the molar mass of the repeat units. A similar equation, eqn (6.1), can be used to describe the decrease in the melt temperature due to the dilution by a solvent or plasticizer, such as glycerol for TPS.[81,82] However, in the case of TPS, as the glycerol content was kept constant at 30% w/w, its effect can be neglected. The value of $\Delta H_f$ observed was almost unchanged with variation in molar mass and eqn (6.3) is applicable to the studied system. The value of $T_m^0$ estimated from the plot of $1/T_m$ *versus* $1/M_n$ was 465 K (192 °C), this being the melting point for the starch with infinite molar mass in the presence of 30% glycerol as plasticizer.

## 6.7 New Challenges and Opportunities

The future of REX of TPS is its use in an intensive form so that starch can be radically modified, generating new materials that can be tuned to a great assortment of uses. REX of TPS has proved to be a green process in that native starch is used without previous modification and the reactions take place at the same time as TPS is produced. The type and degree of modification can be tuned by adjusting not only the process parameters but also the type of reactants used and their concentrations. The greatest challenge in these processes is control of the reaction, since the reactive groups in starch are primary and secondary hydroxyl groups, which are also present in the majority of the plasticizers used, such as glycerol. In addition, water, which is present in starch at a considerable level (around 10–12 wt%) in general reacts with the same reactants that are supposed to react with the starch hydroxyl groups. The use of inhibitors and precise control of the process will be the next frontier to be explored.

## Acknowledgements

The author thanks the State of São Paulo Research Foundation (FAPESP) and the National Council for Science and Technological Development (CNPq) for the financial support provided for the research carried out in this study.

## References

1. R. L. Whistler, Molecular structure of starch, in *Starch: Chemistry and Technology*, ed. R. L. Whistler, J. N. Bemiller and E. F. Paschall, Academic Press, San Diego, CA, 2nd edn, 1984, ch. 6, pp. 153–182.
2. D. French, Organization of starch granules, in *Starch: Chemistry and Technology*, ed. R. L. Whistler, J. N. Bemiller and E. F. Paschall, Academic Press, San Diego, CA, 2nd edn, 1984, ch. 7, pp. 183–247.
3. J. M. V. Blanshard, Starch granule structure and function: a physicochemical approach, in *Starch: Properties and Potential*, ed. T. Galliard, John Wiley & Sons, New York, 1st edn, 1987, ch. 2, pp. 14–54.
4. H. F. Zobel, Starch crystal transformation and their industrial importance, *Starch Stärke*, 1988, **40**, 1–7.
5. W. Jarowenko, Starch based adhesives, in *Handbook of Adhesives*, ed. I. Skeist, Van Nostrand Reinhold, New York, 2nd edn, 1977, ch. 12, pp. 192–211.
6. J. Daniel, R. L. Whistler, A. C. J. Voragen and W. Pilnik, Starch and other polysaccharides, in *Ullmann's Encyclopedia of Industrial Chemistry*, ed. B. Elvers, S. Hawkins and W. Russey, VCH, Weinheim, 5th edn, 1994, vol. A25, pp. 1–62.
7. H. F. Zobel, Molecules to granules: a comprehensive starch review, *Starch Starke*, 1988, **40**, 44–50.
8. A. H. Young, Fractionation of starch, in *Starch: Chemistry and Technology*, ed. R. L. Whistler, J. N. Bemiller and E. F. Paschall, Academic Press, San Diego, CA, 2nd edn, 1984, ch. 8, pp. 249–283.
9. J. P. Robin, C. Mercier, R. Charbonniere and A. Guilbot, Lintnerized starches. Gel filtration and enzymatic studies of soluble residues from prolonged acid treatment of potato starch, *Cereal Chem.*, 1974, **51**, 389–406.
10. J. W. Donovan, Phase transitions of the starch-water system, *Biopolymers*, 1979, **18**, 263–275.
11. P. Colonna, A. Buléon and C. Mercier, Physically modified starches, in *Starch: Properties and Potential*, ed. T. Galliard, New York, John Wiley & Sons, 1st edn, 1987, ch. 4, pp. 79–115.
12. A. Buléon, C. Gérard, C. Riekel, R. Vuong and H. Chanzy, Details of the crystalline ultrastructure of C-starch granules revealed by synchrotron microfocus mapping, *Macromolecules*, 1998, **31**, 6605–6610.
13. N. K. Matheson and R. A. Caldwell, $\alpha$(1–4) Glucan chain disposition in models of $\alpha$(1–4)(1–6) glucans: comparison with structural data for

mammalian glycogen and waxy amylopectin, *Carbohydr. Polym.*, 1999, **40**, 191–209.
14. Z. Nikuni, Studies on starch granules, *Starch Stärke*, 1978, **30**, 105–111.
15. D. Schwartz and R. L. Whistler, *History and Future of Starch in Starch Chemistry and Technology*, ed. J. BeMiler and R. Whistler, Academic Press, Burlington, MA, USA, Elsevier, 3rd edn, 2009, pp. 1–10.
16. R. I. Billmers, B. W. Asplund and D. P. Huang, Edible Hot Melt Composition, *U. S. Pat.* 6,846,502 B1, assigned to National Starch & Chemical Investment Holding Corporation, January 25, 2005.
17. D. Neigel, G. A. Sweey, C. W. Paul, R. L. Billmers and D. C. Rawlins, Process for production of starch based hot melt adhesives, *U. S. Pat.* 5,434,201, assigned to National Starch and Chemical Investment Holding Corporation, July 18, 1995.
18. T. F. Kauffman, J. Wieczorek Jr and S. F. Hattield, Starch based hot melt adhesives for cigarettes, *U. S. Pat.* 5,498,224, assigned to National Starch and Chemical Investment Holding Corporation, March 12, 1996.
19. A. J. F. Carvalho, A. E. Job, N. Alves, A. A. S. Curvelo and A. Gandini, Thermoplastic starch/natural rubber blends, *Carbohydr. Polym.*, 2003, **53**, 95–99.
20. J. L. Willett and R. L. Shogren, Processing and properties of extruded starch/polymer foams, *Polymer*, 2002, **43**, 5935–5947.
21. G. H. Robertson, D. W. S. Wong, C. C. Lee, K. W. Wagschal, M. R. Smith and W. J. Orts, Native or raw starch digestion: a key step in energy efficient biorefining of grain, *J. Agric. Food Chem.*, 2006, **54**, 353–365.
22. A. A. Koutinas, R. Wang and C. Webb, Evaluation of wheat as generic feedstock for chemical production, *Ind. Crops Prod.*, 2004, **20**, 75–88.
23. J. F. Kennedy, J. M. S. Cabral, I. Sá-Correia, C. A. White, Starch biomass: a chemical feedstock for enzyme and fermentation processes, in *Starch: Properties and Potential,* ed. T. Galliard, John Wiley & Sons, New York, 1st edn, 1987, ch. 5, pp. 115–148.
24. F. H. Otey, W. M. Doane, Chemicals from starch, in *Starch: Chemistry and Technology*, ed. R. L. Whistler, J. N. Bemiller and E. F. Paschall, Academic Press, San Diego, CA, 2nd edn, 1984, ch. 11, pp. 389–416.
25. A. J. F. Carvalho, Starch: major sources, properties and applications as thermoplastic materials, in *Monomers, Polymers and Composites from Renewable Resources*, ed. A. Gandini and N. M. Belgacem, Elsevier, Oxford, 1st edn, 2008, vol. 15, pp. 321–342.
26. A. J. F. Carvalho, Starch as source of polymeric materials, in *Biopolymers: Biomedical and Environmental Applications*, ed. S. Kalia and L. Avérous, Wiley, Hoboken, NJ, and Scrivener Publishing, Salem, MA, 2011, pp. 81–98.
27. P. J. Jenkins and A. M. Donald, Gelatinization of starch: a combined SAXS/WAXS/DSC and SANS study, *Carbohydr. Res.*, 1998, **308**, 133–147.
28. D. J. Stevens and G. A. H. Elton, Thermal Properties of the starch/water system. Part I. Measurement of heat of gelatinization by differential scanning calorimetry, *Starch Stärke*, 1971, **23**, 8–11.

29. J. L. Kokini, L. S. Lai and L. L. Chedid, Effects of starch structure on starch rheological properties, *Food Technol.*, 1992, **46**, 124–139.
30. H. Liu, F. Xie, L. Yu, L. Chen and L. Li, Thermal processing of starch-based polymers, *Prog. Polym. Sci.*, 2009, **34**, 1348–1368.
31. P. A. Perry and A. M. Donald, The role of plasticization in starch granule assembly, *Biomacromolecules*, 2000, **1**, 424–432.
32. P. Cassagnau, V. Bounor-Legare and F. Fenouillot, Reactive processing of thermoplastic polymers: a review of the fundamental aspects, *Int. Polym. Process.*, 2007, **22**, 218–258.
33. C. Tzoganakis, Reactive extrusion of polymers: a review, *Adv. Polym. Technol.*, 1989, **9**, 321–330.
34. F. Xie, L. Yu, H. Liu and L. Chen, Starch modification using reactive extrusion, *Starch Stärke*, 2006, **58**, 131–139.
35. S. Kalambur and S. S. H. Rizvi, An overview of starch-based plastic blends from reactive extrusion, *J. Plast. Film Sheeting*, 2006, **22**, 39–58.
36. J. M. Raquez, R. Narayan and P. Dubois, Recent advances in reactive extrusion processing of biodegradable polymer-based compositions, *Macromol. Mater. Eng.*, 2008, **293**, 447–470.
37. G. Moad, Chemical modification of starch by reactive extrusion, *Prog. Polym. Sci.*, 2011, **36**, 218–237.
38. M. A. Huneault and H. Li, Morphology and properties of compatibilized polylactide/thermoplastic starch blends, *Polymer*, 2007, **48**, 270–280.
39. S. Wang, J. Yu and J. Yu, Preparation and characterization of compatible thermoplastic starch/polyethylene blends, *Polym. Degrad. Stab.*, 2005, **87**, 395–401.
40. N. Wang, J. G. Yu and F. Ma, Preparation and characterization of thermoplastic starch/PLA blends by one-step reactive extrusion, *Polym. Int.*, 2007, **56**, 1440–1447.
41. L. Yu, K. Dean, Q. Yuan, L. Cheng and X. Zhang, Effect of compatibilizer distribution on the blends of starch/biodegradable polyesters, *J. Appl. Polym. Sci.*, 2007, **103**, 812–818.
42. F. Xie, T. Xue, L. Yu, L. Chen, X. Li and X. Zhang, Rheological properties of starch-based materials and starch/poly(lactic acid) blends, *Macromol. Symp.*, 2007, **249**, 529–534.
43. H. Wang, X. Sun and P. Seib, Strengthening blends of poly(lactic acid) and starch with methylenediphenyl diisocyanate, *J. Appl. Polym. Sci.*, 2001, **82**, 1761–1767.
44. C. L. Jun, Reactive blending of biodegradable polymers: PLA and starch, *J. Polym. Environ.*, 2000, **8**, 33–37.
45. E. Schwach, J. L. Six and L. Avérous, Biodegradable blends based on starch and poly(lactic acid): comparison of different strategies and estimate of compatibilization, *J. Polym. Environ.*, 2008, **16**, 286–297.
46. H. Wang, X. Sun and P. Seib, Effects of starch moisture on properties of wheat starch/poly(lactic acid) blend containing methylenediphenyl diisocyanate, *J. Polym. Environ.*, 2002, **10**, 133–138.

47. V. D. Miladinov and M. A. Hanna, Starch esterification by reactive extrusion, *Ind. Crops Prod.*, 2000, **11**, 51–57.
48. T. Divers, I. Pillin, J. F. Feller, G. Levesque and Y. Grohens, Starch modification, destructuration and hydrolysis during O-formylation, *Starch Stärke*, 2004, **56**, 389–398.
49. F. Bossard, I. Pillin, T. Aubry and Y. Grohens, Rheological characterization of starch derivatives/polycaprolactone blends processed by reactive extrusion, *Polym. Eng. Sci.*, 2008, **48**, 1862–1870.
50. G. Lammers, E. J. Stamhuis and A. A. C. M. Beenackers, Continuous production of hydroxypropyl starch in a static mixer reactor, *Starch Stärke*, 1993, **45**, 227–232.
51. R. A. De Graaf and L. P. B. M. Janssen, The hydroxypropylation of starch in a self-wiping twin screw extruder, *Adv. Polym. Technol.*, 2003, **22**, 56–68.
52. R. Bhosale and R. Singhal, Process optimization for the synthesis of octenyl succinyl derivative of waxy corn and amaranth starches, *Carbohydr. Polym.*, 2006, **66**, 521–527.
53. R. A. De Graaf, A. Broekroelofs and L. P. B. M. Janssen, The acetylation of starch by reactive extrusion, *Starch Stärke*, 1998, **50**, 198–205.
54. R. A. De Graaf and L. P. B. M. Janssen, The production of a new partially biodegradable starch plastic by reactive extrusion, *Polym. Eng. Sci.*, 2000, **40**, 2086–2094.
55. J. L. Willett and V. L. Finkenstadt, Reactive extrusion of starch–polyacrylamide graft copolymers using various starches, *J. Polym. Environ.*, 2006, **14**, 125–129.
56. V. L. Finkenstadt and J. F. Willett, Reactive extrusion of starch–polyacrylamide graft copolymers: effects of monomer/starch ratio and moisture content, *Macromol. Chem. Phys.*, 2005, **206**, 1648–1652.
57. A. J. F. Carvalho, M. D. Zambon, A. A. S. Curvelo and A. Gandini, Thermoplastic starch modification during melting processing: hydrolysis catalyzed by carboxylic acids, *Carbohydr. Polym.*, 2005, **62**, 387–390.
58. V. R. Miranda and A. J. F. Carvalho, *Blendas de polietileno e amido termoplástico modificado Polímeros*, 2011, **21**, 347–352.
59. A. Taguet, M. A. Huneault and B. D. Favis, Interface/morphology relationships in polymer blends with thermoplastic starch, *Polymer*, 2009, **50**, 5733–5743.
60. N. Wang, J. Yu, X. Ma and Y. Wu, The influence of citric acid on the properties of thermoplastic starch/linear low-density polyethylene blends, *Carbohydr. Polym.*, 2007, **67**, 446–453.
61. J. Jane, Structure features of starch granules II, in *Starch: Chemistry and Technology*, ed. J. N. Bemiller and R. L. Whistler, Elsevier Academic Press, Burlington, MA, 3rd edn, 2009, ch. 6, pp. 193–236.
62. H. E. Grommers and D. A. Van Der Krogt, Potato starch: production, modification and uses, in *Starch: Chemistry and Technology*, ed. J. Bemiller and R. Whistler, Elsevier Academic Press, Burlington, MA, 3rd edn, 2009, vol. 11, pp. 510–539.

63. H. M. Kennedy and A. C. Fischer Jr, Starch and Dextrins in Prepared Adhesives, in *Starch: Chemistry and Technology*, ed. R. L. Whistler, J. N. Bemiller and E. F. Paschall, Academic Press,San Diego, CA, 2nd edn, 1984, ch. 20, pp. 593–610.
64. R. G. Rohwer and R. E. Klem, Acid-Modified Starch: Production and Uses, in *Starch: Chemistry and Technology*, ed. R. L. Whistler, J. N. Bemiller and E. F. Paschall, San Diego, CA, Academic Press, 2nd edn, 1984, ch. 17, pp. 529–541.
65. J. J. G. van Soest, K. Benes and D. de Wit, The influence of acid hydrolysis of potato starch on the stress-strain properties of thermoplastic starch, *Starch Stärke*, 1995, **47**, 429–434.
66. X. Xie and Q. Liu, Development and physicochemical characterization of new resistant citrate starch from different corn starches, *Starch Stärke*, 2004, **56**, 364–370.
67. R. E. Wing, Starch citrate: Preparation and ion exchange properties, *Starch Stärke*, 1996, **48**, 275–279.
68. J. Yu, N. Wang and X. Ma, The effects of citric acid on the properties of thermoplastic starch plasticized by glycerol, *Starch Stärke*, 2005, **57**, 494–504.
69. A. L. Da Róz, M. D. Zambon, A. A. S. Curvelo and A. J. F. Carvalho, *Ind. Crops Prod.*, 2011, **33**, 152–157.
70. S. D. Yoon, S. H. Chough and H. R. Park, Properties of starch-based blend films using citric acid as additive, *J. Appl. Polym. Sci.*, 2006, **100**, 2554–2560.
71. G.-X. Zou, J.-P. Qu and X.-L. Zou, Extruded starch/PVA composites: water resistance, thermal properties and morphology, *J. Elastomers Plast.*, 2008, **40**, 303–316.
72. R. Shi, J. Bi, Z. Zhang, A. Zhu, D. Chen, X. Zhou, L. Zhang and W. Tian, The effect of citric acid on the structural properties and cytotoxicity of the polyvinyl alcohol/starch films when molding at high temperature, *Carbohydr. Polym.*, 2008, **74**, 763–770.
73. N. Wang, J. Yu, P. R. Chang and X. Ma, Influence of citric acid on the properties of glycerol-plasticized dry starch (DTPS) and DTPS/poly(lactic acid) blends, *Starch Stärke*, 2007, **59**, 409–417.
74. M. F. Huang, J. G. Yu and X. F. Ma, High mechanical performance MMT-urea and formamide-plasticized thermoplastic cornstarch biodegradable nanocomposites, *Carbohydr. Polym.*, 2006, **63**, 393–399.
75. K. Majdzadeh-Ardakani, A. H. Navarchian and F. Sadeghi, Optimization of mechanical properties of thermoplastic starch/clay nanocomposites, *Carbohydr. Polym.*, 2010, **79**, 547–554.
76. K. Majdzadeh-Ardakani and B. Nazari, Improving the mechanical properties of thermoplastic starch/poly(vinyl alcohol)/clay nanocomposites, *Compos. Sci. Technol.*, 2010, **70**, 1557–1563.
77. H. Almasi, B. Ghanbarzadeh and A. A. Entezami, Physicochemical properties of starch–CMC–nanoclay biodegradable films, *Int. J. Biol. Macromol.*, 2010, **46**, 1–5.

78. N. Wang, X. Zhang, N. Han and S. Bai, Effect of citric acid and processing on the performance of thermoplastic starch/montmorillonite nanocomposites, *Carbohydr. Polym.*, 2009, **76**, 68–73.
79. R. Shi, Z. Zhang, Q. Liu, Y. Han, L. Zhang, D. Chen and W. Tian, Characterization of citric acid/glycerol co-plasticized thermoplastic starch prepared by melt blending, *Carbohydr. Polym.*, 2007, **69**, 748–755.
80. L. H. Sperling, *Introduction to Physical Polymer Science*, John Willey & Sons, Hoboken, NJ, 4th edn, 2006.
81. P. J. Flory, Thermodynamics of crystallization in high polymers. IV. A theory of crystalline states and fusion in polymers, copolymers and their mixtures with diluents, *J. Chem. Phys.*, 1949, **17**, 223–240.
82. P. J. Flory, *Principles of Polymer Chemistry*, Cornell University Press, Ithaca, NY, 1953.

CHAPTER 7

# Outstanding Features of Starch-based Hydrogel Nanocomposites

ANTONIO G. B. PEREIRA,*[a,b] ANDRÉ R. FAJARDO,[c] ARTUR J. M. VALENTE,[d] ADLEY F. RUBIRA[a] AND EDVANI C. MUNIZ[a,e]

[a] Department of Chemistry, State University of Maringa, Av. Colombo 5790, 87020-900 Maringá, Brazil; [b] Coordenação do Curso de Engenharia de Bioprocessos e Biotecnologia, Federal University of Technology - Paraná (UTFPR), Estrada para Boa Esperança, 85660-000, Dois Vizinhos, PR, Brazil; [c] Centro de Ciências Químicas, Farmacêuticas e de Alimentos, Federal University of Pelotas, 96010-900, Pelotas, Brazil; [d] Department of Chemistry, University of Coimbra, 3004-535 Coimbra, Portugal; [e] Programa de Pós-graduação em Biotecnologia Aplicada a Agricultura, Paranaense University (UNIPAR), 87502-210, Umuarama, Brazil
*Email: guilebasso@hotmail.com

## 7.1 Introduction: Starch Properties

Considered as the energetic reserve of plants, starch is the second most abundant carbohydrate in Nature after cellulose.[1] The potential of this versatile biopolymer has been explored extensively for many years, especially because of its very attractive properties. A brief review of the literature indicates that starch is a non-toxic, biocompatible, biodegradable and abundant biopolymer mainly composed of two homopolymers: amylose and

RSC Green Chemistry No. 37
Starch-based Blends, Composites and Nanocomposites
Edited by Visakh P. M. and Long Yu

Published by the Royal Society of Chemistry, www.rsc.org

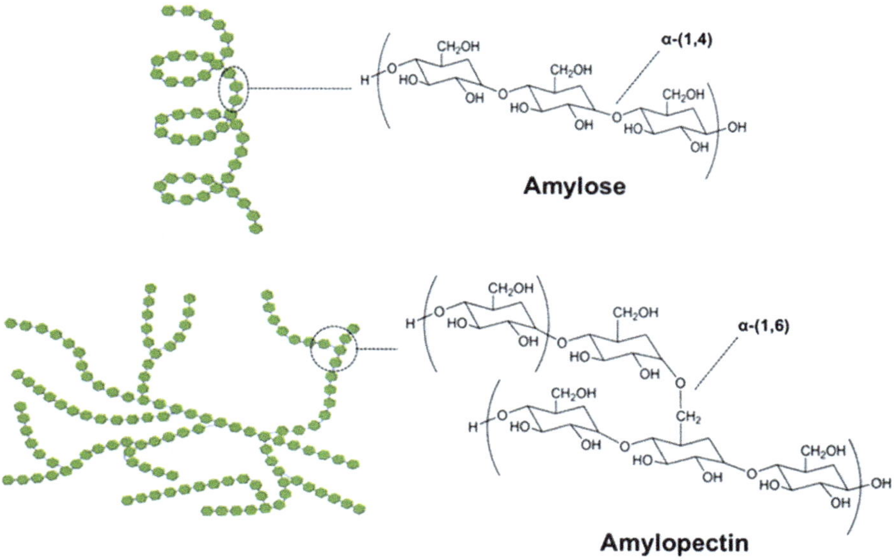


**Figure 7.1** Amylose (linear) and amylopectin (branched) structures.

amylopectin.[2,3] These homopolymers have the same repeating units that are linked in a linear and branched manner, respectively (Figure 7.1).[4]

Amylose is defined as a linear macromolecule of α-(1→4)-D-glucopyranosyl units where some part of the structure could be slightly branched by α-(1→6) linkages, but it is predominantly regarded as linear. Owing to its simple polymeric structure, amylose tends to self-assemble into crystals. In Nature, three crystalline forms of amylose, A, B and C, exist,[1] depending on the source: cereals (A), tubers (B) and certain pea and bean varieties (C). Twisting of the polymer chains into a helix configuration further complicates the so-called linearity of the amylose molecule. Different degrees of hydration of the helix are responsible for the A, B and C forms. In contrast, amylopectin is a highly branched component of starch and is formed mainly by D-glucopyranosyl units linked by α-(1→4) linkages; α-(1→6) linkages can be present at the branch points, corresponding to 5–6% of the linkages.[5] The highly branched structure confers a high molecular weight on amylopectin that is in general 1000 times greater than that of amylose. Indeed, amylopectin is a ''Titan'' natural molecule: one of the largest, with a molecular weight of up to 400 million.[1] The ratio of amylose to amylopectin in any native starch is dependent on the source. In addition to these two main components, starch granules can have other minor components present in its composition, such as some particulate material (*e.g.* cell-wall fragments) and surface and internal components (*e.g.* proteins, enzymes, lipids, amino and nucleic acids).[5,6]

Another important feature of starch is the abundance of hydroxyl groups in the backbone.[5] These groups have a great affinity for other hydroxyl groups (hydrogen bonding), which can act as a driving force to hold the starch chains together in a regular pattern. Where such ordering occurs, crystalline regions are deposited in the starch granules. The remaining regions of unordered starch are denoted amorphous. The crystalline regions are responsible for the granular structure of starch.[1] The size and size distribution of starch granules can be very important for specific applications. In general, the size of granules can vary from 5 to 50 μm depending on the starch source. Furthermore, the hydroxyl groups in the starch backbone form hydrogen bonds with water molecules. The interactions between starch chains and water molecules cause the chains to move apart and, as a consequence, the granules swell. Eventually the granule ruptures and starch polymers are dispersed in the liquid, producing a viscous colloidal state. This is a form of water management that controls structure and texture in food products, for example. "Gelatinization" and "pasting" are, respectively, technical descriptions of the hydration within the granule and the irreversible granule swelling that builds viscosity.[1,5] Beyond these features, the hydroxyl groups allow starch to be used in different applications once physical, chemical or biochemical modifications have been implemented to change the original starch design. Several modification procedures have contributed to the synthesis of numerous highly functional starch derivatives.[7] Such developments have expanded the application frontiers of this multifunctional polymer because now it is possible to tailor raw starch to act with greater specificity in different research fields or to use starch in new processing technologies.[8,9] In this regard, starch and modified starches have been extensively investigated as promising "green" candidates to replace the petroleum-based polymers that have dominated various fields for many years.[9,10] All these efforts contribute to the intention of minimizing the impact of chemicals on the environment and consequently help to alleviate the growing environmental concerns. The completely biodegradable character of starch, its low production cost and renewability[2,6] are other important aspects that encourage the development of sustainable materials from this promising candidate. In the light of this, several processes have been applied to modify starch (*e.g.* stabilization, acid/enzymatic hydrolysis, oxidation, lipophilic substitution, blending, copolymerization and crosslinking).[9–12] As a consequence, the hydrogen bonds are affected in a controllable manner to enhance and extend the applicability of starch or even to create new features. Among all these modification processes, crosslinking is the most important chemical modification. This process involves replacement of the hydrogen bonds between starch chains by stronger covalent bonds. In this manner, the swelling of the starch granule is inhibited, due to preempting disintegration either by chemical attack, mechanical attrition (shear) or cooking. In other words, the starch granule is, in the molecular dimension, "spot welded" at random locations to reinforce it.[1] Another well-established approach in this scenario is the association of starch (in the raw or modified state) with both synthetic and natural macromolecules to form crosslinked networks. In most cases, the crosslinked networks formed between starch and other macromolecules are arranged in a three-dimensional configuration that favors liquid absorption and retention. This feature could be ascribed to the hydrophilic character of starch and might be enhanced or not as a function of the nature of the macromolecule crosslinked with starch chains.

Many research groups have explored in depth the potential of starch-based hydrogels and their derivatives (such as starch-based hydrogel composites) in very varied technological and biological fields, and outstanding advances in starch-based hydrogels have been reported in recent decades. This chapter presents an up-to-date overview of the role of those remarkable hydrogels, with special focus on the development, characterization and application of starch-based hydrogel nanocomposite materials in the biological, engineering and pharmaceutical research fields.

## 7.2 Hydrogels

### 7.2.1 Starch-based Hydrogels

Hydrogels belong to a class of polymeric materials with hydrophilic features that are of great importance in many different areas. The current status achieved by hydrogels, now over 50 years since their first description,[13] can be credited to their interesting and desirable physico-chemical and biological properties. Such properties result from the arrangement of the three-dimensional network that forms the hydrogel structure. The most relevant characteristic of hydrogels is undoubtedly the ability of absorb and retain considerable amounts of aqueous liquids, which results from the hydrophilic crosslinked network that is able to retain the three-dimensional structure even in the swollen state without disintegrating.[14,15] This feature is due to the crosslinked points that hold together the polymeric chains that form the network. In most cases, chemical or physical pathways drive the crosslinking process during the gel formation. The chemical crosslinking process (the products are denoted chemical hydrogels here) conducts the formation of irreversible covalent bonds among the polymeric chains of the hydrogel.[16] On the other hand, in physically crosslinked hydrogels (denoted physical hydrogels here), the polymeric chains are held together by physical interactions (*i.e.* electrostatic interactions, van der Waals forces, hydrogen bonding and physical entanglements).[16] Owing to the reversible character, physical hydrogels can be disintegrated as a result of the impairment of the physical interactions responsible for their crosslinking. Changes in the external medium, such as pH, ionic strength and electric field, can disintegrate the physical hydrogels.

Despite the differences between the pathways that form chemical and physical hydrogels, both of them allow materials to be formed in a great number of formats (films, membranes, spheres, cylinders, *etc.*) and with macro-, micro- and nano-scale dimensions. In the last two decades, the application of hydrogels has no longer focused only on liquid absorption/retention. The potential of this very promising class of materials is now applied in very varied industrial, technological and biotechnological sectors.[17–20] This scenario is due to the formation of a new class of hydrogels – the "smart hydrogels", which are able to provide different responses (*e.g.* volume, porosity and mechanical changes) according to external and internal stimuli.[20,21] In additional to such an evolution, the use of different classes of polymers considerably enhanced their properties and applicability. Biopolymers (such as polysaccharides and proteins) possess unique and desirable physicochemical and biological properties (*i.e.* biocompatibility, biodegradability, non-toxicity and some biological activities) that stimulate their use in preparing different materials.[22,23] For example, hydrogels prepared from biopolymers (mainly polysaccharides) have found great applicability as biomaterials.[24–26] Furthermore, the interesting properties of polysaccharides comes from their structure, which, in general, has a large number of functional groups (–COOH, –OH, $-NH_2$, $-NHOCCH_3$ and $-OSO_3H$) that can be crosslinked by reaction with a coupling agent or that allow the insertion of crosslinkable groups or polymeric chains in the polysaccharide backbone.[27,28] The preparation of polysaccharide-based hydrogels could also be carried out by polyelectrolyte complexation among macromolecules with functional groups having opposite electric charges.[29,30] As discussed in the Introduction, the biopolymer starch shows all the desirable features required for preparing chemical and physical hydrogels. A number of reports and review papers have described the preparation of starch-based hydrogels using several methodologies.[31,32]

In general, the main strategy adopted to prepare starch-based chemical hydrogels is based on the reaction of the hydroxyl groups in the starch backbone with bi- or multifunctional compounds that work as coupling agents. To date, the compounds most commonly used are glutaraldehyde and epichlorohydrin.[33] Despite the production of interesting hydrogels by this strategy, the use of coupling agents is not encouraged when the aim is to prepare hydrogels for biomedical uses because the coupling agents might show some level of toxicity, which decreases the range of applicability of the chemical hydrogels obtained. For this reason, for specific applications such a methodology is far from preferred or ideal.

An efficient strategy usually adopted in order to avoid this undesirable problem of the toxicity of the crosslinker is to perform radical reactions of unsaturated monomers with starch or starch-based macromonomers containing carbon–carbon double bonds. In most cases, superabsorbent hydrogels result from this method. The copolymerization and simultaneous crosslinking of starch with vinyl-functionalized monomers has been reported.[33] Generally, starch is chemically modified in order to add reactive vinyl groups to its backbone, and such groups allow crosslinking among the starch chains or/and crosslinking by grafting with other polymers or monomers.[33] Most of the hydrogels prepared using this methodology show semi-interpenetrating network (semi-IPN) characteristics. IUPAC defines a semi-IPN as a polymeric material comprising at least one network and at least one linear or branched polymer characterized by the penetration of both on a molecular scale.[34] Semi-IPN is distinguished from an interpenetrating network (IPN) because the linear or branched polymers that form the former can, in principle, be separated from the constituent polymer network(s) without breaking chemical bonds; they are, in fact, polymer blends.[34,35]

In general terms, the basic reaction system for radical polymerization of starch is composed of starch (raw or modified), a catalyst (responsible for the radical formation), the monomers with reactive functional groups and a crosslinker molecule [such as *N*,*N*-methylenebisacrylamide (MBA)]. Using raw or chemically modified starch, it is possible to apply different techniques to form hydrogel networks from that reaction system.[8] The technique utilized can be chosen according to the hydrogel destination and it should be taken into account that it affects considerably the final properties of the hydrogel.

A very positive aspect of this methodology is the possibility of preparing starch-based hydrogels in the presence of organic polymers and inorganic compounds. Numerous papers have described the incorporation of inorganic clays,[36] magnetic and metallic nanoparticles,[37] hydroxyapatite,[38] cellulose whiskers[39] and waste residues[40] in the starch-based hydrogel formulation, resulting in composite materials. These hydrogel composites, in the most cases, show superior features compared with the conventional starch-based hydrogels, including applicability and the capacity to respond to external stimuli. Figure 7.2 shows an illustrative scheme of the production of a starch-based hydrogel nanocomposite developed by Spagnol *et al.*[39] In this case, cellulose nanowhiskers were inserted during the hydrogel formulation.

The use of starch to prepare hydrogels is not restricted to chemical crosslinking. In some studies, this versatile polysaccharide was blended with synthetic polymers, such as poly(vinyl alcohol) (PVA).[41] The blend of starch with polymers that have functional groups allowing physical interactions could be a promising strategy to develop hydrogels.[41,42] For example, Liu *et al.*[42] blended starch with PVA by a combination of freeze–thaw and coagulation techniques for artificial blood vessel applications. In this case, the combination of these two techniques promotes physical crosslinking between starch and PVA due to the formation of hydrogen bonds and polymer crystallites and liquid–liquid phase separation during the freeze–thaw cycles and coagulation process.[43] The stable hydrogels obtained without using organic crosslinkers showed significant potential for application in vascular tissue engineering.

From the details provided in this section, just as examples of the many studies on hydrogels and related fields that can be found in the literature, it can be seen that starch provides an excellent starting material for the preparation of both physical and chemical hydrogels, *i.e.* 3D structures based on starch and starch derivatives, namely hydrogels, superabsorbent hydrogels and hydrogel composites.

### 7.2.2 Characterization of Starch-based Hydrogels

Taking into account that the main advantageous properties of hydrogels are their liquid uptake capacity (swelling), morphology, molecular structure and mechanical properties, there are a huge number of qualitative and quantitative techniques that can be applied to characterize these and other properties. For this reason, the characterization and discussion of particular properties of starch-based hydrogels are not considered in this section.

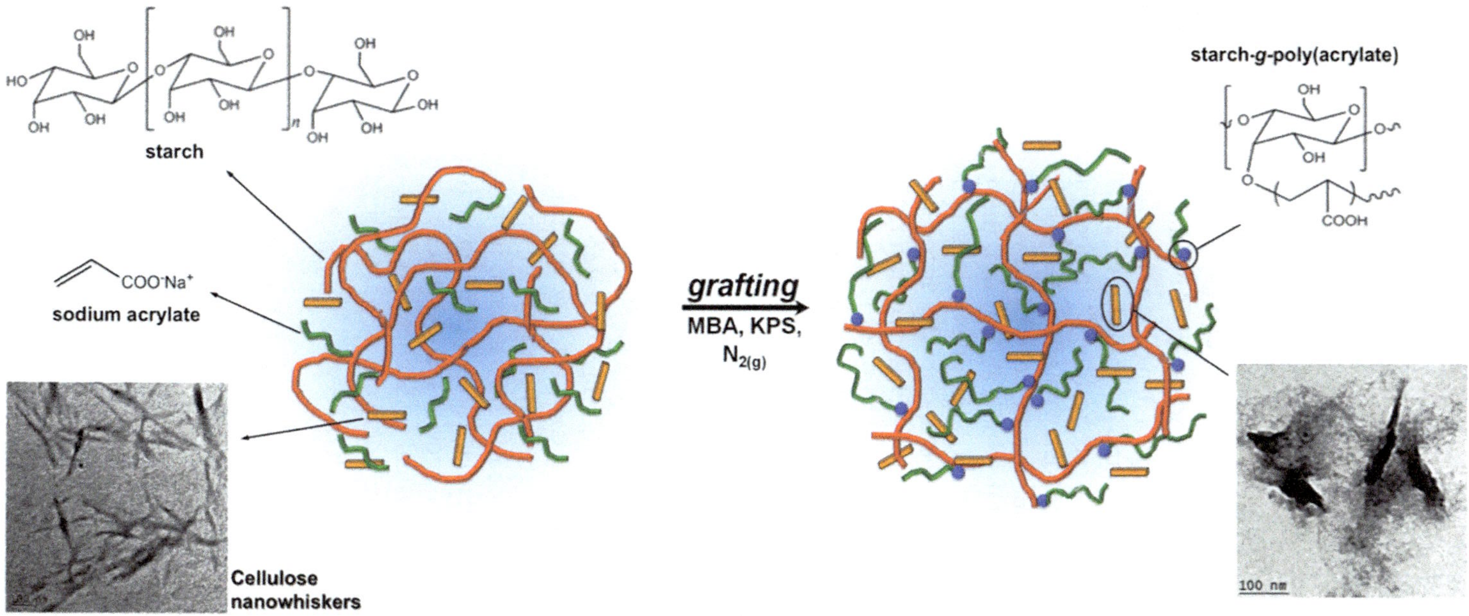


**Figure 7.2** Illustrative scheme of the preparation of a starch-*g*-poly(sodium acrylate)/cellulose nanowhiskers (CNWs) hydrogel composite matrix.

As already mentioned, most of the starch-based hydrogel properties are directly connected with two main factors: (i) the polymers combined with starch and (ii) the method applied to form the hydrogels. These factors affect the number of crosslinking points, the porosity and the distribution of the hydrophilic groups inside and/or at surface of the hydrogel matrix, and other aspects. Generally, these three aspects affect/determine all the hydrogel properties and it is difficult to establish how they are interconnected. The liquid uptake capacity, for example, depends on the hydrophilicity of the starch-based hydrogel matrix. From this characteristic another important aspect arises: the capacity of liquid retention by the hydrogel matrix. Several parameters control the change in either swelling or deswelling, such as temperature, pH, salinity and ionic strength of the medium.[44] The starch-based hydrogels prepared from hydrophilic polymers/monomers, in a general way, have the capacity to absorb and retain large amounts of liquid, which classify these hydrogels as superabsorbent. Raw starch is not so hydrophilic owing to its granular structure and for this reason the association of starch with more hydrophilic polymers is required in order to prepare materials with a high liquid uptake capacity.

The liquid uptake capacity or swelling ($S$) of hydrogels is usually expressed by the following equation:[45]

$$S = \frac{W_s - W_d}{W_d} \tag{7.1}$$

where $W_s$ is the weight of swollen hydrogel and $W_d$ is the weight of dry hydrogel. This relation gives the ratio between the mass of absorbed liquid and the mass of dry hydrogel. Additionally, the parameter $S$ can be evaluated in order to determine the maximum mass of liquid that the hydrogel is able to absorb (at equilibrium conditions) or it can be evaluated in different time intervals to determine the swelling kinetics. The evaluation of the swelling kinetics is important for understanding the hydrogel matrix behavior and the mechanism by which the water is transported into the hydrogel during swelling. Such a mechanism can be either diffusion or relaxation controlled. Different mathematical models have been developed to predict the swelling and solute release behavior from different hydrogel matrices. The semi-empirical equation described by Peppas and co-workers[46,47] is certainly the largest employed even though it is only valid for the initial 60% of the maximum absorbed liquid.

The molecular structure (*e.g.* functional groups present, network type, crosslinking density) and the intermolecular interactions among species (*e.g.* hydrogen bonding, ionic interactions) of both physical and chemical starch-based hydrogels have been studied by Fourier transform infrared (FTIR),[39,48] Raman[49] and nuclear magnetic resonance (NMR)[39] spectroscopic techniques. The thermal behavior and crystallinity of hydrogels have been assessed mostly by differential scanning calorimetry (DSC) and thermogravimetric analysis (TGA)[50] and also by small- and wide-angle X-ray diffraction measurements (SAXS and WAXS).[51,52] Such techniques have great utility in evaluating the changes promoted by insertion of fillers into the starch-based hydrogel matrices. The mechanical properties of starch-based hydrogels/hydrogel composites can be estimated by different means. For instance, the elastic modulus ($E$) can be evaluated using data collected in a texturometer by correlating the necessary force (*stress*) to induce some deformation (*strain*) in a body during compressive tests (*e.g.* compression stress–strain measurements). Additionally, rheology measurements have been reported.[53] Finally, the morphology and surface topography of hydrogels and hydrogel composites have been evaluated by microscopy approaches such as scanning electron microscopy (SEM),[54] transmission electron microscopy (TEM)[51,55] and atomic force microscopy (AFM). The images taken from the hydrogels and hydrogel composites allow us to determine very important parameters such as pore size and distribution and the dispersion of fillers inside the matrix.

## 7.3 Starch Hydrogel Composites

The combination of two or more different classes of materials with very distinct physical and chemical properties forming a composite is a very interesting approach to overcome the unsatisfactory features of the individual components by either having a synergic effect of the constituents improving the already existing properties or by generating new properties.[56] Although the definition of composite materials is broad, a good example of composites with important practical uses is micro/nanoparticle-reinforced polymers.[57,58]

In polymer science, hydrogels have evolved into materials with outstanding features and many potential applications, from soil conditioners and hygienic products to tissue engineering, drug delivery systems and imprinted polymers.[15,59] Therefore, it is not surprising that reports on hydrogels and hydrogel composites are still hot topics in the materials science field. A search (performed in July 2015) on the Scopus database for *hydrogel* revealed 51 511 published papers (almost 80% were published in the last 10 years), of which 510 refer to *starch hydrogels* and 113 papers *starch-containing hydrogel composites*, illustrating the importance of this class of materials in both academia and industry. It is very important to highlight that in many studies reporting hydrogels and/or hydrogel composites, the main matrix is actually composed of synthetic-based monomers (acrylic acid, acrylamide, *etc.*) grafted onto the starch backbone, in which the latter is not the main component. However, those studies are included in this chapter.

In terms of hydrogel composites, regarded as those containing micro- and nano-sized particles in their formulation, the addition of the reinforcement phase is generally performed seeking to improve thermal,[60] mechanical[61] and optical[62] properties, water uptake capacity,[63] release rate of solutes,[64] response to external stimuli[65] and degradation rate,[66] among other properties, tailoring for specific applications. Many different reinforcing phases have been studied, including mineral clays,[67,68] hydroxyapatite,[69] metal and magnetic nanoparticles,[70,71] carbon nanotubes,[72,73] polysaccharide nanocrystals[74] and quantum dots.[75]

Encouraged by the exceptional properties presented by the first micro-powder clay-reinforced starch hydrogel[76] and the first clay-reinforced nanocomposite hydrogel,[77] the majority of starch hydrogel composites reported more recently are those in which the reinforcing phase is based on mineral clays. This is supported by their natural abundance, low cost and nanometric dimensions, with a high aspect ratio of exfoliated sheets providing good interface interactions between layered clays and polymers.[78] Modification of clays can be performed by replacing the metal cations existing in the pristine clay by organic cations, generating modified surfaces that can play a role in the interactions between the polymer and the clay and ultimately in the composite properties.[79]

Different mineral clays have been used as fillers for preparing hydrogel composites, including kaolinite, bentonite, sercite, montmorillonite, attapulgite, halloysite, nontronite and vermiculite in addition to organic-modified clays.

Wu and co-workers demonstrated the effect of micro-powdered (particle size $<1\times10^{-6}$ m) kaolinite, bentonite and sercite on the water uptake capacity of starch-*g*-polyacrylamide hydrogels ($\sim$20 wt% starch, source not stated) prepared by free radical polymerization using MBA as crosslinker and ceric ammonium nitrate as redox initiator.[76,80,81] At 10 wt% clay, the hydrogels containing kaolinite absorbed $\sim$1000 g more water per gram of hydrogel than those containing bentonite or sercite. Bentonite and sercite have higher hydration powers and are more easily exfoliated than kaolinite, forming more particulates in water, which play a role in the crosslinking reaction, contributing to the formation of hydrogels with higher crosslinking density and, as a consequence, with lower water uptake capacity. Moreover, the amount of mineral also affected the absorbance and for formulations at over 20 wt% clay there was a significant reduction in the absorbing properties in relation to the control due to the crosslinking effect of the filler. Li *et al.* studied the effect of attapulgite clay on the water uptake of a hydrogel matrix composed of starch-*g*-polyacrylamide ($\sim$25 wt% corn starch) prepared through radical polymerization using MBA as crosslinker and ammonium persulfate as initiator.[82] They found that the best formulation in terms of water uptake capability was at 10 wt% clay (1317 g $H_2O$ per gram of sample), presenting $\sim$95% water retention after high-speed centrifugation (3000 rpm) for 10 min. The inorganic filler in the network acts as crosslinking points and further addition of attapulgite reduced the water absorption. It has been shown that the optimal amount of attapulgite to provide the highest water absorption is 10 wt%, independent of hydrogel matrix prepared from grafting of acrylic monomers onto a starch backbone such as starch-*g*-poly(acrylic acid)[83] and starch phosphate-*g*-acrylamide.[84]

Al *et al.*[36] prepared superabsorbent hydrogel composites by grafting acrylic acid onto a starch backbone at a monomer-to-starch weight ratio of 1.5 (~40 wt% maize starch) using MBA as crosslinker, ceric ammonium nitrate as initiator and Na/montmorillonite (MMT) as the reinforcing phase. It was showed that increasing the amount of MMT reduced the grafting percentage of acrylic acid from 30% (without MMT) to 26% at 10 wt% MMT as a result of the physical hindrance caused by MMT reducing the effectiveness of collisions between the macro radical and the monomers. The water uptake capacity was dependent on the MMT loading. At 1 and 3 wt% MMT, the water uptake at equilibrium was higher than that for the control whereas at higher clay concentrations (5 and 10 wt%) the absorbance capacity was lower, decreasing about fourfold at 10 wt% in relation to the highest value. The addition of MMT at 1 and 3 wt% did not change the adsorption capacity of a basic dye (Safranine T), which followed the Freundlich isotherm, but the time to reach equilibrium was increased from 24 h (control and 1 wt% MMT) to 48 h (3 wt% MMT), presenting an adsorption capacity of ~2100 mg of dye per gram of hydrogel. Another study performed by the same group reports the removal of heavy metal ions ($Cu^{2+}$ and $Pb^{2+}$) from aqueous solution using a starch-*g*-acrylic acid/MMT hydrogel nanocomposite.[85] The incorporation of MMT did not affect the adsorption capacity of hydrogel for copper and lead, which reached equilibrium conditions in 24 h and was higher for $Cu^{2+}$, but improved the shear modulus from 5370 to 14 778 Pa at 10 wt% of clay, allowing the hydrogel to be regenerated and reused, although the number of cycles was not stated.

Although it has been shown the addition of clay can improve the thermal stability and at low concentration can improve the swelling capacity of hydrogels, Kabiri *et al.*[86] showed that the inclusion of MMT in partially neutralized poly(acrylic acid) hydrogel has a negative effect on the crosslinking process, observed as a decrease in both mechanical strength and gel fraction and also an increase in residual monomer. Such an effect was more pronounced at high clay loadings (>10 wt%) owing to the high viscosity of the solution, which reduced the effectiveness and probability of the free radicals colliding with the monomers, limiting the extension of polymerization.

Ibrahim[87] showed the effect of MMT loading on the tensile strength and barrier properties of maize starch films crosslinked *via* gamma irradiation. The addition of 1 wt% of MMT markedly reduced the water vapor transmission from 1 to 0.85 g $m^2$ $h^{-1}$ and eventually to ~0.75 g $m^2$ $h^{-1}$ at 4 wt% MMT for non-irradiated films and to ~0.67 g $m^2$ $h^{-1}$ for films irradiated at 30 kGy. The tensile strength increased (about sixfold) from 0.625 MPa at 1 wt% MMT to 3.8 MPa at 4 wt% MMT for non-irradiated films, whereas for films irradiated with different doses (10, 20, 30 and 40 kGy) the increase was lower, but the maximum tensile strength was obtained for the film containing 4 wt% MMT at 30 kGy. Moreover, the swelling decreased with increasing clay content owing to physical crosslinking promoted by hydrogen bonds between the starch and MMT. Similarly, increasing radiation dose also promoted more crosslinking points, decreasing the water uptake capacity.

The incorporation of clays has also been used to tune the release properties of solutes. Singh *et al.*[88] incorporated both kaolin and bentonite in starch–alginate (12 : 1 wt%) beads (diameter ~1 mm), crosslinked with 0.1 M $CaCl_2$ aqueous solution, to modify the release features of the fungicide thiram, minimizing agro-environmental pollution. It was found that after 300 h, the amount of thiram released (from 30 mg loaded) was 10 mg from starch–alginate beads and 6.9 and 6.3 mg from beads containing 4 wt% of kaolin and bentonite, respectively, and occurred through non-Fickian diffusion, where both macromolecular relaxation and diffusion are responsible for release.

Aiming at improved mechanical properties and biodegradation rate, Ashamol *et al.*[89] described the preparation of nanocomposites based on crosslinked starch phthalate and various concentrations (up to 10 wt%) of silane-modified nanoclay. It was shown that the relative compressive strength was optimal at 5 wt% clay. Similarly, the flexural strength was a maximum at 4 wt% clay. At high clay concentrations, the clay acts as a third phase, causing agglomeration and saturation of reactive sites, which has a negative effect on mechanical properties. The water uptake increases with increasing amount of clay and reaches a maximum at 6 wt% clay. This is due to the presence of hydrophilic amino groups from the silane-modified clay. For higher clay loadings, the filler induces stiffness, providing rigidity to the chains and, as a consequence, lower water absorption capacity. Biodegradation studies showed that the presence of silane-modified clay retards the onset of degradation. After 18 weeks, the control (crosslinked starch phthalate) was fully degraded whereas the composite containing 10 wt% clay presented a weight loss of 60%, showing the viability of clay incorporation to modulate the degradation rate of composites.

Kamoun and Menzel[90] investigated the effect of organophilic clays, prepared by exchanging metal ions by quaternary alkylammonium ions, such as MMT (modified with hexadecyltrimethylammonium chloride) and laponite-RD (modified with tricaprylylmethylammonium chloride), on the features of a hydrogel composed of hydroxyethyl starch modified with hydroxyethyl methacrylate photopolymerized with visible light (430–490 nm), using camphorquinone–amine as photoinitiator. For all composites studied, the addition of modified MMT (up to 2 wt%) did not improve the thermal stability, but an increase (of ~18 °C) in the glass transition temperature was observed due to the mobility restriction of polymer chains caused by the interactions between polymer and clay sheets. The storage modulus ($G'$) was dependent on the clay used in the preparation of the composites. In the case of laponite, there was an increase in $G'$ with increasing laponite amount whereas for those samples prepared with MMT the effect was only observed at low concentrations (>1.5 wt%). Further, $G'$ was always higher for composites containing laponite than for the counterparts containing MMT. It is important to point out that MMT and laponite were modified with different intercalating agents. In the case of MMT, the sodium ions were exchanged by ammonium ions with a long hydrophobic chain ($C_{16}$), which can reduce the interaction with hydrophilic polymers, whereas in the laponite the metal ions were modified with ammonium ions containing a smaller hydrophobic chain ($C_8$), favoring the interactions with hydrophilic polymers. The swelling capacity decreased progressively as the laponite concentration was increased (up to 5 wt%) due to the formation of cross-linking points. On the other hand, for MMT, the maximum reduction was observed at 1.5 wt% MMT. At higher concentrations, MMT may agglomerate and the crosslinking effect is reduced, allowing the hydrogel to absorb more water.

The addition of active fillers can induce new properties in hydrogels, broadening their range of applications. In this regard, silver nanoparticles have been incorporated into hydrogels owing to their antibacterial features. Vimala *et al.*[55] developed semi-IPN hydrogels composed of starch and polyacrylamide using MBA as crosslinker for producing polyacrylamide hydrogel (PAM) network. Silver nanoparticles were generated using the hydrogel network as template for anchoring the Ag ions followed by *in situ* reduction of $AgNO_3$ with $NaBH_4$ at room temperature. Starch played a role in the anchoring and stabilization of nanoparticles. The particles had varied sizes, from 2 to 20 nm, and were homogeneously distributed throughout the polymer network. The presence of Ag in the oxidized or reduced state improved the swelling capacity of the PAMs due to the electrostatic repulsion of positively charged Ag ions or to the repulsive surface charges of silver nanoparticles. However, there was no significant difference in swelling capacity between hydrogels containing $Ag^+$ or Ag. Moreover, the water up-take of semi-IPN hydrogels was progressively enhanced with increasing the amount of starch, reaching a maximum (about three times higher than that observed for PAM) at an acrylamide-to-starch weight ratio of 2 : 1 due to the extra amount of hydrophilic chains in the final composition. The hydrogels were shown to be effective in the inhibition of *Escherichia coli* microorganisms.

Eid[91] reported the preparation of starch polyelectrolyte hydrogels [starch/polyvinylpyrrolidone, starch/polyacrylamide, starch/poly(acrylic acid)] poly-merized *via* gamma irradiation loaded with silver nanoparticles prepared by *in situ* reduction of silver nitrate with sodium borohydride, at room temperature. The Ag ions homogeneously distributed throughout the network were stabilized by complexation with the nitrogen atoms of polyvinylpyrrolidone, amide groups of polyacrylamide, carboxylic groups of poly(acrylic acid) and hydroxyl groups of starch. When the reduction re-action took place, the hydrogel immediately turned opaque (brown color), proving that the preparation of Ag nanoparticles inside the matrix was efficient. The composite containing poly(acrylic acid) [P(Starch/AAc)–Ag] presented the highest water uptake capability (~6000%) followed by that with polyacrylamide [P(Starch/AAm)–Ag; ~1000%] and then that with poly(*N*-vinylpyrrolidone) [P(Starch/NVP)–Ag; <1000%]. However, the effect of silver nanoparticles on the swelling is not clear since the swelling data for the counterpart hydrogels (those without Ag nanoparticles) were not presented. TEM images showed the nanoparticles had spherical morphology and were randomly distributed in the network, but the size of the particles was dependent on the polymer matrix used, with average diameter ranging from 35.81 nm for P(Starch/NVP)–Ag to 40.45 nm for P(Starch/AAc)–Ag. Although the activity against microorganisms was not tested, the hydrogels have potential to be applied as antibacterial materials.

Xia *et al.*[92] prepared hybrid hydrogels based on dialdehyde starch (from cassava) crosslinked *via* Schiff base formation (reaction between an aldehyde and an amine forming a carbon–nitrogen double bond, $R^1R^2C{=}NR^3$) using chitosan as a natural and biocompatible crosslinker. The excess of aldehyde groups on the modified starch acted as a reducing agent for preparing *in situ* Ag nanoparticles without further addition of toxic reducing chemicals. The swelling capacity of the hydrogels increased gradually with increasing amount of chitosan (from 0.5 to 2.5 wt%) due to the extra hydrophilic chains, reaching at 2.5 wt% chitosan water uptake capacity at equilibrium conditions ($\sim$24 h), $\sim$12 times larger than that of the control. It was shown that the size of the Ag nanoparticles depends on the number of dialdehyde groups and it decreases on increasing the degree of modification of starch owing to the larger number of groups for anchoring and stabilizing the Ag ions.

Hydroxyapatite, a biocompatible calcium phosphate complex with a similar inorganic composition [$Ca_{10}(PO_4)_6(OH)_2$] to natural bone structures, has been introduced into hydrogels to improve biocompatibility and mechanical features for biomedical applications such as bone–tissue repair and growth and adhesives for soft tissue regeneration. Raafat *et al.*[38] prepared hydrogels based on starch and *N*-vinyl-2-pyrrolidone (NVP) crosslinked with $^{60}$Co gamma irradiation followed by hydroxyapatite (HAp) deposition by subsequent immersion in separate $CaCl_2$ and $Na_2HPO_4$ solutions using different deposition cycles. By increasing the amount of NVP from a starch-to-NVP ratio of 1 : 0.09 to 1 : 0.23 wt%, the gel fraction increased from 75 to 94.5%, associated with an increase in swelling capacity from 300 to 500%, but reduced pore sizes. The compressive strength of the hydrogels increased from 8.8 to 25 MPa at one cycle of HAp deposition, but further additional cycles resulted in lower values of 18.8 and 15.7 MPa at five and nine cycles, respectively, possibly owing to a heterogeneous distribution of the particles and phase separation. However, all the samples presented compressive strengths comparable to those of cancellous human bones (2–20 MPa). Moreover, on increasing the number of cycles, the Ca:P ratio increased from 1.30 to 1.56, as verified by energy-dispersive X-ray spectroscopy (EDX), ranging close to the theoretical hydroxyapatite Ca : P ratio of 1.67. *In vitro* bioactivity and blood compatibility tests showed the greater the amount of HAp (number of cycles) in the composites, the higher is the bioactivity and the lower the hemolysis (blood cell rupture, 0.001% at nine cycles) and blood clotting. These results suggested a promising potential bonding ability that facilitates bone in-growth formation and good osteointegration *in vivo*.

Another study reported the effects promoted by the incorporation of micro- (60–180 μm) and nano-powder (100–150 nm) HAp into starch–chitosan (0.38 weight ratio) hydrogels.[93] The elastic modulus (at 1% strain) increased gradually with increase in the amount of HAp from 1.8 GPa for the pristine hydrogel to 5.8 and to 6.1 GPa for composites at 40 wt% of micro- and nano-HAp, respectively. The swelling capacity decreased progressively with increasing HAp loading and this effect was more pronounced in the composites composed of nano-HAp, in which the reduction of about sevenfold at 40 wt% was about double that observed for composites with micro-HAp. This suggested the application of such composites in the hard tissue engineering field.

Magnetic nanoparticles have also been added to starch hydrogels, allowing the modulation of release and absorption features by the use of an external magnetic field and also magnetic separation from aqueous media. Pourjavadi *et al.*[94] introduced silica-coated magnetic nanoparticles ($Fe_3O_4@SiO_2$) into starch-*g*-poly(acrylic acid) hydrogels to be applied as adsorbents for Crystal Violet. Both magnetic and non-magnetic hydrogels reached swelling equilibrium conditions (in distilled water) in 10 min, but the latter presented a higher water uptake capacity (610 g of water per gram of absorbent) because $Fe_3O_4@SiO_2$ can act as crosslinker, increasing the rigidity of the matrix and, as a consequence, the water uptake capacity of the magnetic hydrogel at 20 wt% $Fe_3O_4@SiO_2$ was twofold lower (283 g of water per gram of absorbent). The removal of Crystal Violet from aqueous solution (50 mg $L^{-1}$) was optimal at 25 mg of adsorbent, reaching about 85 and 95% of removal in 24 h for non-magnetic and magnetic hydrogels, respectively. Further, it was shown that the magnetic hydrogels have sufficient superparamagnetic properties for magnetic separation with a conventional magnet and they could be regenerated and reused by several times (five cycles tested) with minimum loss of performance. Therefore, the magnetic hydrogels have potential to be used in the treatment of waste water.

The effect of $Fe_3O_4$ microparticles (<5 μm) in the release of albumin from starch-*co*-DMAAm-*co*-AAc hydrogel was investigated.[51] Non-composite hydrogel presented a higher water uptake capacity ($\sim$150 g $g^{-1}$) than that observed for composites with 0.5, 1 and 2 wt% $Fe_3O_4$ ($\sim$60 g $g^{-1}$) and there was no significant effect of an applied external magnetic field on the water absorption process. The albumin release was governed by macromolecular relaxation in the pristine polymer and in the composites without an external magnetic field. However, on applying a magnetic field, the release features changed, becoming dependent on anomalous transport, as a result of tortuosity effects caused by the presence of $Fe_3O_4$. It was suggested that the prepared composite hydrogels could be used as oral drug vehicles to sustain the drug release for long periods of time.

Polysaccharide nanocrystals, such as those prepared from cellulose (CNC), have appeared as nano-sized fillers in many polymer composites, especially due to some interesting properties including its high crystallinity, which confers an outstanding elastic modulus of 150 GPa, ready availability and high specific surface area.

CNC (from cotton, 300 nm length, 21 nm width, aspect ratio 15) prepared by sulfuric acid hydrolysis was incorporated into a physical hydrogel composed of maize starch and poly(vinyl alcohol) (50 : 50 wt%) prepared by 10 freeze–thaw cycles.[95] The degree of swelling decreased gradually on increasing the amount of CNC from 1715% for the control to 1231% at 30 wt% CNC, but the interconnected pore network promoted fast water absorption, reaching equilibrium in ~20 min. Stress relaxation curves at 25 °C showed a 256% enhancement of initial stress at 30 wt% CNC. The dynamic storage modulus increased by nearly one order of magnitude at 30 wt% CNC, demonstrating the reinforcement effect of CNC. COS-7 cells (biological source: kidney from African green monkey) were able to proliferate and completely cover the porous structure of the hydrogels, indicating biocompatibility and a great potential for the hydrogel composite to be used in the biotechnological field. In another study, HCl-hydrolyzed CNC (from cotton, crystallinity index 90%) was incorporated into a starch-*g*-poly(sodium acrylate) matrix[39] and its effect on the swelling capacity was investigated. The addition of CNC up to 10 wt% improved the water uptake capacity from 90 $g\,g^{-1}$ (water/absorbent) to 110 $g\,g^{-1}$ due to the extra hydrophilic groups of the cellobiose unit. A further increase in CNC loading reduced the swelling ability because some hydroxyl groups from the CNC took part in the crosslinking reaction, increasing the crosslinking density and, as a result, diminishing the water absorption. Both pristine hydrogel and composites were responsive to variations in pH and salts. The Young's modulus was dependent on the CNC content and lower than that observed for the non-composite hydrogel, reaching a minimum at ~10 wt% CNC as a result of the maximum water uptake capacity at that composition favoring the formation of softer material. On the other hand, the Young's modulus reached a maximum for CNC loadings higher than 20 wt% owing to the stiffness of the filler. Therefore, there was a balance between stiffness and the highest water absorption capacity according to the CNC loading.

As demonstrated, many different reinforcing phases composed of micro- and nanoparticles can be introduced into starch-based hydrogels to improve or modify their features, allowing different applications. Although a reasonable number of studies regarding starch hydrogel composites have been reported, the majority were mainly based on copolymers of synthetic monomers and the polysaccharide. Hence the preparation of hydrogel nanocomposites could be a very promising approach to developing fully starch-based hydrogels with consistent features for different applications owing to the biodegradability and ready availability of starch.

## 7.4 Applications of Starch-based Hydrogels

As pointed out previously, the development of new hydrogels based on degradable polymers derived from renewable resources, such as starch and starch-based materials, has been challenging for a significant number of researchers.

Starch is a renewable, biodegradable and low-cost polysaccharide. Starch-based polymers are therefore attractive materials for a wide variety of applications, including implant materials, drug delivery matrices and tissue engineering scaffolds.[96–98] However, starch also presents some drawbacks, such as poor processability, water resistance and high brittleness.[99] These limitations can be overcome by blending starch with hydrophobic polymers to improve the water resistance and mechanical properties. Another method for improving the properties of starch is based on, *e.g.*, adding nanoclay to improve barrier, mechanical and heat resistance properties.[100] Once starch-based blends and composites have become biodegradable, inexpensive and also readily available, they possess an enormous potential for applications ranging from biomedical to environmental areas.[101–103]

It is known that starch-based materials retain their chemical and mechanical integrity in the first implantation stages, starting to degrade thereafter.[104] An approach to improve the material degradability and mechanical properties of (composite) hydrogels is through the incorporation of nanoparticles (NPs).[105] On the other hand, metal NPs aggregate owing to the high active surface area. To stabilize and control the NP structures, various surfactants, polymers and biomacromolecules have been used;[106] a different approach is the use of macroscopic gels as templates or nanopots to prepare composite/hybrid nanoparticles.[107] Both alternatives allow the design of a myriad of blends and composites with suitable features for different applications. One of the first starch-containing nanocomposite (NC) hydrogels was synthesized by Huang *et al.*[105] They reported the synthesis of a novel semi-IPN nanocomposite hydrogel composed of starch-*g*-acrylic acid and sodium bentonite modified with linear poly(dimethyldiallylammonium chloride) (PDMDAAC). The incorporation of sodium bentonite into the semi-IPN network showed better mechanical properties than the corresponding starch-*g*-acrylic acid/PDMDAAC hydrogel. A few months later, the use of polyacrylamide-based semi-IPN hydrogels as effective carriers for Ag nanoparticles and carbohydrates (*e.g.*, starch) was reported (Figure 7.3). The use of carbohydrate polymers was chosen in order to increase the swelling capacity and the biocompatibility of PAAm gels and also to act as a green stabilizer. The NC hydrogel showed mild antibacterial activity against *E. coli*, whereas the counterpart hydrogel did not show any activity. This effect was discussed in terms of the size and charge, and corresponding release, of silver NPs. It is worth noting that silver NPs show unique chemical and physical properties and pronounced antibacterial activity, providing one of the most cost-effective alternatives for the development of new antibacterial agents.[108]

The synthesis of starch–polyelectrolyte (acrylamide, acrylic acid and *N*-vinylpyrrolidone) blend hydrogels using a gamma-radiation polymerization technique also results in appropriate matrices for the formation of silver NPs with a spherical shape and with a uniform distribution throughout the hydrogel matrix.[91]

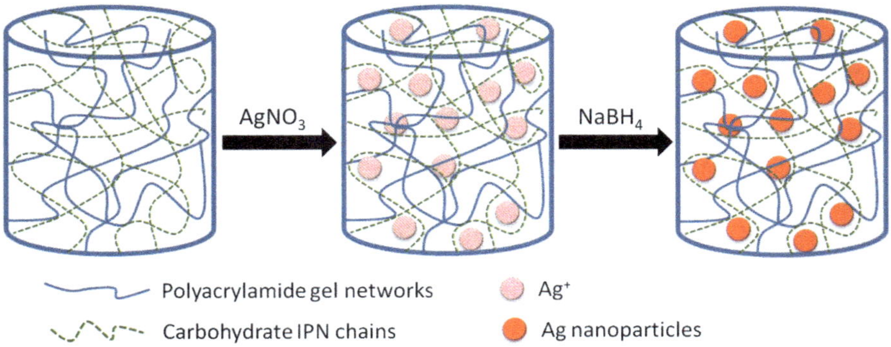


**Figure 7.3** Schematic representation of the synthesis of Ag NPs by using semi-IPN hydrogel networks.

A different route was proposed by Xia *et al.*,[92] consisting of *in situ* reduction of Ag ions in crosslinked tapioca dialdehyde starch–chitosan hydrogels. The synthesis of these composite hydrogels does not require further chemical reductants – resulting in a green approach – they are biocompatible and the swelling and NP size and morphology can be controlled by the dialdehyde starch-to-chitosan ratio and the concentration of silver nitrate, ammonia and dialdehyde starch.

Similar results have been obtained by using other polysaccharide-based nanocomposites. For example, Vimala *et al.*[109] reported the synthesis of blends of non-toxic and biocompatible polycationic chitosan and poly(ethylene glycol) (PEG) for the formation of a porous matrix for incorporation of Ag NPs. The PEG is used as a reducing/stabilizing agent but also acts as a porogenator.[106] The composites obtained exhibited better mechanical and antibacterial properties than the chitosan and chitosan–silver nanocomposites.

Another possible application of starch–chitosan-based materials is for wound dressings and hard-tissue engineering. For the latter, the blend must incorporate a bioactive bone-like compound. One material falling in this category is hydroxyapatite (HA). HA is a calcium phosphate-based compound with a chemical composition very similar to that of bone mineral, being used as a bone implant in a variety of situations, including the filling of osseous defects resulting from inflammatory periodontal diseases.[110] HA also possesses osteoconductive activity, providing a physical matrix suitable for the deposition of new bone,[111] *i.e.* HA tends to bond spontaneously to living bone *via* an apatite layer deposited on the bone surface, improving its mechanical properties.[112] However, the brittleness and the poor mechanical stability of pure HA limit its use for the regeneration of non-load-bearing bone defects and tissue engineering applications. Once starch-based copolymers and composites can be used for orthopedic applications,[113] these HA limitations can be overcome. With this is mind, starch–chitosan composites containing HA micro- and nano-powders have been synthesized and their properties were discussed.[93] The incorporation of HA NPs shows a better distribution throughout the composite, resulting in a higher tensile modulus and lower degree of swelling than for the corresponding HA microparticle composites, suggesting that the former can be used for hard tissue engineering. A different approach in the synthesis of HA-containing starch-based composited was reported by Raafat *et al.*[38] They synthesized a copolymer of starch and NVP by means of gamma radiation-induced graft copolymerization and crosslinking. The starch–NVP hydrogel matrices are an appropriate medium for the deposition and growth of HA layers. *In vitro* blood compatibility studies of the resulting composites confirmed their biocompatibility.

Skin is an important organ in maintaining homeostasis and protects the body from invading microorganisms; consequently, its damage can lead to major disability or even death. Healing is a multistep process involving many processes, such as homeostasis, inflammation, proliferation and remodeling of tissues in order to fill the damaged area and re-establish the skin barrier.[114] Sago starch–chitosan-based composites for wound dressing were synthesized and characterized by Arockianathan *et al.*[115] *In vivo* studies were also carried out. These authors found that the use of Ag NP-containing sago starch–chitosan hydrogel composites with and without an antibiotic (gentamicin) led to faster wound healing in rats than in the control group (sterile cotton gauze with gentamicin was applied in the wound). Furthermore, no significant difference was obtained by using the hydrogel with and without gentamicin. These results were mainly justified by the capacity of these gels to maintain the moisture environment at the wound site for a longer period than occurred with the use of cotton gauze.

In addition to the use of the starch-based materials for tissue engineering and wound dressing, they are also suitable as carrier material for the controlled release of drugs. Starch–MMT clay composites were synthesized to act simultaneously as films for packaging applications and as matrices for release of chlortetracycline, a broad-spectrum antibiotic commonly given to poultry, swine and livestock.[87] MMT – a clay constituted by two fused-silica tetrahedral sheets sandwiching an edge-shared octahedral sheet of either magnesium or aluminum hydroxide – can act as a physical crosslinking agent[116] and thus improve thermal stability and tensile properties and lower the water vapor transmission of pure starch. Those composites were loaded with chlortetracycline, with a loading efficiency of $\sim$14%. The release kinetics and the cumulative release (at equilibrium) of chlortetracycline from the starch–MMT composite depends on the initial amount of the drug loaded.[87] Another study involving the incorporation of MMT was recently published by Kamoun and Menzel.[90] The incorporation of sodium MMT-photopolymerizable hydroxyethyl starch modified with 2-hydroxyethyl methacrylate led to an increase in the thermal and mechanical stability of the nanocomposite hydrogels obtained, which are promising materials for biomedical applications.

A more comprehensive and complete study on the release from hydrogel composites were reported by Guilherme *et al.*[51] They synthesized copolymers of starch, previously vinyl functionalized with glycidyl methacrylate, acrylic acid and *N*,*N*-dimethylacrylamide, in the presence of different amounts of magnetite ($Fe_3O_4$), forming a brain-resembling magnetic responsive hydrogel. Magnetic responsive hydrogels have important advantages because they can be used for different therapeutics where activation for, *e.g.*, drug release can be performed by remote control; furthermore, the use of an appropriate magnetic field does not affect biological tissues and its efficiency does not depend on the clinical status of the patient. The effect of different amounts of $Fe_3O_4$ on the release properties of a model protein, albumin, in the presence and absence of a magnetic field was studied. It was have found that the release profile of albumin from $Fe_3O_4$-containing composite hydrogels can be described by the partitioning-diffusion equation,[117] following first-order release kinetics. Concerning the mechanism of release, the transport of albumin, in the absence of magnetite, follows Super-Case II transport, indicative of coupling of diffusional and relaxational mechanisms. In the presence of magnetite, the release mechanism becomes essentially Fickian. This is justified by the minimization of the macromolecular relaxation due to the incorporation of $Fe_3O_4$ particles, with the consequent attenuation of the anion–anion electrostatic repulsion. It is also believed that the incorporation of mineral into the hydrogel enhances the tortuosity. It is worth noting that in the presence of a magnetic field, the release mechanism of albumin becomes anomalous. This can be explained by a reduction in the motion of the polymeric structure and a corresponding increase in the diffusion pathlength. The rationale is based on the attraction of magnetite, imposed by a magnetic field, bonded to the polymer skeleton and, consequently, the movement of polymer segments becomes slower than in the absence of the magnetic field (see Figure 7.4). These hydrogels seems to be appropriate for oral drug delivery systems because of their capacity to provide a sustained release; furthermore, owing to the acid-resistant properties of starch,[118] these hydrogels are also potential candidates for the transport and storage of acid-responsive drugs.

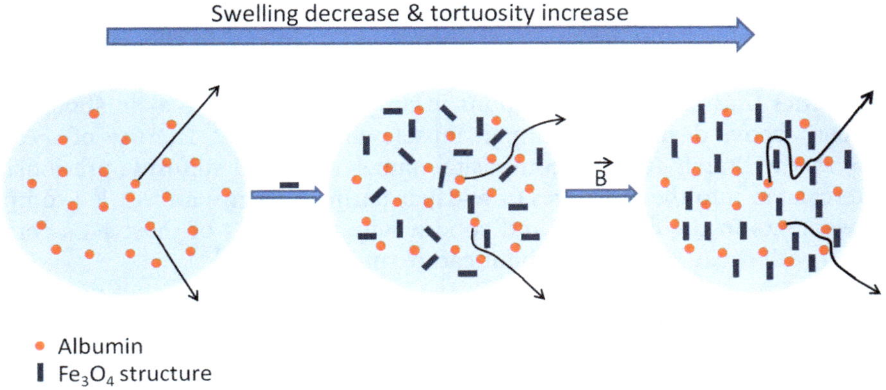


**Figure 7.4** Schematic representation of the effect of magnetite and a magnetic field on the release transport of albumin from starch-based hydrogel.

The same group also developed a novel synthetic strategy for the synthesis of magnetite-nanostructured amylose microspheres.[119] Amylose is the amorphous constituent of starch and among its properties acid resistance is one the most important, as it allows amylose be used as a matrix for transport and preserve acid-sensitive drugs, such as corticoids, often used for colon-specific diseases. Going back to the synthesis of the material, the approach involved the use of ultrasound waves for formation of magnetite microspheres *via* complexation of iron ions with carboxyl groups from carboxyl-functionalized amylose in the interior of the alcohol-confined water droplets. Both the use of ultrasound and the strong interactions contribute to the dispersion of magnetite into small crystals, with a narrow sphere size distribution. The results are promising for the remote delivery of alcohol-soluble drugs acting on, *e.g.*, colon tumors.

The use of magnetite-containing hydrogels has also been successfully applied to dye removal from wastewater. Dyes are widely used in various industries, the including leather, textile and pharmaceutical areas. In general, dyes are toxic and/or mutagenic and they show high stability, remaining in the environmental for long periods. Although their removal from wastewaters is feasible, their decomposition into less toxic compounds is not straightforward using conventional techniques. Magnetic separation can be used for wastewater clarification and purification.[120] For this application, magnetic composite hydrogels were prepared using *in situ* synthesis of starch-*g*-poly(acrylic acid) in the presence of silica-coated iron oxide nanoparticles *via* crosslinking graft copolymerization.[94] The ability of these anionic hydrogels to remove dyes was tested by using the cationic dye Crystal Violet. The magnetic nanocomposite hydrogels showed a slightly lower sorption capacity for Crystal Violet removal than the non-magnetic ones; however, upon application of a magnetic field, the magnetic NPs, with Crystal Violet, can be readily removed and separated from the absorption medium (*e.g.*, wastewater). Furthermore, the sorbent can be regenerated and reused several times.

Starch-based NCs can also be used as electrorheological fluids. These materials can be solidified, in a reversible way, by the application of an external electrical field, with concomitant significant changes in the rheological properties of the fluid (*e.g.* viscosity, yield stress and shear modulus).[121] Such properties find practical applications in different areas, such as in shock absorbers, active devices and human muscle simulators.[122] The use of a carboxymethyl starch–feldspar composite, namely dimethyl sulfoxide previously intercalated into the interlayes of feldspar particles to give an NC, led to improvements in the dielectric and conductivity properties of these NCs compared with those prepared without carboxymethyl starch.[123]

## 7.5 Future Trends in Starch-based Hydrogels

This chapter has demonstrated that starch-based hydrogel nanocomposites are versatile materials owing to their particular properties and can be widely applied in various fields. It is also important to point out that the availability; renewable sources, chemical structure, hydrophilic properties and the possibility of chemical modification render starch a very important starting material for preparing hydrogels and hydrogel nanocomposites. Also, it must be stressed that depending on the methodology used, the as-prepared 3D matrices based on starch and starch derivatives may have different geometries (membranes, hard spheres, hollow spheres, cylinders, *etc.*) and a wide range of sizes, *i.e.*, the macro-, micro- and nano-scales. The possibility of inserting nanoparticles (inorganic and/or organic) during the hydrogel synthesis, conferring specific capabilities (mechanical, magnetic, thermosensitivity, crystallinity, *etc.*) on the 3D nanocomposite matrices gives rise to numerous possibilities for producing starch-based nanostructured hydrogels with a wide variety of properties and applications. Therefore, there are still some topics to be explored that should allow further improvements in this type of material. More comprehensive studies will, of course, broaden the understanding of structure–property relationships. Some future trends in this field can be highlighted:

- Chemical modification of starch, at a specific point in the polymer chain, for further "click" chemistry reactions targeting hydrogel preparation with controlled structure.[124]
- Synthesis of block copolymers by reacting chemically modified starch and other polymers to obtain block copolymers with hydrophilic block and hydrophobic block characteristics targeting self-assembly under certain conditions,[24] allowing, for instance, drug encapsulation/release.
- Development of new and/or improved methodologies for producing starch-based derivatives under controlled conditions allowing control of the chemical structure and molecular mass, for instance, obtaining starch scaffolds and starch nanowhiskers for tissue engineering.[125]
- Although many different nanogels have been prepared, this field is still in the early stages of development, but stands as a very import platform for hydrogel nanomaterials, for instance, in medical applications, mainly those constituted by polysaccharides. In this respect, nanogels based on starch or starch-containing nano-scale moieties are a hot topic in the science and technology of soft materials.[126,127]

## 7.6 Conclusion

Hydrogels are important materials, but the stimuli-responsive hydrogels, as smart materials, occupy a much more important place owing to the versatile and wide applications related to the changes induced by external changes. Starch-based hydrogel nanocomposites are a very important subject in the hydrogel field owing to the set of specific properties discussed in this chapter. Year by year the literature shows increasing numbers of publications claiming novel starch-based hydrogel nanocomposites with new structures and/or properties. As pointed out in this chapter, however, there are challenges to be overcome. More comprehensive studies will broaden the understanding of structure–property relationships, allowing new materials and new applications. Hence the future of starch-based hydrogel nanocomposites should remain an active and continuously growing field in the following decades.

## References


1. G. O. Phillips and P. A. Williams, *Handbook of Hydrocolloids*, CRC Press, Boca Raton, FL, 2000.
2. P. B. Malafaya, C. Elvira, A. Gallardo, J. San Román and R. L. Reis, *J. Biomater. Sci., Polym. Ed.*, 2001, **12**, 1227–1241.
3. L. Hovgaard and H. Brondsted, *Crit. Rev. Ther. Drug Carrier Syst.*, 1996, **13**, 185–223.
4. J.-F. Zhang and X. Sun, *Biomacromolecules*, 2004, **5**, 1446–1451.
5. A. Buleon, P. Colonna, V. Planchot and S. Ball, *Int. J. Biol. Macromol.*, 1998, **23**, 85–112.
6. M. A. Araujo, A. M. Cunha and M. Mota, *Biomaterials*, 2004, **25**, 2687–2693.
7. P. Tomasik and C. H. Schilling, *Adv. Carbohydr. Chem. Biochem.*, 2004, **59**, 175–403.
8. C. Xiao, *Starch Stärke*, 2013, **65**, 82–88.
9. D. R. Lu, C. M. Xiao and S. J. Xu, *Express Polym. Lett.*, 2009, **3**, 366–375.
10. G. Moad, *Prog. Polym. Sci.*, 2011, **36**, 218–237.
11. A. C. Albertsson and S. Karlsson, *Acta Polym.*, 1995, **46**, 114–123.
12. G. J. L. Griffin, Biodegradable synthetic resin sheet material containing starch and a fatty material, *U.S. Pat.* 4016117, 1977.
13. O. Wichterle and D. Lim, *Nature*, 1960, **185**, 117–118.
14. H. Omidian, J. G. Rocca and K. Park, *J. Controlled Release*, 2005, **102**, 3–12.
15. M. J. Zohuriaan-Mehr, H. Omidian, S. Doroudiani and K. Kabiri, *J. Mater. Sci.*, 2010, **45**, 5711–5735.
16. K. Trabbic-Carlson, L. A. Setton and A. Chilkoti, *Biomacromolecules*, 2003, **4**, 572–580.
17. A. Jamshidi, F. A. K. Beigi, K. Kabiri and M. J. Zohuriaan-Mehr, *Polym. Test.*, 2005, **24**, 825–828.
18. D. Buenger, F. Topuz and J. Groll, *Prog. Polym. Sci.*, 2012, **37**, 1678–1719.
19. Y. Qiu and K. Park, *Adv. Drug Delivery Rev.*, 2001, **53**, 321–339.
20. Y. Aizawa, S. C. Owen and M. S. Shoichet, *Prog. Polym. Sci.*, 2012, **37**, 645–658.
21. E. F. S. Vieira, A. R. Cestari, C. Airoldi and W. Loh, *Biomacromolecules*, 2008, **9**, 1195–1199.
22. G. Swift, *Polym. Degrad. Stab.*, 1994, **45**, 215–231.
23. D. L. Kaplan, in *Biopolymers from Renewable Resources*, ed. D. Kaplan, Springer, Berlin, 1998, pp. 1–29.
24. A. R. Fajardo, J. F. Piai, A. F. Rubira and E. C. Muniz, *Carbohydr. Polym.*, 2010, **80**, 934–943.

25. E. G. Crispim, J. F. Piai, A. R. Fajardo, E. R. F. Nakamura, T. U. Nakamura, A. F. Rubira and E. C. Muniz, *Express Polym. Lett.*, 2012, **6**, 383–395.
26. S. Van Vlierberghe, P. Dubruel and E. Schacht, *Biomacromolecules*, 2011, **12**, 1387–1408.
27. J. T. Oliveira and R. L. Reis, *J. Tissue Eng. Regener. Med.*, 2011, **5**, 421–436.
28. M. R. Guilherme, A. V. Reis, A. T. Paulino, A. R. Fajardo, E. C. Muniz and E. B. Tambourgi, *J. Appl. Polym. Sci.*, 2007, **105**, 2903–2909.
29. A. R. Fajardo, M. B. Silva, L. C. Lopes, J. F. Piai, A. F. Rubira and E. C. Muniz, *RSC Adv.*, 2012, **2**, 11095–11103.
30. A. F. Martins, A. G. B. Pereira, A. R. Fajardo, A. F. Rubira and E. C. Muniz, *Carbohydr. Polym.*, 2011, **86**, 1266–1272.
31. G. Crini, *Prog. Polym. Sci.*, 2005, **30**, 38–70.
32. I. Simkovic, *Carbohydr. Polym.*, 2013, **95**, 697–715.
33. V. D. Athawale and V. Lele, *Starch Stärke*, 2001, **53**, 7–13.
34. A. D. Jenkins, P. Kratochvíl, R. F. T. Stepto and U. W. Suter, *Int. J. Pure Appl. Chem.*, 1996, **68**, 2287–2311.
35. Y.-J. Wang, J. Qiao, R. Baker and J. Zhang, *Chem. Soc. Rev.*, 2013, **42**, 5768–5787.
36. E. Al, G. Güçlü, T. B. İyim, S. Emik and S. Özgümüş, *J. Appl. Polym. Sci.*, 2008, **109**, 16–22.
37. G. Bardajee and Z. Hooshyar, *J. Polym. Res.*, 2013, **20**, 1–13.
38. A. I. Raafat, A. A. Saad Eldin, A. A. Salama and N. S. Ali, *J. Appl. Polym. Sci.*, 2013, **128**, 1697–1705.
39. C. Spagnol, F. A. Rodrigues, A. B. Pereira, A. Fajardo, A. Rubira and E. Muniz, *Cellulose*, 2012, **19**, 1225–1237.
40. H. Ismail, M. Irani and Z. Ahmad, *J. Appl. Polym. Sci.*, 2013, **127**, 4195–4202.
41. C. Xiao and M. Yang, *Carbohydr. Polym.*, 2006, **64**, 37–40.
42. Y. Liu, N. E. Vrana, P. A. Cahill and G. B. McGuinness, *J. Biomed. Mater. Res., Part B*, 2009, **90**, 492–502.
43. A. Papancea, A. J. M. Valente, S. Patachia, M. G. Miguel and B. Lindman, *Langmuir*, 2007, **24**, 273–279.
44. P. C. Parvathy and A. N. Jyothi, *Starch Stärke*, 2012, **64**, 803–812.
45. E. S. Dragan and D. F. Apopei, *Chem. Eng. J.*, 2011, **178**, 252–263.
46. N. A. Peppas and P. Colombo, *J. Controlled Release*, 1997, **45**, 35–40.
47. R. W. Korsmeyer, R. Gurny, E. Doelker, P. Buri and N. A. Peppas, *Int. J. Pharm.*, 1983, **15**, 25–35.
48. E. S. Abdel-Halim and S. S. Al-Deyab, *React. Funct. Polym.*, 2014, **75**, 1–8.
49. W. Zou, L. Yu, X. Liu, L. Chen, X. Zhang, D. Qiao and R. Zhang, *Carbohydr. Polym.*, 2012, **87**, 1583–1588.
50. B. S. Kaith, R. Jindal, A. K. Jana and M. Maiti, *Bioresour. Technol.*, 2010, **101**, 6843–6851.
51. M. R. Guilherme, R. S. Oliveira, M. R. Mauricio, T. S. P. Cellet, G. M. Pereira, M. H. Kunita, E. C. Muniz and A. F. Rubira, *Soft Matter*, 2012, **8**, 6629–6637.

52. J. K. Pandey and R. P. Singh, *Starch Stärke*, 2005, **57**, 8–15.
53. A. l. Serrero, S. P. Trombotto, P. Cassagnau, Y. Bayon, P. Gravagna, S. Montanari and L. David, *Biomacromolecules*, 2010, **11**, 1534–1543.
54. J. Kuang, K. Y. Yuk and K. M. Huh, *Carbohydr. Polym.*, 2011, **83**, 284–290.
55. K. Vimala, K. Samba Sivudu, Y. Murali Mohan, B. Sreedhar and K. Mohana Raju, *Carbohydr. Polym.*, 2009, **75**, 463–471.
56. D. D. L. Chung, *Applied Materials Science*, Chapman & Hall/CRC Press, Boca Raton, FL, 2001.
57. A. C. Balazs, T. Emrick and T. P. Russell, *Science*, 2006, **314**, 1107–1110.
58. E. T. Thostenson, C. Li and T.-W. Chou, *Compos. Sci. Technol.*, 2005, **65**, 491–516.
59. K. Deligkaris, T. S. Tadele, W. Olthuis and A. van den Berg, *Sens. Actuators, B*, 2010, **147**, 765–774.
60. M. Sadeghi, *Aust. J. Basic Appl. Sci.*, 2011, **5**, 887–895.
61. D. Zhang, J. Duan, D. Wang and S. Ge, *J. Bionic Eng.*, 2010, **7**, 235–243.
62. C. Chang, J. Peng, L. Zhang and D.-W. Pang, *J. Mater. Chem.*, 2009, **19**, 7771–7776.
63. C. Spagnol, F. H. A. Rodrigues, A. G. B. Pereira, A. R. Fajardo, A. F. Rubira and E. C. Muniz, *Carbohydr. Polym.*, 2012, **87**, 2038–2045.
64. S. R. Sershen, S. L. Westcott, N. J. Halas and J. L. West, *J. Biomed. Mater. Res.*, 2000, **51**, 293–298.
65. X.-L. Luo, J.-J. Xu, Y. Du and H.-Y. Chen, *Anal. Biochem.*, 2004, **334**, 284–289.
66. A. Hawkins, N. Satarkar and J. Z. Hilt, *Pharm. Res.*, 2009, **26**, 667–673.
67. A. Okada and A. Usuki, *Macromol. Mater. Eng.*, 2006, **291**, 1449–1476.
68. K. Kabiri, H. Omidian, M. J. Zohuriaan-Mehr and S. Doroudiani, *Polym. Compos.*, 2011, **32**, 277–289.
69. T. H. Ang, F. S. A. Sultana, D. W. Hutmacher, Y. S. Wong, J. Y. H. Fuh, X. M. Mo, H. T. Loh, E. Burdet and S. H. Teoh, *Mater. Sci. Eng., C*, 2002, **20**, 35–42.
70. N. S. Satarkar and J. Z. Hilt, *J. Controlled Release*, 2008, **130**, 246–251.
71. A. Murugadoss, A. Khan and A. Chattopadhyay, *J. Nanopart. Res.*, 2010, **12**, 1331–1348.
72. S. Bhattacharyya, S. Guillot, H. Dabboue, J.-F. Tranchant and J.-P. Salvetat, *Biomacromolecules*, 2008, **9**, 505–509.
73. S. Chatterjee, M. W. Lee and S. H. Woo, *Bioresour. Technol.*, 2010, **101**, 1800–1806.
74. F. H. A. Rodrigues, C. Spagnol, A. G. B. Pereira, A. F. Martins, A. R. Fajardo, A. F. Rubira and E. C. Muniz, *J. Appl. Polym. Sci.*, 2014, **131**, 15.
75. Y. Liu, J. Yang, P. Zhang, C. Liu, W. Wang and W. Liu, *J. Mater. Chem.*, 2012, **22**, 512–519.
76. J. Wu, J. Lin, M. Zhou and C. Wei, *Macromol. Rapid Commun.*, 2000, **21**, 1032–1034.
77. K. Haraguchi and T. Takehisa, *Adv. Mater.*, 2002, **14**, 1120–1124.

78. A. Olad, *Advances in Diverse Industrial Applications of Nanocomposites*, InTech, Rijeka, Croatia, 2011.
79. P. Maiti, K. Yamada, M. Okamoto, K. Ueda and K. Okamoto, *Chem. Mater.*, 2002, **14**, 4654–4661.
80. J. Wu, Y. Wei, J. Lin and S. Lin, *Polymer*, 2003, **44**, 6513–6520.
81. J. Wu, Y. Wei, J. Lin and S. Lin, *Polym. Int.*, 2003, **52**, 1909–1912.
82. A. Li, R. Liu and A. Wang, *J. Appl. Polym. Sci.*, 2005, **98**, 1351–1357.
83. A. Li, J. Zhang and A. Wang, *Bioresour. Technol.*, 2007, **98**, 327–332.
84. J. Zhang, A. Li and A. Wang, *Carbohydr. Polym.*, 2006, **65**, 150–158.
85. G. Güçlü, E. Al, S. Emik, T. İyim, S. Özgümüş and M. Özyürek, *Polym. Bull.*, 2010, **65**, 333–346.
86. K. Kabiri, S. Hesarian, M. J. Zohuriaan-Mehr, A. Jamshidi, H. Boohendi, M. R. Pourheravi, S. A. Hashemi, H. Omidian and S. Fathollahi, *J. Mater. Sci.*, 2011, **46**, 6718–6725.
87. S. M. Ibrahim, *J. Appl. Polym. Sci.*, 2011, **119**, 685–692.
88. B. Singh, D. K. Sharma, R. Kumar and A. Gupta, *Appl. Clay Sci.*, 2009, **45**, 76–82.
89. A. Ashamol, V. S. Priyambika, G. S. Avadhani and R. R. N. Sailaja, *Starch Stärke*, 2013, **65**, 443–452.
90. E. Kamoun and H. Menzel, *J. Polym. Res.*, 2012, **19**, 1–14.
91. M. Eid, *J. Inorg. Organomet. Polym.*, 2011, **21**, 297–305.
92. B. Xia, Q. Cui, F. He and L. Li, *Langmuir*, 2012, **28**, 11188–11194.
93. J. Ai, M. Rezaei-Tavirani, E. Biazar, K. Saeed Heidari and R. Jahandideh, *J. Nanomater.*, 2011, **2011**, 391596.
94. A. Pourjavadi, S. H. Hosseini, F. Seidi and R. Soleyman, *Polym. Int.*, 2013, **62**, 1038–1044.
95. Y. Wang, C. Chang and L. Zhang, *Macromol. Mater. Eng.*, 2010, **295**, 137–145.
96. N. Teramoto, T. Motoyama, R. Yosomiya and M. Shibata, *Eur. Polym. J.*, 2003, **39**, 255–261.
97. M. Cervera, J. Heinämäki, K. Krogars, A. Jörgensen, M. Karjalainen, A. Colarte and J. Yliruusi, *AAPS PharmSciTech*, 2004, **5**, 109–114.
98. I.-S. Kim and P. N. Kumta, *Mater. Sci. Eng., B*, 2004, **111**, 232–236.
99. A. Dufresne and M. R. Vignon, *Macromolecules*, 1998, **31**, 2693–2696.
100. X. Li, S. Xu, J. Wang, X. Chen and S. Feng, *Carbohydr. Polym.*, 2009, **75**, 688–693.
101. M. E. Gomes, A. S. Ribeiro, P. B. Malafaya, R. L. Reis and A. M. Cunha, *Biomaterials*, 2001, **22**, 883–889.
102. R. C. Thomson, M. C. Wake, M. J. Yaszemski and A. G. Mikos, in *Biopolymers II*, ed. N. Peppas and R. Langer, Springer, Berlin, 1995, vol. 122, pp. 245–274.
103. M. Alexandre and P. Dubois, *Mater. Sci. Eng., R*, 2000, **28**, 1–63.
104. I. Spiridon, M. C. Popescu, R. Bodârlău and C. Vasile, *Polym. Degrad. Stab.*, 2008, **93**, 1884–1890.
105. X. Huang, S. Xu, M. Zhong, J. Wang, S. Feng and R. Shi, *Appl. Clay Sci.*, 2009, **42**, 455–459.

106. S. K. Bajpai, Y. M. Mohan, M. Bajpai, R. Tankhiwale and V. Thomas, *J. Nanosci. Nanotechnol.*, 2007, 7, 2994–3010.
107. Y. Murali Mohan, K. Lee, T. Premkumar and K. E. Geckeler, *Polymer*, 2007, **48**, 158–164.
108. M. Rai, A. Yadav and A. Gade, *Biotechnol. Adv.*, 2009, **27**, 76–83.
109. K. Vimala, Y. M. Mohan, K. S. Sivudu, K. Varaprasad, S. Ravindra, N. N. Reddy, Y. Padma, B. Sreedhar and K. MohanaRaju, *Colloids Surf., B*, 2010, **76**, 248–258.
110. T. J. Gao, T. S. Lindholm, B. Kommonen, P. Ragni, A. Paronzini and T. C. Lindholm, *Biomaterials*, 1995, **16**, 1175–1179.
111. P. T. Sudheesh Kumar, S. Srinivasan, V.-K. Lakshmanan, H. Tamura, S. V. Nair and R. Jayakumar, *Carbohydr. Polym.*, 2011, **85**, 584–591.
112. R. L. Reis, A. M. Cunha, P. S. Allan and M. J. Bevis, *Adv. Polym. Technol.*, 1997, **16**, 263–277.
113. S. C. Mendes, R. L. Reis, Y. P. Bovell, A. M. Cunha, C. A. van Blitterswijk and J. D. de Bruijn, *Biomaterials*, 2001, **22**, 2057–2064.
114. T. T. Irvin, *Arch. Emerg. Med.*, 1985, **2**, 3–10.
115. P. M. Arockianathan, S. Sekar, B. Kumaran and T. P. Sastry, *Int. J. Biol. Macromol.*, 2012, **50**, 939–946.
116. S. Bandi, M. Bell and D. A. Schiraldi, *Macromolecules*, 2005, **38**, 9216–9220.
117. A. V. Reis, M. R. Guilherme, A. F. Rubira and E. C. Muniz, *J. Colloid Interface Sci.*, 2007, **310**, 128–135.
118. V. R. Sinha and R. Kumria, *Int. J. Pharm.*, 2001, **224**, 19–38.
119. M. R. Mauricio, M. R. Guilherme, M. H. Kunita, E. C. Muniz and A. F. Rubira, *Chem. Eng. J.*, 2012, **189–190**, 456–463.
120. V. Rocher, J.-M. Siaugue, V. Cabuil and A. Bee, *Water Res.*, 2008, **42**, 1290–1298.
121. Y. D. Liu and H. J. Choi, *Polym. Int.*, 2013, **62**, 147–151.
122. G. Giammona, G. Pitarresi, G. Cavallaro and G. Spadaro, *J. Biomater. Sci., Polym. Ed.*, 1999, **10**, 969–987.
123. M. R. Saboktakin, R. M. Tabatabaie, A. Maharramov and M. A. Ramazanov, *Carbohydr. Polym.*, 2010, **79**, 1113–1116.
124. D. Kushwaha, P. Dwivedi, S. K. Kuanar and V. K. Tiwari, *Curr. Org. Synth.*, 2013, **10**, 90–135.
125. J. M. Dugan, J. E. Gough and S. J. Eichhorn, *Nanomedicine*, 2013, **8**, 287–298.
126. Z. Y. Qian, S. Z. Fu and S. S. Feng, *Nanomedicine*, 2013, **8**, 161–164.
127. M. Wang and M. Thanou, *Pharmacol. Res.*, 2010, **62**, 90–99.

CHAPTER 8

# Starch-based Blends

JIWEI LI[a,b,c] AND XUEGANG LUO*[a,b]

[a] Engineering Research Center of Biomass Materials, Ministry of Education, Southwest University of Science and Technology, Mianyang 621010, China; [b] School of Material Science and Engineering, Southwest University of Science and Technology, Mianyang 621010, China; [c] School of Chemical Engineering and Technology, Harbin Institute of Technology, Harbin 150001, China

*Email: lxg@swust.edu.cn; lxg-2007@163.com

## 8.1 Introduction

Starch is a natural polymer with the advantages of being renewable, produced in abundance at low cost and totally biodegradable in a wide variety of environments.[1–3] Recently, the development of environmentally friendly and biodegradable materials from starch has attracted increasing attention in an attempt to substitute petroleum-based materials, which present important concerns in terms of pollution and sustainability.[1,2,4,5]

For almost any application, the ordered granular structure of starch is disrupted by heating with a plasticizer or gelatinization agent, and the so-called thermoplastic starch (TPS) was developed by this method.[1,4–7] TPS is economically viable, classified as biodegradable or compostable and can be produced from a range of native sources, including wheat, rice, corn, potato, pea and cassava.[3,4,6,7] However, pure starch normally produces unfavorable final product properties such as poor mechanical and long-term stability characteristics.[1,2,8,9] Furthermore, its mechanical properties are very sensitive to moisture content, which restricts its widespread application.[10–12] In order to overcome these drawbacks, TPS has been blended with other

RSC Green Chemistry No. 37
Starch-based Blends, Composites and Nanocomposites
Edited by Visakh P. M. and Long Yu

Published by the Royal Society of Chemistry, www.rsc.org

polymers, as polymer blends can offer a simple, rapid and cheap method to achieve property combinations that are not generally available in any single polymeric material.[3,4,6,9,13]

## 8.2 Starch-based Blends

The development of biodegradable starch-based blends has attracted increasing attention owing to several advantages, such as (i) improving properties such as its water resistance and mechanical performances, (ii) adjustable rates of degradation and (iii) decreasing the cost of the ensuing blends.[3,4,6,9,13] Starch-based blends can be divided into two main categories according to the source and biodegradation properties of the polymer to be blended with starch, namely partially and completely biodegradable.

### 8.2.1 Partially Biodegradable Starch-based Blends

The initially studied partially biodegradable starch-based polymer blends were based on polyolefins since the 1970s.[14–16] In this case, granular starch was mixed with molten thermoplastic resins without gelatinizing the starch, and starch does not contribute to the mechanical properties of the mixture because it retains its granular form. Basically, starch-filled polymers had remarkably low physical property performance.[17–19] Also, the properties of starch-filled polymers decreased with increasing starch content as the weak starch particles were being "pulled-off" the polymer matrix.[17,19] As usual, a modified blending technique will improve the physical performance of starch-filled polymers. For example, Walker *et al.*[19] used a mechanical process called solid-state shear pulverization (SSSP) to create blends or composites of polyethylene (PE) and starch that resulted in damaged granular structures. Because starch granules are unchanged when polymer–starch blends are made by melt mixing, this is the first time that damage (surface roughening, cracking and clustering) to starch granule morphology was reported in polymer–starch blends or composites (Figure 8.1). These morphological changes result in a 29% reduction in oxygen permeability for a 70 : 30 wt% PE–starch blend made by SSSP relative to neat PE; this compares with a 21% reduction in oxygen permeability when a similar blend is made by melt processing. Recently, Sharif *et al.*[20] performed high-energy ball-milling (HEBM) before melt mixing to fabricate a novel blend containing 85 wt% of a high-density polyethylene (HDPE) and 15 wt% of a maize starch. It was found that the ball-mill-mediated sample showed superior mechanical and biodegradation properties to the HDPE mixtures that contained the same amount of starch but were made by conventional melt blending. These property enhancements were attributed to both the strong HDPE/starch interface and the damaged starch originating from the mechanical action of the milling process.

Although starch in its native form can also be used for the preparation of blends or composites with other thermoplastic polymers, the finer

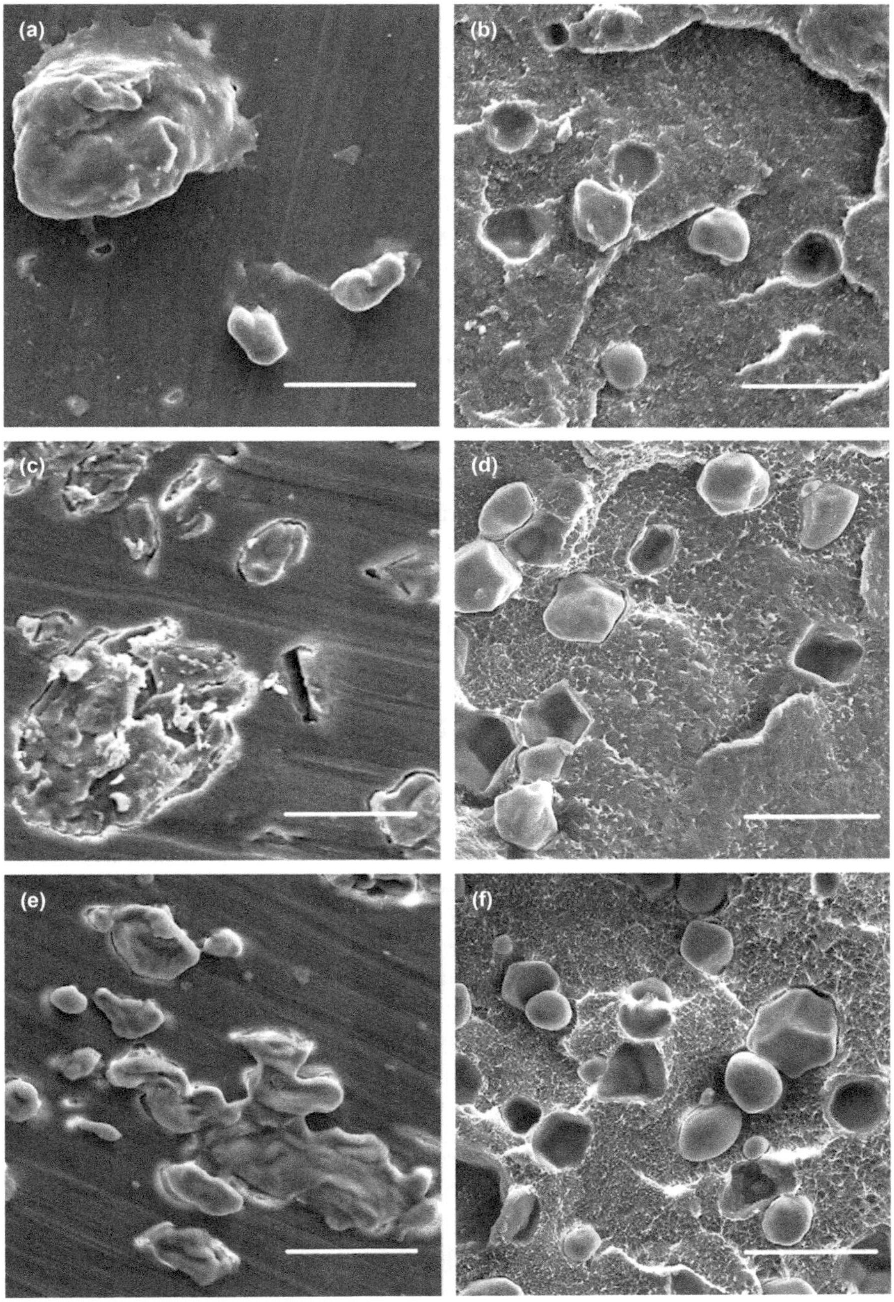


**Figure 8.1** Scanning electron micrographs of a 90 : 10 wt% HDPE–starch blend made by (a) pulverization and (b) melt mixing, an 80 : 20 wt% HDPE–starch blend made by (c) pulverization and (d) melt mixing and a 70 : 30 wt% HDPE–starch blend made by (e) pulverization and (f) melt mixing. Scale bar = 25 mm in all micrographs.[19]

dispersion of TPS or plasticized starch (PLS) in the polymer matrix makes it more suitable.[2,4,5,21,22] Blends of TPS or PLS with traditional non-biodegradable polymers were prepared in efforts to obtain new materials with low cost and high biodegradability.[1,2,4–6,23] Some common non-biodegradable polymers, such as low-density polyethylene (LDPE),[24–26] linear low-density polyethylene (LLDPE),[27,28] high-density polyethylene (HDPE),[29,30] polypropylene (PP),[31–33] polystyrene (PS),[34,35] poly(ethylene–vinyl acetate) (EVA),[36,37] poly(ethylene-*co*-acrylic acid) (EAA),[38,39] poly(vinyl chloride) (PVC),[40,41] polyamide (PA),[42,43] poly(methyl methacrylate) (PMMA)[44,45] and polyaniline (PANI),[46,47] have been used to blend with starch.

The properties of starch-based blends depend not only on the starch content, interface adhesion, the property of each individual polymer and the developed morphology, but also on the process employed.[25,48–50] For instance, Sabetzadeh *et al.*[26] prepared LDPE–thermoplastic corn starch (TPCS) blends containing different amounts of TPCS (0–40 wt%) and a constant amount of LDPE-*g*-maleic anhydride (MA) (3 wt%) by using a single-screw extruder. They found that the tensile properties and relative impact strength (RIS) of the samples decreased with increase in TPCS concentration whereas adding 25 wt% TPCS did not affect the mechanical properties of pure LDPE, which fulfilled the required mechanical properties for application in PE packaging products. Mortazavi *et al.*[51] investigated the effects of phase inversion on the properties of LDPE–TPS blends containing low-density polyethylene grafted with maleic anhydride (LDPE-*g*-MA) as compatibilizer. It was found that the morphology of the blend is matrix droplets until the TPS content exceeds 75 wt%, when phase inversion occurs. Further, the mechanical properties and glass transition temperature of the starch-rich phase decrease with increase in TPS content and these reductions are more significant near the phase inversion point. Moreover, the permeability properties of the blends change gradually at low TPS concentrations and the rate of these variations accelerates around the phase inversion point. Matzinos *et al.*[52] reported that TPS containing 20 wt% glycerol could act as a reinforcing agent in injection molded materials but not in films. It was suggested that the degree of molecular orientation imparted to the LDPE polymer matrix by the two shaping techniques is probably related to this behavior.

However, blends of TPS and polyolefins usually have a relatively high interfacial tension due to the high incompatibility of the non-polar polyolefin and the highly polar TPS.[13,40,44,52] Therefore, the mechanical properties of polyolefin–TPS blends are low because of the high interfacial tension of the blends. In the last two decades, the important drawback of incompatible blending has been realized by researchers and various approaches have been proposed and developed to overcome the problem of incompatibility of starch and synthetic polymer blends.[53–56]

Incorporating chemical compatibilizers in polyolefin–TPS blends is one of the foremost approaches to improve the compatibility of polyolefin and starch,[3–5,13] and polyolefin grafted with maleic anhydride is one of the most commonly used compatibilizers.[3,13,55–58] The esterification reaction between the maleic anhydride groups of the polyolefin-*g*-MA and the hydroxyl groups of the starch results in the *in situ* formation of a grafted copolymer that improves interfacial interactions, leads to a decrease in the dispersed phase size and improves the mechanical properties.[13,25,59–62] In addition, introduction of a third component as a compatibilizer, such as poly(ethylene-*co*-vinyl alcohol) (EVOH),[63] EVA,[64] EAA,[65,66] poly(ethylene-*co*-glycidyl methacrylate) (PEGMA)[67] and poly(LDPE-*g*-dibutyl maleate)[55] has also been found to improve the mechanical properties of the blend. More recently, Da Róz *et al.*[56] developed a compatible blend of TPS and hydrolyzed EVA and reported that the mechanical characteristics and water absorption of the TPS were improved even with a small addition (2.5 wt%) of hydrolyzed EVA. Therefore, the addition of hydrolyzed EVA proved to be an interesting approach to improve the properties of TPS, even when very small quantities were used, such as 2.5 wt%.

However, if too high a concentration of a compatibilizer is used in polymer blends, it could form a distinct phase and no longer contribute to the compatibilization.[53,55,56,67] Alternatively, instead of utilizing commercially available compatibilizers, a compatibilizing effect could also be induced during processing by the reactive extrusion technique. For example, Tanrattanakul and Panwiriyarat[53] enhanced the compatibility of LDPE–cassava starch blends by the addition of potassium persulfate (PPS) and benzoyl peroxide (BPO) to increase the interfacial adhesion between the LDPE matrix and starch granules. Wang *et al.*[57] prepared compatible PE–TPS blends by reactive extrusion and reported that in the presence of dicumyl peroxide (DCP), the thermal plasticization of starch and its compatibilizing modification with polyethylene were accomplished by one-step reactive extrusion in a single-screw extruder at the same time. Furthermore, the efficiency of the reactive compatibilization largely depends on the diffusion of the reactive component migrating to the interface and on the chain entanglement between the modifier and the pure phases in the interfacial region.[3,13,53,57]

Nano/micro inorganic fillers such as organoclay and zeolite could be used to improve the miscibility of TPS-based blends, in addition to enhancing the mechanical properties of the blends.[13,54,68] For example, Thipmanee and Sane[68] reported the compatibilizing effect of zeolite 5A on LLDPE–TPS blends.

Although someone may still be questioning the effectiveness of the biodegradability and practicality of using starch-filled synthetic polymers in reducing environmental impacts, degradation studies on starch-based blends have attracted much attention. It has been found that polyolefin–starch blends degrade faster than pure polyolefins in soil.[69–71] The inclusion of starch enhances the biodegradability of synthetic polymers, mainly due to an increase in the exposed polymer surface created as microorganisms consume the starch phase.[71–73] In practice, pro-degradants and starch are typically used together to develop degradable polyolefin–starch blends. In this case, the starch is consumable by microorganisms and then the remaining synthetic polymer matrix experiences attack by natural elements such as thermal oxidation and ultraviolet photodegradation.[71–75] Recently, new evidence regarding the acceleration of the degradation of polyethylene by starch was reported by Liu *et al.*[71]

Essentially, the efforts made in the development of partially biodegradable polyolefin–TPS blends have contributed positively towards the development of biodegradable materials and life-cycle assessment (LCA) should be used more to quantify the environmental sustainability of starch-based blends in future research.

## 8.2.2 Completely Biodegradable Starch-based Blends

Overcoming the limiting factors of pure starch materials while achieving complete biodegradability in blends can be accomplished with the blending of fully biodegradable polymers.[1,5,6,13,76,77] Both biodegradable synthetic and natural polymers have been blended with starch.

### *8.2.2.1 Blending with Synthetic Biodegradable Polymers*

Fully biodegradable synthetic polymers, such as poly(lactic acid) (PLA),[76,78–80] poly(ε-caprolactone) (PCL),[81–84] poly(vinyl alcohol) (PVA),[85–87] poly(butylene succinate) (PBS),[88–91] poly(butylene adipate-*co*-terephthalate) (PBAT)[92–94] and polyhydroxyalkanoates (PHAs),[95–97] show some interesting and reproducible properties, such as a more hydrophobic character, lower water permeability and some improved mechanical properties, compared with pure starch materials.[1,6,76,98,99] Blending starch with these degradable synthetic polymers can improve their cost competitiveness while maintaining other properties at an acceptable level.

**8.2.2.1.1 PLA–Starch Blends.** PLA is one of the most important biodegradable polyesters, with complete biodegradability and excellent mechanical properties.[76,78,79,100,101] The use of PLA as a component of starch blends has been widely reported in recent years. Early work on PLA–starch blends was focused on the incorporation of dry starch in PLA to reduce the cost of the material while maintaining biodegradability.[102,103] Martin and Avérous[104] first reported the properties of PLA–TPS blends. Previous studies revealed that PLA–starch blends showed excellent biodegradation compared with pure PLA and the biodegradation rates of PLA–starch blends were enhanced with increasing starch content, whereas the mechanical performance decreased.[99,105–107] Further, the crystallization behavior of starch filled PLA composites was examined and it was found that 1% starch effectively increased the crystallization rate of PLA.[108] For PLA–TPS blends, Cai *et al.*[100] also reported that TPS acts as a nucleating agent, improving the crystallinity of PLA.

In general, the blending of starch with PLA reduces the overall cost dramatically, improves the thermal properties and enhances the biodegradability of blends.[80,105] Nonetheless, direct blending of PLA and starch would yield blended products with low mechanical properties owing to the poor compatibility between hydrophilic starch and hydrophobic PLA.[109–111] Some effective means commonly used to improve the compatibility between hydrophilic starch and hydrophobic PLA include (1) the use of gelatinized or plasticized starch alone or combined with plasticizers such as glycerol, formamide, sorbitol and water[13,112,113] and (2) the addition of suitable compatibilizers, such as maleic anhydride,[109] phenylene diisocyanate[111] and methylenediphenyl diisocyanate, [113] during the blending process. Although many compatibilizers have been trialed, the one most commonly used and reported as being successful remains maleic anhydride. The anhydride group on maleic acid could react with the hydroxyl groups present in starch and the interfacial adhesion between starch and PLA was then significantly improved.[105,109,110] The mechanical properties obtained for PLA–starch blends compatibilized with maleic acid are better than those obtained for virgin PLA–starch blends.[109,110] In addition, use of a premade block copolymer containing each of the immiscible phases is also a promising strategy for compatibilizing blends. Recently, biodegradable PLA-grafted amylose copolymer and starch-*g*-PLA have been used as compatibilizer agents in PLA–starch blends.[77,107]

**8.2.2.1.2 PCL–Starch Blends.** PCL is another prominent member of synthetic biodegradable polymer family. It has good water, oil, solvent and chlorine resistance, a low melting point and a low viscosity and is easily processed thermally.[114–116] Blends between starch and PCL have been well documented in the literature. The weakness of pure starch materials, including low impact resistance, low dimensional stability, poor processability and high moisture sensitivity, has been overcome by adding PCL to the starch matrix even at low PCL concentrations.[6,115,117] In addition, blends of starch and PCL are assumed to be completely biodegradable since each component in the blend is readily biodegradable and also compostable.[118,119] Furthermore, blending with PCL modifies the crystalline structure of TPS and the glass transition temperature of the PCL–TPS blend was varied with its composition.[120,121] For example, Cai *et al.*[120] found that TPS functioned as a nucleating agent to improve the crystallinity rate of PCL, which indicated that mobility constraints existed in the PCL chains with increase in TPS, leading to a decline in the crystallization ability of PCL.

However, the mechanical properties of the PCL–starch blends become poorer with increase in the starch content in the blend.[121,122] This may be ascribed to the incompatibility between the hydrophobic PCL and the hydrophilic starch.[117,122] Much effort has been committed to improving the compatibility of the PCL–starch blends, involving the grafting of compounds onto the polymers by using various coupling agents or bonding agents or the pretreatment of polymers with suitable chemicals.[13,123,124] For instance, MA was often used as a coupling agent and it improved the mechanical properties of PCL–starch blends.[125,126] In addition, Chen *et al.*[127] used starch-grafted-poly(L-lactide) copolymer (St-*g*-PLLA) as a compatibilizer to improve the interfacial adhesion between the PCL matrix and starch. Tan *et al.*[124] used synthesized starch-modified polyurethane (St–PCL) as a compatibilizer to compatibilize PCL–starch blends and found that a small amount of St–PCL can effectively improve the compatibility of the blends. Moreover, starch-*g*-PCL has been used as a compatibilizer to enhance the interfacial adhesion between PCL and starch phases, and the properties of the blend were improved considerably.[128,129] It was also confirmed that poly(ethylene glycol) (PEG) with a suitable molecular weight could stabilize PCL–starch blends by locating at the interface and interacting with both the PCL phase and starch phase.[123]

**8.2.2.1.3 Other Polyester–Starch Blends.** In addition to PLA and PCL, other polyesters have also been blended with starch. PHAs are a family of polyesters derived from a variety of microorganisms.[130,131] The use of starch in blends with PHAs could result in completely biodegradable materials with desired physical properties and a concomitant reduction in cost. However, PHA–starch blends showed a lack of success in retarding the inherent brittleness of PHAs.[95,97] For example, Innocentini-Mei *et al.*[95] investigated the blending of poly(3-hydroxybutyrate) (PHB) with three starch derivatives, natural starch, starch adipate and grafted starch–urethane, to try to improve properties, but met with limited success and obtained brittle materials.

PHB is a homopolymer of 3-hydroxybutyrate, partially crystalline, and is one of the most widely investigated members of the PHA family.[132,133] Some attempts have been made to produce PHB–starch blends with improved physical properties. Godbole *et al.*[96] reported that blends with a PHB:TPS ratio of 0.7 : 0.3 showed the best properties and reduced cost compared with other blends. Lai *et al.*[133] used three types of TPS, potato starch, corn starch and soluble potato starch, to blend with PHB and the PHD–TPS (soluble starch) blend gave the highest level of water absorption and weight loss.

PBAT is a biodegradable aliphatic–aromatic copolyester having good thermomechanical properties and has turned out to be an interesting polymeric partner in starch-based melt blends.[92,93,134] Previous studies have shown that the PBAT–starch blend has better mechanical and barrier properties than starch.[93,94] Furthermore, in view of its high toughness and biodegradability, PBAT is considered a good candidate for toughening TPS.[135] Biodegradable films composed of TPS and PBAT with excellent mechanical properties and adequate water vapor barrier properties were obtained by Brandelero and co-workers[136,137] and Olivato *et al.*[138] Nabar and co-workers[139,140] successfully fabricated PBAT–starch blend foams.

PBS is a commercially available aliphatic polyester with many interesting properties, including biodegradability, melt processability and thermal and chemical resistance.[89,91,131,141] Recently, a great deal of work has been carried out on PBS–starch blends.[88–90,140,141]

As expected, melt blending TPS with these polyesters resulted in a significant improvement in the properties of plasticized starch. However, high levels of starch and its poor compatibility with the biodegradable polyesters led to poorer physical and mechanical properties, such as density, compressibility, resilience and flexibility.[88,142,143] The poor interfacial adhesion between the hydrophilic starch and the hydrophobic polyester remains one of the great difficulties in making polyester–starch blends.[90,93,110,142,144] To reduce the tendency for phase separation, various compatibilizers, such as anhydrides,[93,109,110] citric acid (CA),[93,142] PEG,[144] epoxy–styrene–acrylic copolymer[145] and diisocyanate,[111,113] have been added to polyester–starch blends. Another effective way to improve the compatibility between starch and the polyester matrix is to functionalize the starch or polyester component by grafting highly reactive functional groups.[77,92,128] Zeng *et al.*[90] reported a simple and efficient way to develop PBS–starch blends with improved compatibility through reactive extrusion of NCO-terminated PBS with TPS.

**8.2.2.1.4 Starch–PVA Blends.** PVA is also one of the first synthetic polymers to be blended with starch and has excellent mechanical properties and compatibility with starch.[85–87] Moreover, PVA–starch blends are assumed to be biodegradable since both components are biodegradable in various microbial environments.[86,87,146]

In general, blending starch with PVA improves the mechanical properties of starch-based materials and the blend ratio and PVA molecular weight can be adjusted to tailor the mechanical properties of the material for a variety of applications.[85,86,147] Since starch and PVA molecules have a large number of hydroxyl groups, the formation of hydrogen bonds by the hydroxyl functional group (OH) between starch and PVA tends to promote localized stability and subsequently improve the miscibility of starch and PVA.[87,148,149] However, PVA–starch blends have low water resistance, poor water barrier properties and low mechanical properties in high relative humidity environments. For example, Mao *et al.*[150] reported that starch–glycerol–PVA formulations were highly sensitive to relative humidity (RH) and the tensile strength decreased significantly with increase in RH. Attempts have been made to improve the properties of PVA–starch blends further to meet the demands of extensive applications. Some of the effective means commonly used to improve the properties of PVA–starch blends include reinforcement with nanoclays[98,151] and chemically modification of PVA or starch such as by crosslinking,[152–155] grafting[156–158] and surface modification.[159,160]

### 8.2.2.2 *Blending with Natural Biodegradable Polymers*

Natural polymers such as cellulose and their derivatives,[161–163] chitosan,[164–166] protein[167–172] and konjac glucomannan (KGM)[173–175] are inherently biodegradable and exhibit unique properties. A number of investigations have been devoted to study the blend of them with starch.

**8.2.2.2.1 Starch–Cellulose Blends.** Cellulose, a homopolymer of β-(1→4)-linked D-anhydroglucopyranose units, is the major substance obtained from vegetable fibers.[161,163] Since cellulose materials have the advantages of being renewable, biodegradable, abundantly available and inexpensive, cellulose–starch blends have been widely investigated.[161,163,176]

Usually, cellulose–starch blends are more like cellulose-reinforced starch composites. Like many other fiber-reinforced composites, improving the mechanical properties is one of the major driving forces to reinforce starch-based materials using cellulose, and previous work showed that the incorporation of these fibers into TPS led to an increase in tensile strength and elastic modulus but a decrease in elongation.[177–179] For example, Curvelo *et al.*[179] used cellulosic fibers as reinforcement for TPS and the composite showed an increase of 100% in tensile strength and more than 50% in modulus compared with non-reinforced TPS. Woehl *et al.*[177] prepared biocomposites of TPS reinforced with partially hydrolyzed bacterial cellulose (BC) nanofibers and found that the elastic modulus of the bionanocomposite ($575.7 \pm 166.7$ MPa) was 17 times higher than that of the starch matrix ($33.4 \pm 4.3$ MPa). Furthermore, previous studies have shown that cellulose (or its derivatives)–starch blends are biocompatible, the blend showing an increase in the dimensionally and thermomechanical stability, while the rigidity and water sensitivity decreased with increase in the cellulose content.[180–183] Recently, blends of hydroxypropylmethylcellulose (HPMC) with up to 70% of hydroxypropyl starch (HPS) were prepared by casting.[163] According to the authors, the transparency of the films decreased with increase in HPS content up to 70% and then increased with HPS contents higher than 80%, which indicates that the system is immiscible.[163] Furthermore, adding PEG increased the transparency of the blends, indicating that the compatibility between HPMC and HPS had been improved.[163]

**8.2.2.2.2 Protein–Starch Blends.** Proteins are thermoplastic heteropolymers and the primary structure of proteins consists of a sequence of amino acids, linked together by peptide bonds.[167,168,184] Plant proteins that have been used for starch-based blends include zein,[169,170] gluten,[185,186] soy protein isolate (SPI)[167,168,184] and peanut protein.[171,187] Animal proteins such as silk protein[188,189] and gelatin[172,190] have also been used to blend with starch. Compared with animal proteins, plant protein sources were more suitable for the preparation of starch-based blends owing to the renewability of the raw material and ready availability in large quantities and the low cost.[169,184,188,190,191]

Blending of starch with proteins to produce biodegradable plastics is a promising area, since both can be processed in the presence of plasticizers, similarly to most conventional synthetic thermoplastic polymers.[6,169,184] Blends of starch and zein, plasticized by glycerol, were studied by Corradini *et al.*[192] Recently, Leroy *et al.*[193] used glycerol and 1-butyl-3-methylimidazolium chloride ([bmim]Cl) as plasticizers to prepare zein–starch blends. Additionally, blending starch with proteins reduces its sensitivity to water.[194,195] For example, Huang *et al.*[194] reported that the water absorption of soy protein–starch blends decreased with increasing injection molding temperature, up to 130 °C, which could be attributed to increased protein interactions after sufficient denaturing. Furthermore, above 130 °C, the water absorption increased slightly, indicative of protein degradation.

The processability, thermal stability, water vapor permeability and thermodynamic and mechanical properties of protein–starch blends or composites have also been investigated.[194,196–199] For example, Chinma *et al.*[197] found that addition of soy protein concentrate (SPC) to cassava starch resulted in improved functional and pasting properties, which can be desirable in biofilm preparation. Moreover, Chanvrier and co-workers[196,200] showed that the mechanical properties of processed corn flour and zein–starch model blends are poorer than those of both pure starch- and pure zein-based materials. Huang *et al.*[194] reported that the tensile properties of soy protein–starch blends increased with increase in injection molding temperature (up to 130 °C), which can be attributed to increased protein interactions after sufficient denaturing.

However, the thermodynamic incompatibility between starch and proteins is the main problem that limits the significant development of protein–starch blends.[184,193,201] Habeych *et al.*[201] used aldehyde starch as compatibilizer for the *in situ* compatibilization of zein–starch blends and the compatibilized blends showed improved mechanical properties, indicative of good adhesion. Recently, Leroy *et al.*[193] compatibilized starch–zein melt processed blends using [bmim]Cl. The characterization of the materials indicated that, compared with glycerol, the use of [bmim]Cl leads to less hygroscopicity, more efficient plasticization of both the starch and zein phases and compatibilization of starch–zein blends.

Gelatin is prepared by the thermal denaturation of collagen[172,190] and has been blended with starch. Like other plant protein–starch blends, gelatin and starch are immiscible and their morphology and compatibility are affected by various factors, such as processing time, temperature, pH and solids concentration.[172,190,202]

**8.2.2.2.3 Chitosan–Starch Blends.** Chitosan is a natural carbohydrate polymer obtained by the deacetylation of chitin {poly[β-(1→4)-*N*-acetyl-D-glucosamine]} and it is antimicrobial, biodegradable, non-toxic and renewable.[164–166] The blending of starch and chitosan not only improves the mechanical and water vapor barrier properties and antimicrobial attributes of starch film, but also reduces the cost and enhances the biodegradability of chitosan films.[164,166,203–205] There have been a considerable numbers of studies investigating chitosan–starch blend films for food packaging and it was found that the properties (water vapor permeability, mechanical properties and miscibility of biodegradable blend, *etc.*) of chitosan–starch films can be affected by the ratio of starch to chitosan.[164,203,205,206] For example, Zhai *et al.*[205] reported that the incorporation of 20 wt% chitosan in 80 wt% starch film maximized the tensile strength of chitosan–starch film. Chillo *et al.*[206] reported that compared with each individual component, the elongation at break and water vapor transmission rate (WVTR) of chitosan–starch films were greatly improved. The inter- and intramolecular hydrogen bonding that formed between amino groups and hydroxyl groups on the backbone of the two components plays an important role in improving the properties of the blends.[166,203,206] However, a major drawback of chitosan–starch film is its brittleness and sensitivity to water.[164,207] With the addition of plasticizers, the brittleness of the films can be reduced and flexibility and extensibility can be improved.[164,207,208] For instance, Bangyekan *et al.*[164] studied the influence of chitosan and glycerol concentrations in cassava starch films produced by casting and observed that the water vapor permeability decreased with increase in the concentration of chitosan and increased with increase in the concentration of glycerol. This behavior was considered to be due to the hydrophobicity of chitosan.

## 8.3 Modification of Starch-based Blends

### 8.3.1 Gelatinization

For almost any application, the ordered granular structure of starch is disrupted by heating with a plasticizer or gelatinization agent.[5,209,210] This process is known as gelatinization, which is the basis of the conversion of starch to a thermoplastic.[5,18,209] It has been demonstrated that different degrees of starch gelatinization result in different TPS properties.[5,22,209,211] Since the properties of TPS are very sensitive to gelatinization, the properties of starch-based blends can also be adjusted by changing the TPS gelatinization level. A number of researchers have investigated the effects of blending both gelatinized and ungelatinized starch with biodegradable aliphatic polyesters.[21,209,211,212] In Dubois and Narayan's work,[209] high molecular weight PCL and PLA were melt blended with starch, either as corn starch granules or as thermoplastic corn starch, after plasticization with glycerol. They found that biodegradable compositions can be effectively reached by melt blending of granular or thermoplastic corn starch with commercially available PCL or PLA after adequate compatibilization using reactive melt processing. Dean *et al.*[212] investigated ungelatinized and gelatinized starch–PCL blends and light microscopy indicated that blends with gelatinized starch had better interfacial adhesion. Furthermore, the efficacy of gelatinization was demonstrated by a dramatic sixfold reduction in the dispersed TPS phase size and a unimodal TPS phase size distribution in the TPS–PCL blend system.[211] Recently, Hubackova *et al.*[22] assessed the effects of four different types of starch (native Meritena starch, waxy starch from genetically modified maize, Gel-Instant pregelatinized maize starch and amaranth starch) on the degree of biodegradation of polymer blends containing PCL. The results showed that the mixtures of PCL with starches plasticized with glycerol exhibited improved mechanical properties and a higher degree of biodegradation in the anaerobic environment and mixtures of PCL with pure forms of starch were ascertained to be fairly resistant to the anaerobic aqueous environment.

In order to investigate the effect of gelatinization on the properties of starch-based blends, Luo *et al.*[21] prepared ungelatinized and gelatinized corn starch–PVA blend films by a solution casting method. A droplet phase was observed in the blends containing ungelatinized starch and a laminated phase in the blends containing gelatinized starch. Additionally, thermogravimetric analysis results showed that over the first 300 °C, the weight loss for ungelatinized starch–PVA blends was higher than that for gelatinized starch–PVA blends; however, the gelatinized starch–PVA blends showed a greater weight loss after scanning up to 400 °C. These various morphologies and thermal behaviors were attributed to the different hydrogen bonding interaction characteristics between starch and PVA under different conditions (Figure 8.2). It was suggested that hydroxyl groups are insufficient for the formation of hydrogen bonds and the interaction intensity is low as most of the hydroxyl groups are embedded in the internal structure of the starch granules. When casting solutions were prepared at a higher temperature, the starch granules eventually collapsed. As a result, the number of free hydroxyl groups in this system increased, which contributed greatly to the interaction between PVA and gelatinized starch.

The gelatinization of starch using plasticizers can significantly improve processing, and thus also positively influence the mechanical properties of starch-based blends.[18,21,22,209,211,212] Therefore, before blending with other polymers, application of a process for gelatinization of starch is desirable.[4,5,18,21,209]

## 8.3.2 Amylose/Amylopectin Ratio

Starch consists of two major components, linear amylose and highly branched amylopectin, at ratios that vary with the starch source.[3–5,108]

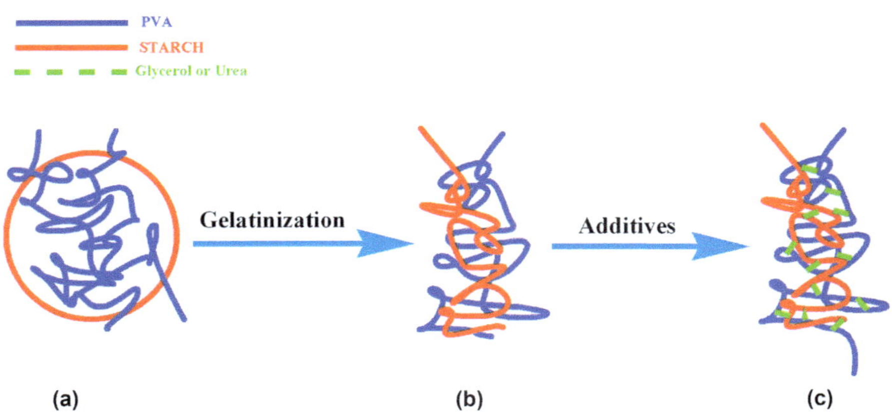


**Figure 8.2** Schematic presentation of preferential interactions between starch and poly(vinyl alcohol) in various PVA–starch blend films.[21]

This variation provides a natural mechanism for regulating material properties of starch. Previous studies reported that starches with different amylose:amylopectin ratios have different phase transition behaviors and rheological properties.[3,4,7,213–218] Further, TPS made from high-amylose starch has better thermal and mechanical properties.[219,220] However, the processing (especially extrusion) of high-amylose starch is much more difficult than that of normal starch, because a higher die pressure is required owing to the higher melt viscosity and unstable flow under some conditions.[218,220,221]

Also, the type of starch used in the polymer blend contributes significantly to the properties of the blend. Previous studies have demonstrated that altering the types of starch may be a promising and effectively approach to adjust the morphology and mechanical, thermal and dynamic mechanical properties of starch-based blends.[7,108,222–226] For example, Wang *et al.*[223] reported that the morphology and tensile properties of LDPE–starch blends were dependent not only on granule size but also on the intrinsic properties of starch. In another study, dynamic mechanical analysis (DMA) results indicated that the glass transition temperature of starch decreased as the amylose content increased in the extrusion and injection-molded ethylene–vinyl acetate–maleic anhydride (EVAMA)–starch, low-density polyethylene–maleic anhydride (EMA)–starch and high-density polyethylene–maleic anhydride (HDPEMA)–starch blends.[227] Additionally, blends containing high amylopectin contents in starch showed a co-continuous phase formation with synthetic polymers and high-amylose blends showed phase separation between starch and synthetic polymers.[227] Taghvaei-Ganjali *et al.*[226] reported that the physico-mechanical properties of natural rubber (NR) compounds filled by various starches are mainly correlated with the amylose-to-amylopectin ratio in the starch samples and the higher the content of amylopectin, the greater is the possible rubberization of filler (because of the highly branched and networking structure of amylopectin). Moreover, Yun and Yoon[222] reported that with higher amylose contents of starch used in preparing PVA–starch blend film, the tensile strength and degree of swelling of films increased, whereas elongation at break and solubility decreased.

The amylose content of starch can also affect the water sensitivity of starch-based blends.[2,4,5,11,108,224] In Ke *et al.*'s study,[108] four corn starches with amylose contents of approximately 0, 28, 50 and 70 wt% were blended at 185 °C with PLA at various starch-to-PLA ratios using an extruder. At starch contents <40%, differences in water absorption were not pronounced because the continuous PLA phase inhibited water penetration into the blend. However, as the starch content increased, the water absorption of blends with gelatinized starch increased as the amylopectin content increased, because the crystalline region of starch melted upon gelatinization and the swelling of amylopectin played a major role in water absorption.[108] In addition, Le Bolay and Boisseau[225] reported that the presence of starch with a high amylopectin content favors the water uptake of poly(vinyl acetate) (PVAc)–starch blends if the starch is well dispersed in the matrix under the effect of an optimized co-grinding treatment.

Several studies have established how the amylose content alters the properties of starch-based blends.[97,224,228] For example, Zhang and Thomas[97] reported that PHB–starch blends containing 70% amylose have better thermal, rheological and mechanical properties than a blend containing 72% amylopectin, which can be attributed to the enhanced hydrogen bonding between PHB and starch with a high amylose content. Phetwarotai *et al.*[228] reported that the tensile properties of gelatinized corn starch (GCS)–plasticized PLA blend films were slightly better than those of gelatinized tapioca starch (GTS)–plasticized PLA blend films. However, the water absorption isotherms of GTS–plasticized PLA blend films were slightly higher than those of GCS–plasticized PLA blend films. This was attributed to the difference in amylose content between tapioca and corn starches (30.79 and 32.35 wt%, respectively), as the starch with a higher amylose content (corn starch) gave higher crystallinity, which led to better mechanical properties compared with the lower amylose starch (tapioca starch). Furthermore, blend films with GTS exhibited higher water absorption than its counterpart owing to its lower amylose content (higher amylopectin). More recently, Li *et al.*[224] compared PBS–waxy thermoplastic starch (WTPS) and PBS–normal thermoplastic corn starch (NTPS) blends and found that the moisture absorption isotherms of PBS–WTPS blends are slightly lower than those of the corresponding PBS–NTPS blends (Figure 8.3). This behavior was attributed

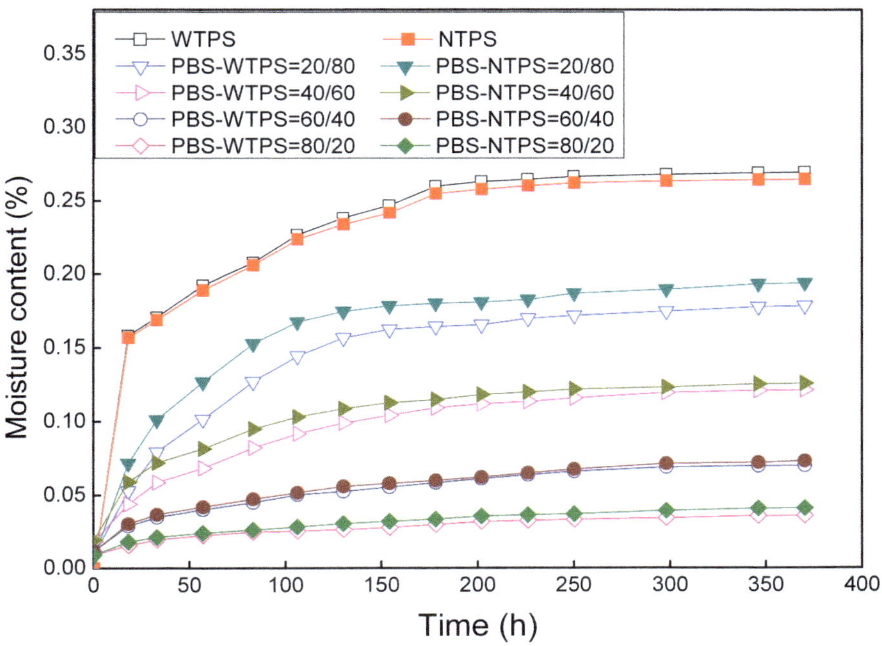


**Figure 8.3** Moisture absorption isotherms of different PBS–TPS blends with different PBS contents.[224]

to a finer dispersion of WTPS phase in the PBS matrix owing to its lower melt viscosity, which leaves fewer free hydroxyl groups for water binding. Additionally, the dense and homogeneous microstructure can hinder the diffusion of water molecules in the PBS–WTPS blends. Furthermore, compared with the PBS–NTPS blends, the combination of PBS with WTPS could result in some greatly improved performances such as excellent processability, excellent mechanical properties and higher water resistance. WTPS has a higher degree of plasticization and easier processability than NTPS, which can explain the resulting better performance of PBS–WTPS blends.

Generally, to account for the relationship between starch type (or amylose:amylopectin ratio) and properties of starch-based blends, complicated reasons including the degree of gelatinization and the microstructure of the blends should be considered.[97,213,224,226]

### 8.3.3 Additives

Various additives have been evaluated and used in different formulations of starch-based blends to improve the properties of the materials.[4,5,21,208,229] Compatibilizers have been mentioned earlier, and here additives such as plasticizers, crosslinking agents and chain extenders are considered.

#### *8.3.3.1 Plasticizer*

If dry granular starch is blended with other polymers directly, the resulting materials can be considered more as composites rather than as blends and they often possess very poor properties as the strong hydrogen bonds in native starch hinder the dispersion of these particles in a polymer matrix on both microscopic and molecular scales.[17,19,102,103] Various low molecular weight and polymeric compounds that have been used in TPS include water,[160,207,215] polyols (such as glycerol,[24,61,112,150,207,214,215] glycol,[144,207,230,231] sorbitol,[112,207,232,233] xylitol,[234,235] maltitol,[236,237] ethylene glycol,[207,238] propylene glycol[239,240] and butanediol[239,241]), amide-functionalized chemicals (such as urea,[242–244] formamide[243–245] and ethanolamine[242,246]), glucose,[247] sucrose,[247,248] fructose,[247] mannose[248] and acids (such as citric acid[155] and myristate[249,250]).

Water is a strong plasticizer for starch and is much more effective than glycerol.[207,215,251] However, the use of water alone as a plasticizer is not advantageous because the resulting product will be brittle when equilibrated with ambient humidity, which is a major drawback in many applications.[4,5,215,251] In addition, water would result in serious degradation of biodegradable polyesters (such as PCL and PLA) during processing at elevated temperatures, which results in the degradation of mechanical properties.[252–255] In order to overcome the effect of water, dry starch-based blends were developed in numerous studies. For instance, in order to achieve high performance of PLA–TPS blends, Wang *et al.*[252] used CA as an additive instead of water. In another study, the same group[254] used formamide and glycerol as plasticizers for dry starch alone or combined in different ratios to prepare dry PLA–dried thermoplastic starch (DTPS) blends by one-step extrusion. Further, Ma *et al.*[253] prepared biodegradable poly(propylene carbonate)–dried thermoplastic starch (PPC–DTPS) composites. Additionally, Lawton *et al.*[256] reported that the tensile properties of the starch–poly(hydroxyester ether) (PHEE) (40 : 60) composites were not significantly affected when the total moisture content was 6% or less. Furthermore, above 6% moisture content, the tensile strength and modulus decreased rapidly and composites with moisture contents greater than 7% showed extensive debonding of the PHEE from the starch granules.

Glycerol is typically used as a non-volatile plasticizer and is especially used in combination with water, which could improve the plasticization ability.[112,206,214] Furthermore, glycerol appears to be a less effective plasticizer than water, but it is a classical plasticizer of starch, and is perhaps the most widely studied and used polyol plasticizer of TPS.[5,32,61,247,252] This is due to its ready availability, low cost, non-toxicity (for food and biomedical applications) and high boiling point.[112,150,234,238] However, the use of glycerol also has some difficulties, including high hydrophilicity, low thermal stability, surface migration, TPS recrystallization and decreased mechanical properties because of the plasticization effect.[112,207,215,232,237,257] Sorbitol is another plasticizer that has frequently been used in TPS.[4,5,257] The main advantage of sorbitol in addition to its effective plasticization of starch is its higher thermal stability and lower water sensitivity compared with glycerol.[5,112,232,250]

In order to prevent retrogradation, some amide-based coplasticizers such as formamide and acetamide, which can form stable hydrogen bonds with starch, have been used.[243,244,254] However, these chemicals are toxic and would not be allowed in many food-contact and biomedical applications.[4,5,254]

Citric acid (CA) is usually utilized as a coplasticizer in TPS, since it forms stronger hydrogen bonds with starch than glycerol and, as a result, it provides aging resistance even at very low contents.[237,258,259] Furthermore, it was reported to increase the melt processing properties of TPS, which can be explained by acid hydrolysis of starch.[142,155,252,255] CA is composed of carboxyl and hydroxyl groups which increase the different interactions between the components of the blend.[27,155,258] For example, when added to TPS–LDPE blends, CA improved starch plasticization and the mechanical properties of the blend.[27] In another study, CA was used in PBAT–TPS blends to improve the interaction between the polymeric phases and produce films with better properties.[258] Moreover, CA has also been used as an additive to PVA–starch films.[260] It was reported that adding CA improved the compatibility of the polymeric phases and the mechanical stability of the materials, which can be attributed to better hydrogen bonding between CA and PVA–starch molecules than with glycerol.[260] In addition, the effect of water on the properties of CA-modified TPS–PLA blends was investigated.[255] It was found that when 10 wt% water was used as plasticizer in TPS–PLA blends, the fluidity and plasticization of TPS could be increased dramatically, but the serious depolymerization of PLA and starch degraded the mechanical properties of the TPS–PLA blend.[255]

In addition to conventional plasticizers, new kinds of plasticizers were investigated in several studies.[193,237,261] For instance, PVA–starch films were prepared with calcium chloride as the plasticizer.[261] The results indicated that $CaCl_2$ could interact with starch and PVA molecules and then effectively destroy the crystals of starch and PVA.[261] Furthermore, PVA–starch films plasticized with $CaCl_2$ became soft and ductile, with lower tensile strength and higher elongation at break compared with pure PVA–starch film, which is an important cause of the plasticization of $CaCl_2$ on PVA–starch film.[261]

Reactive plasticizers have been developed and used in starch-based blends[3,13,262,263] For example, Liu *et al.*[262] successfully prepared PP–starch blends with greatly enhanced mechanical properties by a one-step reactive compatibilization technique using maleated polypropylene (MAPP) as the interfacial agent and diethanolamine (DEA) as a reactive plasticizer for starch. It was found that the addition of MAPP can effectively reduce the size of the starch domains in the PP matrix and improve the interfacial bonding between starch and the polymer matrix because of *in situ* reactions among MAPP, starch and DEA. More recently, bio-based plasticizers have attracted attention owing to their renewability and they have been used to plasticize starch.[237,264] For example, epoxidized soybean oil (ESO) has been used as a plasticizer in a PLA–starch system.[264] It was found that the elongation at break and impact strength of PLA–ESO–starch increased markedly compared to virgin composites of PLA–starch blends, which indicated that ESO could be used as a bio-based reactive plasticizer for PLA and the starch compounds.[264]

Recently, the combination of different plasticizers was investigated as a means of overcoming or reducing problems such as the starch retrogradation, long-term plasticizer migration, water sensitivity, poor tensile properties and TPS embrittlement.[4,5,215,241,247] Combinations such as glycerol–water,[215,241] glycerol–sorbitol,[265,266] glycerol–CA,[267,268] sugar–water,[247] formamide–urea[269] and ethylenebisformamide–sorbitol[270,271] have been investigated. For instance, coplasticizing starch with glycerol–CA mixture is promoted as it increases starch plasticization, partial esterification can occur and chains with lower molecular weight are obtained.[267] In addition, it has been found that even without water, plasticization of starch with glycerol was possible when using a small amount of CA (0–4%).[255] Furthermore, when water was present, the blend was more homogeneous but thermal stability decreased.[255]

In general, PLA–starch blends are rigid and brittle owing to the brittle nature of both PLA and starch.[76,108–110] Therefore, the choice of suitable plasticizers to increase the ductility of PLA-starch blends is an important issue. The use of plasticizers such as glycerol,[112,272,273] sorbitol,[101,112,274] adipate and citrate esters[101,275,276] to increase flexibility and improve the processability of PLA–starch blends has been studied. The incorporation of plasticizers, which have a different molecular structure and show significantly different gelatinization temperatures, can potentially influence the state of the morphology of the starch-based blends.[112,237,257,277] For example, Li and Huneault[112] investigated the structures and properties of PLA–TPS blends where the TPS phase was plasticized by sorbitol, glycerol and glycerol–sorbitol mixtures. It was found that the sorbitol-plasticized TPS phase can be more finely dispersed and more uniformly distributed in the PLA matrix even in absence of any compatibilization. Further, the sorbitol-plasticized PLA–TPS blends exhibited much higher tensile strength and modulus. In addition, CA-coplasticized PLA–TPS blends have been reported in different studies.[252,255,278] For example, Wang *et al.*[252] investigated the influence of CA addition on a glycerol-plasticized PLA–starch system. The rheological investigations showed that CA decreased the viscosity both of TPS and of PLA–TPS blends, which resulted in a better dispersion by decreasing the interfacial tension between TPS and PLA, as also proved by scanning electron microscopy (SEM), and enhanced mechanical properties.[252] Additionally, Yokesahachart and Yoksan[229] investigated the effect of three different types of amphiphilic molecules – Tween 60, linoleic acid and zein – on the characteristics and tensile properties of TPS and its blends with PLA. It was found that amphiphiles facilitated processability, flowability and extensibility of PLA–TPS blends and linoleic acid was a more effective additive in enhancing the flowability of the PLA–TPS blends than were Tween 60 and zein.

When PVA–starch blend polymers are prepared, plasticizers generally have to be combined because of their high rigidity and low workability.[147,149,155,156] The conventional plasticizers used for PVA–starch blend polymers are water and polyols (such as glycerol,[279,280] sorbitol[260,281,282] and ethylene glycol[281]), which can improve the flexibility and workability, but several other chemicals, such as urea,[21] organic acids (*e.g.* malic acid,[98,282,283] tartaric acid, ascorbic acid, citric acid[260,284]), PEG[281,285,286] and glycerol–urea complex plasticizers[287] have also been successfully employed. More recently, the plasticizing mechanism and effect of calcium chloride on PVA–starch films were investigated by Jiang *et al.*[261] They found that the addition of $CaCl_2$ can remarkably improve the mechanical properties of PVA–starch film and overcome its brittleness, which effectively demonstrated that $CaCl_2$ can be used as a plasticizer for PVA–starch film.[261] In addition, the type of plasticizer plays a key role in determining the physical properties, such as mechanical properties, elongation at break and water resistance, of PVA–starch blend polymers.[98,260,288] For example, the strength and flexibility of the films with an added organic acid are superior to those of the films with an added polyol because those with the organic acid contain carboxyl groups as functional groups.[98,260,288]

However, there are some problems with using plasticizers in starch-based blends. First, a high plasticizer concentration is usually required in order to obtain a sufficient plasticization effect, which leads to instability and migration to the film surface.[232,240,245,257,260] Second, some small plasticizes only react with starch particles through hydrogen bonding and easily migrate from the matrix of TPS and result in the retrogradation of crystalline starch in TPS, which could cause deterioration of the mechanical properties of the starch-based blends.[232,246,249,259] In addition, some small-molecule plasticizers (*e.g.* water) may accelerate the hydrolytic degradation of polyesters during the melt blending process.[234,237,252,255] Moreover, plasticizers such as glycerol and sorbitol are hydrophilic compounds, which increase the sensitivity of blend films and their instability under different conditions of RH.[232–234,238]

### *8.3.3.2 Crosslinking Agents*

Crosslinking is an efficient and commonly used technique to increase the compatibility, water resistance and physical and mechanical properties of PVA–starch blends.[289–292] Crosslinking agents, such as EAA,[98,279] formaldehyde,[293] glutaraldehyde,[153,294] sodium hexametaphosphate,[292,295] epichlorohydrin,[293,296,297] hexamethoxymethylmelamine,[98,298,299] zinc oxide,[293] borax,[285,293] maleic acid,[282,283] boric acid,[288] CA,[260,284] ascorbic acid[289] and linear polyols,[98,282,283] have been used to react with the hydroxyl groups in starch or PVA to introduce modifications. For instance, Liu *et al.*[291] investigated the effect of post-crosslinking modification with glutaraldehyde on the properties of PVA–TPS blend films. It was found that the crosslinking modification significantly reduced the moisture sensitivity of the PVA–TPS blend films and increased the tensile strength and Young's modulus but decreased the elongation at break of the films. Yoon[289] investigated crosslinked potato starch-based blend films using ascorbic acid as a plasticizer and the results indicated that the cured films possess enhanced mechanical, thermal and optical properties compared with non-cured films.

In addition to crosslinking agents, crosslinking modifications of TPS and PVA can also be implemented by irradiation,[157,286,295,300,301] and irradiation seems easier and more convenient for practical applications. For example, previous studies have shown that surface photocrosslinking modification of TPS sheets and PVA–TPS films significantly reduces the surface hydrophilic characteristics and improves the water resistance of the materials.[295]

Although crosslinking improved the mechanical properties and the reduced moisture absorption of starch-based blends, some studies also reported that crosslinking reduced the rate of biodegradation.[293,302,303] For instance, Maiti *et al.*[293] investigated the effect of crosslinkers on the biodegradation behavior of PVA–starch blend films by a weight loss study, SEM and differential scanning calorimetry (DSC). The weight loss study showed that crosslinking retarded the biodegradation of the films in the first 15 days, but after that there was a significant increase in weight loss. Furthermore, the DSC analysis revealed that the consumption of starch and the consequent rearrangement of the PVA molecules were distinctly different in the crosslinked films due to the effect of different crosslinking agents.

Garg and Jana[304] observed that the addition of a crosslinked starch to LDPE increased the tensile strength and elongation of LDPE–starch blends in comparison with blends containing glycerol-modified starch. In addition, LLDPE–starch blends compatibilized with crosslinked and glycerol-modified starch produced smoother films.

#### *8.3.3.3 Chain Extenders*

PLA is highly vulnerable to thermal and oxidative degradation, especially when heated in the presence of water, which will result in decrease in molecular weight and deterioration of rheological properties.[145,305–308] Furthermore, because of the linear structure of PLA, it has relatively low melt strength.[305,308] These degenerative effects result in poor properties of TPS–PLA blends.[145,305–307] This limitation can be improved by the use of a polymeric chain extender through chain-end reactions. For instance, Li and Huneault[145] studied the effects of a multifunctional, epoxy-based chain extender on the properties of PLA–TPS blends. They found that the carboxyl end groups of PLA could potentially react with the epoxy groups in the chain extender and this led to an increase in the molecular weight through the chain-extension reaction. In addition, Zhang *et al.*[305] investigated the effect of a commercial polymer chain extender, Joncryl ADR 4370S (at a level of 1% w/w) on the physical properties of TPS and plasticized poly(lactic acid) (p-PLA) blends. The found that addition of Joncryl significantly improved the film tensile strength, 0.2% offset yield strength and elongation, especially evident with 250% elongation of 70 : 30 TPS–p-PLA film. Further, they also reported that addition of Joncryl had limited effects on the moisture content, water solubility, glass transition temperature and moisture sorption behavior of TPS–p-PLA blend films.

Considering that both PLA and starch are relatively brittle, Zhu *et al.*[264] used epoxidized soybean oil as a toughening modifier of PLA–starch blends to improve the toughness. They found that the elongation at break and impact strength of PLA–ESO–starch increased markedly compared with virgin composites of PLA–starch blends.

### 8.3.4 Radiation

Radiation technology has frequently been used to improve the properties of starch-based products owing to the chemical reactions among polymer molecules under irradiation.[157,205,309,310]

The poor processability and mechanical properties of PE–starch blends due to poor interfacial compatibility can be improved by radiation modification.[311–315] For example, Senna *et al.*[311] reported that the compatibility and tensile strength of LDPE–TPS foams was improved by the addition of a compatibilizing agent and electron beam irradiation. In another study, potato starch was radiolytically degraded by irradiation with $^{60}$Co gamma radiation and then plasticized using glycerol and water to obtain radiation-processed thermoplastic starch (RTPS).[314] The RTPS showed better miscibility with LDPE and the use of RTPS led to improved processability of the blends.[314] However, Ferreira *et al.*[312] obtained inconsistent results and found that ionizing irradiation of LDPE–modified starch blends (80 : 20 m/m) resulted in reductions of 11% in the tensile strength and 35% in percentage elongation. The inconsistent results of radiation modification of PE–starch blends can be attributed to the different processing methods and additives used in different studies. In addition, Ferreira *et al.*[313] sterilized LDPE–modified starch blends by gamma irradiation and they found that gamma radiation increased the fragility of the material and thus the biodegradation process was accelerated.

The irradiation modification of PVA–starch blend films has been reported by several groups. For example, PVA–starch blends have been prepared by using gamma,[316,317] plasma,[300,318] ultraviolet[87,157,319] and electron beam irradiation.[316,320] It was shown that irradiation treatment can improve the thermal stability and mechanical properties of PVA–starch blends, as they can cause a chemical reaction or graft functional groups on to the PVA backbone without adding an initiator or chemical substances.[316,318–320] Furthermore, results showed that surface photo-crosslinking modification of TPS sheets and PVA–TPS films significantly reduced the surface hydrophilic characteristics and improved the water resistance of the materials.[98,318,321] In addition, PCL–starch,[309,322] EVOH–starch[323,324] and chitosan–starch[205,310] blends showed improved properties. For example, Vignon *et al.*[322] reported the use of $\gamma$-irradiation and reactive extrusion in extensive process engineering to obtain PCL–TPS blends. The irradiated blends were thermally more stable than the non-irradiated materials and mechanical tests also showed that the efficiency of the irradiation depended greatly on the dose applied to the initial materials. Additionally, Zhai *et al.*[205] reported that after irradiation, there was no obvious change in the structure of chitosan–starch blend films, but antibacterial activity was induced even when the content of chitosan was only 5% owing to the degradation of chitosan in the blend films under the action of irradiation.

In addition to starch-based blends, starch-based blends containing fillers or fibers have also been modified by irradiation.[309,315] Recently, Yasin *et al.*[315] investigated the effect of electron beam irradiation on HDPE–starch blends containing functionalized sepiolite. It was found that the radiation imparted a crosslinked network, which embedded the additive sepiolite particles. Furthermore, the ultimate tensile strength, Young's modulus and thermal distortion temperature values of the nanocomposite were significantly improved after irradiation. In addition, Campos *et al.*[309] reported that after ultraviolet-C (UV-C) irradiation, the tensile properties of a TPS and PCL biocomposite with sisal bleached fibers decreased owing to chain scission. Moreover, samples of irradiated PCL and TPS showed an increase in crystallinity, whereas the blend and composites showed a decrease.

Microwave-assisted modification of starch was used for compatibilizing LLDPE–starch blends and it was observed that the addition of 10 wt% of compatibilizer with respect to the dispersed phase content led to a decrease in the starch phase size and to an improvement of the mechanical properties of the blends.[325] In addition, Le Bolay *et al.*[326] used co-grinding to improve the interface between the PLA matrix and the starch without adding any agent. It was shown that the properties of composite materials could be significantly improved in comparison with blends owing to better interactions between the components.

### 8.3.5 Filling Modification

Preparations of blends or composites using inorganic or natural fibers are among the routes to improve some of the properties of biodegradable polymers.[327,328] Lately, the application of organo-modified clays or natural fibers in polymer–starch blends has attracted great interest.[98,328,329]

#### *8.3.5.1 Nanoclays*

To improve further the moisture barrier and mechanical properties of starch-based blends, attentions has been directed to producing nanocomposites by adding nanoscale particles.[330–336] For example, previous studies have shown that whatever the clay type, the incorporation of nanoclays into PVA–starch, PCL–starch,[330,337] PLA–starch,[333,338] PE–starch[332] and PP–starch[334] blends led to an increase in the material rigidity, mechanical properties, water resistance and thermal stability. Moreover, since starch is only partially compatible with chitosan, the tensile strength and elongation at break of chitosan–starch blends were not significantly improved.[336,339,340] Ashori[336] first reported on the feasibility of biopolymer films made from blends of native starch and chitosan with graphene nanoparticles. The incorporation of a tiny amount of graphene presented a promising method for improving the mechanical and barrier properties of chitosan–starch biocomposite films. Further, Deepthi *et al.*[340] reported chitosan–TPS/multi-walled carbon nanotube (MWCNT) nanocomposites for electroactive polymer applications. The results showed that the incorporation of MWCNTs improved the flexural and wear properties and electrical conductivity of the nanocomposites.

The incorporation of nanoclays can improve the compatibility between starch and other polymers. For example, Ikeo *et al.*[337] suggested that addition of nanoclay could improve the compatibility between starch and PCL. Vertuccio *et al.*[330] found that the addition of inorganic materials dispersed at a nanometric level was an effective method to overcome the weak adhesion between starch and PCL. Thipmanee and Sane[68] used zeolite 5A as an inorganic phase to reinforce PCL–starch blends and reported that this improved the physical properties of the composites. Recently, Ferreira *et al.*[341] found that addition of clay results in dispersed TPS, which improved compatibility with PCL. Liao and Wu[332] investigated the effects of organoclay-modified compatibilizer–maleic anhydride grafted polyethylene (mPE-*g*-MA) on the properties of mPE–TPS/nanocompatibilizer composites. They showed that the average size of the starch dispersed phase in the mPE–TPS/nanocompatibilizer system was smaller than that of the mPE–pristine starch/nanocompatibilizer system. Moreover, it was demonstrated that the nanoclay-reinforced compatibilizer was more efficient in enhancing the thermal properties, elastic modulus and tensile strength of both mPE–pristine starch and mPE–TPS blends than the neat compatibilizer.

As usual, increased properties of starch-based blend composites are linked to the clay stiffness and dispersion state. In DeLeo *et al.*'s study,[334] transmission electron microscopy (TEM) indicated that natural sodium montmorillonite (MMT) was well dispersed in the plasticized starch (PLS) phase of the blends, whereas Cloisite 30B (which is MMT that has been organically modified with methyl groups) was located both within the PLS phase and at the interface between PLS and PP. In starch–PLA nanocomposites, Arroyo *et al.*[338] also found that MMT-$Na^+$ (Cloisite $Na^+$) was preferentially located in the starch phase or at the blend interface. Bocchini *et al.*[335] reported that by blending the TPS–nanoclay with 20 wt% of poly(butylenesuccinate-*co*-adipate) (PBSA), exfoliation of polymer chains into the Cloisite30B gallery was observed, which can be ascribed to the high affinity between Cloisite 30B and PBSA. Furthermore, Mondragón *et al.*[342] revealed that, in starch–natural rubber nanocomposites, MMT-$Na^+$ nanolayers were mainly dispersed in the natural rubber domains, forming a well-ordered intercalated structure. In these cases, MMT-$Na^+$ might have preferential polarity matching with one of the two polymers in the matrix and, as a result, interactions and stress transfer between the phases could be modified.[342] Further, the effect of clay type on the properties of starch-based blend composites has also been studied.[12,98,343] For example, Boonprasith *et al.*[343] used sodium montmorillonite (Cloisite $Na^+$) and organo-modified MMT (Cloisite 30B) as fillers in TPS–PBS/montmorillonite nanocomposites and found that the use of Cloisite 30B led to a nanocomposite with a higher tensile modulus value, whereas the use of Cloisite $Na^+$ slightly enhanced the thermal stability of the material.

Several studies investigated the enhancement mechanism of nanoclays.[331,339,344] For example, Tang and co-workers[331,344] developed several PVA–starch/nano-silicon dioxide ($SiO_2$) biodegradable blend films and discovered that the mechanism responsible for the improved tensile strength and water resistance of the composites was the strong chemical bond formation between nano-$SiO_2$ and PVA–starch blend. In another study, Ma *et al.*[339] reported that the addition of graphene oxide (GO) improved the miscibility of chitosan and oxidized starch, which resulted in improved mechanical properties, thermal stability and water resistance. They suggested that the carboxyl groups introduced into the starch and the incorporation of GO can improve the properties of the starch-based composites due to the synergistic interaction and hydrogen bonding between GO, chitosan and oxidized starch.

In addition to starch-based blend films, the incorporation of nanoclay-layered silicates has also been used to improve the properties of starch-based blend foams.[333,345,346] For example, Lee *et al.*[346] compared the addition of different MMTs (Cloisite 10A, Cloisite 25A, Cloisite 93A and Cloisite 15A) in the melt processing of tapioca starch–PLA nanocomposite foams. Their results showed that Cloisite 30B, instead of MMT-$Na^+$, resulted in the greatest extent of intercalation and also gave better functional properties. This was caused by strong hydrogen bonding between the –OH groups of the matrix and those of the Cloisite 30B organo modifier. Furthermore, Lee and coworkers[333,345,346] reported that the addition of nanoclay to starch–PLA foams can improve the thermal stability, mechanical and foaming properties of the blends.

However, addition of nanoclay to starch-based blends had the effect of drastically reducing the elongation and relative slowness of the degree of biodegradation.[98,327–329] Moreover, when starch-based blends are prepared by the blending and casting method, agglomeration occurs between inorganic nanomaterials and polymers.[329,347,348] Therefore, to solve the problems of reduced elongation, slow biodegradation and agglomeration, it is necessary to develop new nano-sized particles or pretreatment methods.[23,98,328,348]

### 8.3.5.2 Natural Fibers

Many studies have investigated starch-based blends filled with natural fiber (such as flax,[349,350] ramie,[351,352] jute,[353,354] kenaf core[355,356] and sisal[357,358]). The incorporation of natural fibers into starch-based blends can lead to enhanced biodegradability, higher strength and modulus and lower density of the composites at lower cost and without compromising biodegradability.[355,359–362]

One interesting approach to improve the mechanical properties and impart degradability is through the use of natural fibers to reinforce the LDPE–TPS blend. For example, Sarifuddin *et al.*[356] evaluated the effect of the incorporation of plasticized sago starch (TPSS) on the properties of LDPE reinforced with kenaf core fiber (KCF) and found that there was a gradual decrease in tensile strength, modulus and elongation at break with increase in TPSS loading. Moreover, the thermal stability of the composite was clearly improved with the addition of TPSS. Meanwhile, water uptake and the hydrophilic character of the composite system tended to increase as the TPSS was added to the composites. In another study, Sarifuddin *et al.*[355] reported that the incorporation of KCF into the blend caused considerable improvements in tensile strength and Young's modulus. Furthermore, the decomposition temperature was improved because of the higher thermal stability of the fibers, and DSC results illustrated the phase compatibility between components in the composite system. Meanwhile, water uptake tended to increase as the hydrophilic character of KCF was imparted to the composites.

Rosa *et al.*[363] reported that starch–EVOH blends reinforced with the treated fibers exhibited superior properties to neat starch–EVOH, which can be attributed to the chemical similarity of starch and fibers, providing good compatibility between fibers and starch.

Alvarez and co-workers[180,358,364] carried out systematic studies on the properties of cellulose derivative–starch blends with different sisal short fiber contents. They found that the addition of the sisal fibers did not have a great effect on the thermal degradation of composite materials but promoted a significant improvement of the composite creep resistance in comparison with the matrix alone. Furthermore, they investigated the effects of the moisture and fiber content on the mechanical properties of the composites and found that when the fiber content increased, the loss of flexural modulus slightly increased, but the effect of RH was greater than that of the fiber content.

The incorporation of sisal fibers into TPS and PCL can be a good alternative to improve the properties and decrease the density of TPS–PCL composites.[309,365,366] The crystallization, relationship between processing and properties and effects of fiber content and processing conditions on flexural creep behavior and also the effect of chemical treatment on the mechanical properties PCL–starch–sisal fiber composites were systematically studied by Cyras and co-workers.[357,367–369] In addition, Campos and co-workers also investigated the development of TPS–PCL–sisal fiber composites.[309,359,365] They reported that the improved properties of composites with bleached sisal fibers incorporated in TPS–PCL blends was due to the efficient dispersibility of the fibers in the matrix.[309,359,365] Moreover, Fabunmi *et al.*[360] prepared TPS–PCL–flax fiber composites and found that the maximum tensile strength of 11.55 MPa was observed at 10% flax fiber and 40% PCL reinforcement. They also reported that both flax fiber and PCL acted as independent reinforcements for plasticized pea starch as no interfacial bonding between the two was observed. However, in spite of the above-mentioned studies, there is a lack of information about the biodegradation and hydrolytic stability of PCL–starch blends reinforced with natural fibers. Therefore, di Franco *et al.*[366] investigated the degradation of PCL–starch blends and composites with sisal fiber (SF) and found that in soil burial, PCL–starch and 15% SF–PCL–starch were degraded to about 50% of the initial mass. Furthermore, the weight loss pattern showed by the composite was associated with the presence of strong fiber–fiber and fiber–matrix interactions, which are absent in the neat matrix. Recently, Taghizadeh and co-workers[362,370] investigated the biodegradation behavior and water adsorption of PVA–starch–carboxymethylcellulose–clay nanocomposites.

Several lignocellulosic fibers have been used in TPS–PLA blends to ensure better properties.[361,371,372] For example, Smitthipong *et al.*[372] reported that TPS with pineapple leaf fiber (PALF)–PLA composite gave better mechanical properties and water resistance than the TPS–PLA blend, but it showed the same flow behavior (based on the power law index) as the PLA alone. Further, Iovino *et al.*[371] investigated the biodegradability of PLA–TPS–short coir fiber composites and found that the fibers seemed to play a secondary role in the process, as confirmed by the slight differences in carbon dioxide produced. Furthermore, lignocellulosic fiber-reinforced materials have a short survival time in biotic environments such as compost and therefore after their use they are suitable for disposal in landfills.[371,372] Recently, Bocz *et al.*[361] reported that the combined application of surface-treated reinforcing flax fibers and plasticizer in biodegradable, flame-retardant PLA–TPS–chopped flax fiber biocomposites resulted in well-balanced strength and stiffness.

### 8.3.6 Ternary Blending

Blending with a third polymer can lead to ternary starch-based blends with excellent properties.[66,144,373,374]

Yang *et al.*[375] first studied the properties of ternary blends [starch–styrene–maleic anhydride (SMA)/ethylene–propylene–maleic anhydride (EPMA)] containing starch. They found that ternary blends of starch–SMA–EPMA gave products with a range of physical properties that depended on the ratio of SMA to EPMA in the blend, and as the SMA content increased, the samples became more rigid and had higher tensile strengths. In order to increase the biodegradability and also the mechanical properties of LDPE–starch blends, Matzinos *et al.*[373] added PCL to LDPE–starch blends. In addition, Prinos *et al.*[64] used EVA copolymer as a compatibilizer in order to increase the amount of plasticized starch that can be incorporated in the LDPE matrix. It was found that the elongation of the LDPE–starch–EVA blends was superior even to that of LDPE, which meant that higher amounts of plasticized starch (20–30 wt%) could be used for blending preparation with satisfactory mechanical properties. Moreover, the rheological and thermal behavior and thermo-oxidative degradation of starch–LDPE blends containing EAA have also been investigated.[66,376]

Considering that both PLA and starch are relatively brittle, the introduction of PCL, a ductile biodegradable polymer, to PLA–starch blends may improve the toughness of the blends.[117,377] Binary and ternary blends of PLA, TPS and PCL have been reported by Sarazin *et al.*,[117] who found that the addition of PCL greatly increased the ductility of PLA–TPS blends. Liao and Wu[377] also used PCL to moderate the brittle behavior of PLA in PLA–starch blends. In addition, PBAT is a flexible copolyester and can also be considered as a good candidate for the toughening of rigid polymers such as TPS–PLA binary blends.[137,378] For example, Ren *et al.*[378] reported that the mechanical properties of the TPS–PLA–PBAT blends exhibited a dramatic improvement in elongation at break with increasing PBAT content. Recently, Shirai *et al.*[379] evaluated the effect of the addition of PLA on the functional properties of TPS–PBAT films produced using the blown extrusion method. They found that the addition of PLA decreased the elongation, blow-up ratio and opacity and increased the elastic modulus, tensile strength and viscoelastic parameters of the films. Moreover, films with 20 wt% of PLA exhibited a lower water vapor permeability due to the hydrophobic nature of this polymer.

Some polymers were blended with PLA–starch blends to improve the weak interfacial affinity between PLA and starch.[380,381] For instance, since both starch and PVA are polyols and starch will form a continuous phase with PVA during blending, PVA can be used to enhance the compatibility of starch–PLA blends and the mechanical properties of the starch–PLA blends are improved.[380] In addition, Shi *et al.*[381] used glycidyl methacrylate grafted poly(ethylene octane) (GPOE) as a toughening agent and compatibilizer for TPS–PLA blends. They found that the elongation at break and impact strength of the ternary blends were greatly increased by the filling of GPOE. Moreover, compared with non-GPOE-containing binary blends, the morphology of ternary blends with GPOE indicated that starch granules melted and there was good compatibility between the PLA matrix and TPS. Additionally, ternary blends of PLA–PEG–starch have also been developed to improve the compatibility and mechanical properties of PLA–starch blends.[144,374,382]

In order to improve the antimicrobial activity of PLA–starch blend films, Bie *et al.*[383] used chitosan as the antimicrobial agent and reported the suitability of using ternary PLA–starch–chitosan films for antimicrobial packaging applications. Further, Gomes *et al.*[384] prepared starch–chitosan–PVA blend films and found that the addition of PVA led to an increase in film flexibility, as characterized by mechanical tests.

The lack of nitrogen needed for microbial metabolism will limit the biodegradation rate of PCL–starch blends.[385] In order to solve this problem, SPI was added to the blends.[385] It was found that increasing amounts of SPI and decreasing amounts of starch in PCL–starch blends showed a tendency to recover partially the elastic modulus of PCL, despite the decrease in tensile strength and elongation.[385] Furthermore, the elongation of the PCL–starch decreased owing to the stiffness imposed on the system by the presence of SPI.[385]

Applying a coating of another biodegradable polymer on a product's surface is another way to control the properties of starch-based materials.[386–388] Several studies have shown that coatings can improve the mechanical properties and water vapor barrier of starch films or sheets.[386,389,390] However, the coating method also needs to enhance the interfacial adhesion between the surface layer and TPS substrate.[164,388,389] Recently, this method was also used to modify the properties of starch-based blends. For example, Soares and co-workers[386,387] reported that the coating of TPS–PLA sheets with crosslinked chitosan improved the mechanical and water vapor barrier properties and reduced the water solubility owing to its hydrophobic characteristics.

Unfortunately, the third polymer usually only affects one component in the ternary blends and cannot produce perfect interfacial adhesion among all components.[391–394] Some methods have been developed in order to improve the properties of ternary starch-based blends.[393–395] For instance, Kahar *et al.*[394] investigated the properties of *N*,*N*′-*m*-phenylenebismaleimide (HVA-2)-vulcanized HDPE–natural rubber–TPS (HDPE–NR–TPS) blends and found that tensile strength, Young's modulus and elongation at break exhibited significant improvements when the blends were subjected to the HVA-2 crosslinker. It was suggested that the improvement in tensile properties and modulus could be correlated with the formation of crosslinks within the natural rubber phase. In addition, Kahar *et al.*[393] also investigated the effects of PE-*g*-MA as a compatibilizer and the effect of citric acid-modified starch on the properties of ternary HDPE–NR–TPS blends. Also recently, Ma *et al.*[395] tailored the morphology and properties of PLA–EVA–starch blends *via* reactive compatibilization in the presence of MA, benzoyl peroxide and glycerol.

## 8.4 Preparation of Starch-based Blends

The development of starch-based blends should include the knowledge of processing techniques to ensure workability of materials and the effects that different processes can have on various end material properties.[3–5,7] In addition, some studies have demonstrated that the blend production method and the processing technique can influence the properties of starch-based blends.[4,5,7,396–398] Various processing techniques, such as casting,[174,399–401] extrusion,[402–406] film blowing,[175,258,373,407,408] foaming,[140,409–411] injection and compression molding[102,227,412–418] and spinning,[419–422] have been adapted for processing starch-based blends.

### 8.4.1 Casting

Casting is essentially involves heat application but low mechanical input processing.[96,174,399,401,423] It first involves the dispersion or solubilization of the material in a solvent and in a second step the suspension or film-forming solution is placed in a suitable mold and the solvent is evaporated under controlled conditions.[4,399,423] Casting has been the most widely used technique for the laboratory production of starch-based films.[4,96,174,401,424] Most starch-blended films obtained by casting from a solution are based on starch and PVA.[399,401,423,425] Since the 1980s, the PVA–starch blends have been studied primarily for producing films by means of solution casting.[98,149,152] Polymers such as PHB,[96,423] chitosan,[165,205] KGM[174,400] and PS have been blended with starch by the casting method. For example, Godbole *et al.*[96] prepared PHB–TPS blend films by solvent casting and found that blends with a PHB:TPS ratio of 0.7 : 0.3 showed the best properties and had lower costs compared with other blends. In addition, Nair and Jyothi[400] prepared cassava starch–KGM blend films by casting and found that the blend films exhibited lower crystallinity, better mechanical properties and lower water vapor transmission rate and solubility in comparison with the neat starch film.

Casting has also been used to prepare filler-containing starch-based blends.[424,425] For example, jute micro/nanofibril (JNF)-reinforced PVA–starch based biocomposite films have been prepared by the solution casting method.[424] In another study, composite films of chitosan and sago starch

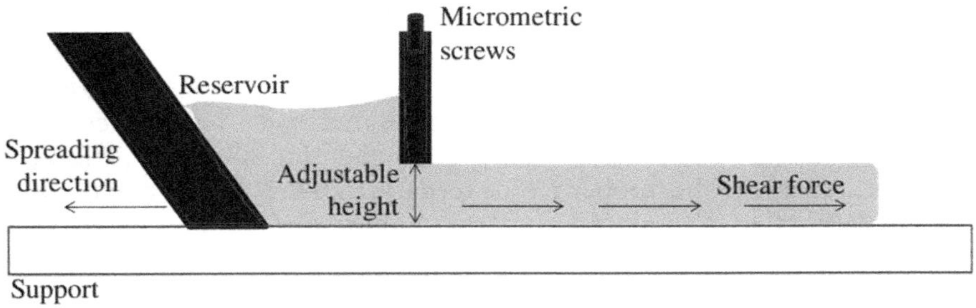


**Figure 8.4** Sketch of the tape casting process.[401]

impregnated with silver nanoparticles (AgNPs) with and without the antibiotic gentamicin were prepared by the solvent casting method.[425]

The properties of the casting blend films depend mainly on the casting conditions: temperature, type and concentration of solvent and the drying relative humidity and temperature.[4,165,423] Even though the casting method is simple and has been used in most of the research on films based on starch and its blends in most laboratories, the drying step makes this method slow, energy consuming and discontinuous, so it is not practical on an industrial scale.[4,96,423,424] Recently, the tape-casting technique has been used to prepare starch-based films. It involves the spreading of a suspension on large supports, with control of the thickness by an adjustable blade at the bottom of the spreading device, and the drying of the film can be carried out on the support itself, under controlled conditions.[399,401] Figure 8.4 presents a sketch of the tape-casting process. By repeating the tape-casting steps, it is possible to prepare multilayer starch-based films with different suspensions.[399,401]

## 8.4.2 Extrusion

Extrusion has the advantage of being a low-cost, continuous, energy efficient and versatile production system with easy industrial scale-up.[402,403,426,427] Thus, extrusion seems definitely to have advantages over casting, being a more attractive process for producing starch-based blends.[4,5,396,402,403,428] Various studies have been reported on the extrusion processing of starch-based blends. Zhu *et al.*[402] investigated the soy protein–high-amylose corn starch (SPC–HACS) blend for the first time and determined the effects of screw speed (230 and 330 rpm) on the physical properties of the blends. Compared with 230 rpm, a screw speed at 330 rpm resulted in increased specific mechanical energy, expansion ratio, water absorption and water solubility indices and decreased bulk density and piece density. In addition, varying the screw speed did not significantly affect the mechanical strength of extrudates or the starch molecular weight distribution. In another study, polyamide 12-plasticized starch blends were prepared by high-shear extrusion and the mechanical properties of the blends were found to increase with increasing shear rate of blend processing.[403]

In the extrusion processing of starch-based blends, melt blending, plasticization of starch and chemical reactions may occur at the same time, which will affect the blend morphology and properties of the blends.[4,106,402,403] Furthermore, processing pure TPS using a twin-screw extruder is difficult because of the very high viscosity of starch, and blending with other polymers is expected to improve the flow behavior of starch, ease the extrusion process and improve the poor mechanical properties of TPS.[99,108,111,140,141] However, in the processing of starch-based blends by extrusion, the temperature and shear stress are higher compared with those employed in a casting technique, which promotes the efficient destruction of starch chains.[5,111,134,381]

By simultaneous extrusion of a feed rod assembled from multiple materials, biphasic or multilayered structures are easily fabricated.[427,429,430] To combine the properties of two or more polymers into a single multilayered structure, a co-extrusion process has been widely used in the preparation of starch-based blends.[397,426,427,430] For instance, TPS and PLA have been processed by co-extrusion into laminated films and this multilayer arrangement structure decreases the water sensitivity of starch and improves the mechanical properties of the film.[397,426] Dole *et al.*[427] reported that the gas barrier properties of PE–starch blends were improved by using a multilayer structure (PE–PEg–starch–PEg–PE) (PEg = maleic anhydride-functionalized polyethylene). Recently, Alix *et al.*[430] prepared different wheat starch- and wheat flour-based blends with PLC by extrusion and co-extrusion (Figure 8.5). It was found that the water absorption of the blend films could be modulated as a function of the material surface composition imposed by the chosen extrusion mode (extrusion or co-extrusion). Co-extrusion seems to be preferred for manufacturing multilayered, functionally graded or textured

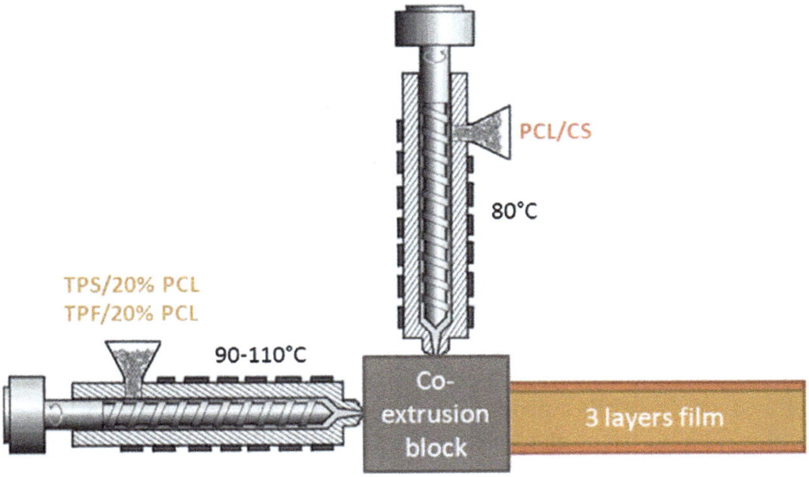


**Figure 8.5** Co-extrusion device used to produce three-layer films composed of two materials.[430]

structures since no solvents are used and the entire process is completed in a single step.[397,426,427] In addition, the preparation of starch-based multilayers can also be carried out through compression molding of plasticized starch and polyesters.[427,429]

Compared with the blending process mentioned above, the important advantages of reactive extrusion (REX) are that it combines several chemical process operations (polymer synthesis and/or modification) and extrusion (melting, blending, structuring, devolatilization and shaping) into a single process carried out in the extruder.[4,5,396,404,428] In addition, REX is a highly effective process and there are no by-products (or with the ability to remove or recycle excess reagents and by-products simply and efficiently).[3,4,209,262,263]

REX has been widely used in the past to modify starch, such as starch graft copolymers,[3,431] hydroxypropylated starch,[396,432] starch succinates,[433] starch phosphates,[404,405] starch nanoparticles[434,435] and other functional starches.[406,428,436] Recently, REX has been studied widely for preparing starch–polymer blends.[143,263,322,395] In order to enhance the compatibility of starch and other polymers, a reactive functional agent and an initiator were usually added to the blend.[143,209,395,436,437] Previous studies have shown that the reactive blending resulted in high interfacial adhesion between starch and other polymers and also improved mechanical properties (such as high tensile strength, elongation at break and impact strength) compared with the virgin composites of starch-based blends.[57,93,143,263,431] More recently, Frost *et al.*[437] successfully created TPS–silica–PVA composite films by reactive extrusion.

However, REX is not powerful enough for free-radical generation in some reactions.[57,126,262,263] Furthermore, unused reagents that remain in the extrudates cannot be removed economically.[3,4]

### 8.4.3 Film Blowing

Film blowing is a commonly used method for producing self-supporting plastic films.[4,5,11,407,438] During film blowing, a hollow tube is extruded and then expanded into a bubble by increasing the pressure inside the tube.[4,5,11] In order to form a good bubble, the materials used for film blowing are required to have sufficiently high melt strength and extensibility.[4,5,407] Because of this, film blowing of only starch is restricted by the poor melt tenacity of TPS and the film blowing preparation of pure starch films is still a challenging technology.[407,438,439] Therefore, from blends of starch and other biodegradable polymers, *e.g.* PLA,[379] PVA,[175] PCL,[440] PBAT[137,258] and chitosan,[408] is possible to obtain biodegradable blown films with better mechanical and processability properties than those prepared only with TPS. In particular, TPS has been blended with PCL or PBAT to adjust the rheological properties of the melt before the film blowing process.[258,440,441] In addition, the production of films of blends of TPS–PLA using the blown extrusion method is difficult because PLA is rigid and brittle and this problem can be improved by adding PCL or PBAT.[117,377,379] Further, with the aim of increasing the mechanical properties of PCL–starch and making the LDPE films partially biodegradable, Matzinos *et al.*[373] prepared blends of PCL–starch–LDPE. More recently, in order to improve the mechanical properties of starch–PBAT films produced by blown extrusion, CA was used as the compatibilizer for the blend.[142]

Considering the industrial need for large-scale production, film blowing is preferred to casting owing to its energy efficiency combined with high productivity.[4,373,407,440]

### 8.4.4 Injection Molding

Injection molding is a widely accepted processing technique in the plastics industry.[4,257,415,417] It is characterized by low cost, high production rates, design flexibility and dimensional accuracy of the product.[257,416,418,442] Furthermore, during the injection molding process, effectively no shear is involved and as a result there are no residual stresses in the final product.[257,418,442,443] The production of starch-based blends has also been conducted by injection molding.[257,415,417,418,442,444] For example, a series of studies by the same group systematically investigated the injection molding process parameters (melt temperature, screw speed, injection and backpressure) and starch type on the mechanical properties (tensile and flexural properties), water absorption, and dynamic mechanical properties of injection-molded corn starch blended with ethylene–vinyl acetate–maleic anhydride (EVAMA) copolymer and with ethylene–maleic anhydride (EMA) copolymer in a weight ratio of 70 : 30.[227,417,442,445] Mold injection was also achieved in a PE–starch blend manufacturing process, even for a 70% starch-containing sample.[443] In addition, PCL,[416,444] PBS,[416] soy protein,[194] chitin[446] and natural rubber[342] have been blended with starch by injection molding. In particular, Avérous and Fringant[447] showed that the less viscous PCL migrated to the surface during injection molding, in effect encapsulating the starch, and potentially reducing water sorption. Moreover, Rosa and Andrade[446] found that the addition of chitin contributes to the improvement of water resistance of composite samples compared with injection-molded starch alone.

If high water contents are used in the processing of starch-based materials, distortion and shrinkage will occur as the equilibrium water content is naturally achieved after processing.[253,255] Therefore, moisture control is very important for injection molding starch-based blends and industrial injection molding resins are often dried using a desiccant to reduce resin decomposition during processing.[418,448] For example, Kovács and Tábi[418] found that with adequate drying of starch and PLA, strong adhesion can be created between the two phases, which resulted in significantly higher mechanical parameters. In another study, Abdillahi *et al.*[448] used citric acid to improve the mechanical and barrier properties of injection-molded wheat flour–PLA–glycerol blends.

### 8.4.5 Compression Molding

Compression molding is useful as a processing method because of its simplicity and capability of producing materials with high melt viscosity that cannot be produced by extrusion or injection.[4,102,412,449] Furthermore, compression molding can also be used to prepare starch-based multilayers based on plasticized starch and other polymers.[4,427,429,430] Like injection molding, compression molding typically requires the elimination of water and a dry process is often included in the compression molding of starch-based blends.[255,427,449]

Polymers such as PCL,[360] polyurethane,[398,412] PLA,[102] PHB,[132] EVA,[413] cellulose[449] and zein[169,450] have been used to blend with starch by thermocompression. For example, Lopez *et al.*[451] prepared films based on TPS and chitosan/chitin by thermocompression and the thermocompression effectiveness was demonstrated by the presence of homogeneous cross-sections with the absence of unmelted starch granules and no visible agglomerates of chitin and chitosan. Moreover, Thiré *et al.*[132] processed PHB–starch blends by compression molding and observed that the addition of starch decreased the mechanical properties and degree of crystallinity of PHB without affecting its crystalline lattice.

The different properties of starch-based blends processed by compression, extrusion, casting and injection have been investigated. Yang *et al.*[375] studied the properties of ternary blends of starch and maleated polymers of styrene and ethylene–propylene rubber. The tests showed that injection molding gave higher tensile strengths than compression molding and the tensile strength increased with increase in SMA content. Ke and Sun[102] studied the effect of processing procedures on the properties of TPS–PLA blends, and the blends prepared by compression molding had higher crystallinities than those prepared by injection molding. However, the blends prepared by injection molding had higher tensile strengths and elongations and lower water absorption values than those made by compression molding. In addition, González-Gutiérrez *et al.*[414] compared the effects of compression molding, extrusion and their combination on the mechanical and optical properties of albumen–starch-based blends. It was found that materials with good mechanical properties and acceptable degree of transparency could be obtained by compression molding, which was concluded to be a more effective way of preparing improved albumen–starch-based blends than extrusion. Furthermore, an appropriate combination of extrusion and compression molding, with plasticization of starch prior to mixing with albumen, was demonstrated to improve further the bioplastics properties of albumen–starch blends.

### 8.4.6 Foaming

Starch-based foams have been widely investigated and produced by means of baking technology and direct extrusion foaming.[409,410,452] Because of the inherent long cycle time, the baking process cannot be as readily and economically used for large-scale production as can continuous processes such as extrusion.[409,453,454] However, starch-based foams prepared by direct extrusion foaming are difficult to mold into shaped products, which limits the commercial application of the materials.[410,411,454–456] Recently, a novel technology has been developed for making starch-based foams by microwave heating of extruded starch pellets.[457–459]

Starch-based foams made from starch alone have limited applications owing to their brittleness, poor water resistance, poor mechanical properties and dimensional instability.[409,455,460] These problems could be overcome by blending starch with other polymers such as PS,[410] natural rubber,[456] PLA,[455,461] PVA,[462,463] PCL[139] and PBAT.[140,452] For example, extruded starch foams with PVA were first patented by Altieri and Lacourse in 1989.[453] Some studies have also indicated that the resistance of starch foams to direct contact with water showed an improvement on addition of a high proportion of PVA.[462,463] Further, by blending starch with PLA, the resulting foams retained the important biodegradability characteristic and showed improved mechanical properties and water resistance.[411,461] Moreover, Tanrattanakul and Chumeka[456] found that the mechanical properties of TPS foams can be improved by blending starch with natural rubber latex. More recently, Kaisangsri *et al.*[464] prepared various compositions of cassava starch-based foam blended with chitosan and kraft fiber. The results showed that foam produced from cassava starch with 30% kraft fiber (w/w of starch) and 4% chitosan (w/w of starch) had properties similar to those of polystyrene foam.

Furthermore, many efforts have been made to improve the compatibility of starch–polyester-based foams.[140,454] For instance, Nabar *et al.*[140] prepared starch foams in the presence of PBAT with better physicomechanical and hydrophobic properties; maleic anhydride grafted PBAT was used as a compatibilizer between pristine PBAT and starch.

### 8.4.7 Spinning

Melt spinning, wet spinning and electrospinning are some of the techniques that have been used to generate fibers of starch at the micro- and nanoscales.[465–469]

Melt spinning is a process in which a viscous melt is extruded through a contraction nozzle (normally with a circular cross-section) and then further strained during cooling to align the material along the axis of the fiber.[466,467,470,471] SPCL (starch with PCL, 30 : 70%) and SPLA (starch with PLA, 30 : 70%) can be processed by melt spinning to create fibers that can be combined, under the influence of temperature and pressure, into physically bonded fiber meshes.[466,467,470–472] Previous studies showed that these fiber meshes have outstanding potential for providing adequate porous structure and mechanical properties to be used in a range of tissue-engineering strategies.[467,471,472] Although melt spinning can avoid the use of solvents, it usually cannot be used to make fibers on the nanoscale.[466,467,470]

The wet spinning technique is based on the extrusion of polymer solutions and is a viable alternative of the commonly used melt spinning for obtaining fine fibers from thermosensitive/degradable polymers.[468,469,473] Blends of starch and PCL have been used to produce three-dimensional fiber mesh scaffolds by the wet spinning technique.[468,469] It was found that the addition of starch to the PCL improves the mechanical properties, providing higher resistance of the blend to tensile forces and therefore closer similarity to those found in native bone.[468,469] Additionally, previous studies showed that these functionalized materials have the potential for future applications in bone tissue engineering.[468,469] Pashkuleva *et al.*[421] reported a poly(ethylene-*co*-vinyl alcohol) (SEVA-C) scaffold that presented an adequate pore size and high interconnectivity, allowing better transport inside the scaffold.

The electrospinning processing technique is a simple, versatile and efficient method to produce high-performance polymeric fibers with diameters ranging from the micro- to the nanoscale.[420,474,475] Its benefits include cost-effectiveness, adjustable membrane pore sizes, simplicity of the process, controllability of fiber diameter and versatility.[420,474–476] Many attempts have been made to fabricate starch-based blend fibers by electrospinning. For example, starch–PCL based fibrous scaffolds have been prepared *via* the electrospinning technique.[474–476] These scaffolds can provide an ideal structure for cell deposition and organization and their unique architecture which supports and guides the cells, making them a suitable candidate for bone tissue engineering applications.[474–477] In addition, Oktay *et al.*[478] prepared starch–poly(ethylene-*alt*-maleic anhydride) (PE-*alt*-MA) membranes with both a large surface area and nano-uniform porosity, which made them suitable for many uses, ranging from tissue engineering to filtration. Moreover, oxidized starch (OS)–PVA nanofibers produced by electrospinning were reported by Wang *et al.*[420] Although they focused on the fabrication and characterization of OS–PVA fibers, these fibers may be used in tissue engineering, wound dressing and drug delivery as a new potential material because of its non-toxicity, solubility and biodegradability. Furthermore, although electrospinning is a convenient method to produce fibers in nano or micro dimensions, it often requires the use of a large amount of non-starch components, including plasticizers, crosslinker resins or other polymers.[420,422,476,477] Recently, Kong and Ziegler[422,465,479] developed pure starch fibers with a modified electrospinning setup and the resulting starch fibers had diameters in the order of microns, with potentials use in various applications, *e.g.* in the food, textile and biomedical industries.

## 8.5 Applications of Starch-based Blends

Starch-based blends show great versatility of processing, good physical properties and commercial availability, hence they have been widely used in different application areas such as agriculture,[480–483] packaging,[454,484–487] pharmacy and biomedicine[488–492] and wastewater treatment.[493–496]

### 8.5.1 Agricultural Applications

Owing to the increased economic and environmental concerns over the removal and disposal of mulch films, much research has been focused on using starch-based blends as mulch films.[497–499] Starch-based blends are useful for making agricultural mulch films because of the cost-effectiveness (starch is an inexpensive and abundant natural polymer), ease of manufacture and satisfactory mechanical properties.[5,497,498] Furthermore, they meet the biodegradability criteria and degrade into harmless products when placed in contact with soil microorganisms without interfering with the next cultivation.[5,499,500]

Photo-biodegradable polyethylene films containing starch have been developed and used in agriculture.[481,497,499,500] They are better able to raise the temperature, preserve moisture and increase yields than common polyethylene films and can be degraded environmentally after use.[499,501–503] Although starch–polyethylene blends can easily pass through harvesting machinery without difficulty, they are actually not 100% biodegradable and the rate of degradation depends on the starch content and is very sensitive to the environmental conditions and other ingredients in the formulation.[480,499,500] Biodegradation of starch–polyethylene is limited with respect to accessibility by microorganisms to part of the biodegradable compound (starch).[497,499,501,502] The films break down into small, brittle pieces (disintegrate), but do not actually biodegrade.[497,499,501,502] Real 100% biodegradable mulch films have also been developed based on starch and other biodegradable polymers, since they are broken down by the action of humidity and microorganisms, decomposing completely into $CO_2$ and water.[5,483,504] For example, Halley *et al.*[483] developed a biodegradable film from a blend of starch and biodegradable polyester polymers, and acceptable behavior of the films was achieved for general use. In addition, PVA–starch blends have been developed and tested for use as agricultural mulch film.[298,503] However, the high cost of these fully biodegradable starch-based mulch films prevents their commercialization.[298,498,499]

Starch-based agricultural films can be prepared using conventional plastics processing technology.[5,483,500,501] Products currently on the market that are derived from or blended with plant starch include Biosafe (Xinfu Pharmaceutical, China), Eastar Bio (Novamont, Italy), Eco-Flex (BASF, Germany), Ingeo (NatureWorks, USA), Mater-Bi (Novamont, Italy) and Paragon (Avebe, The Netherlands).[482,500]

Recently, starch-based sustained-release fertilizers have attracted considerable attention. The hydrophilic matrix provides effective control over the fertilizer encapsulated in the starch for only a relatively short period after rainfall or irrigation, which reduces the survival life in field use, especially in heavy water environments.[505–509] Therefore, modified starch or starch-based blends have prominent application in fertilizer development. For instance, Chen *et al.*[509] prepared both native starch–urea and PLLA-grafted starch–urea composite films by solution casting and washing rapidly with water.

It was found that starch-*g*-PLLA is an effective material for encapsulating water-soluble fertilizers such as urea for controlled release as it exhibits a relatively low swelling ability, high encapsulating capacity and slow release rate. Han *et al.*[508] used formaldehyde-crosslinked PVA–starch films for controlling the release of fertilizers and found a 50% weight loss after 50 days of biodegradation in compost.

Starch-based slow- or controlled-release fertilizers could effectively avoid or decrease the loss of normal fertilizers and environmental pollution.[506,508,509] However, many of these sustained-release materials only achieve slow release of fertilizers, and not water retention. Recently, Zhong *et al.*[505] prepared a novel agricultural superabsorbent polymer (SAP) based on sulfonated corn starch–PLA embedding phosphate rock (SCS–PAA–PHR) to integrate the retention and sustained release of water and fertilizers. The results revealed that the system with good sustained fertilizer release and excellent water retention had several significant advantages over conventional slow-release fertilizers and agricultural SAP. This combined system is potentially useful in crop cultivation. In addition, Jin *et al.*[506] reported a novel and degradable coated slow-release compound in which the nutrient was entrapped in a crosslinked starch matrix granule (SF) and starch–poly(acrylic acid-*co*-acrylamide) [poly(AA-*co*-AM)] interpenetrating polymer networks (IPNs) were used as an outer coating [starch–poly(acrylic acid-*co*-acrylamide) superabsorbent (SAAmF)]. This material proved to be an efficient and superior source of nitrogen and phosphorus and could be applied to improve the utilization efficiency of fertilizers and water and it has potential practical applications in the remediation of saline soil.

### 8.5.2 Packaging

Starch-based materials not only retain the advantages of traditional packaging materials but also do not have the environmental pollution and disposal problems associated with traditional food packaging materials such as PE and PP.[487,510,511] For packaging applications, starch-based materials are usually used for films, loose fill and single-use trays.[430,454,511] Furthermore, starch-based packaging materials display very low permeability to oxygen at low RH and have been proposed for food product protection to improve quality and shelf life without impairing consumer acceptability.[430,485,487]

Biodegradable films based on LDPE, wheat starch and soluble starch for food packaging applications have been developed and it was reported that the gas permeability and water vapor transmission rate increased proportionally with the decrease in starch content in the blend.[484,487] Berruezo *et al.*[486] prepared PS–TPS blends by melt blending and found that although the mechanical properties of the blend were inferior to those of neat polystyrene, they were suitable for use in manufacturing packaging products for food and disposable beverage containers (trays, cups, bottles, *etc.*), with the advantage of being biodegradable and thus promoting environmental care. In addition, PVA–starch blends have developed rapidly in recent decades and are now widely applied as packaging films. For example, Guo *et al.*[512] investigated the anaerobic digestion (AD) of starch–poly(vinyl alcohol) biopolymer packaging and reported that with highly active inocula, 58–62% biodegradation of PVA–starch-based biopolymers can be achieved under AD conditions. Recently, Kuchaiyaphum *et al.*[485] prepared PVA–rice starch–silk fibroin-blended (PVA–RS–SF) films by solution casting. Results from film characterization indicated that the optimal composition for packaging properties was 60 : 40 w/w PVA–RS with 2% w/v of SF and under these conditions the film had a high content of biopolymer and still had high strength, flexibility and transparency and low water swelling and solubility.

Starch-based foam is another major starch-based packaging material and has been developed and used by manufacturers to protect and preserve a wide array of food and industrial products.[453,460,464,510,511] For example, it can be pressed into trays or disposable dishes, which can dissolve in water and leave a non-toxic solution, and subsequently can be consumed in the microbial environment.[460,510,513] In addition, this product has been developed as a replacement for PS, which is used to produce loose fillers and other expanded items.[410,460,510] The use of fillers as reinforcement has been explored to improve starch-based foams.[513,514]

Unfortunately, starch-based packaging materials present some drawbacks, such as a strong hydrophilic behavior (poor moisture barrier) and poorer mechanical properties than the conventional non-biodegradable plastic films used in the food packaging industries.[453,510,515] Furthermore, components in the conventional starch-based polymer packaging materials are not completely inert and if they migrate into foods they may be harmful to the human body.[5,11,510,515] This limits the commercial applications of starch-based packaging materials. The application of the nanocomposite concept has proven to be a promising option in order to improve the mechanical and barrier properties of starch-based packaging materials.[98,515,516] For instance, a PVA–starch–clay nanocomposite food packaging material has been developed that can offer better mechanical properties and lower migration of polymer and additives.[516] Further, in active packaging, some additives can be incorporated in the formulation to induce an antimicrobial or bacterial adhesion inhibitory effect.[430,510,516]

### 8.5.3 Pharmacy and Biomedicine

Starch-based blends have some advantages for use in the biomedical area: (a) good biocompatibility; (b) degradation as a requirement and its degradation products are non-toxic; (c) no tissue irritation and do not induce a severe immune response; (d) good mechanical properties; (e) no risk of viral transmission from animal or human components contained in other hemostats.[183,489,491,517–519] Starch-based blends have been widely used in several biomedical applications, ranging from drug delivery carriers and screws to tissue engineering scaffolds.[183,489–492,517,520] Also, starch-based blends have been used in wound dressing and anti-adhesion applications.[425,518,519,521,522]

### 8.5.3.1 Drug Delivery

Starch and starch derivatives have been used in formulations of tablets, microparticles and coating films for controlled drug delivery and release rates.[163,183,202,419,523] For example, many studies have demonstrated the successful use of high-amylose starches crosslinked by different chemicals, such as epichlorohydrin and sodium trimetaphosphate, in the development of controlled drug-delivery systems.[524–527] Furthermore, to inhibit or reduce enzymatic degradation taking place in the stomach in order to allow an adequate amount of the therapeutic agent to be absorbed, starch is often compounded with other polymers in the field of controlled drug release.[163,183,524,526,527] For example, Pereira *et al.*[528] reported a new biodegradable hydrogel based on corn starch–cellulose for use as bone cement or drug-delivery carrier. Soares *et al.*[524] evaluated the influence of drug loading and polymer ratio on the physicochemical properties of microparticles of crosslinked high-amylose starch–pectin blends loaded with diclofenac for further application in controlled drug-delivery systems. Moreover, Bagliotti *et al.*[529] prepared films from resistant starch-pectin dispersions intended for colonic drug delivery. The release behavior of starch–alginate capsules has been investigated by Wang *et al.*[419] and Córdoba *et al.*[530] In addition, starch has been also blended with other polymers such as gelatin,[202] KGM,[400] PCL[531,532] and EVOH[533] to develop starch-based blends for controlled drug delivery.

More recently, Zhang *et al.*[163] developed hydroxypropylmethylcellulose–hydroxypropyl starch (HPMC–HPS) blends for use as capsule materials. Photographs of capsules made from solutions with different HPMC–HPS contents are shown in Figure 8.6. It can be seen that the transparencies of

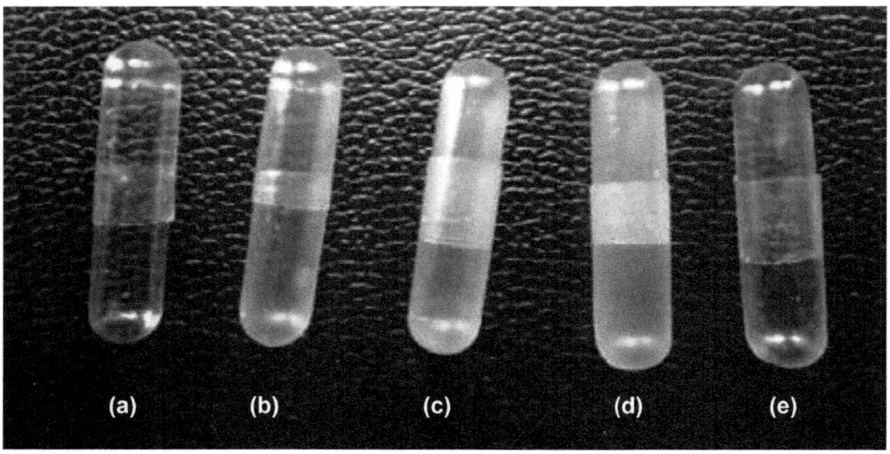


**Figure 8.6** Photographs of capsules made from solutions with different hydroxypropylmethylcellulose–hydroxypropyl starch (HPMC–HPS) contents: (a) 100 : 0; (b) 70 : 30; (c) 50 : 50; (d) 30 : 70; (e) 0 : 100.[163]

both pure HPMC and HPS capsules are reasonably good; the transparency of the capsules decreased with increasing HPS content, which indicates that there is phase separation for the blends. Furthermore, the higher viscosity of the starch HPS at lower temperature improved the viscosity balance of the system, which improved the processability of the capsule by the well-established dipping–drying method. It was also reported that the lower price of the raw materials was another advantage of this product.

Starch-based capsules have been reported recently but the small number of products in the market are mainly soft capsules produced by extrusion with large amounts of plasticizer and gelling agent.[488,489,523] In addition, although starch is generally regarded as safe, its derivatives and in fact at submicron levels may pose some safety challenges, especially as carriers in drug delivery systems.[163,488,491,531,533]

### *8.5.3.2 Tissue Engineering*

Owing to the advantages of (a) natural origin, (b) good biocompatibility and adequate mechanical properties, (c) showing no harmful cytotoxic behavior when in contact with cells and (d) being processable by various methodologies, the use of starch-based blends and materials has attracted considerable attention in tissue engineering applications.[183,490,492,520]

Starch-based scaffolds obtained from a blend of corn starch and PCL have been proposed as candidates for bone tissue engineering applications. For instance, Oliveira *et al.*[534] originally tested the suitability of corn starch–PCL (SPCL) scaffolds for pursuing a cartilage tissue engineering approach. In addition, Martins *et al.*[476] prepared SPCL scaffolds by a combination of starch–PCL micro- and PCL nano-motifs, produced by rapid prototyping and electrospinning techniques, respectively. SEM and micro-computed tomographic analysis showed the successful fabrication of a multilayer scaffold composed of parallel aligned microfibers in a grid-like arrangement, intercalated by a mesh-like structure with randomly distributed nanofibers.[476] In addition, starch–EVOH (SEVA-C),[490,520,535] starch–PLA (SPLA)[520] and starch–chitosan[492] scaffolds have been developed for use in bone tissue engineering applications. Compounding of starch with other polymers would allow improved mechanical properties of the scaffold to be obtained, which is very relevant for load-bearing applications.[183,520,534,535] Furthermore, previous studies demonstrated that biodegradable starch-based scaffolds were able to support the attachment, proliferation and differentiation of cells.[183,476,490,492,536] For instance, the ability of outgrowth endothelial cells (OECs) to form perfused vascular structures as early as 48 h following subcutaneous implantation of starch–PCL biomaterial *in vivo* was studied by Ghanaati *et al.*[536]

Various processing techniques have been and are being developed to fabricate starch-based blend scaffolds, such as melt-based technology, injection molding, spinning, supercritical immersion–precipitation and selective laser sintering.[183,535,537] For example, Gomes *et al.*[535] developed a new approach based on conventional injection molding with the addition of a solid blowing agent to generate foaming of the bulk of the molded part. The proposed route allows the production of scaffolds with a compact skin and a porous core, with promising mechanical properties. Duarte *et al.*[537] used the supercritical immersion–precipitation technique to prepare scaffolds of a polymeric blend of starch and PLA for tissue engineering purposes. Additionally, Salmoria *et al.*[183] reported the rapid fabrication of starch–cellulose and cellulose acetate scaffolds by selective laser sintering and evaluated the effects of the laser power, laser scan speed and polymer particle size on the properties of the scaffold. The major problem with the scaffolds produced by the methods developed so far is their mechanical weakness, which does not allow for their use in hard-tissue regeneration where high-strength scaffolds are required.[183,491,520,535,537]

Crosslinked PVA–starch films have been employed as dialysis membranes and the films showed transport properties comparable to, and in some cases better than, those of currently used commercial membranes.[538] In another study, polymer blends of PEG–polypropylene glycol (PPG)–PEG copolymer mixtures with pregelatinized starch at various compositions ranging from 0 to 3 wt% were prepared and evaluated for potential use as novel resorbable bone hemostatic wax.[522] The prepared samples had adequate properties for use as a bone wax. Starch-based materials have also been suggested for incorporation in orthopedic implants or as bone replacement. For example, Pereira *et al.*[491] reported an implant composed of corn starch and poly(vinylidene fluoride) or poly(vinylidene fluoride–trifluoroethylene) obtained by compressing/annealing processing and an *in vivo* study in animals showed that the blends, regardless of composition, were tolerated by cancellous bone.

Recently, starch based materials have been employed as hemostatic wound dressings since they effectively absorb water and low molecular weight compounds from blood, concentrate platelets and clotting proteins and thereby enhance the endogenous clotting mechanism.[425,517,519,521,522,539] For example, Arockianathan *et al.*[425] prepared biocomposite films containing chitosan and sago starch impregnated with silver nanoparticles. An evaluation of the films as temporary biological wound dressing materials for experimental wounds of rats revealed that the experimental wounds healed faster than control wounds. In addition, in several studies starch–PVA blend films or hydrogels were developed as wound dressings. For instance, Kenawy *et al.*[521] prepared physically crosslinked hydroxyethyl starch (HES)–PVA blend hydrogel membranes and showed that the addition of HES–PVA hydrogels slightly improved the utility as hydrogel membranes for biomedical applications, specifically for wound dressing application. Moreover, Park *et al.*[518] designed a novel liquid-type adhesion barrier using anti-coagulating HES blended with hyaluronic acid (HA). Test results in Sprague–Dawley rats showed that the HES–HA blend solution effectively reduced the adhesion area to 6.5% of the control, which might be due to the inhibited blood coagulation by HES and the highly effective barrier properties of hygroscopic high molecular weight HA.

### 8.5.4 Absorbent Materials

It is well known that starch, which is an abundant, renewable and biodegradable resource, has the capacity to associate by physical and chemical interactions with a wide variety of molecules.[494,495,540] Hence starch-based sorbents have been proposed as chelating agents to remove compounds or heavy metal ions from wastewater and as a useful tool for protecting the environment.[493–496,540] For example, the adsorption of Pb(II) and Cd(II) ions with crosslinked carboxymethyl starch (CCS) was investigated as a function of the solution pH, contact time, initial metal ion concentration and temperature by Chen *et al.*[495] In addition, Dong *et al.*[494] reported the preparation of amino starch and the removal of Cu(II) and Cr(VI) from aqueous solution by amino starch. More recently, Zhang *et al.*[540] used a crosslinked starch-based polymer as a solid-phase extraction material for the determination of nitrophenols at trace levels in environmental water.

Recently, starch-based blends have been developed as useful low-cost sorbents for heavy metals. In particular, Zhou *et al.*[541] prepared a porous PBS–corn starch blend adsorbent using α-amylase at pH 5 and50 °C, with a 2 wt% enzyme dosage (based on dry starch) and an 8 h hydrolysis time. The optimum ratio of starch to PBS was 100 : 25. SEM images showed that the surface of the composite particles treated with α-amylase was full of micropores with pore sizes of about 1–4μm, which may be advantageous fro its adsorption ability. They also reported that the enzyme-treated starch–PBS particles had a good adsorption efficiency for Cu(II) ions compared with the untreated starch–PBS particles.

## 8.6 Conclusion

This chapter has demonstrated that starch has been widely studied and successfully applied at an industrial level in blends with other polymers to reduce the final cost of the product and to improve biodegradability. It has been shown that after modification, the physical and mechanical properties of starch-based blends are similar to those of some conventional plastics, but they have the advantage of being biodegradable in different environments. In addition, starch-based blends can easily be adapted to different processes, such as casting, extrusion, blowing, foaming, injection and compression molding, spinning and reactive extrusion, which enhances the potential commercial applications. Although starch-based blends have been widely used in different application fields such as agriculture, packaging, pharmacy and biomedicine and wastewater treatment, further scientific and technological developments are still needed to adapt to the wide range of potential applications of starch-based blends.

## References

1. D. R. Lu, C. M. Xiao and S. J. Xu, *Express Poly. Lett*, 2009, **3**, 366.
2. M. Kaseem, K. Hamad and F. Deri, *Polym. Sci., A*, 2012, **54**, 165.

3. G. Moad, *Prog. Polym. Sci.*, 2011, **36**, 218.
4. H. S. Liu, F. W. Xie, L. Yu, L. Chen and L. Li, *Prog. Polym. Sci.*, 2009, **34**, 1348.
5. L. Janssen, L. Moscicki, *Thermoplastic Starch*, Wiley-VCH, Weinheim, 2009.
6. L. Yu, K. Dean and L. Li, *Prog. Polym. Sci.*, 2006, **31**, 576.
7. F. Xie, P. J. Halley and L. Avérous, *Prog. Polym. Sci.*, 2012, **37**, 595.
8. S. Y. Lee and M. A. Hanna, *Polym. Compos.*, 2009, **30**, 665.
9. S. Thomas, P. M. Visakh and A. P. Mathew, *Adv. Nat. Polym*, 2013, 1.
10. M. A. L. Russo, C. O'Sullivan, B. Rounsefell, P. Halley, R. Truss and W. P. Clarke, *Bioresour. Technol.*, 2009, **100**, 1705.
11. N. Peelman, P. Ragaert, B. De Meulenaer, D. Adons, R. Peeters, L. Cardon, F. Van Impe and F. Devlieghere, *Trends Food Sci. Technol.*, 2013, **32**, 128.
12. M. P. M. Dicker, P. F. Duckworth, A. B. Baker, G. Francois, M. K. Hazzard and P. M. Weaver, *Composites, Part A*, 2014, **56**, 280.
13. A. R. Rahmat, W. A. W. A. Rahman, L. T. Sin and A. A. Yussuf, *Mater. Sci. Eng., C*, 2009, **29**, 2370.
14. G. J. L. Griffin, Biodegradable synthetic resin sheet material containing starch and a fatty material, *U.S. Pat.* 4016117 A, 1977.
15. G. J. L. Griffin, Synthetic resin sheet material, *U.S. Pat.* 4021388 A, 1977.
16. G. J. L. Griffin, Synthetic/resin based compositions, *U.S. Pat.* 4125495 A, 1978.
17. J. L. Willett, *J. Appl. Polym. Sci.*, 1994, **54**, 1685.
18. J. H. Lin, W. T. Kaob, Y. C. Tsaib and Y. H. Chan, *Carbohydr. Polym.*, 2013, **98**, 1553.
19. A. M. Walker, Y. Tao and J. M. Torkelson, *Polymer*, 2007, **48**, 1066.
20. A. Sharif, J. Aalaie, H. Shariatpanahi, H. Hosseinkhanli and A. Khoshniyat, *J. Appl. Polym. Sci.*, 2013, **128**, 145.
21. X. Luo, J. Li and X. Lin, *Carbohydr. Polym.*, 2012, **9**, 1595.
22. J. Hubackova, M. Dvorackova, P. Svoboda, P. Mokrejs, J. Kupec, I. Pozarova, P. Alexy, P. Bugaj, M. Machovsky and M. Koutny, *Polym. Test.*, 2013, **32**, 1011.
23. S. O. Adeosun, G. I. Lawal, A. B. Sambo and E. I. Akpan, *J. Miner. Mater. Charact. Eng.*, 2012, **11**, 385.
24. H. A. Pushpadass, P. Bhandari and M. A. Hanna, *Carbohydr. Polym.*, 2010, **82**, 1082.
25. C. Cercle, P. Sarazin and B. D. Favis, *Carbohydr. Polym.*, 2013, **92**, 138.
26. M. Sabetzadeh, R. Bagheri and M. Masoomi, *J. Appl. Polym. Sci.*, 2012, **126**, E63.
27. N. Wang, J. G. Yu, X. F. Ma and Y. Wu, *Carbohydr. Polym.*, 2007, **67**, 446.
28. R. Chandra and R. Rustg, *Polym. Degrad. Stab.*, 1997, **56**, 85.
29. M. L. Méndez-Hernández, C. S. Tena-Salcido, Z. Sandoval-Arellano, M. C. González-Cantú, M. Mondragón and F. J. Rodríguez-González, *Polym. Bull.*, 2011, **67**, 903.
30. T. Yasin, M. Nisar, M. Shafiq, Y. C. Nho and R. Ahmad, *Polym. Compos.*, 2013, **34**, 408.

31. X. Ramis, A. Cadenato, J. M. Salla, J. M. Morancho, A. Vallés, L. Contat and A. Ribes, *Polym. Degrad. Stab.*, 2004, **86**, 48.
32. D. S. Rosa, M. A. G. Bardi, L. D. B. Machado, D. B. Dias, L. G. A. Silva and Y. Kodama, *J. Therm. Anal. Calorim.*, 2010, **102**, 181.
33. M. Kaseem, K. Hamad and F. Deri, *Polym. Bull.*, 2011, **68**, 1079.
34. H. A. Pushpadass, R. W. Weber, J. J. Dumais and M. A. Hann, *Bioresour. Technol.*, 2010, **101**, 7258.
35. M. Berruezo, L. N. Ludueña, E. Rodriguez and V. A. Alvarez, *J. Plast. Film Sheeting*, 2013, 1.
36. Y. Zheng, Y. F. Zhang, L. Zhang, Y. Liu and Q. Zhou, *Starch Stärke*, 2013, **65**, 840.
37. S. Lustig, J. M. Schuetz, *Eur. Pat.* 0292894 A2, 1988.
38. R. L. Shogren, R. V. Greene and Y. V. Wu, *J. Appl. Polym. Sci.*, 1991, **42**, 2279.
39. R. L. Shogren, R. V. Greene and Y. V. Wu, *J. Appl. Polym. Sci.*, 1991, **42**, 1701.
40. I. M. Thakore, S. Desai, B. D. Sarwade and S. Devi, *J. Appl. Polym. Sci.*, 1999, **71**, 1851.
41. R. P. Westhoff, F. H. Otey, C. L. Mehltretter and C. R. Russell, *Ind. Eng.Chem.Prod .Res. Dev.*, 1974, **13**, 123.
42. E. Landreau, L. Tighzerta, C. Bliardb, F. Berzina and C. Lacostea, *Eur. Polym. J.*, 2009, **45**, 2609.
43. F. Teyssandier, P. Cassagnau, J. F. Gérard, N. Mignard and F. Mélis, *Mater. Chem. Phys.*, 2012, **133**, 913.
44. I. M. Thakoore, S. Desai and S. Devi, *J. Appl. Polym. Sci.*, 2001, **79**, 48.
45. S. K. Kisku and S. K. Swain, *Polym. Compos.*, 2012, **33**, 79.
46. M. Lukasiewicz, P. Ptaszek, A. Ptaszek and S. Bednarz, *Starch Stärke*, 2014, **66**, 1.
47. E. N. Zareh, P. N. Moghadam, E. Azariyan and I. Sharifian, *Iran. Polym. J.*, 2011, **20**, 319.
48. A. Sharif, J. Aalaie, H. Shariatpanahi, H. Hosseinkhanli and A. Khoshniyat, *J. Polym. Res.*, 2011, **18**, 1955.
49. M. M. Pang, M. Y. Pun and Z. A. M. Ishak, *Polym. Eng. Sci.*, 2014, **54**, 1357.
50. D. S. Rosa, C. G. F. Guedes and C. L. Carvalho, *J. Mater. Sci.*, 2007, **42**, 551.
51. S. Mortazavi, I. Ghasemi and A. Oromiehie, *Polym. Test.*, 2013, **32**, 482.
52. P. Matzinos, D. Bikiaris, S. Kokkou and C. Panayiotou, *J. Appl. Polym. Sci.*, 2001, **79**, 2548.
53. V. Tanrattanakul and W. Panwiriyarat, *J. Appl. Polym. Sci.*, 2009, **114**, 742.
54. S. Kalambur and S. S. Rizvi, *J. Appl. Polym. Sci.*, 2005, **96**, 1072.
55. B. G. Girija and R. R. N. Sailaja, *J. Appl. Polym. Sci.*, 2006, **101**, 1109.
56. A. L. Da Róz, A. M. Ferreira, F. M. Yamaji and A. J. F. Carvalho, *Carbohydr. Polym.*, 2012, **90**, 34.
57. S. Wang, J. Yu and J. Yu, *Polym. Int.*, 2005, **54**, 279.
58. R. R. N. Sailaja, A. P. Reddy and M. Chanda, *Polym. Int.*, 2001, **50**, 1352.

59. D. Bikiaris, J. Prinos, K. Koutsopoulos, N. Vouroutzis, E. Pavlidou, N. Frangis and C. Panayiotou, *Polym. Degrad. Stab.*, 1998, **59**, 287.
60. W. Liu, Y. J. Wang and Z. Sun, *J. Appl. Polym. Sci.*, 2003, **88**, 2904.
61. Y. J. Wang, W. Liu and Z. Sun, *J. Appl. Polym. Sci.*, 2004, **92**, 344.
62. D. Bikiaris and C. Panayiotou, *J. Appl. Polym. Sci.*, 1998, **8**, 1503.
63. R. R. N. Sailaja and M. Chanda, *J. Appl. Polym. Sci.*, 2002, **86**, 3126.
64. J. Prinos, D. Bikiaris, S. Theologidis and C. Panayiotou, *Polym. Eng. Sci.*, 1998, **38**, 954.
65. I. Arvanitoyannis, E. Psomiadou, C. G. Biliaderis, H. Ogawa, N. Kawasaki and A. Nakayama, *Starch Stärke*, 1997, **49**, 306.
66. Q. Yin, A. Q. Dong, J. L. Wang and Y. P. Yin, *Polym. Compos.*, 2008, **29**, 745.
67. R. R. N. Sailaja, *Polym. Int.*, 2005, **54**, 286.
68. R. Thipmanee and A. Sane, *J. Appl. Polym. Sci.*, 2012, **126**, E252.
69. M. Elanmugilan, P. A. Sreekumar, N. K. Singha, M. A. Al-Harthi and S. K. De, *Polym. Eng. Sci.*, 2013, **53**, 2389.
70. T. A. Hottle, M. M. Bilec and A. E. Landis, *Polym. Degrad. Stab.*, 2013, **98**, 1898.
71. X. Liu, L. Yu, F. W. Xie, E. Petinakis, P. Sangwan, S. Shen, K. Dean, A. Ammala and S. Wong-Holmes, *J. Appl. Polym. Sci.*, 2013, **130**, 2282.
72. H. Vieyra, M. A. Aguilar-Méndez and E. S. Martín-Martínez, *J. Appl. Polym. Sci.*, 2013, **127**, 845.
73. M. M. Pang, M. Y. Pun and Z. A. M. Ishak, *J. Appl. Polym. Sci.*, 2013, **129**, 3656.
74. D. M. Whitacre, F. A. Gunther, *Reviews of Environmental Contamination and Toxicology*, Springer, 2008.
75. M. Elanmugilan, P. A. Sreekumar, N. K. Singha, M. A. Al-Harthi and S. K. De, *J. Appl. Polym. Sci.*, 2013, **129**, 449.
76. M. A. Huneault and H. Li, *Polymer*, 2007, **48**, 270.
77. J. Wootthikanokkhan, P. Kasemwananimit, N. Sombatsompop, A. Kositchaiyong, S. I. Ayutthaya and N. Kaabbuathong, *J. Appl. Polym. Sci.*, 2012, **126**, E389.
78. M. Mihai, M. A. Huneault, B. D. Favis and H. B. Li, *Macromol. Biosci.*, 2007, **7**, 907.
79. B. Imre, *Express Polym. Lett.*, 2013, **8**, 2.
80. O. H. Arroyo, M. A. Huneault, B. D. Favis and M. N. Bureau, *Polym. Compos.*, 2009, **31**, 114.
81. J. Cai, Z. Xiong, M. Zhou, J. Tan, F. Zeng, S. Lin and H. Xiong, *Carbohydr. Polym.*, 2014, **102**, 746.
82. E. Perez, C. J. Pérezb, V. A. Alvarezc and C. Bernal, *Carbohydr. Polym.*, 2013, **97**, 269.
83. V. B. Carmona, A. de Campos, J. M. Marconcini and L. H. C. Mattoso, *J. Therm. Anal. Calorim.*, 2013, **115**, 153.
84. Z. Shen and J. Wang, *Bioresour. Technol.*, 2011, **102**, 8835.
85. L. T. Sin, W. A. W. A. Rahman, A. R. Rahmat and A. A. Samad, *Polymer*, 2010, **51**, 1206.

86. S. Y. Yang, C. I. Liu, J. Y. Wu, J. C. Kuo and C. Y. Huang, *Macromol. Symp.*, 2008, **272**, 150.
87. M. Khan, S. K. Bhattacharia, M. A. Kader and K. Bahari, *Carbohydr. Polym.*, 2006, **63**, 500.
88. K. Suchao-In, P. Koombhongse and S. Chirachanchai, *Carbohydr. Polym.*, 2014, **102**, 95.
89. W. Wang, G. X. Zhang, W. Z. Zhang, W. H. Guo and J. K. Wang, *J. Polym. Environ.*, 2012, **21**, 46.
90. J. B. Zeng, L. Jiao, Y. D. Li, M. Srinivasan, T. Li and Y. Z. Wang, *Carbohydr. Polym.*, 2011, **83**, 762.
91. S. M. Lai, C. K. Huang and H. F. Shen, *J. Appl. Polym. Sci.*, 2005, **97**, 257.
92. J. A. Stagner, V. D. Alves and R. Narayan, *J. Appl. Polym. Sci.*, 2012, **126**, E135.
93. J. B. Olivato, M. V. E. Grossmann, F. Yamashita, D. Eiras and L. A. Pessan, *Carbohydr. Polym.*, 2012, **87**, 2614.
94. J. M. Raquez, Y. Nabar, R. Narayan and P. Dubois, *Polym. Eng. Sci.*, 2008, **48**, 1747.
95. L. H. Innocentini-Mei, J. R. Bartoli and R. C. Baltieri, *Macromol. Symp.*, 2003, **1**, 77.
96. S. Godbole, S. Gote, M. Latkar and T. Chakrabarti, *Bioresour. Technol.*, 2003, **86**, 33.
97. M. Zhang and N. L. Thomas, *J. Appl. Polym. Sci.*, 2010, **116**, 688.
98. X. Tang and S. Alavi, *Carbohydr. Polym.*, 2011, **85**, 7.
99. W. Phetwarotai, P. Potiyaraj and D. Aht-Ong, *J. Polym. Environ.*, 2013, **21**, 95.
100. J. Cai, M. Liu, L. Wang, K. H. Yao, S. Li and H. G. Xiong, *Carbohydr. Polym.*, 2011, **86**, 941.
101. M. Nampoothiri, K. N. R. Nair and R. P. John, *Bioresour. Technol.*, 2010, **101**, 8493.
102. T. Ke and X. Sun, *J. Appl. Polym. Sci.*, 2001, **81**, 3069.
103. S. Jacobsen and H. Fritz, *Polym. Eng. Sci.*, 1996, **36**, 2799.
104. O. Martin and L. Avérous, *Polymer*, 2001, **42**, 6209.
105. H. Gao, S. Hu, F. H. Su, J. Zhang and G. N. Tang, *Polym. Compos.*, 2011, **32**, 2093.
106. G. Li, P. Sarazin, W. J. Orts, S. H. Imam and B. D. Favis, *Macromol. Chem. Phys.*, 2011, **212**, 1147.
107. E. Schwach, J. L. Six and L. Avérous, *J. Polym. Environ.*, 2008, **16**, 286.
108. T. Ke, S. X. Sun and P. Seib, *J. Appl. Polym. Sci.*, 2003, **89**, 3639.
109. B. M. S. Fiber, *Macromol. Mater. Eng.*, 2013, **298**, 603.
110. J. F. Zhang and X. Sun, *Biomacromolecules*, 2004, **5**, 1446.
111. S. Karagoz and G. Ozkoc, *Polym. Eng. Sci.*, 2013, **53**, 2183.
112. H. Li and M. A. Huneault, *J. Appl. Polym. Sci.*, 2011, **119**, 2439.
113. H. Wang, X. Sun and P. Seib, *J. Polym. Environ.*, 2002, **10**, 133.
114. M. Labet and W. Thielemans, *Chem. Soc. Rev.*, 2009, **38**, 3484.
115. I. Pillin, T. Divers, J. F. Feller and Y. Grohens, *Macromol. Symp.*, 2005, **222**, 233.

116. B. Y. Shin, S. Lee, Y. S. Shin, S. Balakrishnan and R. Narayan, *Polym. Eng. Sci.*, 2004, **44**, 1429.
117. P. Sarazin, G. Li, W. J. Orts and B. D. Favis, *Polymer*, 2008, **49**, 599.
118. M. Vikman, S. H. D. Hulleman, M. V. D. Zee, P. Myllärinen and H. Feil, *J. Appl. Polym. Sci.*, 1999, **74**, 2594.
119. H. Cho, H. S. Moon, M. Kim, Nam and J. Y. Kim, *Waste Manage.*, 2011, **31**, 475.
120. J. Cai, Z. Y. Xiong, M. Zhou, J. Tan, F. B. Zeng, M. H. Ma, S. Lin and H. G. Xiong, *Carbohydr. Polym.*, 2013, **102**, 746.
121. A. Mahieu, C. Terrié, A. Agoulon, N. Leblanc and B. Youssef, *J. Polym. Res.*, 2013, **20**, 1.
122. L. L. Avérous, M. P. Doleb and C. Fringant, *Polymer*, 2000, **41**, 4157.
123. C. H. Kim, E. J. Choi and J. K. Park, *J. Appl. Polym. Sci.*, 2000, **77**, 2049.
124. Y. Tan, P. X. Wang, K. Xu, H. Y. An and L. Dong, *Chem. Res. Chin. Univ.*, 2010, **26**, 483.
125. C. S. Wu, *Polym. Degrad. Stab.*, 2003, **80**, 127.
126. S. Kalambur and S. S. Rizvi, *J. Plast. Film Sheeting*, 2006, **22**, 39.
127. L. Chen, Z. Zhang, X. L. Zhuang, X. S. Chen and X. B. Jing, *J. Appl. Polym. Sci.*, 2010, **117**, 2724.
128. E. J. Choi, C. H. Kim and J. K. Park, *J. Polym. Sci., Part B: Polym.*, 1999, **37**, 2430.
129. R. Mani, J. Tang and M. Bhattacharya, *Macromol. Rapid Commun.*, 1998, **19**, 283.
130. Y. Parulekar and K. Mohanty, *Macromol. Mater. Eng.*, 2007, **292**, 1218.
131. P. Pan and Y. Inoue, *Prog. Polym. Sci.*, 2009, **34**, 605.
132. R. M. Thiré, T. A. Ribeiro and C. T. Andrade, *J. Appl. Polym. Sci.*, 2006, **100**, 4338.
133. S. M. Lai, T. M. Don and Y. C. Huang, *J. Appl. Polym. Sci.*, 2006, **100**, 2371.
134. S. Chiangga, S. Suwannasopon and N. Trivijitkasem, *Kasetsart J.: Nat. Sci.*, 2012, **46**, 653.
135. E. Schwach and L. Avérous, *Polym. Int.*, 2004, **53**, 2115.
136. R. P. H. Brandelero, M. V. E. Grossmann and F. Yamashita, *Carbohydr. Polym.*, 2011, **86**, 1344.
137. R. P. H. Brandelero, F. Yamashita and M. V. E. Grossmann, *Carbohydr. Polym.*, 2010, **82**, 1102.
138. J. B. Olivato, M. V. E. Grossmann, A. P. Bilck, F. Yamashita and L. M. Oliveira, *Polímeros*, 2013, **23**, 32.
139. Y. U. Nabar, D. Draybuck and R. Narayan, *J. Appl. Polym. Sci.*, 2006, **102**, 58–68.
140. Y. Nabar, J. M. Raquez, P. Dubois and R. Narayan, *Biomacromolecules*, 2005, **6**, 807.
141. T. Ohkita and S. H. Lee, *J. Appl. Polym. Sci.*, 2005, **97**, 1107.
142. P. S. Garcia, M. V. E. Grossmann, M. A. Shirai, M. M. Lazaretti, F. Yamashita, C. M. O. Muller and S. Mali, *Ind. Crop.Prod.*, 2014, **52**, 305.

143. R. Maliger, S. A. McGlashan, P. J. Halley and L. G. Matthew, *Polym. Eng. Sci.*, 2006, **46**, 248.
144. M. M. F. Ferrarezi, M. Taipina, O. de, L. C. E. da Silva, M. Gonçalves and C. do, *J. Polym. Environ.*, 2013, **21**, 151–159.
145. H. Li and M. A. Huneault, *J. Appl. Polym. Sci.*, 2011, **122**, 134.
146. W. L. Chai, J. D. Chow and C. C. Chen, *J. Polym. Environ.*, 2012, **20**, 550.
147. K. Katerinopoulou, A. Giannakas, K. Grigoriadi, N. M. Barkoula and A. Ladavos, *Carbohydr. Polym.*, 2014, **102**, 216.
148. L. T. Sin, W. A. W. A. Rahman, A. R. Rahmat and M. I. Khan, *Carbohydr. Polym.*, 2010, **79**, 224.
149. L. T. Sin, W. A. W. A. Rahman, A. R. Rahmat and A. A. Samad, *Polymer*, 2010, **51**, 1206.
150. L. Mao, S. Imam, S. Gordon, P. Cinelli and E. Chiellini, *J. Polym. Environ.*, 2000, **8**, 205.
151. S. D. Yoon, M. H. Park and H. S. Byun, *Carbohydr. Polym.*, 2012, **87**, 676.
152. S. D. Yoon, S. H. Chough and H. R. Park, *J. Appl. Polym. Sci.*, 2007, **106**, 2485.
153. B. Ramaraj, *J. Appl. Polym. Sci.*, 2007, **103**, 909.
154. B. Ramaraj, *J. Appl. Polym. Sci.*, 2007, **103**, 1127.
155. S. D. Yoon, S. H. Chough and H. R. Park, *J. Appl. Polym. Sci.*, 2006, **100**, 2554.
156. G. H. Zhao, Y. Liu, C. Fang, M. Zhang, C. Q. Zhou and Z. D. Chen, *Polym. Degrad. Stab.*, 2006, **91**, 703.
157. M. A. Khan, S. K. Bhattacharia, M. A. Kader and K. Baharic, *Carbohydr. Polym.*, 2006, **63**, 500.
158. M. K. Beliakova, A. A. Aly and F. A. Abdel-Mohdy, *Starch Stärke*, 2004, **56**, 407.
159. Y. L. Chung and H. M. Lai, *Carbohydr. Polym.*, 2006, **63**, 527.
160. R. Jayasekara, I. Harding, I. Bowater, G. B. Y. Christie and G. T. Lonergan, *Polym. Test.*, 2004, **23**, 17.
161. B. Ghanbarzadeh, H. Almasi and A. A. Entezami, *Innov. Food Sci. Emerging*, 2010, **11**, 697.
162. M. A. El-Kader and H. Ragab, *Ionics*, 2013, **19**, 361.
163. L. Zhang, Y. F. Wang, H. S. Liu, L. Yu, X. X. Liu, L. Chen and N. Z. Zhang, *Carbohydr. Polym.*, 2013, **98**, 73.
164. C. Bangyekan, D. Aht-Ong and K. Srikulkit, *Carbohydr. Polym.*, 2006, **63**, 61.
165. F. Liu, B. Qin, L. H. He and R. Song, *Carbohydr. Polym.*, 2009, **78**, 146.
166. F. M. Pelissari, M. V. E. Grossmann, F. Yamashita and E. A. G. Pineda, *J. Agric. Food Chem.*, 2009, **57**, 7499.
167. R. Soares, F. F. Scremin and V. Soldi, *Macromol. Symp.*, 2005, **229**, 258.
168. J. Gonzalez-Gutierrez, P. Partal, M. Garcia-Morales and C. Gallegos, *Bioresour. Technol.*, 2010, **101**, 2007.
169. E. Corradini, A. J. F. de Carvalho, A. A. Curvelo, S. da, J. A. M. Agnelli and L. H. C. Mattoso, *Mater. Res.*, 2007, **10**, 227.

170. M. Zeng, Y. W. Huang, L. Y. Lu, L. Fan and D. Lourdin, *Carbohydr. Polym.*, 2011, **84**, 323.
171. Q. Sun and C. S. L. Xiong, *Carbohydr. Polym.*, 2014, **101**, 1134.
172. X. X. Liu, Y. F. Wang, N. Z. Zhang, R. A. Shanks, H. S. Liu, Z. Tong, L. Chen and L. Yu, *Chin. J. Polym. Sci.*, 2014, **32**, 108.
173. J. Chen, C. H. Liu, Y. Q. Chen, Y. Chen and P. R. Chang, *Carbohydr. Polym.*, 2008, **74**, 946.
174. S. B. Nair, A. N. Jyothi, M. S. Sajeev and R. Misra, *Starch Stärke*, 2011, **63**, 728.
175. K. S. Mikkonen, M. P. Yadav, P. Cooke, S. Willför, K. B. Hicks and M. Tenkanen, *BioResources*, 2008, **3**, 178.
176. S. Yoo and J. M. Krochta, *Int. J. Food Sci. Technol.*, 2012, **47**, 255.
177. M. A. Woehl, C. D. Canestraro, A. Mikowski, M. R. Sierakowski, L. P. Ramos and F. Wypych, *Carbohydr. Polym.*, 2010, **80**, 866.
178. Y. Wan, H. L. Luo, F. He, H. Liang, Y. Huang and X. L. Li, *Compos. Sci. Technol.*, 2009, **69**, 1212.
179. A. Curvelo, A. D. Carvalho and J. Agnelli, *Carbohydr. Polym.*, 2001, **45**, 183.
180. V. Alvarez and A. Vázquez, *Polym. Degrad. Stab.*, 2004, **84**, 13.
181. V. A. Alvarez and A. Vazquez, *J. Compos.Mater.*, 2004, **38**, 1165.
182. A. Marques, R. Reis and J. Hunt, *Biomaterials*, 2002, **23**, 1471.
183. G. V. Salmoria, H. Luo, F. He, H. Liang, Y. Huang and X. L. Li, *Polym. Test.*, 2009, **28**, 648.
184. F. Song, D. L. Tang, X. L. Wang and Y. Z. Wang, *Biomacromolecules*, 2011, **12**, 3369.
185. F. Koksel and M. Scanlon, *J. Cereal Sci.*, 2012, **56**, 445.
186. E. Corradini, J. M. Marconcini, J. A. M. Agnellia and L. H. C. Mattoso, *Carbohydr. Polym.*, 2011, **83**, 959.
187. Q. Sun, C. Sun and L. Xiong, *Carbohydr. Polym.*, 2013, **98**, 630.
188. P. Srihanam, *Biotechnology*, 2011, **10**, 114.
189. J. Prachayawarakorn and W. Hwansanoet, *Fiber. Polym.*, 2012, **13**, 606.
190. F. M. Fakhouri, D. Costa, F. Yamashita, S. M. Martelli, R. C. Jesus, K. Alganer, F. P. Collares-Queiroz and L. H. Innocentini-Mei, *Carbohydr. Polym.*, 2013, **95**, 681.
191. C. J. Verbeek and L. E. van den Berg, *Macromol. Mater. Eng.*, 2010, **295**, 10.
192. E. Corradini, E. S. de Medeiros, A. J. F. Carvalho, A. A. S. Curvelo and L. H. C. Mattoso, *J. Appl. Polym. Sci.*, 2006, **101**, 4133.
193. E. Leroy, P. Jacquet, G. Coativy, A. L. Reguerre and D. Lourdin, *Carbohydr. Polym.*, 2012, **89**, 955.
194. H. Huang, T. Chang and J. Jane, *J. Am. Oil Chem. Soc.*, 1999, **76**, 1101.
195. J. U. Otaigbe, H. Goel, T. Babcock and J. Jane, *J. Elastomers Plast.*, 1999, **31**, 56.
196. H. Chanvrier, P. Colonna, G. D. Valle and D. Lourdin, *Carbohydr. Polym.*, 2005, **59**, 109.

197. C. E. Chinma, C. C. Ariahu and J. O. Abu, *J. Food Sci. Technol.*, 2013, **50**, 1179.
198. C. C. Enyinnaya, C. C. Ariahu and J. Alakali, *Int. J. Food Sci. Technol.*, 2013, **48**, 2400.
199. C. Chinma, C. Ariahu and J. Alakali, *J. Food Sci. Technol.*, 2013, 1.
200. H. Chanvrier, G. D. Valle and D. Lourdin, *Carbohydr. Polym.*, 2006, **65**, 346.
201. E. Habeych, A. J. van der Goot and R. Boom, *Chem. Eng. Sci.*, 2009, **64**, 3516.
202. N. Zhang, P. Colonna, G. D. Valle and D. Lourdin, *Carbohydr. Polym.*, 2013, **92**, 455.
203. Y. Zhong, Y. Li and Y. Zhao, *J. Food Sci.*, 2012, 77, E280.
204. F. M. Pelissari, F. Yamashita and M. V. E. Grossmann, *Int. J. Food Sci. Technol.*, 2011, **46**, 702.
205. M. Zhai, L. Zhao, F. Yoshii and T. Kume, *Carbohydr. Polym.*, 2004, **57**, 83.
206. S. Chillo, S. Flores, M. Mastromatteo, A. Conte, L. Gerschenson and M. A. Del Nobile, *J. Food Eng.*, 2008, **88**, 159.
207. T. Bourtoom, *J. Sci. Technol.*, 2008, **30**, 149.
208. S. Mathew and T. E. Abraham, *Food Hydrocolloid*, 2008, **22**, 826.
209. P. Dubois and R. Narayan, *Macromol. Symp.*, 2003, **198**, 233.
210. M. A. Huneault and H. Li, *J. Appl. Polym. Sci.*, 2012, **126**, E96.
211. G. Li, P. Sarazin and B. D. Favis, *Macromol. Chem. Phys.*, 2008, **209**, 991.
212. K. Dean, L. Yu, S. Bateman and D. Y. Wu, *J. Appl. Polym. Sci.*, 2007, **103**, 802.
213. W. Zou, L. Yu, X. X. Liu, L. Chen, X. Q. Zhang, D. L. Qiao and R. Zhang, *Carbohydr. Polym.*, 2012, **87**, 1583.
214. V. D. Alves, S. Mali, A. Beléia and M. V. E. Grossmann, *J. Food Eng.*, 2007, **78**, 941.
215. P. Liu, F. W. Xie, M. Li, X. X. Liu, L. Yu, P. J. Halley and L. Chen, *Carbohydr. Polym.*, 2011, **85**, 180.
216. X. Liu, L. Yu, F. W. Xie, M. Li, L. Chen and X. X. Li, *Starch Stärke*, 2010, **62**, 139.
217. P. Liu, L. Yu, X. Y. Wang, D. Li, L. Chen and X. X. Li, *J. Cereal Sci.*, 2010, **51**, 388.
218. J. Wang, L. Yu, F. W. Xie, L. Chen, X. X. Li and H. S. Liu, *Starch Stärke*, 2010, **62**, 667.
219. S. J. Van and P. Essers, *J. Macromol. Sci., Part A: Pure Appl.Chem.*, 1997, **34**, 1665.
220. L. Yu and G. Christie, *J. Mater.Sci.*, 2005, **40**, 111.
221. M. Li, P. Liu, W. Zou, L. Yu, F. W. Xie, H. Pu, H. S. Liu and L. Chen, *J. Food Eng.*, 2011, **106**, 95.
222. Y. H. Yun and S. D. Yoon, *Polym. Bull.*, 2010, **64**, 553.
223. Y. J. Wang, W. Liu and Z. Sun, *J. Mater. Sci. Lett.*, 2003, **22**, 57.
224. J. Li, X. G. Luo, X. Y. Lin and Y. Zhou, *Starch Stärke*, 2013, **65**, 831.
225. N. Le Bolay and S. Molina-Boisseau, *Powder Technol.*, 2014, **255**, 36.

226. S. Taghvaei-Ganjali, S. Taghvaei-Ganjali, F. Motiee, E. Shakeri and A. Abbasian, *J. Appl. Chem. Res*, 2010, **4**, 53.
227. R. Mani and M. Bhattacharya, *Eur. Polym. J.*, 1998, **34**, 1467.
228. W. Phetwarotai, P. Potiyaraj and D. Aht-Ong, *J. Appl. Polym. Sci.*, 2012, **126**, E162.
229. C. Yokesahachart and R. Yoksan, *Carbohydr. Polym.*, 2011, **83**, 22.
230. H. S. Cho, H. S. Moon, M. Kim, K. Nam and J. Y. Kim, *Waste Manage.*, 2011, **31**, 475.
231. X. Ma and J. Yu, *Starch Stärke*, 2004, **56**, 545.
232. C. M. Müller, F. Yamashita and J. B. Laurindo, *Carbohydr. Polym.*, 2008, **72**, 82.
233. S. Gaudin, D. Lourdinb, D. L. Botlanc, J. L. llaria and P. Colonna, *J. Cereal Sci.*, 1999, **29**, 273.
234. D. Muscat, B. Adhikari, R. Adhikari and D. S. Chaudhary, *J. Food Eng.*, 2012, **109**, 189.
235. D. S. Chaudhary, B. P. Adhikari and S. Kasapis, *Food Hydrocolloid.*, 2011, **25**, 114.
236. X. Qiao, Z. Tang and K. Sun, *Carbohydr. Polym.*, 2011, **83**, 659.
237. M. G. A. Vieira, M. A. da Silva, L. O. dos Santos and M. M. Beppu, *Eur. Polym. J.*, 2011, **47**, 254.
238. A. Smits, P. H. Kruiskamp, J. J. G. Soest and J. F. G. Vliegenthart, *Carbohydr. Polym.*, 2003, **53**, 409.
239. A. Da Róz, A. J. F. Carvalho, A. Gandini and A. A. S. Curvelo, *Carbohydr. Polym.*, 2006, **63**, 417.
240. C. Bastioli, V. Bellotti, R. Lombi, *U.S. Pat.* 5,412,005, 1995.
241. J. M. Forssell, M. Moates, K. Graham and R. Parker, *Carbohydr. Polym.*, 1997, **34**, 275.
242. X. Ma, J. Yu and J. Wan, *Carbohydr. Polym.*, 2006, **64**, 267.
243. X. F. Ma, J. G. Yu and J. Feng, *Polym. Int.*, 2004, **53**, 1780.
244. X. Ma and J. Yu, *Acta Polym. Sin.*, 2004, 483.
245. X. Ma and J. Yu, *Carbohydr. Polym.*, 2004, **57**, 197.
246. M. Huang, J. Yu and X. Ma, *Polym. Degrad. Stab.*, 2005, **90**, 501.
247. E. Teixeira, A. L. D. Róz, A. J. F. Carvalho and A. A. S. Curvelo, *Carbohydr. Polym.*, 2007, **69**, 619.
248. Y. Zhou, D. F. Wang, L. Zhang, X. F. Du and X. L. Zhou, *Food Hydrocolloid*, 2008, **22**, 505.
249. S. N. Raphaelides and N. Georgiadis, *Food Res. Int.*, 2008, **41**, 75.
250. G. Mantzari, S. N. Raphaelides and S. Exarhopoulos, *Carbohydr. Polym.*, 2010, **79**, 154.
251. P. Lescher, K. Jayaraman and D. Bhattacharyya, *Mater. Sci. Eng., A*, 2012, **532**, 178.
252. N. Wang, J. G. Yu, P. R. Chang and X. F. Ma, *Starch Stärke*, 2007, **59**, 409.
253. X. F. Ma, P. R. Chang, J. G. Yu and N. Wang, *Carbohydr. Polym.*, 2008, **71**, 229.
254. N. Wang, J. G. Yu, P. R. Chang and X. F. Ma, *Carbohydr. Polym.*, 2008, **71**, 109.

255. N. Wang, X. X. Zhang, N. Han and J. M. Fang, *J. Thermoplast. Compos. Mater.*, 2010, **23**, 19.
256. J. Lawton, W. Doane and J. Willett, *J. Appl. Polym. Sci.*, 2006, **100**, 3332.
257. T. Mekonnen, P. Mussone, H. Khalil and D. Bressler, *J. Mater. Chem. A*, 2013, **1**, 13379.
258. J. Olivato, M. V. E. Grossmann, A. P. Bilck and F. Yamashita, *Carbohydr. Polym.*, 2012, **90**, 159.
259. R. Shi, Z. Z. Zhang, Q. Y. Liu, M. HanY, L. Q. Zhang and W. Tian, *Carbohydr. Polym.*, 2007, **69**, 748.
260. H. R. Park, S. H. Chough, Y. H. Yun and S. D. Yoon, *J. Polym. Environ.*, 2005, **13**, 375.
261. X. Jiang, T. Jiang, L. L. Gan, X. F. Zhang, H. Dai and X. Zhang, *Carbohydr. Polym.*, 2012, **90**, 1677.
262. L. Liu, Y. Yu and P. Song, *Ind. Eng. Chem. Res.*, 2013, **52**, 16232.
263. C. L. Jun, *J. Polym. Environ.*, 2000, **8**, 33.
264. X. Zhu, Y. Yang, J. X. Feng, X. M. Zhang, C. Z. Zhang, Z. B. Tang and J. Zhu, *Carbohydr. Polym.*, 2013, **92**, 810.
265. S. Mali, L. S. Sakanaka, F. Yamashita and M. V. E. Grossmann, *Carbohydr. Polym.*, 2005, **60**, 283.
266. E d M Teixeira, D. Pasquini, A. A. S. Curvelo, E. Corradini, M. N. Belgacem and A. Dufresne, *Carbohydr. Polym.*, 2009, **78**, 422.
267. J. G. Yu, N. Wang and X. F. Ma, *Starch Stärke*, 2005, **57**, 494.
268. W. Rymowicz, A. Rywińska and W. Gładkowski, *Chem. Pap.*, 2008, **62**, 239.
269. X. F. Ma, J. G. Yu and Y. B. Ma, *Carbohydr. Polym.*, 2005, **60**, 111.
270. J. H. Yang, J. G. Yu and X. F. Ma, *Carbohydr. Polym.*, 2006, **66**, 110.
271. J. H. Yang, J. G. Yu and X. F. Ma, *Starch Stärke*, 2006, **58**, 580.
272. J. W. Park, S. S. Im, S. H. Kim and Y. H. Kim, *Polym. Eng. Sci.*, 2000, **40**, 2539.
273. N. Wang, J. G. Yu and X. F. Ma, *Polym. Int.*, 2007, **56**, 1440.
274. T. Ke and X. Sun, *Trans. ASAE*, 2001, **44**, 945.
275. T. Ke and X. S. Sun, *J. Appl. Polym. Sci.*, 2003, **88**, 2947.
276. M. Shirai, M. V. E. Grossmann, S. Mali, F. Yamashita, P. S. Garcia and C. M. O. Müller, *Carbohydr. Polym.*, 2013, **92**, 19.
277. A. Taghizadeh, P. Sarazin and B. D. Favis, *J. Mater. Sci.*, 2013, **48**, 1799.
278. E. d. M. Teixeira, A. A. S. Curvelo, A. C. Corrêa, J. M. Marconcini, G. M. Glenn and L. H. C. Mattoso, *Ind. Crop. Prod.*, 2012, **37**, 61.
279. J. Lawton and G. Fanta, *Carbohydr. Polym.*, 1994, **23**, 275.
280. K. Das, D. Ray, N. R. Bandyopadhyay, A. Gupta, S. Sengupta, S. Sahoo, A. Mohanty and M. Misra, *Ind. Eng. Chem. Res.*, 2010, **49**, 2176.
281. R. Westhoff, W. Kwolek and F. Otey, *Starch Stärke*, 1979, **31**, 163.
282. Y. H. Yun, Y. H. Na and S. D. Yoon, *J. Polym. Environ.*, 2006, **14**, 71–78.
283. S. D. Yoon, S. H. Chough and H. R. Park, *J. Appl. Polym. Sci.*, 2006, **100**, 3733.
284. R. Shi, J. L. Bi, Z. Z. Zhang, A. C. Zhu, D. F. Chen, X. H. Zhou, L. Q. Zhang and W. Tian, *Carbohydr. Polym.*, 2008, **74**, 763.

285. B. Sreedhar, M. Sairam, D. K. Chattopadhyay, P. A. S. Rathnam and D. V. M. Rao, *J. Appl. Polym. Sci.*, 2005, **96**, 1313.
286. M. Zhai, F. Yoshii and T. Kume, *Carbohydr. Polym.*, 2003, **52**, 311.
287. X. Y. Zhou, Y. F. Cui, D. M. Jia and D. Xie, *Polym.-Plast. Technol. Eng.*, 2009, **48**, 489.
288. Y. Yin, J. F. Li, Y. C. Liu and L. Zhong, *J. Appl. Polym. Sci.*, 2005, **96**, 1394.
289. S. D. Yoon, *J. Agric. Food. Chem.*, 2013, **62**, 1755.
290. P. Chopra, S. Garg, A. K. Jana and D. Kohli, *Ultrason. Sonochem.*, 2013, **9**, 129.
291. Z. Liu, M. Jiang, X. Bai, X. G. Dong, J. Tong and J. Zhou, *J. Appl. Polym. Sci.*, 2012, **124**, 3774.
292. Z. Q. Liu, Y. Dong, H. Men, M. Jiang, J. Tong and J. Zhou, *Carbohydr. Polym.*, 2012, **89**, 473.
293. S. Maiti, D. Ray and D. Mitra, *J. Polym. Environ.*, 2012, **20**, 749.
294. H. Ismail and N. Zaaba, *Polym.-Plast. Technol. Eng.*, 2011, **50**, 1214.
295. P. Chopra, *Ultrason. Sonochem.*, 2013, **9**, 129.
296. B. Sreedhar, D. K. Chattopadhyay, M. S. H. Karunakar and A. R. K. Sastry, *J. Appl. Polym. Sci.*, 2006, **101**, 25.
297. C. M. Xiao, *Starch Stärke*, 2013, **65**, 82.
298. P. Cinelli, E. Chiellini, S. H. Gordon and S. H. Imam, *Macromol. Symp.*, 2003, **197**, 143.
299. N. Reddy and Y. Yang, *Food Chem.*, 2010, **118**, 702.
300. S. Y. Yang and C. Y. Huang, *J. Appl. Polym. Sci.*, 2008, **109**, 2452.
301. J. Zhou, Y. H. Ma, L. L. Ren, J. Tong, Z. Q. Liu and L. Xie, *Carbohydr. Polym.*, 2009, **76**, 632.
302. R. Jayasekara, I. Harding, I. Bowater, G. B. Y. Christie and G. T. Lonergan, *J. Polym. Environ.*, 2003, **11**, 49.
303. J. S. Park, J. H. Yang, D. H. Kim and D. H. Lee, *J. Appl. Polym. Sci.*, 2004, **93**, 911.
304. S. Garg and A. K. Jana, *Eur. Polym. J.*, 2007, **43**, 3976.
305. Y. Zhang, X. Yuan, Q. Liu and A. Hrymak, *J. Polym. Environ.*, 2012, **20**, 315.
306. H. Oza, M. R. Thompson, A. N. Hrymak and Q. Liu, *Starch Stärke*, 2012, **64**, 944.
307. H. Eslami and M. R. Kamal, *J. Appl. Polym. Sci.*, 2013, **129**, 2418.
308. A. J. Lasprilla, G. A. R. Martinez, B. H. Lunelli, A. L. Jardini and R. Maciel Filho, *Biotechnol. Adv.*, 2012, **30**, 321.
309. A. Campos, J. M. Marconcini, S. M. Martins-Franchetti and L. H. C. Mattoso, *Polym. Degrad. Stab.*, 2012, **97**, 1948.
310. M. M. Senna, Y. K. A. Moneam, A. A. A. Hakiem and H. M. Said, *J. Polym. Res.*, 2012, **19**, 1.
311. M. M. Senna, A. M. Yossef, F. M. Hossam and A. W. M. El-Naggar, *J. Appl. Polym. Sci.*, 2007, **106**, 3273.
312. F. G. D. Ferreira, M. A. G. Lima, A. de and Y. M. B. de Almeida, *Quím. Nova*, 2008, **31**, 1043.

313. F. G. D. Ferreira, M. A. G. Lima, A. de, Y. M. B. de Almeida and G. M. Vinhas, *J. Polym. Environ.*, 2010, **18**, 196.
314. K. Dubey, C. V. Chaudhari, N. Raje, L. Panickar, Y. K. Bhardwaj and S. Sabharwal, *J. Appl. Polym. Sci.*, 2012, **124**, 3501.
315. T. Yasin, M. Nisar, M. Shafiq, Y. C. Nho and R. Ahmad, *Polym. Compos.*, 2013, **34**, 408.
316. M. Zhai, F. Yoshii, T. Kume and K. Hashim, *Carbohydr. Polym.*, 2002, **50**, 295.
317. M. M. Senna, H. A. El-Shahat and A. W. M. El-Naggar, *J. Polym. Res.*, 2011, **18**, 763.
318. S. Y. Yang, C. I. Liu, J. Y. Wu, J. C. Kuo and C. Y. Huang, *Macromol. Symp.*, 2008, **272**, 150.
319. N. Follain, C. Joly, P. Dole and C. Bliard, *Carbohydr. Polym.*, 2005, **60**, 185.
320. B. Wongsuban, K. Muhammad, Z. Ghazali, K. Hashim and M. A. Hassan, *Nucl. Instrum. Methods Phys. Res., Sect. B*, 2003, **211**, 244.
321. J. Zhou, J. Zhang, Y. H. Ma and J. Tong, *Carbohydr. Polym.*, 2008, **74**, 405.
322. A. Vignon, A. Ayoub and V. Massardier, *J. Appl. Polym. Sci.*, 2013, **127**, 4168.
323. A. D. Sagar, M. A. Villar, E. L. Thomas, R. C. Armstrong and E. W. Merrill, *J. Appl. Polym. Sci.*, 1996, **61**, 139.
324. A. D. Sagar, M. A. Villar, E. L. Thomas, R. C. Armstrong and E. W. Merrill, *J. Appl. Polym. Sci.*, 1996, **61**, 157.
325. I. E. Rivero, V. Balsamo and A. J. Müller, *Carbohydr. Polym.*, 2009, **75**, 343.
326. N. Le Bolay, A. Lamure, L. N. Gallego and A. Subhani, *Chem. Eng. Process.*, 2012, **56**, 1.
327. F. Xie, E. Pollet, P. J. Halley and L. Avérous, *Prog. Polym. Sci.*, 2013, **38**, 1590.
328. L. Avérous, E. Pollet, Green nano-biocomposites., in *Environmental Silicate Nano-Biocomposites*, ed. L. Avérous and E. Pollet, Springer, London, 2012, pp. 1–11.
329. J. K. Pandey, *Polymer Nanocomposites Handbook*, CRC Press, Boca Raton, FL, 2010.
330. L. Vertuccio, G. Gorrasi, A. Sorrentino and V. Vittoria, *Carbohydr. Polym.*, 2009, **75**, 172.
331. H. Tang, H. G. Xiong, S. W. Tang and P. Zou, *J. Appl. Polym. Sci.*, 2009, **113**, 34.
332. H. T. Liao and C. S. Wu, *J. Appl. Polym. Sci.*, 2005, **97**, 397.
333. S. Y. Lee and M. A. Hanna, *Polym. Compos.*, 2009, **30**, 665.
334. C. DeLeo, C. A. Pinotti, M. Goncalves, C. do and S. Velankar, *J. Polym. Environ.*, 2011, **19**, 689.
335. S. Bocchini, D. Battegazzore and A. Frache, *Carbohydr. Polym.*, 2010, **82**, 802.
336. A. Ashori, *Polym. Eng. Sci.*, 2014, **54**, 2258.

337. Y. Ikeo, K. Aoki, H. Kishi, S. Matsuda and A. Murakami, *Polym. Adv. Technol.*, 2006, **17**, 940.
338. O. Arroyo, M. A. Huneault, B. D. Favis and M. N. Bureau, *Polym. Compos.*, 2010, **31**, 114.
339. J. Ma, C. H. Liu, R. Li and J. Wang, *J. Appl. Polym. Sci.*, 2012, **123**, 2933.
340. M. Deepthi, G. S. Ananthapadmanabha, P. Sampathkumaran, S. Seetharamu, V. V. Pattenshetti, S. Ganga, V. Asai Thambi, R. R. N. Sailaja, *Properties and Applications of Dielectric Materials (ICPADM), 2012 IEEE 10th International Conference on the*, IEEE, 2012, 1.
341. W. H. Ferreira, R. R. Khalili, M. J. Figueira and C. T. Andrade, *Ind. Crops. Prod.*, 2014, **52**, 38.
342. M. Mondragón, E. M. Hernández, J. L. Rivera-Armenta and F. J. Rodríguez-González, *Carbohydr. Polym.*, 2009, **77**, 80.
343. P. Boonprasith, J. Wootthikanokkhan and N. Nimitsiriwat, *J. Appl. Polym. Sci.*, 2013, **130**, 1114.
344. S. Tang, P. Zou, H. G. Xiong and H. L. Tang, *Carbohydr. Polym.*, 2008, **72**, 521.
345. S. Y. Lee, M. A. Hanna and D. D. Jones, *Starch Stärke*, 2008, **60**, 159.
346. S. Y. Lee, H. Chen and M. A. Hanna, *Ind. Crops. Prod.*, 2008, **28**, 95.
347. B. R. Mose and S. M. Maranga, *J. Mater. Sci. Eng. B*, 2011, **1**, 239.
348. F. Chivrac, E. Pollet and L. Avérous, *Mat. Sci. Eng.: R*, 2009, **67**, 1.
349. N. Cañigueral, F. Vilaseca, J. A. Méndez, J. P. López, L. Barberà, J. Puig, M. A. Pèlach and P. Mutjé, *Chem. Eng. Sci.*, 2009, **64**, 2651.
350. N. Soykeabkaew, P. Supaphol and R. Rujiravanit, *Carbohydr. Polym.*, 2004, **58**, 53.
351. M. Sreekala, K. Goda and P. Devi, *Compos. Interface.*, 2008, **15**, 281.
352. Y. Lu, L. Weng and X. Cao, *Carbohydr. Polym.*, 2006, **63**, 198.
353. M. Iman and T. K. Maji, *Carbohydr. Polym.*, 2012, **89**, 290.
354. F. Vilaseca, J. A. Mendez, A. Pèlach, M. Llop, N. Cañigueral, J. Gironès, X. Turon and P. Mutjé, *Process Biochem.*, 2007, **42**, 329.
355. N. Sarifuddin, H. Ismail and Z. Ahmad, *J. Phys. Sci.*, 2013, **24**, 4294.
356. N. Sarifuddin, H. Ismail and Z. Ahmad, *J. Thermoplast. Compos. Mater.*, 2013, 0892705713486125.
357. C. Vallo, J. M. Kenny, A. Vazquez and V. P. Cyras, *J. Compos. Mater.*, 2004, **38**, 1387.
358. V. A. Alvarez, J. M. Kenny and A. Vázquez, *Polym. Compos.*, 2004, **25**, 280.
359. A. Campos, K. B. R. Teodoro, E. M. Teixeira, A. C. Corrêa, J. M. Marconcini, D. F. Wood, T. G. Williams and L. H. C. Mattoso, *Polym. Eng. Sci.*, 2013, **53**, 800.
360. O. O. Fabunmi, L. G. Tabil, S. Panigrahi and P. R. Chang, *J. Polym. Environ.*, 2011, **19**, 841.
361. K. Bocz, B. Szolnoki, A. Marosi, T. Tábi, M. Wladyka-Przybylak, G. Marosi, *Polym. Degrad. Stab.*, 2013.
362. M. T. Taghizadeh, Z. Abbasi and Z. Nasrollahzade, *J. Taiwan Inst. Chem. E.*, 2012, **43**, 120.

363. M. F. Rosa, B. Chiou, E. S. Medeiros, D. F. Wood, T. G. Williams, L. H. C. Mattoso, W. J. Orts and S. H. Imam, *Bioresour. Technol.*, 2009, **100**, 5196.
364. V. A. Alvarez, A. N. Fraga and A. Vázquez, *J. Appl. Polym. Sci.*, 2004, **91**, 4007.
365. A. de Campos, H. D. T. Gustavo, J. M. Marconcini, L. H. C. Mattoso, A. Klamczynski, K. S. Gregorski, D. Wood, T. Williams, B. Chiou and S. H. Imam, *J. Polym. Environ.*, 2013, **21**, 1.
366. C. R. di Franco, V. P. Cyras, J. P. Busalmen, R. A. Ruseckaite and A. Vázquez, *Polym. Degrad. Stab.*, 2004, **86**, 95.
367. V. P. Cyras, S. Iannace, J. M. Kenny and A. Vázquez, *Polym. Compos.*, 2001, **22**, 104.
368. V. P. Cyras, A. Vázquez and J. M. Kenny, *Polym. Eng. Sci.*, 2001, **41**, 1521.
369. V. P. Cyras, J. F. Martucci, S. Iannace and A. Vázquez, *J. Thermoplast. Compos. Mater.*, 2002, **15**, 253.
370. M. T. Taghizadeh and N. Sabouri, *Int. Nano Lett.*, 2013, **3**, 51.
371. R. Iovino, R. Zullo, M. A. Rao, L. Cassar and L. Gianfreda, *Polym. Degrad. Stab.*, 2008, **93**, 147.
372. W. Smitthipong, R. Tantatherdtam and R. Chollakup, *J. Thermoplast. Compos. Mater.*, 2013, 0892705713489701.
373. P. Matzinos, V. Tserki, C. Gianikouris, E. Pavlidou and C. Panayiotou, *Eur. Polym. J.*, 2002, **38**, 1713.
374. K. Wang and M. Yang, *J. Wuhan Univ. Technol.*, 2013, **28**, 157.
375. Z. Yang, M. Bhattacharya and U. R. Vaidya, *Polymer*, 1996, **37**, 2137.
376. D. Bikiaris, J. Prinos, C. Perrier and C. Panayiotou, *Polym. Degrad. Stab.*, 1997, **57**, 313.
377. H. T. Liao and C. S. Wu, *Mater. Sci. Eng., A*, 2009, **515**, 207.
378. J. Ren, H. Fu, T. Ren and W. Yuan, *Carbohydr. Polym.*, 2009, **77**, 576.
379. M. A. Shirai, J. B. Olivato, P. S. Garcia, C. M. O. Müller, M. V. E. Grossmann and F. Yamashita, *Mater. Sci. Eng.:* C, 2013, **33**, 4112.
380. T. Ke and X. S. Sun, *J. Polym. Environ.*, 2003, **11**, 7.
381. Q. Shi, C. Chen, L. Gao, L. Jiao, H. Xu and W. Guo, *Polym. Degrad. Stab.*, 2011, **96**, 175.
382. J. Wang, W. Zhai and W. Zheng, *J. Polym. Environ.*, 2012, **20**, 528.
383. P. Bie, P. Liu, L. Yu, X. Li, L. Chen and F. Xie, *Carbohydr. Polym.*, 2013, **98**, 959.
384. A. M. M. Gomes, P. L. da Silva, C. E. M. da Silva and N. M. P. Ricardo, *Macromol. Symp.*, 2011, **299**, 220.
385. P. Mariani, K. Allganer, F. B. Oliveira, E. J. B. Cardoso and L. H. Innocentini-Mei, *Polym. Test.*, 2009, **28**, 824.
386. F. C. Soares, F. Yamashita, C. M. O. Müller and A. T. N. Pires, *Polym. Test.*, 2014, **33**, 34.
387. F. C. Soares, F. Yamashita, C. M. O. Müller and A. T. N. Pires, *Polym. Test.*, 2013, **32**, 94.
388. S. Arun, K. A. A. Kumar and M. S. Sreekala, *Int. J. Plast. Technol*, 2012, **16**, 50.

389. S. Y. Ryu, J. W. Rhim, H. J. Roh and S. S. Kim, *LWT—Food Sci. Technol.*, 2002, **35**, 680.
390. J. Zhou, L. Ren, J. Tong and Y. Ma, *J. Appl. Polym. Sci.*, 2009, **114**, 940.
391. M. K. Abdul Wahab, H. Ismail and N. Othman, *Polym.-Plast Technol.*, 2012, **51**, 298.
392. M. K. Abdul Wahab, H. Ismail and N. Othman, *Polym.-Plast Technol*, 2011, **50**, 748.
393. A. W. M. Kahar, H. Ismail and N. Othman, *J. Vinyl Addit. Technol.*, 2012, **18**, 65.
394. A. W. M. Kahar, H. Ismail and N. Othman, *J. Appl. Polym. Sci.*, 2012, **125**, 768.
395. P. Ma, D. G. Hristova-Bogaerds, P. Schmit, J. G. P. Goossens and P. J. Lemstra, *Polym. Int.*, 2012, **61**, 1284.
396. R. Kavitha and J. N. BeMiller, *Carbohydr. Polym.*, 1998, **37**, 115.
397. A. Copinet, C. Bertrand, A. Longieras, V. Coma and Y. Couturier, *J. Polym. Environ.*, 2003, **11**, 169.
398. Q. Wu and L. Zhang, *Ind. Eng. Chem. Res.*, 2001, **40**, 558.
399. A. Wonisch, P. Polfer, T. Kraft, A. Dellert, A. Heunisch and A. Roosen, *J. Am. Ceram. Soc.*, 2011, **94**, 2053.
400. S. B. Nair and A. N. Jyothi, *Starch Stärke*, 2013, **65**, 273.
401. J. O. de Moraes, A. S. Scheibe, A. Sereno and J. B. Laurindo, *J. Food Eng.*, 2013, **119**, 800.
402. L. J. Zhu, R. Shukri, N. J. de Mesa-Stonestreet, S. Alavi, H. Dogan and Y. C. Shi, *J. Food Eng.*, 2010, **100**, 232.
403. F. Teyssandier, P. Cassagnau, J. F. Gérard, N. Mignard and F. Mélis, *Mater. Chem. Phys.*, 2012, **133**, 913.
404. S. O'Brien, Y. J. Wang, C. Vervaet and J. P. Remon, *Carbohydr. Polym.*, 2009, **76**, 557.
405. N. A. Landerito and Y. J. Wang, *Cereal Chem.*, 2005, **82**, 271.
406. N. Gimmler, F. Lawn and F. Meuser, *Starch Stärke*, 1995, **47**, 268.
407. M. Thunwall, V. Kuthanová, A. Boldizar and M. Rigdahl, *Carbohydr. Polym.*, 2008, **71**, 583.
408. F. M. Pelissari, F. Yamashita, M. A. Garcia, M. N. Martino, N. E. Zaritzky and M. V. E. Grossmann, *J. Food Eng.*, 2012, **108**, 262.
409. M. Sauceau, J. Fages, A. Common, C. Nikitine and E. Rodier, *Prog. Polym. Sci.*, 2011, **36**, 749.
410. M. Mihai, M. A. Huneault and B. D. Favis, *Macromo. Biosci.*, 2007, **43**, 215.
411. J. Guan and M. A. Hanna, *Ind. Eng. Chem.Res.*, 2005, **44**, 3106.
412. Q. Wu and L. Zhang, *J. Appl. Polym. Sci.*, 2001, **79**, 2006.
413. M. A. Rodriguez-Perez, R. D. Simoes, C. J. L. Constantino and J. A. de Saja, *J. Appl. Polym. Sci.*, 2011, **121**, 2324.
414. J. González-Gutiérrez, P. Partal, M. García-Morales and C. Gallegos, *Carbohydr. Polym.*, 2011, **84**, 308.
415. H. Vieyra, M. Aguilar-Méndez and S. Martín-Martínez, *J. Appl. Polym. Sci.*, 2013, **127**, 845.

416. A. Sen and M. Bhattacharya, *Polymer*, 2000, **41**, 9177.
417. D. Ramkumar, U. R. Vaidya, M. Bhattacharya, M. Hakkarainen, A. C. Albertsson and S. Karlsson, *Eur. Polym. J.*, 1996, **32**, 999.
418. J. G. Kovács and T. Tábi, *Polym. Eng. Sci.*, 2011, **51**, 843.
419. Q. Wang, X. Hu, Y. Du and J. F. Kennedy, *Carbohydr. Polym.*, 2010, **82**, 842.
420. H. Wang, W. Wang, S. Jiang, S. Jiang, L. Zhai and Q. Jiang, *Iran. Polym. J.*, 2011, **20**, 551.
421. I. Pashkuleva, P. M. López-Pérez, H. S. Azevedo and R. L. Reis, *Mater. Sci. Eng., C*, 2010, **30**, 981.
422. L. Kong and G. R. Ziegler, *Carbohydr. Polym.*, 2013, **92**, 1416.
423. K. C. Reis, J. Pereira, A. C. Smith, C. W. P. Carvalho, N. Wellner and I. Yakimets, *J. Food Eng.*, 2008, **89**, 361.
424. K. Das, D. Ray, N. R. Bandyopadhyay, S. Sahoo, A. K. Mohanty and M. Misra, *Composites, Part B*, 2011, **42**, 376.
425. P. M. Arockianathan, S. Sekar, B. Kumaran and T. P. Sastry, *Int. J. Biol. Macromol.*, 2012, **50**, 939.
426. R. Gattin, A. Copinet, C. Bertrand and Y. Couturier, *J. Appl. Polym. Sci.*, 2003, **88**, 825.
427. P. Dole, L. Avérous, C. Joly, G. Della Valle and C. Bliard, *Polym. Eng. Sci.*, 2005, **45**, 217.
428. V. D. Miladinov and M. A. Hanna, *Ind. Crops Prod.*, 2000, **11**, 51.
429. O. Martin, E. Schwach and Y. Couturier, *Starch Stärke*, 2001, **53**, 372.
430. S. Alix, A. Mahieu, C. Terrie, J. Soulestin, E. Gerault, M. G. Feuilloley, R. Gattin, V. Edon, T. Ait-Younes and N. Leblanc, *Eur. Polym. J.*, 2013, **49**, 1234.
431. E. Hablot, S. Dewasthale, Y. Zhao, Y. Zhiguan, X. Shi, D. Graiver and R. Narayan, *Eur. Polym. J.*, 2013, **49**, 873.
432. A. Xu and P. A. Seib, *J. Cereal Sci.*, 1997, **25**, 17.
433. L. Wang, R. L. Shogren and J. L. Willett, *Starch Stärke*, 1997, **49**, 116.
434. D. Song, Y. S. Thio and Y. Deng, *Carbohydr. Polym.*, 2011, **85**, 208.
435. S. B. Kalambur and S. S. Rizvi, *Polym. Int.*, 2004, **53**, 1413.
436. R. E. Wing and J. L. Willett, *Ind. Crops. Prod.*, 1997, **7**, 45.
437. K. Frost, J. Barthes, D. Kaminski, E. Lascaris, J. Niere and R. Shanks, *Carbohydr. Polym.*, 2011, **84**, 343.
438. R. Zullo and S. Iannace, *Carbohydr. Polym.*, 2009, **77**, 376.
439. O. V. López, N. E. Zaritzky, M. V. Grossmann and M. A. García, *J. Food Eng.*, 2013, **116**, 286.
440. P. Matzinos, V. Tserki, A. Kontoyiannis and C. Panayiotou, *Polym. Degrad. Stab.*, 2002, **77**, 17.
441. J. B. Olivato, M. V. Grossmann, F. Yamashita, M. M. Nobrega, M. R. Scapin, D. Eiras and L. A. Pessan, *Int. J. Food Sci. Tech.*, 2011, **46**, 1934.
442. R. Mani and M. Bhattacharya, *Eur. Polym. J.*, 1998, **34**, 1477.
443. H. Vieyra Ruiz, E. S. M. Martínez and M. A. A. Mendez, *Starch Stärke*, 2011, **63**, 42.

444. A. Sen, M. Bhattacharya, K. A. Stelson and V. R. Voller, *Mater. Sci. Eng., A*, 2002, **338**, 60.
445. D. H. S. Ramkumar, M. Bhattacharya and U. R. Vaidya, *Eur. Polym. J.*, 1997, **33**, 729.
446. R. C. R. Rosa and C. T. Andrade, *J. Appl. Polym. Sci.*, 2004, **92**, 2706.
447. L. U. C. Avérous and C. Fringant, *Polym. Eng. Sci.*, 2001, **41**, 727.
448. H. Abdillahi, E. Chabrat, A. Rouilly and L. Rigal, *Ind. Crops Prod.*, 2013, **50**, 104.
449. S. Srithongkham, L. Vivitchanont and C. Krongtaew, *J. Mater. Sci. Eng. B*, 2012, **2**, 213.
450. E. Habeych, B. Dekkers, A. J. van der Goot and R. Boom, *Chem. Eng. Sci.*, 2008, **63**, 5229.
451. O. Lopez, M. A. Garcia, M. A. Villar, A. Gentili, M. S. Rodriguez, L. Albertengo, *LWT-Food Sci. Technol.*, 2014, **57**, 106.
452. Y. Xu and M. A. Hanna, *Carbohydr. Polym.*, 2005, **59**, 521.
453. P. A. Altieri, N. L. Lacourse, 1989, *U.S. Pat.* 4863655.
454. M. Mihai, M. A. Huneault, B. D. Favis and H. Li, *Macromol. Biosci.*, 2007, **7**, 907.
455. J. L. Willett and R. L. Shogren, *Polymer*, 2002, **43**, 5935.
456. V. Tanrattanakul and W. Chumeka, *J. Appl. Polym. Sci.*, 2010, **116**, 93.
457. X. Peng, J. Song, A. Nesbitt and R. Day, *J. Cell. Plast.*, 2013, **49**, 245.
458. X. Peng, J. Song, A. Nesbitt and R. Day, *J. Cell. Plast.*, 2013, **49**, 147.
459. J. Zhou, J. Song and R. Parker, *Carbohydr. Polym.*, 2006, **63**, 466.
460. J. Y. Cha, D. S. Chung, P. A. Seib, R. A. Flores and M. A. Hanna, *Ind. Crops Prod.*, 2001, **14**, 23.
461. J. F. Zhang and X. Sun, *J. Appl. Polym. Sci.*, 2007, **106**, 3058.
462. S. Mali, F. Debiagi, M. V. Grossmann and F. Yamashita, *Ind. Crops Prod.*, 2010, **32**, 353.
463. P. Cinelli, E. Chiellini, J. W. Lawton and S. H. Imam, *Polym. Degrad. Stab.*, 2006, **91**, 1147.
464. N. Kaisangsri, O. Kerdchoechuen and N. Laohakunjit, *Ind. Crops Prod.*, 2012, **37**, 542.
465. L. Kong and G. R. Ziegler, *Biomcromolecules*, 2012, **13**, 2247.
466. M. E. Gomes, H. S. Azevedo, A. R. Moreira, V. Ellä, M. Kellomäki and R. L. Reis, *J. Tissue Eng. Regener. Med.*, 2008, **2**, 243.
467. J. Pal, N. Kankariya, S. Sanwaria, B. Nandan and R. K. Srivastava, *Mater. Sci. Eng., C*, 2013, **33**, 4213.
468. A. I. Rodrigues, M. E. Gomes, I. B. Leonor and R. L. Reis, *Acta Biomater.*, 2012, **8**, 3765.
469. T. C. Santos, T. J. Morton, M. Moritz, S. Pfeifer, K. Reise, A. P. Marques, A. G. Castro, R. L. Reis and M. van Griensven, *Tissue Eng. Part A*, 2013, **19**, 834.
470. M. P. Pavlov, J. F. Mano, N. M. Neves and R. L. Reis, *Macromol. Biosci.*, 2004, **4**, 776.
471. M. E. Gomes, H. L. Holtorf, R. L. Reis and A. G. Mikos, *Tissue Eng.*, 2006, **12**, 801.

472. D. P. Link, L. S. Gardel, V. M. Correlo, M. E. Gomes and R. L. Reis, *J. Biomed. Mater. Res., Part A*, 2013, **101**, 3059.
473. Q. Wang, N. Zhang, X. Hu, J. Yang and Y. Du, *Eur. J. Pharm. Biopharm.*, 2007, **66**, 398.
474. M. I. Santos, K. Tuzlakoglu, S. Fuchs, M. E. Gomes, K. Peters, R. E. Unger, E. Piskin, R. L. Reis and C. J. Kirkpatrick, *Biomaterials*, 2008, **29**, 4306.
475. K. Tuzlakoglu, N. Bolgen, A. J. Salgado, M. E. Gomes, E. Piskin and R. L. Reis, *J. Mater. Sci. - Mater. Med.*, 2005, **16**, 1099.
476. A. Martins, S. Chung, A. J. Pedro, R. A. Sousa, A. P. Marques, R. L. Reis and N. M. Neves, *J. Tissue Eng. Regener. Med.*, 2009, **3**, 37.
477. M. A. da Silva, A. Crawford, J. Mundy, A. Martins, J. V. Araújo, P. V. Hatton, R. L. Reis and N. M. Neves, *Tissue Eng. Part A*, 2008, **15**, 377.
478. B. Oktay, E. Baştürk, N. Kayaman-Apohan and M. V. Kahraman, *Polym. Compos.*, 2013, **34**, 1321.
479. L. Kong and G. R. Ziegler, *Food Hydrocolloids*, 2014, **36**, 20.
480. N. Sharma, L. P. Chang, Y. L. Chu, H. Ismail, U. S. Ishiaku and Z. A. Mohd Ishak, *Polym. Degrad. Stab.*, 2001, **71**, 381.
481. F. H. Otey, A. M. Mark, C. L. Mehltretter and C. R. Russell, *Ind. Eng. Chem. Prod. Res. Dev.*, 1974, **13**, 90.
482. D. G. Hayes, S. Dharmalingam, L. C. Wadsworth, K. K. Leonas, C. Miles, & D. A. Inglis, Biodegradable agricultural mulches derived from biopolymers, in *Degradable Polymers and Materials, Principles and Practice*, ed. K. C. Khemani and C. Scholz, ACS Symposium Series, American Chemical Society, Washington, DC, 2nd edn, 2012, Vol. 1114, pp. 201–223.
483. P. Halley, R. Rutgers, S. Coombs, J. Kettels, J. Gralton, G. Christie, M. Jenkins, H. Beh, K. Griffin, R. Jayasekara and G. Lonergan, *Starch Stärke*, 2001, **53**, 362.
484. E. Psomiadou, I. Arvanitoyannis, C. G. Biliaderis, H. Ogawa and N. Kawasaki, *Carbohydr. Polym.*, 1997, **33**, 227.
485. P. Kuchaiyaphum, W. Punyodom, S. Watanesk and R. Watanesk, *J. Appl. Polym. Sci.*, 2013, **129**, 2614.
486. M. Berruezo, L. N. Ludueña, E. Rodriguez and V. A. Alvarez, *J. Plast. Film Sheeting*, 2013, 8756087913504581.
487. I. Arvanitoyannis, C. G. Biliaderis, H. Ogawa and N. Kawasaki, *Carbohydr. Polym.*, 1998, **36**, 89.
488. M. J. Santander-Ortega, T. Stauner, B. Loretz, J. L. Ortega-Vinuesa, D. Bastos-González, G. Wenz, U. F. Schaefer and C. M. Lehr, *J. Controlled Release*, 2010, **141**, 85.
489. A. Rodrigues and M. Emeje, *Carbohydr. Polym.*, 2012, **87**, 987.
490. A. J. Salgado, O. P. Coutinho and R. L. Reis, *Tissue Eng.*, 2004, **10**, 465.
491. J. D. Pereira, R. C. Camargo, N. Alves, M. A. Rodriguez-Perez and C. J. Constantino, *Mater. Sci. Eng., C*, 2014, **36**, 226.
492. A. M. Martins, J. D. Kretlow, A. R. Costa-Pinto, P. B. Malafaya, E. M. Fernandes, N. M. Neves, C. M. Alves, A. G. Mikos, F. K. Kasper and R. L. Reis, *J. Biomed. Mater. Res. A*, 2012, **100**, 599.

493. B. R. Pant, H. J. Jeon and H. H. Song, *Macromol. Res.*, 2011, **19**, 307.
494. A. Dong, J. Xie, W. Wang, L. Yu, Q. Liu and Y. Yin, *J. Hazard. Mater.*, 2010, **181**, 448.
495. Y. X. Chen, B. H. Zhong and W. M. Fang, *J. Appl. Polym. Sci.*, 2012, **124**, 5010.
496. S. E. Abdel-Aal, Y. H. Gad and A. M. Dessouki, *J. Hazard. Mater.*, 2006, **129**, 204.
497. S. Kasirajan and M. Ngouajio, *Agron. Sustainable Dev.*, 2012, **32**, 501.
498. G. Scarascia-Mugnozza, C. Sica and G. Russo, *J. Agric. Eng. Res.*, 2011, **42**, 15.
499. D. Briassoulis, *J. Polym. Environ.*, 2004, **12**, 65.
500. A. Corbin, J. Cowan, C. Miles, D. Hayes, J. Dorgan, D. Inglis, *Washington State University Extension Fact Sheet*, 2013, 1.
501. Y. Z. Wang, K. K. Yang, X. L. Wang, Q. Zhou, C. Y. Zheng and Z. F. Chen, *J. Polym. Environ.*, 2004, **12**, 7.
502. F. Khabbaz, A. C. Albertsson and S. Karlsson, *Polym. Degrad. Stab.*, 1999, **63**, 127.
503. S. H. Imam, P. Cinelli, S. H. Gordon and E. Chiellini, *J. Polym. Environ.*, 2005, **13**, 47.
504. G. Scarascia-Mugnozza, E. Schettini, G. Vox, M. Malinconico, B. Immirzi and S. Pagliara, *Polym. Degrad. Stab.*, 2006, **91**, 2801.
505. K. Zhong, Z. T. Lin, X. L. Zheng, G. B. Jiang, Y. S. Fang, X. Y. Mao and Z. W. Liao, *Carbohydr. Polym.*, 2013, **92**, 1367.
506. S. Jin, Y. Wang, J. He, Y. Yang, X. Yu and G. Yue, *J. Appl. Polym. Sci.*, 2013, **128**, 407.
507. M. L. H. Tiing, Master Dissertation, 2013.
508. X. Han, S. Chen and X. Hu, *Desalination*, 2009, **240**, 21.
509. L. Chen, Z. Xie, X. Zhuang, X. Chen and X. Jing, *Carbohydr. Polym.*, 2008, 72, 42.
510. *Developments in Sustainable Chemical and Bioprocess Technology*, ed. R. Pogaku, A. Bono and C. Chu, Springer, New York, 2013.
511. I. S. Arvanitoyannis, *J. Macromol Sci., Rev. Macromol. Chem. Phys.*, 1999, **39**, 205.
512. M. Guo, A. P. Trzcinski, D. C. Stuckey and R. J. Murphy, *Bioresour. Technol.*, 2011, **102**, 11137.
513. E. Teixeira, M. de, A. de Campos, J. M. Marconcini, T. J. Bondancia, D. Wood, A. Klamczynski and G. M. Glenn, *RSC Adv.*, 2014, **4**, 6616.
514. Y. Han, S. O. Manolach, F. Denes and R. M. Rowell, *Carbohydr. Polym.*, 2011, **86**, 1031.
515. M. Avella, J. J. de Vlieger, M. E. Errico, S. Fischer, P. Vacca and M. G. Volpe, *Food Chem.*, 2005, **93**, 467.
516. X. Tang and S. Alavi, *J. Agric. Food Chem.*, 2012, **60**, 1954.
517. K. Björses, L. Faxälv, C. Montan, K. Wildt-Persson, P. Fyhr, J. Holst and T. L. Lindahl, *Acta Biomater.*, 2011, **7**, 2558.
518. J. K. Park, J. Yeom, S. K. Hahn, E. J. Hwang, J. S. Shin, I. H. Cho and B. S. Kim, *Macromol. Res.*, 2010, **18**, 1076.

519. K. Pal, A. Banthia and D. Majumdar, *Trends Biomater. Artif. Organs*, 2006, **20**, 59.
520. N. M. Neves, A. Kouyumdzhiev and R. L. Reis, *Mater. Sci. Eng., C*, 2005, **25**, 195.
521. E. R. Kenawy, E. A. Kamoun, E. M. S. Mohy and M. A. El-Meligy, *Arab. J. Chem*, 2013.
522. J. Suwanprateeb, W. Suvannapruk, F. Thammarakcharoen, W. Chokevivat and P. Rukskul, *J. Mater. Sci.: Mater. Med.*, 2013, **24**, 2881.
523. V. D. Vilivalam, L. Illum and K. Iqbal, *Pharm. Sci. Technol. Today*, 2000, **3**, 64.
524. G. A. Soares, A. D. de Castro, B. S. Cury and R. C. Evangelista, *Carbohydr. Polym.*, 2013, **91**(1), 135–142.
525. G. A. Soares, F. M. Carbinatto, B. S. Cury and R. C. Evangelista, *Drug Dev. Ind. Pharm.*, 2013, **39**, 284.
526. F. M. Carbinatto, A. D. de Castro, R. C. Evangelista and B. S. Cury, *Asian J. Pharm. Sci.*, 2014, **54**, 2258.
527. F. M. Carbinatto, A. D. de Castro, B. S. Cury, A. Magalhães and R. C. Evangelista, *Int. J. Pharm.*, 2012, **423**, 281.
528. C. S. Pereira, A. M. Cunha, R. L. Reis, B. Vazquez and J. San Roman, *J. Mater. Sci.: Mater. Med.*, 1998, **9**, 825.
529. M. A. Bagliotti, B. S. F. Cury and R. C. Evangelista, *Carbohydr. Polym.*, 2014, **99**, 140.
530. A. L. Córdoba, L. Deladino and M. Martino, *Carbohydr. Polym.*, 2014, **99**, 150.
531. T. K. Dash and V. B. Konkimalla, *J. Controlled Release*, 2012, **158**, 15.
532. E. R. Balmayor, K. Tuzlakoglu, A. P. Marques, H. S. Azevedo and R. L. Reis, *J. Mater. Sci.: Mater. Med.*, 2008, **19**, 1617.
533. C. Elvira, J. F. Mano, J. San Roman and R. L. Reis, *Biomater*, 2002, **23**, 1955.
534. J. T. Oliveira, A. Crawford, J. M. Mundy, A. R. Moreira, M. E. Gomes, P. V. Hatton and R. L. Reis, *J. Mater. Sci.: Mater. Med.*, 2007, **18**, 295.
535. M. E. Gomes, A. S. Ribeiro, P. B. Malafaya, R. L. Reis and A. M. Cunha, *Biomaterials*, 2001, **22**, 883.
536. S. Ghanaati, S. Fuchs, M. J. Webber, C. Orth, M. Barbeck, M. E. Gomes and K. C. James, *J. Tissue Eng. Regener. Med.*, 2011, **5**, e136.
537. A. R. C. Duarte, J. F. Mano and R. L. Reis, *J. Supercrit. Fluid.*, 2009, **49**, 279.
538. M. Cascone, N. Barbanib, C. Cristallini, P. Giustic, G. Ciardellid and L. Lazzeri, *J. Biomater. Sci., Polym. Ed.*, 2001, **12**, 267.
539. Y. Wang, M. Xu, H. Y. Dong, Y. Liu, P. T. Zhao, W. Niu, D. Q. Xu, X. Ji, C. Xing, D. D. Lu and Z. C. Li, *Acta Histochem.*, 2012, **114**, 311.
540. Q. Zhang, T. Liang, L. Wang, Y. Cai and C. P. Okoli, *J. Sep. Sci.*, 2014, **37**, 257.
541. H. Zhou, X. G. Luo, X. Y. Lin, J. W. Li and H. P. Zhang, *Mater. Sci. Forum.*, 2012, **724**, 77.

CHAPTER 9

# Fracture and Failure of Starch-based Composites

CELINA R. BERNAL

Instituto de Tecnología en Polímeros y Nanotecnología (ITPN) (UBA-CONICET), Universidad de Buenos Aires, Av. Las Heras 2214 (CP 1127AAQ), Ciudad Autónoma de Buenos Aires, Argentina
Email: cbernal@fi.uba.ar; celinabernal55@gmail.com

## 9.1 Introduction

In recent years, increasing environmental concerns have led to much attention being focused on the development of biodegradable polymers for applications in which there is a high level of residual materials, such as packaging and in agriculture, with the aim of reducing environmental pollution derived from plastic wastes.[1–4] However, in order to compete with common synthetic plastics, biodegradable polymers must also be inexpensive and have comparable mechanical properties.

Among biodegradable polymers, starch is an agro-sourced polymer having strong advantages such as low cost (compared with other biodegradable polymers), wide availability and compostability without toxic residues.[5] Starch-based blends have enormous potential to be widely used in the areas of biomedical applications and environmentally friendly materials.[6,7] Blends of starch with biodegradable polymers such as polyhydroxybutyrate (PHB), poly(ε-caprolactone) (PCL) and poly(lactic acid) (PLA) have been extensively investigated[8–13] in order to lower the cost and to enhance their properties and biodegradability.[14]

RSC Green Chemistry No. 37
Starch-based Blends, Composites and Nanocomposites
Edited by Visakh P. M. and Long Yu

Published by the Royal Society of Chemistry, www.rsc.org

In many applications of polymers and their blends and composites, adequate fracture toughness is usually required. Several fracture mechanics approaches have been reported in order to characterize fracture in polymers and composites depending on the fracture behavior of the materials.

Linear elastic fracture mechanics (LEFM) is applicable to brittle materials that fail by catastrophic crack growth after reaching a threshold load value. The related criteria rely either on the stress field ahead of the crack tip through the stress intensity factor ($K_c$) or the energy release during crack extension through the strain energy release rate ($G_c$). These fracture parameters find application in polymers that behave in a brittle manner (*i.e.* undergoing little plastic deformation).[15]

When no significant crack growth resistance is observed, the value of the *J*-integral at initiation ($J_{Ic}$) is a good measure of toughness. The *J*-integral is conventionally defined for non-linear elastic materials as a path-independent line integral. In fact, the single-specimen *J* formulation has been extensively used to characterize fracture in ductile polymers.[16,17] Although ASTM E813 and ASTM E1152 apply only to ductile fracture, more recent standards allow *J*-integral testing of materials that fail by cleavage.[18,19]

The *J*-integral approach is a natural extension of LEFM and works best for not too ductile fractures. In the case of very ductile fracture, the essential work of fracture (EWF) approach works best. This approach was first proposed for plane stress ductile metal fractures[20] and later applied to polymers.[21] The aim of the EWF approach is to separate the work performed in the fracture process zone, $W_e$, from the total work of fracture, $W_f$, which in ductile polymers is often dominated by the work of plastic deformation, $W_p$:

$$W_f = W_e + W_p \tag{9.1}$$

The work performed in the fracture process zone, termed the essential work of fracture, is a quasi-material property dependent only on the specimen thickness. The EWF method makes use of the fact that the essential work and the plastic work scale differently:

$$W_f = w_e lB + \beta w_p l^2 B \tag{9.2}$$

where $B$ is the plate thickness, $l$ is the uncracked ligament length, $\beta$ is the shape factor and $w_e$ and $w_p$ are the specific essential and non-essential work of fracture, respectively (Figure 9.1). By dividing $W_f$ by the ligament area $lB$, one obtains the specific total work $w_f$, which can be expressed as

$$w_f = w_e + \beta w_p l \tag{9.3}$$

If the entire specimen ligament deforms plastically before fracture initiation, then the specific essential work can be found by testing different ligament lengths and extrapolating the specific total work of fracture to zero ligament length (Figure 9.2). In recent years, the EWF method has been

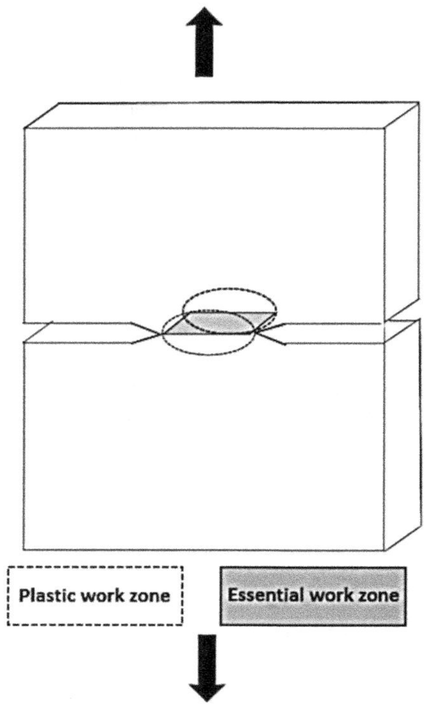


**Figure 9.1** Development of plastic zone ($w_p$) and process zone ($w_e$) in a double-edge notched tensile DENT specimen.

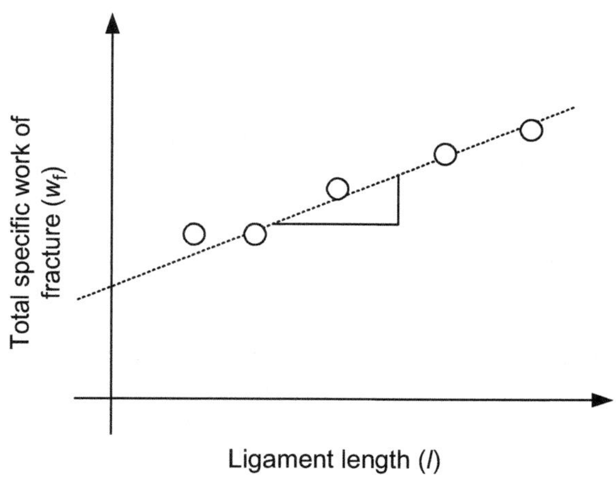


**Figure 9.2** Data reduction procedure for the EWF method.

extensively applied to polymers and there is a draft standard of the European Structural Integrity Society (ESIS).[22] The EWF method delivers a single fracture parameter that is representative of crack propagation.

It is important to note that $J_{Ic}$ and $w_e$ can be compared since they should be identical or very similar, as has been corroborated by several authors.[23–26] The EWF approach can also be used for plane strain fracture $J_{Ic}$, obtained either from slow strain rate tests[27,28] or from impact tests.[29]

The EWF concept was used by Wang *et al.*[14] to characterize the fracture behavior of poly(propylene carbonate) (PPC) modified with starch. During the EWF testing, they found that a significant amount of plastic deformation took place around the initial ligament region (*i.e.* outer plastic zone) for pure PPC and the PPC–starch blends containing 10 and 20 wt% of starch. However, for the PPC–starch system with 30 wt% of starch, they observed brittle fracture and consequently the EWF analysis could not be applied. Plots of the specific total fracture work against ligament length for pure PPC and PPC–starch blends containing 10 and 20 wt% of starch showed that a good linear relationship existed between specific total fracture work ($w_f$) and ligament length ($l$). Hence the $y$-intercepts for the $w_f$ *versus* $l$ plots gave the specific essential fracture work $w_e$ (*i.e.* fracture toughness) for the respective materials. The specific non-essential work of fracture $\beta w_p$ could also be obtained from the slopes of the $w_f$ *versus* $l$ plots. The $w_e$ and $\beta w_p$ values for pure PPC and PPC–starch were found to decrease with increase in the starch content, which indicated that the fracture toughness of PPC might have been affected by adding starch. The ductility of the samples was maintained with the incorporation of up to 20 wt% of starch particles. However, when 30 wt% of starch was added to PPC, the material exhibited brittle fracture. From the EWF analysis, the fracture toughness (which was expressed as the specific essential fracture work $w_e$) of the PPC–starch materials decreased with the incorporation of starch into PPC. In addition, morphology observations indicated that PPC and starch had weak interfacial adhesion.

Although a large number of studies have been already published on biodegradable materials and particularly on starch-based materials, most of them have not been specifically focused on the mechanical performance and the fracture behavior is usually disregarded.[30] In order to extend the applications of this type of polymers and their blends and composites, a full understanding of their performance is required. Their ultimate performance properties such as their fracture properties should be deeply analyzed and the different toughening mechanisms operative in these materials and responsible for this behavior should be elucidated.

In addition, there are two main approaches to study the fracture behavior of the materials. One is to develop a quantitative calculation of fracture toughness values by fracture mechanics as described above. The other is to determine and analyze the fracture morphology directly by optical or electron microscopy.[31] However, most of the literature on biodegradable polymers and their composites disregards one of these approaches or even both. The fracture behavior of these materials affects their resistance during processing, handling and transport. Hence it determines the extent to which biopolymers can replace synthetic polymers in many applications, such as packaging films, and consequently is of major economic concern.[32]

Furthermore, the moisture content greatly influences the fracture behavior of biopolymers. This behavior can even change from brittle to ductile due to water plasticization, which complicates the scenario.[32]

In this chapter, recent results on the fracture and failure of different starch-based composites are discussed. These composites include starch reinforced with natural fibers, carbon nanotubes or nanoclays. The emphasis is placed on those studies that not only analyzed the fracture behavior of the starch-based composites but also identified the failure mechanisms responsible for this behavior. Finally, a summary of the results and future trends on this topic are presented.

## 9.2 Fracture and Failure of Starch Reinforced with Natural Fibers

Discontinuous fibers have been widely used as reinforcement for thermoplastic matrices[33] in order to improve their properties and to allow better processing. Several cellulosic products and wastes were preferred as the reinforcement mainly to achieve cost savings.[34] Natural fibers also have many advantages compared with artificial fibers.[35] They are inexpensive, abundant and renewable, light weight, biodegradable and non-abrasive to processing equipment. In the case of biodegradable matrices, the use of these fibers is primarily due to environmental concerns.[36–40]

In short fiber-reinforced polymer composites, the mechanical properties are determined by many factors, such as fiber–matrix interactions, fiber volume fraction, fiber aspect ratio (length-to-diameter ratio) and fiber orientation. The fiber aspect ratio should be above a critical value for maximum stress in the fiber before composite failure. In addition, dissipation mechanisms are also affected by the fiber aspect ratio: fibers with a subcritical aspect ratio will be pulled out rather than broken. On the other hand, the fiber orientation has a significant influence on the mechanical properties of the composites in that the stress value is the highest along the axis of orientation of the fiber. Processing conditions have an important effect on the mechanical properties of fiber-reinforced composites due not only to the fiber orientation, but also to fiber breakage.[41] The combined effect of fiber content and microstructure on the fracture toughness of short fiber-reinforced composites has already been widely investigated in the case of different polymers reinforced with inorganic fibers.[42]

A significant number of studies have been published on the effect of natural fibers on the properties of thermoplastic starch (TPS),[43–49] but the literature regarding the fracture and failure behavior of these composites is still limited.

Recently, Muller *et al.*[50] presented a comprehensive investigation of the interfacial interactions in TPS reinforced with wood particles and a study of the mechanical properties including the failure mechanisms. They observed that modification of TPS with wood led to a very strong reinforcement, especially for fibers with a larger aspect ratio. However, they found an initial steep increase in mechanical properties that subsequently slowed at higher fiber contents, which they attributed to structural effects such as physical contact of wood particles or the formation of a network in the case of fibers with a large aspect ratio. They pointed out that various deformation processes may take place around the inclusions in composites. They suggested that in addition to debonding, *i.e.* the separation of the interface between the matrix and the reinforcement, several other processes such as matrix yielding, fiber pullout or fiber fracture or cavitation may also take place during the deformation of composites. From scanning electron microscopy (SEM) analysis, they observed that most of the fibers fractured perpendicular to their axis owing to the strong interactions between the wood fibers and the starch matrix. The mechanism of debonding was absent and fiber fracture was the dominant process during deformation of their composites.

The quasi-static in-plane fracture properties of a commercial starch blend (MaterBi-Y) reinforced with short sisal fibers was studied by Alvarez *et al.*[51] They observed that the matrix displayed almost linear load-displacement traces with an abrupt decrease in load to zero immediately after crack initiation, revealing the brittle-like behavior of the material. Sisal-reinforced composites, in contrast, showed greater stiffness and a higher degree of non-linearity before maximum load as a result of irreversible events.[52] A gradual decrease in load from the maximum point was observed thereafter and the samples exhibited stable crack propagation. However, no signs of macroscopic yielding of the matrix were observed on the tested samples. LEFM through the stress intensity factor ($K_{IQ}$) and the strain energy release rate ($G_{IQ}$) parameters, calculated from the maximum in the load displacement curves,[53] was adopted to characterize the resistance to crack initiation. A significant increase in $K_{IQ}$ and $G_{IQ}$ was observed as a result of the addition of sisal fibers to the plain matrix. This result was attributed to the development of new failure mechanisms induced by the presence of the fibers. A slight increasing trend of fracture parameters with fiber content was also found as a result of the increase of the dissipation mechanisms.

The work of fracture approach that mainly describes the stage of crack propagation was also applied by Alvarez *et al.*[51] in order to detect any difference in the fracture propagation mode that was not reflected by the fracture initiation parameters.[54,55] Higher values of the work of fracture were exhibited by the MaterBi-Y–sisal composites compared with the unreinforced matrix, indicating that the incorporation of sisal fibers in the matrix also improved the fracture propagation behavior in these composites. On the other hand, the work of fracture was found to be independent of sisal fiber content, suggesting that no significant differences in the fracture propagation mode existed among these biocomposites.

In addition, Alvarez *et al.*[51] applied the Pukánszky and Maurer model[56] for the composition dependence of $K_{IQ}$ and $G_{IQ}$ on their MaterBi-Y–sisal composites. A substantial linear correlation was obtained for both fracture parameters, suggesting that the model successfully fitted their experimental

**Table 9.1** Interaction parameter values obtained from the application of the Pukánszky and Maurer model to fracture toughness data.[51]

| Material | Loading conditions | $B_{GC}$ | $r^2$ |
|---|---|---|---|
| MaterBi-Y–sisal composite | Quasi-static (2 mm min$^{-1}$) | 12.2 | 0.9530 |
| MaterBi-Y–sisal composite | Impact (1 m s$^{-1}$) | 13.8 | 0.9776 |
| MaterBi-Y–sisal composite with treated fibers | Impact (1 m s$^{-1}$) | 14.3 | 0.9590 |

data. Hence the corresponding parameters $B_{Kc}$ and $B_{Gc}$ could also be used to study the effect of interaction on the fracture resistance in these composites. Based on their results on quasi-static fracture parameters, Alvarez *et al.*[51] also used the Pukánszky and Maurer model[56] to fit the impact fracture data they had previously reported for the same composites containing treated and untreated fibers.[56] They confirmed that the Pukánszky and Maurer model[56] could adequately describe the composition dependence of fracture parameters in their composites. However, the difference in the interaction parameter $B_{GC}$ values was within the experimental scatter of impact data and was unable to quantify the small improvement in interfacial properties that they had obtained with an alkali treatment of sisal fibers[57] (Table 9.1).

Furthermore, Alvarez *et al.*[51] also studied the impact out-of-plane fracture properties of their different MaterBi-Y–sisal composites. They observed a decrease in the disk strength in comparison with the plain matrix for the composite containing up to 15 wt% of sisal fibers. They hypothesized that the fibers spread through the matrix would have acted as crack initiation points during impact for those compositions.[33] In addition, a regular increasing trend of disk strength with fiber content was reported, as there were sufficient fibers at 20 wt% of sisal to enhance load transfer.[33] Conversely, the addition of sisal fibers to the MaterBi-Y matrix led to a sharp increase in the ductility index as new energy dissipation mechanisms resulting from the presence of the fibers became active. For the composites investigated by Alvarez *et al.*,[51] the ductility index was found to increase with fiber loading, especially for the highest fiber content they used (20 wt%). A similar trend for the energy at initiation and the fracture energy with increase in fiber content was also observed. This result suggested that the incorporation of sisal fibers in the biodegradable matrix was also an efficient way to improve impact fracture properties. In macrophotographs of the opposite side of the impacted areas, many radial cracks were observed around the hole area of matrix samples. In the composites, in contrast, circumferential cracking became more dominant as the fiber content increased. Mouzakis *et al.*[33] had previously pointed out that the circumferential shear-cracking phenomenon was more effective in toughening than the radial cracking mechanism. Hence Alvarez *et al.*[51] explained the increasing trend of fracture energy with fiber content that they observed in terms of an increase in the circumferential shear-cracking mechanism as the fiber content increased. The failure mechanisms observed by Alvarez *et al.*[51] in biaxial impact samples are illustrated in Figure 9.3.

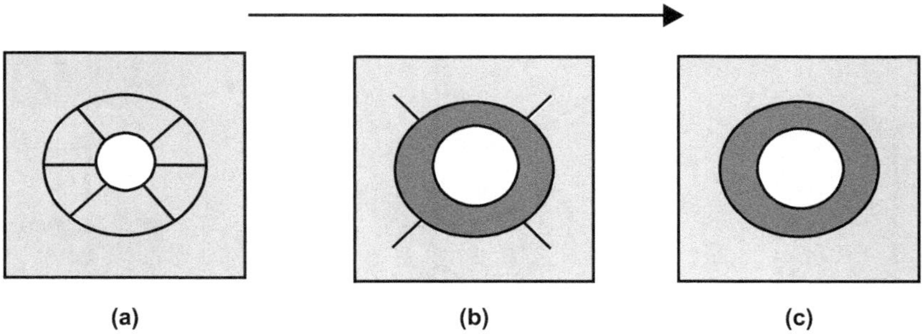


**Figure 9.3** (a) Radial cracking, (b) combined circumferential shear and radial cracking and (c) circumferential shear cracking. The arrow indicates increasing fiber content.

Finally, from fracture surface analysis by SEM, Alvarez *et al.*[51] observed that the matrix displayed flat surfaces due to the absence of any plastic damage. Furthermore, sisal fibers did not hold the matrix material, which suggested a relatively poor fiber–matrix interface. The presence of fiber pullout was predicted by Alvarez *et al.*[51] by considering the critical fiber length concept ($L_c$).[58] $L_c$ estimated by applying the Kelly–Tyson model[59] was found to be about 4.9 mm, which was longer than the mean fiber length of the composites after processing ($L = 2.42$ mm). Therefore, the fibers were expected to be pulled out rather than broken as they were not able to reach their strength. They observed fiber pullout and also other energy dissipation mechanisms that were active and produced the increase in toughening. Such mechanisms were identified as modest fiber–matrix debonding, separation of sisal fibers into ultimate cells by axial splitting of the boundary layer, uncoiling of these spirally arranged microfibrils and microcracking of ultimate cells (Figure 9.4). These failure mechanisms had been identified previously by Romhány *et al.*[60] for TPS-based composites containing flax fibers in unidirectional and crossed-ply arrangements. They determined the tensile behavior of their composites with different fiber contents and arrangements and also identified the failure mechanisms by using the acoustic emission technique. In addition, it had been previously reported by Lu *et al.*[61] that even uncoiling of microfibrils inside the plant fibers consumes substantial energy and consequently imparts high mechanical performance to the composite. On the other hand, surface roughness was also clearly observed by Alvarez *et al.*[51] in SEM fractographs of samples tested in fracture. This was due to the organic matrix surrounding the primary cell wall of fibers, which might have been responsible for additional energy-absorbing mechanisms such us friction and pullout of ultimate cells.[62]

Alvarez *et al.*[63] also investigated the effect of microstructure on the tensile and fracture properties of biodegradable composites obtained by injection molding. As they did not observe significant differences in fiber length and fiber aspect ratio between the composites with different fiber contents, they mainly associated the changes in microstructure with changes in fiber

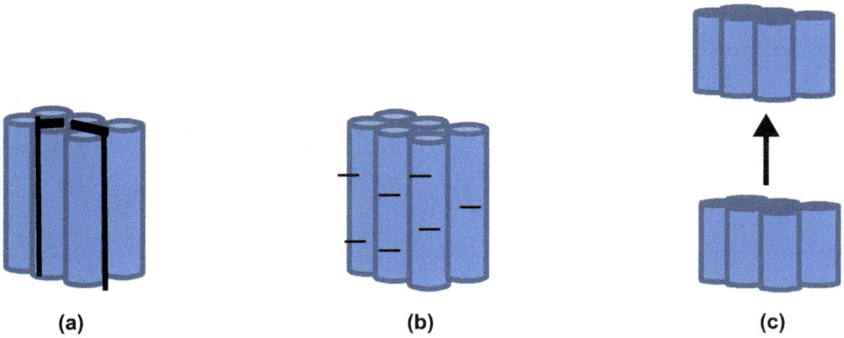


**Figure 9.4** Some typical energy dissipation mechanisms of natural fibers. (a) Splitting of microfibrils, (b) transversal microcracking and (c) multiple fracture of elementary fibers.
Adapted from Romhány *et al.*[60]

alignment. The tensile properties of the composites were found to be dependent on microstructure: longitudinally oriented samples (L samples) with respect to the mold filling direction displayed much higher values of Young's modulus and tensile strength than transversally oriented samples (T samples) as a consequence of fiber alignment during processing. In addition, irrespective of sample orientation, a trend of increasing Young's modulus and tensile strength with fiber content was found, suggesting that some kind of interaction between the reinforcement and the polymer matrix existed in their composites.[63] The polysaccharide and most likely starch blends appear to be compatible matrices for vegetable fibers owing to the intrinsic adhesion of the fiber–matrix interface as a result of the chemical similarity of such thermoplastic and plant fibers.[64] On the other hand, the observed trend in tensile properties was found to be more pronounced for L samples and also much higher values of tensile properties were displayed by these samples than transversally oriented samples. It had been well established in the literature[65] that injection-molded short fiber-reinforced polymer composites usually have a processing-induced three-layer structure. In the skin layers, fibers are mainly oriented parallel to the mold filling direction, whereas in the core layer, fibers are mainly oriented perpendicular to the mold filling direction. Hence L tensile samples were found to have many fibers mainly oriented longitudinal to the loading direction, whereas T tensile samples were found to have many fibers aligned preferentially transversal to the loading direction, as reported previously.[66] Other workers had also found that mechanical properties improve when the axis of orientation of the fibers is parallel to the loading direction compared with the other two mutually perpendicular directions. The maximum strength and reinforcement is virtually along the direction of fiber alignment, whereas reinforcement is low in the transverse direction because the fibers act as barriers that prevent the distribution of stresses throughout the matrix.[2,67] Intermediate values were found by Alvarez *et al.*[63] for randomly oriented fiber composites. In the case of natural fiber composites, additional factors contribute to the low reinforcement in the transverse direction: the boundary layer between elementary sisal fibers is easily split, hence fiber–matrix and elementary fiber interfaces are created.[60]

Regarding the fracture properties, Alvarez *et al.*[63] found that irrespective of sample orientation, $K_{IQ}$ values increased with the addition of sisal fibers to the neat matrix owing to the different failure mechanisms derived from the presence of the fibers, in agreement with their previous findings for compression-molded composites described above.[51] By comparing their actual fiber length values after processing (about 3.9) with the critical length value for similar composites, they found that irrespective of the fiber content, most fibers were shorter than the critical fiber length. Hence properly oriented fibers were expected to be pulled out rather than broken. Furthermore, the fiber pullout energy ($W_{po}$) was determined by Alvarez *et al.*[63] by assuming that the fibers were perpendicular to the crack growth direction and that the interfacial shear stress was half of the matrix tensile strength.[51] This was a rough approximation because for the low fiber volume fractions they examined, although the more random core layer can vanish,[42] a number of fibers were expected still to be aligned oblique to the crack propagation direction. An increasing trend of fiber pullout energy with fiber content was observed.

It had been already established in the literature[42] that the fiber pullout energy was higher for composites with oblique fibers to the crack propagation direction than those with transverse fibers. Hence the contribution of transverse fibers to the total composite toughness was smaller than that of oblique fibers. On the other hand, fibers parallel to the crack propagation direction did not contribute this type of energy and therefore their contribution to the total composite fracture toughness was much lower. The fracture toughness values obtained by Alvarez *et al.*[63] showed the expected result that the best improvement was achieved by a random alignment of fibers (larger number of oblique fibers) and the worst by the samples with many fibers aligned preferentially parallel to the crack propagation direction (L fracture samples). Hence the results obtained suggested that the fiber pullout mechanism made the largest contribution to the fracture toughness in their composites and also that this mechanism was able to explain the trends of fracture toughness with fiber content and microstructure. In addition, from fracture surface observations they confirmed the above results. L tensile samples had many fibers mainly oriented parallel to the loading direction and also fiber pullout was the main failure mechanism in those samples. In contrast, T tensile samples had many fibers preferentially aligned perpendicular to the loading direction and therefore fiber pullout appeared to be suppressed.

Alvarez *et al.*[63] also studied the effect of microstructure on the out-of-plane impact fracture properties of their injection-molded starch blend–sisal composites. Under impact loads parallel to the thickness direction, the incorporation of sisal fibers in the matrix led to an improvement in ductility

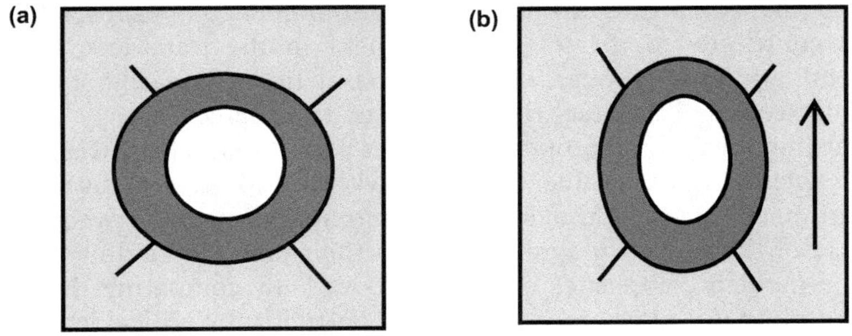


**Figure 9.5** Morphology of holes on tested samples in biaxial impact tests as a function of microstructure. (a) Compression-molded composites and (b) injection-molded composites. The arrow indicates the melt flow direction.

and fracture energy. In addition, the fiber orientation induced different patterns in the damage zone. From measurements of hole diameters in parallel and perpendicular directions with respect to the mold filling direction, it was observed that for the matrix these diameters did not differ markedly, indicating that crack propagation took place to a similar extent in both directions. When fibers were incorporated in the matrix, in contrast, the hole diameter became larger in the melt flow direction (Figure 9.5), *i.e.* crack propagation occurred mainly in the injection direction. On the other hand, the damage area was reduced as the fiber content increased. The results obtained in this study differed from the previously reported results for randomly oriented composites mentioned earlier.[51] For compression-molded composites, both diameters were similar irrespective of the fiber content as a consequence of the absence of any preferential fiber alignment and, therefore, cracks propagated to a similar extent in all directions (Figure 9.5).

On the other hand, Romhány *et al.*[68] pointed out that in thermoplastic composites reinforced with textile fabrics, the fracture behavior is very complex. These composites fail *via* crack propagation and, consequently, the application of linear elastic fracture mechanics is very limited. Elastic–plastic and post-yield fracture mechanics such as the *J*-integral and the essential work of fracture approaches have to be used. In addition, a large damage zone that may even change in shape during crack growth develops in those composites. In order to characterize the fracture behavior, the determination of the corresponding resistance curves is needed. For this purpose, the crack growth should be traceable during specimen loading, but this is a problem when crack growth can hardly be resolved by visual inspection and compliance calibration techniques are not suitable.[68] Other techniques, such as acoustic emission and infrared thermography, were successfully used by Romhány *et al.*[68] to trace the crack growth in TPS composites containing crossed quasi-unidirectional flax fiber reinforcement that failed in a ductile manner with pronounced crack growth. They found that the resistance to crack growth, $J_0$, and the resistance to crack propagation, $\Delta J/\Delta a$, as a function of flax content passed through a minimum. In addition, for flax contents higher than 40 wt%, the reinforcement increased the initiation value but decreased the tearing modulus, as is usual in thermoplastic composites having a very ductile polymer matrix. The main goal of this work was to characterize the complex fracture behavior of TPS reinforced with natural textile fabrics. This was achieved by using alternative experimental techniques such as acoustic emission and infrared thermography.

## 9.3 Fracture and Failure of Starch-based Nanocomposites

Although many commercial products made of starch-based materials are currently available, these materials are known to have limitations, such as poor processability and properties (*e.g.* weak mechanical properties, poor long-term stability and high water sensitivity). Hence significant efforts are still needed in order to improve the properties of starch-based materials to satisfy industrial needs and make them truly competitive with traditional petroleum-based plastics over a wider range of applications.[69] One of the most promising approaches has been the development of nanocomposites, *i.e.* the dispersion of nano-sized fillers (at least one dimension in the nanometer range, *i.e.* 1–100 nm) into a starch matrix. Different types of nanofillers (with different size, shape and surface chemistry) have been incorporated into plasticized starch such as phyllosilicates (montmorillonite, hectorite, sepiolite, *etc.*), polysaccharide nanofillers (nanowhiskers/nanoparticles from cellulose, starch, chitin and chitosan) and carbonaceous nanofillers [carbon nanotubes (CNTs), graphite oxide and carbon black], among others.[5] In the following, a discussion of some relevant results on the fracture and failure behavior of two special kinds of starch-based nanocomposites is presented: starch reinforced with carbon nanotubes and starch reinforced with nanoclays.

### 9.3.1 Starch Reinforced with Carbon Nanotubes

Different formulations of biodegradable composites based on the incorporation of CNTs into a starch matrix have been reported. Nevertheless, many difficulties have arisen in attempts to obtain well-dispersed nanotubes in starch without appropriate filler functionalization and/or matrix modification.[70–77] The homogeneous dispersion of the filler throughout the polymer matrix and adequate interfacial adhesion between phases are determining factors for the efficient use of composite materials.

The structures of CNTs, along with their low density, high aspect ratio and remarkable mechanical properties, make them particularly attractive for the reinforcement of polymers.[78–81] Several studies on starch–CNT composites have been reported in which different methods were used to disperse the filler into the matrix.[71–75] Recently, Famá and co-workers reported significant improvements in the mechanical properties of cassava starch-based nanocomposites with extremely small amounts of CNTs previously wrapped with a starch–iodine complex.[76,82] A significant improvement in Young's modulus and an enhancement of tensile strength values were found following the incorporation of very small amounts of multi-walled CNTs (MWCNTs) to a cassava starch matrix, maintaining deformations higher than 80%. They also observed that stiffness increased by almost 70% and ultimate tensile strength by ~35% with the addition of only 0.055 wt% of MWCNTs. In addition, they reported that all tensile parameters increased with the incorporation of wrapped CNTs and, consequently, tensile toughness also showed the same trend with filler loading.[82]

The results of Famá *et al.*[82] are not very frequently reported in the literature, especially with such small proportions of filler. They attributed their results to the good dispersion of the MWCNTs in the starch matrix and also to the strong interfacial adhesion between components. Hence efficient load transfer from the matrix to the filler was achieved and the high tensile strength and strain at break of the CNTs was transferred to the composites. Consequently, the good interfacial adhesion prevented the formation of holes and maintained the high deformations.

It is also important to note that at the time of writing, with the exception of the works of Famá and co-workers,[76,82] in all previous investigations the concentrations of fillers used were higher than 0.1 wt% and/or covalent functionalization was applied to the filler.[71,72,74]

Famá *et al.*[82] also investigated the biaxial impact fracture behavior of cassava starch-based films plasticized by glycerol and reinforced with different concentrations of MWCNTs. In agreement with their quasi-static tensile results, the features of the impact load–displacement curves of glycerol–starch films did not change with the incorporation of CNTs. All samples exhibited semi-brittle fracture behavior characterized by non-linear force–displacement curves that dropped to zero immediately upon reaching maximum load. In addition, typical radial cracks were observed around the hole area of impacted samples, which were responsible for the final failure of the samples. These cracks and the small deformed area around the hole also confirmed the semi-brittle fracture behavior reflected by load–displacement records.

Famá *et al.*[82] observed that thickness-related perforation energy, disc maximum strength and maximum displacement values increased significantly with increase in filler content, reaching improvements as high as ~100%. Their results demonstrated that the addition of even a very small proportion of CNTs to a starch matrix was also an efficient way to improve the material impact fracture behavior. They explained that in agreement with their tensile results, the improvements observed in impact properties were also the result of efficient load transfer from the matrix to the filler. This was determined by the extremely high area-to-volume ratio (which is about 1000 times that of spheres of similar radius[83]) and the high aspect ratio of CNTs. In addition, the surface-modified CNTs developed by Famá *et al.*[82] were highly compatible with the starch matrix, as they were wrapped with a starch–iodine complex composed of the same starch as the matrix. Hence excellent adhesion between phases and good dispersion of the CNTs in the matrix were achieved. In summary, they concluded that the excellent filler–matrix adhesion of their composites induced a cohesive fracture of the matrix and the good dispersion of nanotubes gave place to a tortuous crack path, resulting in an increase in toughness with increase in filler content.

## 9.3.2 Starch Reinforced with Nanoclays

Polymer–clay nanocomposites are particularly interesting because they are able to lead to a large reinforcing effect at low filler contents.[84] Phyllosilicates or layered silicates have strong advantages over other nanofillers such as wide availability, low cost, versatility, eco-friendliness and low toxicity.[5] Phyllosilicates are an important group of minerals that include the clay minerals, micas, chlorite, serpentine, talc, *etc.* They have different structures, textures and/or morphology.

Starch–PCL blend-based composites reinforced with nanoclays have been extensively investigated.[85–87] The creep and tensile properties and the effect of water uptake on these properties for starch–PCL blends reinforced with different types of clays have been analyzed.[84,88,89] Further, the influence of content and type of clay on the crystallization behavior of these materials has been reported.[90,91] However, in order to extend the applications of these materials, as mentioned earlier, knowledge of their tensile, fracture, impact and damage behavior, including the different toughening mechanisms, is also needed. Unfortunately, these properties are usually disregarded.[30]

Cyras *et al.*[92] prepared glycerol-plasticized starch–clay nanocomposites films from potato starch and three different loadings of montmorillonite (MMT) aqueous suspensions by casting, to study the effect of the nanoclay on the properties of starch. They carried out micro-tensile tests on dog-bone samples cut from the films and observed an important increase of five- and sixfold in the modulus when 3 and 5 wt% of MMT were added to the starch, respectively. This was in agreement with previous results of other workers for plasticized starch–clay systems.[6,93] They also found an increase of 57% in the maximum stress for the composite containing 5 wt% of MMT compared with neat starch. Huang *et al.*[93] also observed that the Young's modulus increased with increase in MMT content. However, this increment was greater when up to 5 wt% of clay was added to the starch. Then the values of Young's modulus did not follow the same trend with the addition of MMT. The slope was lower than that observed with less MMT, which was expected and was attributed by Cyras *et al.* to the resistance exerted by the clay itself and to the orientation and aspect ratio of the intercalated silicate layers. In addition, the stretching resistance of the oriented backbone of the polymer chain in the gallery bonded by hydrogen interactions was also considered to be a contributing factor to enhancing the modulus and the stress. The deformation decreased with increase in the MMT content. The layered silicate acted as a mechanical reinforcement of starch, reducing the flexibility of the polymer. The nanocomposites exhibited a remarkable improvement in mechanical properties, especially in stiffness, which was mainly attributed by Cyras *et al.* to the stronger interfacial interaction between the matrix and layered silicate due to the vast surface of the clay layers that was exposed. SEM images of the fracture surface after tensile testing were obtained for the films. The resulting micrographs revealed a smooth surface for the native starch. It had been reported previously[94] that in intercalated nanocomposites the clay platelets were poorly dispersed and formed aggregates that broke upon loading, as was observed in the fracture surface of the starch composites of Cyras *et al.*

Recently, Pérez *et al.*[95] determined the effects of different types of nanoclay (MMT, C30B and C10A) on the quasi-static fracture behavior of a commercial starch–PCL blend (MaterBi-Z). Some theoretical and experimental analyses were applied to study the fracture behavior of the composites. It was reported that most composites exhibited non-linear quasi-brittle fracture behavior characterized by the presence of unstable crack growth before maximum load or near it. However, they observed that the composites with 1 or 2.5 wt% C30B displayed a greater extent of stable crack propagation followed by a precipitous decrease in load to zero at a certain point after the maximum. Significantly higher displacement levels were also observed in these samples. As all materials exhibited non-linear fracture behavior until maximum load and unstable crack growth at a certain point on the load–displacement curve coincident with the point of fracture, the *J*-integral parameter at instability ($J_c$) was adopted by Pérez *et al.*[95] to characterize the material's fracture behavior.[96]

The integral $J$ was evaluated at the instability point ($J_c$) by calculating the total fracture energy of pr-cracked specimens having a crack depth-to-width ratio of $0.45 \leq a/W \leq 0.55$ as

$$J_{\max} = \frac{\eta U_{\text{tot}}}{B(W-a)} \tag{9.4}$$

where $U_{\text{tot}}$ is the fracture energy up to the point of fracture, *i.e.* the area under the load–deflection curve up to that point, $B$ the thickness of the tested specimen and $\eta$ a geometry factor that for DENT specimens is expressed as[97]

$$\eta = -0.06 + 5.99\left(\frac{a}{W}\right) - 7.42\left(\frac{a}{W}\right)^2 + 3.29\left(\frac{a}{W}\right)^3 \tag{9.5}$$

$J_c$ represents the material's resistance to unstable fracture.

Pérez *et al.*[95] observed that the fillers they used affected the fracture response of the starch–PCL blend differently. MMT and C10A did not significantly influence the material fracture toughness with up to 2.5 wt% of filler (1.06 and 1.6 vol.% of filler, respectively). Then, further increases in clay content were found to be beneficial and detrimental to the $J_c$ value in the case of MMT and C10A, respectively. Small amounts of C30B (1 and 2.5 wt%, 0.6 and 1.5 vol.%, respectively), on the other hand, led to significant improvements in the material fracture toughness but amounts of clay higher than 5 wt% (3.1 vol.%) led to slightly reduced $J_c$ values with respect to the matrix. The dispersion of the different types of clay in MaterBi-Z obtained from X-ray diffraction analysis had been reported previously in another investigation.[88] In the case of the composites with MMT and C10A, the position of the (0 0 1) peak shifted to lower angles, indicating that the polymer chains were intercalated into the clay platelets. On the other hand, although the peak was weaker, it remained at the same position for the composite with C30B, which suggested that no polymer intercalation took place in this case.[88] The composites with C10A displayed the highest values of Young's modulus and tensile strength among the three kind of fillers investigated by Pérez *et al.*,[88] which they attributed to their higher compatibility with the hydrophobic matrix. The observed differences in the dispersion of modified clays, and reflected in tensile properties, were explained in terms of the organic group of each modifier, as mentioned before. The benzyl group present in C10A had the largest surface area, producing a strong interaction with the polymer matrix. In addition, the hydroxyethyl group of C30B had hydroxyl (OH) groups that were not highly compatible with the hydrophobic matrix, leading to a lesser interaction and allowing the debonding process to occur. On the other hand, the intermediate values of tensile stiffness and strength for the MaterBi-Z–MMT composites were related to a better filler–matrix interphase, as had been reported previously.[88] The observed fracture behavior for the different composites was in agreement with previously reported tensile results. The composites with C30B, which had initial intermediate compatibility with the starch–PCL blend but the worst dispersion among the three nanoclays investigated, displayed the best fracture behavior as a result of the matrix ductile tearing favored by filler-induced voiding. On the other hand, the composites with C10A and MMT showed the lowest and intermediate values of the $J_c$ parameter, respectively.

On the other hand, the analysis of fracture surfaces represents an important approach to identifying the different toughening mechanisms and the effects of different parameters on the fracture behavior of composites.[96] Pérez *et al.*[95] also analyzed SEM fractographs of the specimens broken in fracture tests at different magnifications for the starch–PCL blend matrix and the composites with 2.5 wt% of MMT, C30B and C10A. They observed generalized ductile tearing of the starch–PCL blend matrix and that the composite with MMT displayed ductile tearing in a similar way to the matrix. Composite with 2.5 wt% of C30B also exhibited, as the matrix, generalized ductile tearing. This composite tended to induce even greater plastic deformation than the neat matrix, in accordance with the corresponding load–displacement curves. The composite with C10A displayed regions with significantly different topographies. Zones of localized matrix ductile tearing and also other regions exhibiting smooth surfaces were observed. These different kinds of topography were related to ductile and brittle fracture mechanisms, respectively. Their presence was in agreement with the low fracture toughness displayed by this composite. A really complex topography for all of the systems analyzed was recognized. Especially the voids necessary to activate ductile tearing of the matrix were not clearly observed, possibly due to the large plastic deformation of the matrix.

As the composites with C30B clay displayed the largest variations in fracture toughness, Pérez *et al.*[95] also applied the multifractal theory to these composites with the aim of correlating the topography of the fracture surfaces with the characteristic parameter spectra. The fractal and multifractal theories had previously been successfully applied to describe the morphology of fracture surfaces by variation of the characteristic parameters of the multifractal spectra.[98–102] The contents of C30B that corresponded to the minimum and maximum fracture toughness values (2.5 and 7.5 wt%, respectively) and the matrix were considered for the multifractal analysis by Pérez *et al.*[95] In order to apply the multifractal theory to experimental results, the $\log N(\varepsilon)$ *versus* $\log \varepsilon$ curve should exhibit a linear regression, called the scaling range.[103] As the data of Pérez *et al.*[95] verified this condition, multifractal theory was applied to study the fracture surfaces of starch–PCL blend. Similar scaling ranges had been reported previously for fracture surfaces analysis.[99] The corresponding curves for the other composites were also linearly fitted. Multifractal spectra for the starch–PCL blend and its composites with C30B showed that the shapes and widths of the spectra were different. The width of the curve increased for the composite with 2.5 wt% of C30B and decreased with increase in filler content. These variations suggested that the fracture surfaces tended to be more irregular with the initial incorporation of C30B and more regular with increasing content of clay.[103] The presence of brittle and ductile mechanisms is related to more regular or irregular surfaces, respectively. Pérez *et al.*[95] qualitatively confirmed their multifractal results by analyzing SEM micrographs. Moreover, observed topographical variations were detected by the mean gray value distribution. Hence the observed trend of fracture toughness with C30B content was also confirmed from the results of multifractal analysis.

The dependence of fracture toughness on filler content for the different composites with the model proposed by Pukánszky and Maurer[56] was also analyzed by Pérez *et al.*[95] For all nanoclays considered, a linear regression for low filler contents was found. On the other hand, deviation from a linear fitting was observed at high clay contents, suggesting the presence of some structural effect, probably clay agglomerates. Therefore, the model proposed by Pukánszky and Maurer could be applied to low clay contents. The lowest value of the $B$ parameter was obtained for the composites with C10A, indicating a strong interfacial interaction between the matrix and filler in this case. In contrast, the highest value corresponded to the composite with

C30B, which had the weakest filler–matrix interaction. An intermediate value of *B* was found for the composite with MMT, suggesting that although it had the lowest initial compatibility with the matrix, a better interphase existed between this filler and the blend of plasticized starch and PCL within the composites, as mentioned before. The results obtained from the application of Pukánszky and Maurer model were in good agreement with the analysis of fracture surfaces.

It has been established that surface modification of nanoclays can facilitate their exfoliation, normally considered as the best way to activate the filler effect. The potential toughening mechanisms for nanoclays are based on their large surface area-to-volume ratio and are assumed to be related to debonding for exfoliated nanoclays or to splitting for intercalated nanoclays. It should be pointed out that increased toughness was reported with not well exfoliated nanoclays. In addition, to activate the toughening mechanisms, voids are necessary to initiate crazes, but the structure or chemistry of nanoclays can inhibit multiple crazing of the polymer matrix.[96]

The composites with C10A exhibited the strongest filler–matrix interaction among the materials analyzed by Pérez *et al.*,[95] which would have hindered the void formation necessary to promote multiple crazing of the polymer matrix, leading to more brittle composites. This behavior was in accordance with fracture surface observations. In addition, for higher filler contents, clay agglomerates were detected, which could be considered as critical-size flaws.[96] Hence these agglomerates would have contributed also to the embrittlement observed in the composites with C10A. The toughening effect of C30B, in contrast, was ascribed to easy debonding of the clay due to the low compatibility between this filler and the matrix. This assumption was in agreement with fracture surface analysis, where intensive ductile tearing was observed for composites with C30B for filler contents lower than 5 wt%. The presence of agglomerates was detected at high filler contents. Hence a maximum was observed for fracture toughness above a certain value of filler fraction in this composite. Composites with MMT did not shown any significant toughening effect independently of filler content. Based on the intermediate value of the interfacial interaction in this case, Pérez *et al.* suggested that the voiding necessary to initiate crazes could not be fully developed, hence no significant toughening was achieved for these composites. Finally, for the different composites investigated, variations in fracture toughness values related to the filler type and content could be observed. These changes were explained in terms of two counteracting effects: the debonding process for a weak filler–matrix interaction, leading to subsequent matrix plastic deformation (increased toughness), and the presence of agglomerates. An increase in filler content induced larger filler agglomerates, which were detrimental to the fracture behavior of the material. Filler agglomeration was confirmed by Pérez *et al.* from deviations from a linear fitting in the model of Pukánszky and Maurer for the dependence of fracture toughness on composition.

## 9.4 Summary and Future Trends

Most of the literature on biodegradable polymers and their blends and composites is mainly focused on the analysis of the mechanical behavior, without a deep insight into the failure mechanisms responsible for this behavior. In particular, the fracture properties of starch-based composites, which mainly determine the ability of these materials to replace synthetic polymer-based composites in many applications, are usually disregarded.

Several studies have been already published regarding the fracture and failure of starch-based composites such as starch reinforced with natural fibers, with CNTs or with clay. The results of these studies are discussed in this chapter. However, at the time of writing, the number of papers on this issue is still very limited.

From the literature considered, it is clear that adequate methodology derived from fracture mechanics for starch-based composites can vary greatly depending on the fracture behavior of the materials, similarly to what happens with other polymer-based composites. However, in the case of starch matrices, additional factors also influence the ultimate performance of the materials, the most relevant being the moisture content due to water plasticization. Hence the scenario is even more complicated than in the case of synthetic polymer-based composites.

Much further work is still needed in order to take full advantage of the beneficial properties of starch-based composites and make them truly competitive with synthetic polymer-based composites. The fracture and failure behavior of these materials should be more deeply investigated under different loading and environmental conditions. The effect of time on this behavior should be also analyzed.

## References

1. J. Sahari, S. M. Sapuan, E. S. Zainudin and M. A. Maleque, Thermo-mechanical behaviors of thermoplastic starch derived from sugar palm tree (*Arenga pinnata*), *Carbohydr. Polym.*, 2013, **92**, 1711–1716.
2. G. A. Davis and J. H. Song, Biodegradable packaging based on raw materials from crops and their impact on waste management, *Ind. Crops Prod.*, 2006, **23**, 147–161.
3. G. Scott, 'Green' polymers, *Polym. Degrad. Stab.*, 2000, **68**, 1–7.
4. V. Siracusa, P. Rocculi, S. Romani and M. Dalla Rosa, Biodegradable polymers for food packaging: a review, *Trends Food Sci. Technol.*, 2008, **19**, 634–643.
5. F. Xiea, E. Pollet, P. J. Halleya and L. Avérous, Starch-based nano-biocomposites, *Prog. Polym. Sci.*, 2013, **38**, 1590–1628.
6. B. Chen and J. R. G. Evans, Thermoplastic starch–clay nanocomposites and their characteristics, *Carbohydr. Polym.*, 2005, **61**, 455–463.

7. A. Sorrentino, G. Gorrasi and V. Vittoria, Potential perspectives of bio-nanocomposites for food packaging applications, *Trends Food Sci. Technol.*, 2007, **18**, 84–95.
8. K. C. Reis, J. Pereira, A. C. Smith, C. W. P. Carvalho, N. Wellner and I. Yakimets, Characterization of polyhydroxybutyrate–hydroxyvalerate (PHB-HV)/maize starch blend films, *J. Food Eng.*, 2008, **89**, 361–369.
9. E. Corradini, J. M. Marconcini, J. A. M. Agnelli and H. C. L. Mattoso, Thermoplastic blends of corn gluten meal/starch (CGM/starch) and corn gluten meal/polyvinyl alcohol and corn gluten meal/poly-(hydroxybutyrate-*co*-hydroxyvalerate) (CGM/PHB-V), *Carbohydr. Polym.*, 2011, **83**, 959–965.
10. D. S. Rosa, D. R. Lopes and M. R. Calil, Thermal properties and enzymatic degradation of blends of poly(3-caprolactone) with starches, *Polym. Test.*, 2005, **24**, 756–761.
11. C. M. Chaléat, P. J. Halley and R. W. Truss, Properties of a plasticised starch blend. Part 1: Influence of moisture content on fracture properties, *Carbohydr. Polym.*, 2008, **71**, 535–543.
12. S. Zhang, X. Feng, S. Zhu, Q. Huan, K. Han, Y. Ma and M. Yu, Novel toughening mechanism for polylactic acid (PLA)/starch blends with layer-like microstructure via pressure-induced flow (PIF) processing, *Mater. Lett.*, 2013, **98**, 238–241.
13. Z. Xiong, C. Li, S. Ma, J. Feng, Y. Yang, R. Zhang and J. Zhu, The properties of poly(lactic acid)/starch blends with a functionalized plant oil: tung oil anhydride, *Carbohydr. Polym.*, 2013, **95**, 77–84.
14. X. L. Wang, R. K. Y. Li, Y. X. Cao and Y. Z. Meng, Short communication. Essential work of fracture analysis for starch filled poly(propylene carbonate) composites, *Mater. Des.*, 2007, **28**, 1934–1939.
15. T. Bárány, T. Czigány and J. Karger-Kocsis, Application of the essential work of fracture (EWF) concept for polymers, related blends and composites: a review, *Prog. Polym. Sci.*, 2010, **35**, 1257–1287.
16. E. Plati and J. G. Williams, The determination of the fracture parameters for polymers in impact, *Polym. Eng. Sci.*, 1975, **15**, 6.
17. C. Bernal, M. Rink and P. M. Frontini, Load separation principle in determination of *J–R* curve for ductile polymers, *Macromol. Symp.*, 1999, **147**, 235.
18. L. Fasce, V. Pettarin, R. Seltzer and P. Frontini, , Evaluation of impact fracture toughness of polymeric materials by means of the *J*-integral approach, *Polym. Eng. Sci.*, 2003, **43**, 1081–1095.
19. ASTM E1820-13, *Standard Test Method for Measurement of Fracture Toughness*, American Society for Testing and Materials, Philadelphia, PA, 2013.
20. B. Cotterell and J. K. Reddel, The essential work of plane stress fracture, *Int. J. Fract.*, 1977, **13**, 267–277.
21. Y. W. Mai and B. Cotterell, On the essential work of ductile fracture in polymers, *Int. J. Fract.*, 1986, **32**, 105–125.

22. E. Clutton, Essential work of fracture, in *Fracture Mechanics Testing for Polymers* ed. D. R. Moore, A. Pavan and J. G. Williams, ESIS Pub. 28, Elsevier, Amsterdam, 2001, pp. 177–195.
23. Y. W. Mai and P. J. Powell, Essential work of fracture and *J*-integral measurements for ductile polymers, *Polym. Sci., Part B: Phys*, 1991, **29**, 785–793.
24. C. A. Paton and S. J. Hashemi, Plane-stress essential work of ductile fracture for polycarbonate, *Mater. Sci.*, 1992, **27**, 2279–2290.
25. A. S. Saleemi and J. A. Nairn, The plane-strain essential work of fracture as a measure of the fracture toughness of ductile polymers, *Polym. Eng. Sci.*, 1990, **30**, 211–218.
26. G. Levita, L. Parisi and A. Marchetti, The work of fracture in semi ductile polymers, *J. Mater. Sci.*, 1994, **29**, 4545–4553.
27. J. S. Wu and Y. W. Mai, The essential fracture work concept for toughness measurement of ductile polymers, *Polym. Eng. Sci.*, 1996, **36**, 2275–2288.
28. P. Luna, C. Bernal, A. Cisilino, B. Frontini and Y.-W. Cotterell, Mai, The application of the essential work of fracture methodology to the plane strain fracture of ABS 3-point bend specimens, *Polymer*, 2003, **44**, 145–1150.
29. J.-S. Wu, Y.-W. Mai and B. Cotterell, Fracture-toughness and fracture mechanisms of PBT/PC/IM blend. Part 1. Fracture properties, *J. Mater. Sci.*, 1993, **28**, 3373–3384.
30. F. Tuba, L. Oláh and P. Nagy, Characterization of reactively compatibilized poly(d,l-lactide)/poly(ε-caprolactone) biodegradable blends by essential work of fracture method, *Eng. Fract. Mech.*, 2011, **78**, 3123–3133.
31. H. Zhao and R. K. Y. Li, Crystallization, mechanical and fracture behaviors of spherical alumina-filled polypropylene nanocomposites, *J. Polym. Sci. Polym. Phys.*, 2005, **43**, 3652–3664.
32. S. S. Paes, I. Yakimets, N. Wellner, S. E. Hill, R. H. Wilson and J. R. Mitchell, Fracture mechanisms in biopolymer films using coupling of mechanical analysis and high speed visualization technique, *Eur. Polym. J.*, 2010, **46**, 2300–2309.
33. D. E. Mouzakis, T. Harmia and J. Karger-Kocsis, Fracture behaviour of discontinuous long glass fiber reinforced injection molded polypropylene, *Polym. Polym. Comp*, 2000, **8**, 167–175.
34. K. Joseph, S. Thomas, C. Pavithran and M. Brahmakumar, Tensile properties of short sisal polyethylene composites, *J. Appl. Polym. Sci.*, 1993, **47**, 1731–1739.
35. H. D. Rozman, G. B. Peng and Z. A. Mohd Ishak, The effect of compounding techniques on the mechanical properties of oil palm empty fruit bunch–polypropylene composites, *J. Appl. Polym. Sci.*, 1998, **70**, 2647–2655.
36. S. Iannace, L. Nocilla and L. Nicolais, Biocomposites based on sea algae fibers and biodegradable thermoplastic matrices, *J. Appl. Polym. Sci.*, 1999, **73**, 583–592.

37. H. Hanselka, A. S. Herrman, Fiber properties and characterization, in: 7 Internationales Techtexil Symposium, 1994, pp. 20–22.
38. A. Dufresne and M. R. Vignon, Improvement of starch film performances using cellulose microfibrils, *Macromolecules*, 1998, **31**, 2693.
39. J. Kuruvilla and L. H. C. Mattoso, Sisal fibre reinforced polymer composites: status and future, *Natural Polymers and Composites*, 1998, p. 333.
40. C. Biastioli, Starch polymer composites, in *Degradable Polymers: Principles and Applications*, ed. G. Scott and D. Gilead, Chapman & Hall, London, 2nd edn, 1995.
41. P. V. Joseph, K. Joseph and S. Thomas, Effect of processing variables on the mechanical properties of sisal-fiber-reinforced polypropylene composites, *Compos. Sci. Technol.*, 1999, **59**, 1625–1640.
42. S.-Y. Fu, Y.-W. Mai, B. Lauke, G. Xu and C.-Y. Yue, Combined effect of fiber content and microstructure on the fracture toughness of SGF and SCF reinforced polypropylene composites, *J. Mater. Sci.*, 2002, **37**, 3067–3074.
43. H. Ibrahim, M. Farag, H. Megahed and S. Mehanny, Characteristics of starch-based biodegradable composites reinforced with date palm and flax fibers, *Carbohydr. Polym.*, 2014, **101**, 11–19.
44. J. L. Guimarães, F. Wypych, C. K. Saul, L. P. Ramos and K. G. Satyanarayana, Studies of the processing and characterization of corn starch and its composites with banana and sugarcane fibers from Brazil, *Carbohydr. Polym.*, 2010, **80**, 130–138.
45. X. Ma, P. R. Chang and J. Yu, Properties of biodegradable thermoplastic pea starch/carboxymethyl cellulose and pea starch/microcrystalline cellulose composites, *Carbohydr. Polym.*, 2008, **72**, 369–375.
46. W. N. Gilfillan, D. M. T. Nguyen, P. A. Sopade and W. O. S. Doherty, Preparation and characterisation of composites from starch and sugar cane fibre, *Ind. Crops Prod.*, 2012, **40**, 45–54.
47. A. P. Kumar and R. P. Singh, Biocomposites of cellulose reinforced starch: Improvement of properties by photo-induced crosslinking, *Bioresour. Technol.*, 2008, **99**, 8803–8809.
48. L. Avérous, C. Fringant and L. Moro, Plasticized starch–cellulose interactions in polysaccharide composites, *Polymer*, 2001, **42**, 6565–6572.
49. M. G. Ivo Martins, S. P. Magina, L. Oliveira, C. S. R. Freire, A. J. D. Silvestre, C. Pascoal Neto and A. Gandini, New biocomposites based on thermoplastic starch and bacterial cellulose, *Compos. Sci. Technol.*, 2009, **69**, 2163–2168.
50. P. Muller, K. Renner, J. Móczó, E. Fekete and B. Pukánszky, Thermoplastic starch/wood composites: interfacial interactions and functional properties, *Carbohydr. Polym.*, 2014, **102**, 821–829.
51. V. Alvarez, C. Bernal and A. Vázquez, Fracture characterization of sisal-reinforced starch based composites, *Polym. Compos.*, 2005, **26**, 316–323.
52. M. Hughes, C. A. S. Hill and J. R. B. Hague, Fracture toughness of bast fiber reinforced polyester composites, *J. Mater. Sci.*, 2002, **37**, 4669.

53. ASTM D5045-93, Standard Test Methods for Plane-Strain Fracture Toughness and Energy Release Rate Determination of Plastics Materials, American Society for Testing and Materials, Philadelphia, PA, 1993.
54. B. Lauke and W. Pompe, Relation between work of fracture and fracture-toughness of short-fibre reinforced polymers, *Compos. Sci. Technol.*, 1986, **26**, 37.
55. P. Montemartini, T. Cuadrado and P. Frontini, Fracture evaluation of acrylic bone cements modified with hydroxyapatite: influence of the storage conditions, *J. Mater. Sci.*, 1999, **10**, 309.
56. B. Pukánszky and F. H. Maurer, Composition dependence of the fracture toughness of heterogeneous polymer systems, *Polymer*, 1995, **36**, 1617.
57. V. Alvarez, R. Ruseckaite and A. Vázquez, Kinetic analysis of thermal degradation of pvoh-co-ethylene blends, *J. Compos. Mater.*, 2003, **37**, 1575.
58. S. C. Tjong, S. A. Xu and Y.-W. Mai, Impact fracture toughness of short glass fiber-reinforced polyamide 6,6 hybrid composites containing elastomer particles using essential work of fracture concept, *Mater. Sci. Eng.*, 2003, **A347**, 338.
59. S. G. Wong, G. X. Sui, Y. Yue and Y.-W. Mai, Characterization of microstructures and toughening behavior of fiber-containing toughened nylon 6,6, *J. Mater. Sci.*, 2002, **37**, 2659.
60. G. Romhány, J. Karger-Kocsis and T. Czigány, Tensile fracture and failure behavior of thermoplastic starch with unidirectional and cross-ply flax fiber reinforcements, *Macromol. Mater. Eng.*, 2003, **288**, 699–707.
61. X. Lu, M. Q. Zhang, M. Z. Rong, G. Shi and G. C. Yang, All-plant fiber composites. I: Unidirectional sisal fiber reinforced benzylated wood, *Polym. Compos.*, 2002, **23**, 624.
62. A. Stamboulis, C. Baillie and E. Schulz, Interfacial characterization of flax fibre-thermoplastic polymer composites by the pull-out test, *Angew. Makromol. Chem.*, 1999, **272**, 117.
63. V. Alvarez, A. Vázquez and C. Bernal, Effect of microstructure on the tensile and fracture properties of sisal fiber/starch based composites, *J. Compos. Mater.*, 2006, **40**, 21–35.
64. L. Lundquist, B. Marque, P.-O. Hagstrand, Y. Leterrier and J.-A. E. Manson, Novel pulp fibre reinforced thermoplastic composites, *Comp. Sci. Technol.*, 2003, **63**, 152.
65. M. Wollerdorfer and H. Bader, Influence of natural fibres on the mechanical properties of biodegradable polymers, *Ind. Crops Prod.*, 1998, 105–112.
66. T. Harmia and K. Friedrich, Fracture toughness and failure mechanisms in unreinforced and long-glass-fiber-reinforced PA6/PP blends, *Compos. Sci. Technol.*, 1995, **53**, 423–430.
67. R. P. Kumar, M. L. G. Amma and S. Thomas, Short sisal fiber reinforced styrene–butadiene rubber composites, *J. Appl. Polym. Sci.*, 1995, **58**, 597–612.

68. G. Romhány, T. Czigány and J. Karger-Kocsis, Determination of J-R curves of thermoplastic starch composites containing crossed quasi-unidirectional flax fiber reinforcement, *Compos. Sci. Technol.*, 2006, **66**, 3179–3187.
69. Y. Zhang, C. Rempel, D. McLaren, Thermoplastic starch, in *Innovations in Food Packaging*, ed. J. Han, Elsevier Academic Press, London, 2nd edn, 2014, ch. 16, pp. 391–412.
70. A. Casey, G. F. Farrell, M. McNamara, H. J. Byrne and G. Chambers, Interaction of carbon nanotubes with sugar complexes, *Synth. Met.*, 2005, **153**, 357.
71. X. Cao, Y. Chen, P. R. Chang and M. A. Huneault, Preparation and properties of plasticized starch/multiwalled carbon nanotubes composites, *J. Appl. Polym. Sci.*, 2007, **106**, 1431.
72. X. Ma, P. R. Chang, J. Yu and P. Lu, Properties of biodegradable citric acid-modified granular starch/thermoplastic pea starch composites, *Carbohydr. Polym.*, 2008, **74**, 895.
73. P. R. Chang, P. Zheng, B. Liu, D. P. Anderson, J. Yu and X. Ma, Characterization of magnetic soluble starch-functionalized carbon nanotubes and its application for the adsorption of the dyes, *J. Hazard. Mater.*, 2011, **18**, 6.
74. Z. Liu, L. Zhao, M. Chen and J. Yu, Effect of carboxylate multi-walled carbon nanotubes on the performance of thermoplastic starch nanocomposites, *Carbohydr. Polym.*, 2011, **83**, 447.
75. L. Yan, P. R. Chang and P. Zheng, Preparation and characterization of starch grafted multiwall carbon nanotube composites, *Carbohydr. Polym.*, 2011, **84**, 1378.
76. L. Famá, P. G. Gañán Rojo, C. Bernal and S. Goyanes, Biodegradable starch based nanocomposites with low water vapor permeability and high storage, *Carbohydr. Polym.*, 2012, **87**, 1989.
77. S. L. Ruan, P. Gao, X. G. Yang and T. X. Yu, Toughening high performance ultrahigh molecular weight polyethylene using multiwalled carbon nanotubes, *Polymer*, 2003, **44**, 5643.
78. M. Moniruzzaman and K. Winey, Polymer nanocomposites containing carbon nanotubes, *Macromolecules*, 2006, **39**, 5194.
79. W. Jiang, W. Qiao and K. Sun, Mechanical and thermal properties of thermoplastic acetylated starch/poly(ethylene-*co*-vinyl alcohol) blends, *Carbohydr. Polym.*, 2006, **65**, 139.
80. D. Zilli, S. Goyanes, M. M. Escobar, C. Chiliotte, V. Bekeris, A. L. Cukierman and G. H. Rubiolo, Comparative analysis of electric, magnetic and mechanical properties of epoxy matrix composites with different contents of multiple walled carbon nanotubes, *Polym. Compos.*, 2007, **28**, 612.
81. A. De Falco, A. J. Marzocca, M. A. Corcuera, A. Ecesiza, I. Mondragon, G. H. Rubiolo and S. Goyanes, Accelerator adsorption onto carbon nanotubes surface affects the vulcanization process of styrene–butadiene rubber composites, *J. Appl.d Polym. Sci.*, 2009, **113**, 2851.

82. L. M. Famá, V. Pettarin, S. N. Goyanes and C. R. Bernal, Starch/multi walled carbon nanotubes composites with improved mechanical properties, *Carbohydr. Polym.*, 2011, **83**, 1226.
83. J. Coleman, U. Khan, W. Blau and Y. Gunko, Small but strong: a review of the mechanical properties of carbon nanotube–polymer composites, *Carbon*, 2006, **44**, 1624.
84. C. J. Pérez, V. A. Alvarez, I. Mondragón and A. Vázquez, Water uptake behavior of layered silicate/starch–polycaprolactone blend nanocomposites, *Polym. Int.*, 2008, **57**, 247–253.
85. B. Q. Chen and J. R. G. Evans, On the driving force for polymer intercalation in smectite clays, *Philos. Mag.*, 2005, **85**(14), 1519–1538.
86. L. Vertuccio, G. Gorrasi and A. Sorrentino, Nano clay reinforced PCL/starch blends obtained by high energy ball milling, *Carbohydr. Polym.*, 2009, **75**, 172–179.
87. Q. X. Zhang, Z. Z. Yu, X. L. Xie, K. Naito and Y. Kagawa, Preparation and crystalline morphology of biodegradable starch/clay nanocomposites, *Polymer*, 2007, **48**, 7193–7200.
88. C. J. Pérez, V. A. Alvarez, I. Mondragón and A. Vázquez, Mechanical properties of layered silicate/starch–polycaprolactone blend nanocomposites, *Polym. Int.*, 2007, **56**, 686–693.
89. C. J. Pérez, V. A. Alvarez and A. Vázquez, Creep behavior of layered silicate/starch–polycaprolactone blends nanocomposites, *Mater. Sci. Eng. A*, 2008, **480**, 259–265.
90. C. J. Pérez and V. A. Alvarez, Isothermal crystallization of layered silicate/starch–polycaprolactone blend nanocomposites, *J. Therm. Anal. Calorim.*, 2008, **91**, 749–757.
91. C. J. Pérez, V. A. Alvarez, P. M. Stefani and A. Vázquez, Non-isothermal crystallization of MaterBi-Z/clay nanocomposites, *J. Therm. Anal. Calorim.*, 2006, **88**, 825–832.
92. V. Cyras, L. Manfredi, M. Ton-That and A. Vázquez, Physical and mechanical properties of thermoplastic starch/montmorillonite nanocomposite films, *Carbohydr. Polym.*, 2008, **73**, 55–63.
93. M. Huang, J. Yu, X. Ma and J. Peng, High performance biodegradable thermoplastic starch–EMMT nanoplastics, *Polymer*, 2005, **46**(9), 3157.94.
94. S. C. Tjong, Structural and mechanical properties of polymer nanocomposites, *Mater. Sci. Eng.*, 2006, **53**, 73–197.
95. E. Pérez, C. J. Pérez, V. Alvarez and C. Bernal, Fracture behavior of a commercial starch/polycaprolactone blend reinforced with different layered silicates, *Carbohydr. Polym.*, **97**, 2013, 267–276.
96. B. Cotterell, J. Y. H. Chia and K. Hbaieb, Fracture mechanisms and fracture toughness in semicrystalline polymer nanocomposites, *Eng. Fract. Mech.*, 2007, **74**, 1054–1078.
97. W. Grellmann and K. Reincke, Quality improvement of elastomers – application of instrumented notched tensile-impact testing for assessment of toughness, *Material prüfung Jahrgang*, 2004, **46**, 168–175.

98. C. Liu, X. L. Jiang, T. Liu, L. Zhao, W. X. Zhou and W. K. Yuan, Multifractal analysis of the fracture surfaces of foamed polypropylene/polyethylene blends, *Appl. Surf. Sci.*, 2009, **255**, 4239–4245.
99. E. Pérez, C. Bernal and M. Piacquadio, Multi fractal analysis of tensile toughness and filler dispersion for polypropylene–$CaCO_3$ composites, *Appl. Surf. Sci.*, 2012, **258**, 8940–8945.
100. M. Tarafder, S. K. Das, I. Chattoraj, M. Nasipuri and S. Tarafder, Fractal-based quantification of crack paths for determination of effective microstructural length scales and fracture toughness, *Scr. Mater.*, 2010, **62**, 109–112.
101. B. Venkatesh, D. L. Chen and S. D. Bhole, Three-dimensional fractal analysis of fracture surfaces in a titanium alloy for biomedical applications, *Scr. Mater.*, 2008, **59**, 391–394.
102. Y. H. Zhang, B. F. Bai, J. Q. Li, J. B. Chen and C. Y. Shen, Multifractal analysis of the tensile fracture morphology of polyvinylidene chloride/glass fiber composite, *Appl. Surf. Sci.*, 2011, **257**, 2984–2989.
103. O. Malcai, D. A. Lidar and O. Biham, Scaling range and cutoffs in empirical fractals, *Phys. Rev. E: Stat. Phys., Plasmas, Fluids, Relat. Interdiscip. Top.*, 1997, **56**, 2817–2828.

CHAPTER 10

# Application of Starch Nanocomposites in the Food Industry

KHONGSAK SRIKAEO

Faculty of Food and Agricultural Technology, Pibulsongkram Rajabhat University, Muang, Phitsanulok 65000, Thailand
Email: khongsak@live.psru.ac.th

## 10.1 Introduction

Nanotechnology is opening up new horizons in almost all scientific and technological fields. Among these, applications of nanotechnologies are expected to bring large benefits and add value to the food and food-related industries through the whole food chain, from production to processing, safety, packaging, transportation, storage and delivery.[1]

Nanotechnology is a multidisciplinary field that covers a vast range of processes, materials and applications encompassing physical, chemical, biological, engineering and electronic sciences. It focuses on the characterization, fabrication and manipulation of substances at sizes in the nanoscale range, between approximately 1 and 100 nm. The smaller particle size, in combination with an increased surface area, confers unique and novel properties, thus creating vast potential for applications.[2–4] A nanomaterial is defined as any material that has one or more dimensions in the nanoscale range, whereas a nanoparticle is a discrete entity that has all three dimensions at the nanoscale.[5] Nanomaterials and nanoparticles can encompass various nanoforms, which derive their names from their

RSC Green Chemistry No. 37
Starch-based Blends, Composites and Nanocomposites
Edited by Visakh P. M. and Long Yu

Published by the Royal Society of Chemistry, www.rsc.org

individual shapes and dimensions, *e.g.* nanotubes, nanofibres, nanorods, nanofilms, nanolayers, nanocoatings and nanosheets.[6,7]

Starch as an agro-sourced polymer has attracted much attention recently owing to its strong advantages such as low cost, wide availability and total compostability without toxic residues. However, despite a considerable number of commercial products being available, the fundamental properties (mechanical properties, moisture sensitivity, *etc.*) of plasticized starch-based materials have to be enhanced to enable such materials to be truly competitive with traditional petroleum-based plastics over a wider range of applications.[8] In this respect, one of the most promising technical advances has been the development of starch nanocomposites. Starch is mostly converted into starch nanocomposites to enhance its properties and to overcome its drawbacks. Starch can be used as the matrix either singly or in combination with other polymers and with the incorporation of different types of nanofillers, such as nanoclay, cellulose nanoparticles, carbon nanotubes (CNTs) and other inorganic nanofillers. In another approach, starch granules are transformed into nanoparticles and are introduced as fillers in various polymer matrices.[9] Moreover, nanoparticles produced from starches may be used directly as food ingredients.

This chapter provides an overview of the implications of nanotechnology for the food sector, with particular focus on starch nanocomposites.

## 10.2 Nanotechnology in Foods

The phenomena that take place at the nanometre scale offer multiple opportunities for innovation that have the potential to impact the food industry substantially worldwide, as nanotechnologies can be applied to the whole food chain, from production to processing, safety, packaging, transportation, storage and delivery.[4,6,10] The impacts of and needs for nanotechnological applications in food and related sectors are depicted in Figure 10.1.

The potential applications of nanotechnologies in food-related fields are almost endless and the design and production of nanomaterials and nano-based structures, devices and systems provide new ways of producing and processing foods and also protecting the safety and authenticity of food products in the supply chain. Despite the projected benefits, this exciting technological frontier is raising regulatory issues and public concern regarding safety and environmental effects. For regulatory purposes, various countries have adopted different and sometimes diverging approaches.[6,11] Table 10.1 summarizes the current and projected applications of nanotechnology in the food production chain.

### 10.2.1 Application of Nanotechnology in Food Production and Nutrition

The intentions of applying nanotechnology in foods are to improve the quality, safety and nutritional value of food, and also to reduce costs.

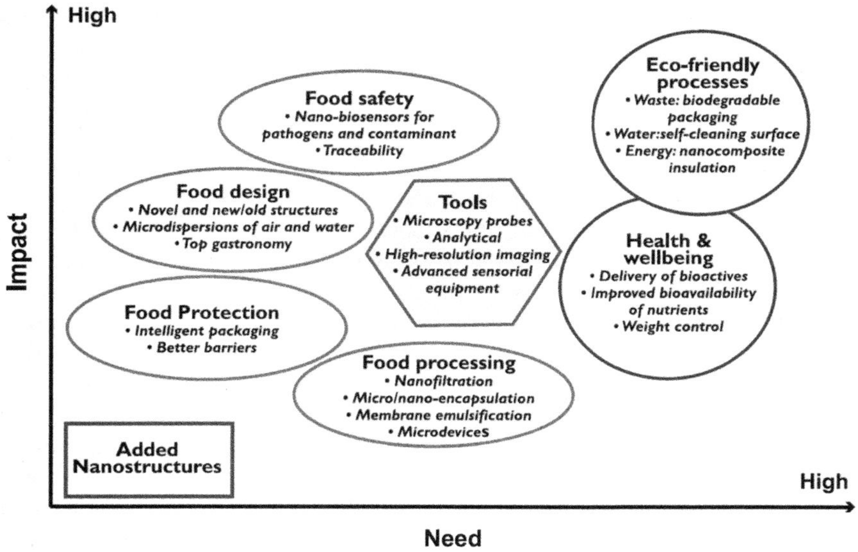


**Figure 10.1** Impacts of and needs for nanotechnological applications in foods and food processing.
Reproduced from Rossi *et al.*[1] with permission of Elsevier.

**Table 10.1** Current and projected applications of nanotechnology in the food production chain. Reproduced from Rossi *et al.*[1] with permission of Elsevier.

| Nanotechnology | Applications | Functions |
|---|---|---|
| *Agricultural production* | | |
| Nanosensors | Nanosprays | Binding and colouring microorganisms |
| Hand-held devices | Detection of contaminants, mycotoxins and microorganisms | |
| Nano-sized additives | Organic or inorganic nano-sized additives for feed applications, including removal of toxins in feed | Various (including nutritional additives) with lesser amounts needed for a specific function |
| Nanopesticides | Nanoemulsions, nanoencapsulates | Increased efficacy and water solubility |
| Triggered-release nanoencapsulates | Triggered (local) release | |
| Other nano-sized agrochemicals | Nano-sized fertilizers, biocides, veterinary medicines | Improved delivery of agrochemicals in the field, better efficacy, better control of application/dose, less use of solvents in agricultural spraying |

**Table 10.1** (*Continued*)

| Nanotechnology | Applications | Functions |
|---|---|---|
| *Food processing* | | |
| Processed nano-structured or nano-textured food products | Nanoemulsions, micelles | Use of less fat and emulsifiers, stable emulsions, better tasting food products |
| Nano-sized additives | Organic or inorganic nano-sized additives for food and health-food applications | Various, but lesser amounts would be needed for a function or a taste attribute, better dispersibility may also occur |
| Nanoencapsulates | Nanocarrier systems in the form of liposomes or biopolymer-based nanoencapsulated substances | Providing protective barriers, flavour and taste masking, controlled release and better dispersibility for water-insoluble food ingredients and additives |
| Nutritional supplements and nutraceuticals | Nanoingredients and additives | Enhanced absorption and bioavailability of nano-sized ingredients in the body |
| *Nano-carrier systems for delivery of nutrients* | | |
| Nanofilters | Nanofiltration (*e.g.* porous silica, regenerated cellulose membranes) | Filtration of water and removal of some undesired components in food, such as bitter taste in some plant extracts |
| *Food contact materials, food storage and handling* | | |
| Nanocomposites | Incorporating nanoparticles in a polymer to form a composite | Improving strength of materials, durability, barrier properties, biodegradation |
| Nanocoatings | Incorporating nanomaterials on the packaging surface | Improving barrier properties |
| "Active" food contact materials | Incorporating active nanoparticles with intentional release into and consequent effect on the packaged food | Oxygen scavenging, prevention of growth of pathogens |
| "Intelligent" packaging materials | Incorporating nanosensors for food labelling | Detection of food deterioration, monitoring storage conditions |
| Surface biocides | Incorporating nanoparticles (Ag, ZnO, MgO) on surfaces | Antimicrobial coatings for refrigerators, storage containers, equipment for food processing, handling and preparation |

Consumers can benefit from this application in terms of meeting individual dietary and health requirements or taste preferences, and benefits to food companies include product differentiation, new market opportunities and economic gains.

Several reviews have identified that the emerging applications of nanotechnology in food processing are focused on the development of nano-sized food ingredients and additives and delivery systems for nutrients and supplements in the form of nutraceuticals. A diverse range of processes are being utilized to aid this, such as nanoemulsions, surfactant micelles, emulsion layers, reverse micelles and functionally designed nanocapsules.[3,7,12–20]

There have been major advances in the design and fabrication of food-grade nanoparticles that can be used by food manufacturers to develop effective micronutrient delivery systems.[21–24]

A key application of nanotechnology in food processing involves the development of nano-structured food ingredients that offer improvements in consistency, taste and texture.[19] Nanoemulsion technology is frequently used to create low-fat mayonnaise, spreads and ice cream, which manufacturers claim to be as creamy as the full-fat alternatives, thus offering consumers healthier options.[6] For example, mayonnaise can be nano-textured using oil in water emulsion containing nanodroplets of water inside oil droplets. The mayonnaise offers taste and texture attributes that are similar to those of the full-fat equivalent but with significant reductions in the fat content. Ice cream with reductions in fat from 8–16 to 1% while not compromising on the flavour has also been produced. Consumers can also benefit from more rapid and simpler thawing of frozen foods in the microwave oven using nanoemulsion technology.[17,18]

Food companies can benefit greatly from adding nanoparticles to their food and beverage products in terms of improvements to flavour, colour, flow properties and stability during processing or extension of shelf life. For instance, aluminosilicate materials are commonly used as anticaking agents in granular or powdered processed foods, while anatase titanium dioxide is a food whitening and brightening additive that is commonly used in confectionery, some cheeses and sauces. Titanium dioxide is widely used in commercially available food and beverages, including chocolate, cheese, ready prepared mashed potato, coffee creamer, yoghurts, pop tarts, mints, sports drinks and salad dressing.[7]

Nanotechnology offers opportunities to alter and manipulate food and beverage products to allow for more effective delivery of nutrients such as proteins, vitamins and minerals, in addition to antioxidants, to specifically target nutritional and health benefits to consumers. This application also enables food companies to gain a competitive advantage by satisfying individual dietary requirements and consumers' varied demands for foods. An important current nanotechnology application is nanoencapsulation of food ingredients and additives. Nanocarrier systems including emulsions, micelles, liposomes, biopolymer matrices and association colloids have been developed for use in food and beverage products.

Nanoencapsulation can control the release of certain active ingredients (*e.g.* proteins, vitamins, minerals, enzymes and preservatives), mask undesirable odours and flavours such as fish oils, enhance the shelf life and stability of the ingredients and the finished food product and also improve the uptake of encapsulated nutrients and supplements.[15,18] The modified characteristics of nanocarriers permit their use in a vast array of food and beverage products. For example, nanocapsules containing tuna fish oil have been successfully incorporated into bread for additional health benefits. The nanocapsules are designed to be secreted once they enter the stomach, thereby avoiding the unpleasant taste of the fish oil. Another example is the use of minute compressed micelles or nanodrops in the development of canola active oil. The micelles work as a liquid carrier, allowing the penetration of vitamins, minerals and phenolic compounds that are insoluble in water or fats. The micelles are added to food products and so pass through the digestive system efficiently, without breaking up, to the absorption site.[7]

A number of nanoparticle-based delivery systems built of different ingredients suitable for encapsulation of micronutrients are illustrated in Figure 10.2.

Nanoencapsulation technologies are being employed to protect bioactive compounds, including vitamins, minerals, proteins, lipids, carbohydrates and antioxidants, for the manufacture of functional foods with improved functionality and stability, thus offering great potential for improvements to public health and nutrition. Novel carriers for nutraceuticals to be incorporated into food systems have been developed, thereby enhancing the bioavailability of the product. Lycopene, β-carotenes and phytosterols are some of the nutraceuticals incorporated in the carriers and used in the production of healthy foods, especially to prevent the accumulation of cholesterol.[18,25]

It has been reported that micelles are capable of encapsulating non-polar molecules, including flavourants, lipids, antimicrobials, vitamins and antioxidants.[3] Various nanomicelle-based carrier systems have been developed for nutraceuticals and nutritional supplements and are commercially available. For example, liposomes as a supplement delivery system were used in vitamin C; it can produce serum levels of vitamin C that are nearly twice the level of any other oral form of vitamin C. The nanocochleates ($\sim$50 nm in size) are based on a phosphatidylserine carrier derived from soya bean, generally regarded as safe (GRAS). The nanocochleate system appears to prevent the degradation of micronutrients and antioxidants during manufacture and storage.[17,18]

A recent trend being investigated is the nanoencapsulation of live probiotic microbes for the promotion of gastrointestinal health. They can be incorporated into various food and drink products, including fermented milk, yoghurts, cheese, puddings and fruit-based drinks. Nanoencapsulation technology is applied to aid in the development of designer probiotic bacterial preparations that can be transited to the gastrointestinal tract, where they interact with specific receptors and can improve intestinal microflora.[26,27]

Numerous products based on nanocluster delivery systems are available commercially worldwide. This technology has been used to create a slimming product based on cocoa nanoclusters, which are coated on the surface

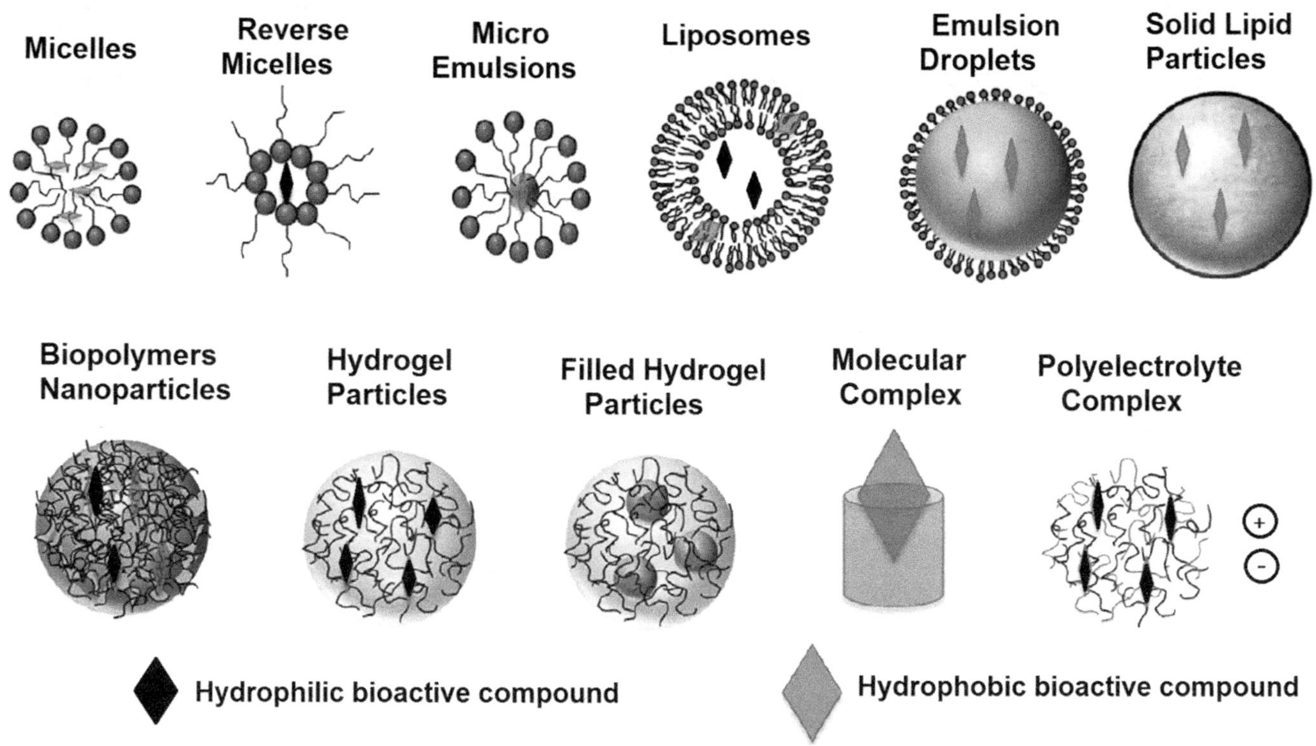


**Figure 10.2** Examples of colloidal delivery systems that can be used to encapsulate, protect and deliver functional food ingredients with the most likely location of the bioactive compound based on its hydrophilicity of hydrophobicity (not drawn to scale). Reproduced from Joye *et al.*[20] with permission of Elsevier.

of engineered nanomaterials to enhance the chocolate flavour through the increase in surface area that targets the taste buds. This product offers consumers an effective solution to weight loss, while appealing to their taste preferences.[18]

A nanotube is a discrete hollow-fibre entity, which has two dimensions at the nanoscale. Self-assembly of hydrolysed calcium-binding milk protein α-lactalbumin into nanotubes is another recent development. These food protein-derived nanotubes show good stability and offer potential applications in foods, nutrients and pharmaceutics. The α-lactalbumin plays an important role in lactose formation, which is essential for milk production; it is already used as a food ingredient in infant formula. Infant formula is designed to bear a close resemblance to human breast milk and so extensive research has been dedicated to improving its protein quality, so that it is more like human milk. The relatively high content of essential amino acids in α-lactalbumin makes it desirable for improved infant formula protein systems by offering a similar protein content to that of human milk, thus helping to meet the nutritional needs of infants.[17,28]

### 10.2.2 Application of Nanotechnology in Food Packaging

Nanopackaging applications of food contact materials are growing rapidly and these are now considered to be one of the most active areas of nanotechnology in the food sector.[18] However, industrial applications of food packaging based on nanomaterials are strongly dependent on exploration of regulatory aspects, with consideration of their efficiency in preserving the physical, chemical, microbiological and sensorial quality of food but also their possible and unanticipated risks for the environment and human health.[29] Manufacturers claim that nanopackaging can extend product shelf life, and also maintain, improve or monitor the quality and safety of foods. For instance, the use of nanoparticles in food packaging can improve the mechanical and heat resistance properties, thereby affecting gas or water vapour permeability and thus increasing shelf life. Several reviews have reported three main categories of nanopackaging: improved packaging, active and intelligent packaging and biodegradable nanocomposites food packaging.[3,10,12,14,17,30–33]

Improved packaging has been described as incorporating nanoparticles in the polymer matrix materials with improved packaging properties in terms of temperature/moisture stability, flexibility, durability and gas barrier properties (*e.g.* nanocomposites, silicate nanoparticles and nanosilver) and also antibacterial properties.[10,30]

Active and intelligent food packaging is a novel concept of packaging compared with traditional materials. Polymer nanocomposites integrating metal or metal oxide nanoparticles have been developed for active packaging. These include silver, gold, zinc oxide, silica, titanium dioxide and iron oxides.[13] Active packaging also has the ability to remove undesirable tastes and flavours and improve the colour or odour of the packed food.[30]

For example, carbon black nanoparticles incorporated into polymer packaging can absorb odours released from the food or packaging. An emerging active packaging application integrates nanoparticles with antimicrobial or oxygen scavenging properties; this packaging is designed to stop microbial growth once the package is opened by the consumer and rewrapped with an active-film portion of the package.[17] A number of food contact materials have been developed using nanosilver, and are claimed to preserve the food for longer and inhibit the growth of microorganisms.[34]

Intelligent or smart food packaging incorporates a nanobiosensor for sensing and signalling microbial and biochemical changes and release of antimicrobials, antioxidants, enzymes, flavours and nutraceuticals to extend shelf life. A diverse range of devices have been developed to detect food spoilage organisms in food packaging (*e.g.* nanowires and antibodies), thus enabling versatility and much cheaper production.[14] For example, an electronic tongue was incorporated into food packaging as a nanosensor that is highly sensitive to the gases released when food is spoiled. These gases result in a colour change of a sensor strip to warn the consumer that the food is unfit for consumption.[17]

Figure 10.3 illustrates the key components of a typical active packaging system enabled by nanotechnology.[35] The active components are assembled using nanotechnology to produce an active component carrier that can be integrated into a food package. The carrier can interact with internal and/or external factors, thereby activating an intended action that enhances the shelf life, food quality, and/or safety of the product.

The key elements of a nanotechnology-based intelligent packaging system are summarized in Figure 10.4. The indicator/sensor, fabricated using suitable nanotechnology, can interact with internal factors (food components and headspace species) and/or external environmental factors. As a result of this interaction, the indicator/sensor will generate a response (*e.g.* visual cue, electrical signal) that correlates with the state of the food product. The information generated not only is useful for communication with consumers by informing them about the safety and quality of the products, but can also be utilized by the producers in their decision support systems to determine when and what actions are to be taken throughout the entire product distribution channel and production process.[36]

Biodegradable nanocomposites food packaging incorporates inorganic particles, such as clay, in the biopolymeric matrix, which can improve the delivery of micronutrients.[32] The nanolayered structure also restricts the access of gases and offers considerable improvements in terms of gas barrier properties of nanocomposites. Biodegradable materials have potential use in a wide range of food packaging applications, including processed meats, cheese, confectionery, cereals and boil-in-the-bag foods, and also extrusion-coating applications for fruit juices and dairy products or co-extrusion processes for the production of bottles for beer and carbonated drinks.[13] The plastic incorporates clay supplied by Nanocor to produce a film that is lighter, stronger and more heat resistant than traditional packaging

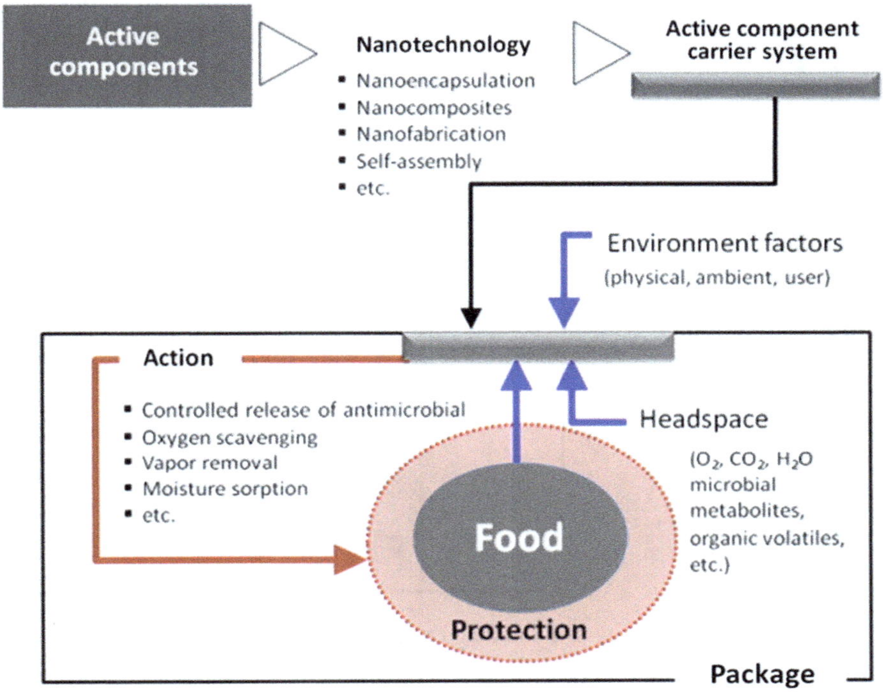


**Figure 10.3** Schematic diagram showing the active packaging concept and its nanotechnology involvement.
Reproduced from Mihindukulasuriya and Lim[35] with permission of Elsevier.

materials. The film is intended to prevent the entry of oxygen and other gases and the exit of moisture, thus preventing food spoilage.[25,30] Conversely, some workers have suggested that biodegradable materials demonstrate poor barrier and mechanical properties and require substantial improvements before they can replace traditional packaging materials.[14]

Nanopackaging has the potential to provide manufacturers with a wide range of benefits, including the ability to keep packaged food fresher for longer.[32] This may enable food to travel further and remain in storage for an extended period, thus resulting in a more reliable food supply. By increasing the shelf life of food products, manufacturers will also be able to sell food that would otherwise have been discarded owing to spoilage and hence contributing to reductions in food waste. Innovative and novel packaging that is light weight, stronger and functional can also significantly reduce transportation costs and packaging materials in the environment. Smart labels on food packaging are likely to appeal to manufacturers owing to the ability to monitor effectively the safety, quality and security of food and beverage products during transportation and storage, thereby

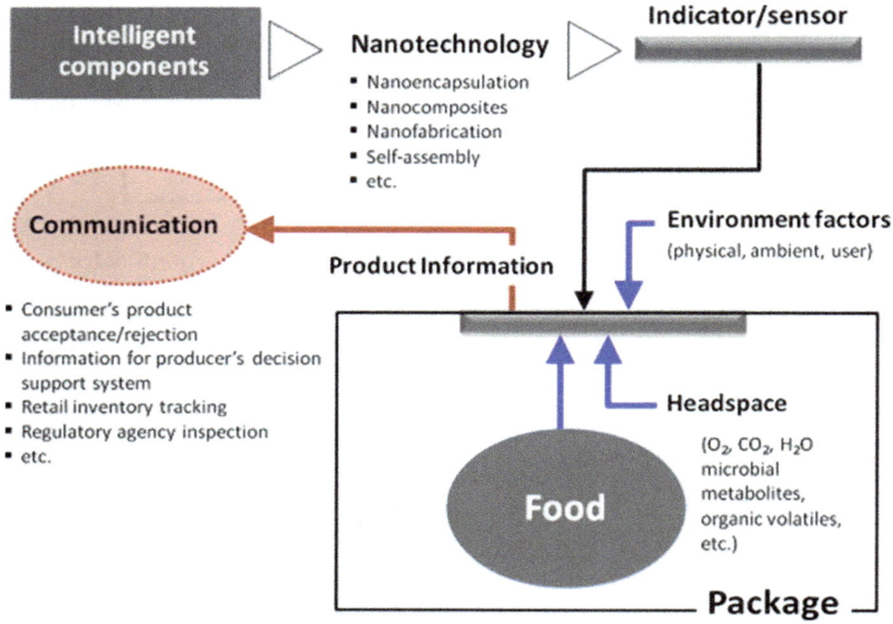


**Figure 10.4** Schematic diagram showing the intelligent packaging concept and its nanotechnology involvement.
Reproduced from Mihindukulasuriya and Lim[35] with permission of Elsevier.

reducing the risks of food-borne illness. Consumers can also benefit from attractive new products on the market, which are safer and of better quality.[7]

### 10.2.3 Other Applications

There are a number of other emerging applications of nanotechnology that could offer innovative solutions in the food and related sectors. Examples include the use of nanoporous materials for water filtration and for removal of undesirable tastes, flavours or allergens from food products. The use of nanomaterials such as zerovalent iron is finding increasing applications in water decontamination. Other developments nearing the market include nanocoatings (*e.g.* of titanium dioxide) for the photocatalytic sterilization of surfaces and water, nano(bio)sensors for food safety and nanobarcodes for food authenticity. Nanotechnology applications in water treatment, filtration and desalination offer many benefits in terms of safe use/reuse of potable water. The nanosensors are expected to allow the multi-analyte detection of pathogens and food contaminants. They should also be inexpensive and usable in the field by operators with relatively little training. The potential benefits of these sensors will include microbial and chemical safety of foods to protect consumer health.[12,37]

Available nanosensor types and their potential applications in the food sector are varied. They include, but are not limited to, array biosensors for the detection of food-borne contaminants, nanoelectromechanical systems for the detection of pathogens and electronic nose nanosensors for wine discrimination. Taking into account the crucial importance of time in food microbiology, the main aim of nanosensors is to reduce the time for pathogen detection from days to hours or even minutes.[38] Such nanosensors could be placed directly in the packaging material, where they would serve as an "electronic tongue" or "electronic nose" by detecting chemicals released during food spoilage.[39,40] Other types of nanosensors are based on microfluidic devices and can also be used to detect pathogens efficiently in real time and with high sensitivity. A major advantage of microfluidic sensors is their miniature format and their ability to detect compounds of interest rapidly in only microlitres of required sample volumes, which has already led to widespread applications in medical, biological and chemical analysis.[33,41–43]

## 10.2.4 Safety Assessment of Nanotechnology in Foods

Nanoparticles are small in size and consequently have a much larger surface-to-mass ratio than larger sized materials. Because of this, discussions were initiated on how to assess the safety of nanoparticles in general and within food and feed products specifically. This resulted in various (draft) opinions issued by the US Food and Drug Administration (FDA) and the European Food Safety Authority (EFSA) on the adequacy of the risk assessment paradigm and guidelines for industry regarding the safety evaluation of nanoparticles.[44–47]

So far, only for a few nanoparticles are safety assessments available in the scientific literature.[34,48,49] These safety assessments are surrounded by uncertainties, mainly due to the lack of reliable characterization data on the nanomaterial in the product and inadequate material characterization in the toxicological studies performed. The clear need to develop, implement and validate routine analytical methods for the characterization of nanoparticles in foods has long been recognized. In addition, experimental (tiered) approaches for the hazard assessment of nanoparticles are being developed.[47]

In order to assess the safety of nanoparticles in foods, it is important to understand the analytical techniques for the detection of nanoparticles. There is a large array of analytical techniques for characterizing pristine nanoparticles, as powder or in clean aqueous liquids. However, food, biological and agricultural samples are heterogeneous mixtures and therefore require separation or pretreatment to isolate the target analyte from the interfering matrix components. Owing to their high reactivity, nanoparticles can change in composition and size in response to changes in their environment. Results obtained after sample preparation do not, or only partially, reflect the original situation in the original sample. Ideally, sample preparation is kept to the minimum.[50–53] Roughly three categories of detection and characterization techniques can be considered: imaging, separation and spectroscopic techniques. A drawback of imaging techniques such as dynamic light scattering (DLS) and electron microscopy (EM) is that often they are not quantitative.

The most widely used separation techniques for nanoparticles in food and agricultural samples are hydrodynamic chromatography (HDC) and field flow fractionation (FFF), often in combination with detection methods such as multiple angle light scattering (MALS) and inductively coupled plasma mass spectrometry (ICP-MS).[54–56]

The detection limits of DLS and EM techniques are typically in the mg $L^{-1}$ range and those of combinations of HDC with MALS and ICP-MS are in the mg $L^{-1}$ and μg $L^{-1}$ range, respectively.

Whereas there are many techniques for the detection, quantification and measurement of the properties of inorganic nanoparticles, this is not the case for organic nanoparticles composed of polymers, lipids, proteins and polysaccharides. Imaging techniques are limited and EM is generally useful only if staining techniques are applied. Separation techniques such as HDC and FFF can be used but care has to be taken not disrupt the micelle-like structures.[56,57]

There is no single analytical method that can be used to detect all types and sizes of nanoparticles in all matrices. In particular, the requirement for measuring the constituent particles inside aggregates, the difficulty in converting experimentally measured signals into accurate number-based size distributions and the detection and counting of particles in the lower size range, *e.g.* smaller than 10 nm, can be problematic. Analytical methods able to identify materials as nanoparticles according to the recommended nanomaterial definition are under development. However, most current methods have a detection limit higher than 1 nm or a lower sensitivity for smaller particles. As a consequence, they can only be used for a positive test to prove that a material is a nanomaterial, and not for a negative test to prove that a material is not a nanomaterial. At present, none of the currently available methods is able to determine for all kinds of potential nanoparticles whether they fulfil the definition or not. Therefore, a range of measurement methods is required to investigate whether nanoparticles fulfil the regulatory definition.[58,59]

In terms of toxicological assessment of nanoparticles in food, the (regulatory) safety assessment of nanoparticles still relies mainly on *in vivo* data generated according to government guidelines. For these studies, the need for additional nano-specific requirements is frequently debated. In addition to the long lists of required parameters for an adequate material characterization, requirements for nano-specific biological endpoints have been discussed.[60] Chronic (or sub-chronic) oral studies are best performed by administering the nanoparticles directly *via* the diet, as this is most realistic compared with human oral exposure. It is very likely that the nanoparticles interact with the feed matrix or gavage solutions. Clearly, during transit through the digestive tract, the physicochemical properties of the nanoparticles change, as has been shown for silver and silica nanoparticles using an *in vitro* digestion model.[61–63]

Given the enormous diversity of nanoparticles applied in foods and feeds, it is scientifically and ethically not defensible to rely only on animal experiments in future safety assessments. The incorporation of alternative testing approaches in a tiered toxicological safety assessment has been suggested. In the first tier of assessment, physicochemical properties of the pristine nanomaterial and its stability in suspension are assessed. As mentioned earlier, a broad range of analytical procedures is available. The second tier of assessment of nanoparticles following oral exposure is proposed to consist of an assessment of solubility and agglomeration using *in vitro* digestion models. Several *in vitro* digestion models of increasing complexity are available, ranging from static to dynamic models that have been used for conventional chemicals.[53,64,65]

At present, there is no combination of *in vitro* approaches that can truly replace chronic oral *in vivo* studies. While more knowledge is rapidly being obtained on the applicability and robustness of the *in vitro* methods using nanoparticles, none of the methods have yet been validated. Hence *in vivo* studies might still be required. The number of available oral studies evaluating the uptake, distribution and potential effects of nanoparticles is increasing; however, only a few studies can be used to assess the contribution of nanoparticle-related properties to the observed toxicokinetic or toxicodynamic effects. For nanoparticles that are prone to dissolution, such as zinc oxide and silver, it is important to be able to distinguish between potential ionic or nanoparticle effects. Two 28-day oral exposure studies using silver in nanoparticle and ionic forms showed highly comparable organ distribution and biological effects of both forms of silver and indicated up to 2 months' retention of silver in some tissues.[66,67]

Oral uptake of organic nanoparticles has been studied in relation to nanomedicine and targeted release of bioactive compounds; these are not covered here and may be found elsewhere.[68–70]

In spite of the studies that are being undertaken, there still is not a good understanding of the health effects that might arise from exposure to nanoparticles. Knowledge gaps exist in key areas that are essential for predicting potential health risks, such as routes of exposure, the mechanism of uptake of nanoparticles into the body, the parameters that drive the distribution of nanoparticles once they are inside the body (*e.g.* toxicokinetics) and the ways in which nanoparticles interact with the body's biological systems (*e.g.* toxicodynamics). Although more and more evidence has appeared that some nanoparticles might have negative effects, the range of nanoparticles for which comprehensive hazard data are available is small. Given the wide diversity of nanoparticles and observations that different nanoforms with the same chemical composition can have different toxicological properties, it is likely that new approaches that do not rely on conventional toxicity testing will need to be found to assess potential hazards of nanoparticles.[47]

A number of concepts and approaches have been developed to estimate the risks of nanoparticles. Examples include the Control Banding Nanotool,[71] the Precautionary Matrix for Synthetic Nanomaterials[72] and the Stoffenmanager Nano.[73] These methods are categorization methods to determine an overall risk level for working with pure nanoparticles in the laboratory or the workplace and do not estimate risks for consumers or the environment. Alternative methods use multiple criteria decision analysis, which were originally developed for traditional chemicals but which are now being revised for nanoparticles.[74–78] In such a model, nanoparticles are clustered in various risk classes using not only criteria for the physicochemical properties of nanoparticles but also the bioavailability, bioaccumulation and toxic potential of the nanoparticles. In addition, value of information analysis has been proposed to further direct potential additional studies, to estimate in advance the potential added value of this information and to link the outcome of research with decision-making by producers and regulators.[79]

More recently, the NanoRiskCat tool was developed as a systematic approach to assess the hazard and exposure potentials for a nanomaterial-based product in a regulatory perspective.[80] The procedure results in five colour-coded ranks, three for exposure and two for hazard. The first three dots indicate the potential exposure for professional end-users, for consumers and for the environment, and the other two dots indicate the potential hazard for humans and the environment. To evaluate the exposure potential for a specific nanomaterial in a nanoproduct, a substantial amount of information is needed. As this information is rarely available for nanoparticles, four potential exposure categories (high, medium, low, unknown) are used, based on the location of the nanomaterial in the product or application. The potential human health hazard is evaluated using a number of criteria, including the aspect ratio of the nanomaterial, adverse effects of the bulk form of the nanomaterial, acute toxicity of the nanomaterial and indications that the nanomaterial causes genotoxic, mutagenic, carcinogenic, respiratory, cardiovascular, neurotoxic or reproductive effects in humans or (laboratory) animals or for which organ-specific accumulation has been documented. For each of these endpoints, the user has to review the literature and interpret data from scientific results to guide the tool. For a number of these criteria, cut-off values are given to aid the interpretation of literature results. Combining all information finally produces the outcome of using NanoRiskCat, which can primarily be used to understand and categorize what is known about the hazard and exposure potential of using a given nanomaterial in a given application.[47,80]

The above analysis is strongly dependent on active expert involvement. Inclusion of the newest data in the analysis is a continuous challenge. A system that potentially can overcome these constraints is a decision support system (DSS) that mimics human reasoning and is able to utilize scientific knowledge directly as it becomes available from multiple sources. A DSS thus helps to identify those nanoparticles and applications that should receive priority in the safety assessment.[81]

## 10.3 Starch Nanocomposites in the Food Industry

"Composites" are materials that are composed of two types of components: the matrix, whose role is to support and protect the filler materials and transmit and distribute the applied load to them, and the mentioned fillers, which are the stronger and stiffer components that reinforce the matrix. Furthermore, nanocomposites mean polymeric composite materials that are filled with nano-sized rigid particles. The advantages of these nanocomposite materials, compared with conventional composites, are their superior mechanical, barrier and thermal properties at low levels (*e.g.* $\leq$5 wt%) and also their recyclability, transparency and low weight.[82–84]

Starch nanocomposites are mainly used as biopolymers or sources of packaging materials in the food industry. Recently, fossil energy depletion and growing environmental concerns have generated increasing interest in bio-based eco-efficient and high-technology materials. In this context, there is an increasing demand for products made from renewable and sustainable non-petroleum-based resources. It has brought two scientific fields together: nanotechnologies, which allow the development of innovative and efficient materials, and biomaterial processing, with the use of renewable raw materials for more environmentally friendly and sustainable solutions.[82]

A number of biopolymers based on renewable resources, such as starch, cellulose, soya, poly(lactic acid) and polyhydroxyalkanoates, have been investigated as alternative materials to conventional synthetic polymers. This is because some synthetic polymers are less environmental friendly.[85] Of these, starch is the most promising material for the production of biodegradable plastics and their composites, because it is natural, abundant, sustainable and biodegradable, in addition to being capable of exhibiting thermoplastic behaviour under the action of high temperature and shear stress.[86] Materials produced using such material are characterized by their high solubility in water, high hygroscopicity, poor melting point, faster ageing due to retrogradation (a process in which the amylose and amylopectin chains realign themselves, making the liquid starch turn to gel) and poorer mechanical properties in comparison with the materials based on synthetic polymers. In addition to these aspects, starch is one of the most difficult materials to process, because of its brittleness. Therefore, it requires the addition of a plasticizer to make it suitable for engineering applications.[87,88]

As mentioned earlier, starch can be used as the matrix either singly or in combination with other polymers and with the incorporation of different types of nanofillers such as nanoclay, cellulose nanoparticles, CNTs and other inorganic nanofillers. In another approach, starch granules are transformed into nanoparticles and are introduced as fillers in various polymer matrices.

Prior to other processes, native granular starch is usually converted into a molten state, which is known as "plasticized starch" or "thermoplastic starch", by gelatinization/melting processes. When native starch granules are heated in water, their semicrystalline nature and three-dimensional architecture are gradually disrupted, resulting in a phase transition from the ordered granular structure into a disordered state in water, which is known as "gelatinization".[89] Full gelatinization of starch under shearless conditions requires excess water. However, if the water content is limited, the swelling forces due to the water will be much less significant and the steric hindrance is high. In this case, a higher temperature is required to facilitate the mobility of starch molecules and the destructuration of the crystalline regions.[90] The process of gelatinization with low water content could more accurately be defined as the "melting" of starch.[91] The gelatinization/melting behaviour of starch is completely different when shear treatment is imposed.[92] It has been shown that shear can enhance the destructuration of starch granules in abundant water and the melting of crystallites with limited water.[93–95] The significance of such studies is that most processing techniques for starch involve shear treatment.

The techniques that have been used to process starch, such as solution casting, internal mixing, extrusion, injection moulding and compression moulding, are similar to those widely used for conventional synthetic thermoplastics.[96]

### 10.3.1 Starch Matrix with Different Nanofillers

Various nanofillers have been examined with plasticized starch, including phyllosilicates, polysaccharide nanofillers and carbonaceous nanofillers, and these are summarized in Table 10.2. They have different geometries (size and shape) and surface chemistries. Regarding their shapes, three

**Table 10.2** Overview of the nanofillers used for starch-based nanocomposites. Reproduced from Xie *et al.*[8] with permission of Elsevier.

| Group | Nanofiller | Dimensional type | Remarks |
|---|---|---|---|
| *Phyllosilicates* | | | |
| 1. Clay minerals | | | |
| 1.1 Kaolinite group | Kaolinite, halloysite | Nanolayer | Non-expandable |
| 1.2 Smectite group | Montmorillonite | Nanolayer | Most expandable; most frequently used; natural sodium montmorillonite being a preferred choice owing to the matching of polarity with starch |
| | Hectorite | Nanolayer | Expandable |
| 1.3 Sepiolite group | Sepiolite | Nanolayer (actually needle shaped) | – |

**Table 10.2** (*Continued*)

| Group | Nanofiller | Dimensional type | Remarks |
|---|---|---|---|
| 2. Synthetic clays | Somasif ME 100 fluorohectorite/ fluoromica | Nanolayer | Montmorillonite- or hectorite-type synthetic clay |
| | Laponite B, Laponite RD | Nanolayer | Hectorite-type synthetic clay |
| 3. Mica group | Muscovite | Nanolayer | Non-expandable |
| | Paragonite | Nanolayer | Non-expandable |
| | Illite (hydrous mica) | Nanolayer | Non-expandable |
| *Polysaccharides* | Nanowhiskers/ nanofibrils/ nanofibres from cellulose | Nanotube/ nanoparticle | Size and shape dependent on the preparation method and conditions |
| | Nanoparticles/ nanocrystals from starch | Nanoparticle | – |
| | Nanoparticles from chitin/chitosan | Nanoparticle | – |
| *Carbonaceous materials* | Carbon nanotubes | Nanotube | – |
| | Graphite oxide | Nanolayer | – |
| | Carbon black | Nanoparticle | – |
| *Metalloid oxides* | Silicon dioxide (silica) ($SiO_2$) | Nanoparticle | – |
| | Antimony trioxide ($Sb_2O_3$) | Nanoparticles | – |
| *Metal oxides and chalcogenides* | Zinc oxide (ZnO) | Nanoparticle | – |
| | Hydrous zirconium dioxide (zirconia) ($ZrO_2 \cdot nH_2O$) | Nanoparticle | – |
| | Titanium dioxide (titania) ($TiO_2$) | Nanoparticle | – |
| | Cadmium sulfide (CdS) | Nanoparticle | – |
| | Cadmium selenide (CdSe) | Nanoparticle | – |
| *Metal phosphates* | α-Zirconium phosphate $[Zr(HPO_4) \cdot H_2O]$ | Nanolayer | – |
| *Layered double hydroxides* | $[M^{II}{}_{1-x}M^{III}x(OH)_2]_{intra}$, $[A^{m-}{}_{x/m} \cdot nH_2O]_{inter}$[a] | Nanolayer | – |
| *Non-silicate minerals* | Brucite $[Mg(OH)_2]$ | Nanolayer | – |
| | Hydroxyapatite $[Ca_{10}(PO_4)_6(OH)_2]$ | Nanoparticle | – |

[a]$M^{II}$ and $M^{III}$ denote metal cations, A is the anion and intra and inter denote the intralayer domain and the interlayer space, respectively.

distinct types of nanofillers can be observed; nanoparticles, nanotubes and nanolayers.[97] Nevertheless, the term "nanoparticles" is also frequently used in a broad sense to describe a nanofiller regardless of its shape.

Unlike most conventional synthetic polymers, starch has a unique chemical structure and processing behaviour. Therefore, the preparation and properties of starch-based nanocomposites are inherently dissimilar to those of other polymer nanocomposite systems. Furthermore, the incorporation of appropriately tuned nanoparticles in starch as a biopolymer with complex structures and special properties will provide a rich new class of polymer nanocomposites able to be designed for a wide range of both conventional and emerging applications.[8]

Phyllosilicates, or layered silicates, are an important group of minerals that includes the clay minerals, micas, chlorite, serpentine, talc, *etc.* They have different structures, textures, and/or morphologies. The layered structure of phyllosilicates can be found in detail elsewhere.[98] Some phyllosilicates do not display a normal layered structure, including sepiolite, which displays a kind of fibrous structure, and halloysite, which has the usual layered structure but forms spheroidal aggregates.[99] Cation exchange is the most common technique for chemical modification of the phyllosilicate surface, which increases the interlayer spacing. Using this technique, various organomodified montmorillonites and hectorites used for starch-based nanocomposites can be produced, which differ mainly in the nature of the counter-cation and the cation-exchange capacity. Nevertheless, most conventional organomodifiers increase the hydrophobicity of the phyllosilicate, resulting in reduced compatibility with the hydrophilic starch matrix. This is one of the key points to consider when developing starch nanocomposites.[8]

The dispersion state of a typical phyllosilicate (except sepiolite and halloysite) in a polymer, which depends on the preparation conditions and the matrix–nanolayer affinity, determines the structure of the resulting composite, which can be either a phase-separated composite (microcomposite), intercalated nanocomposite or exfoliated nanocomposite.[100–105] Normally, incorporation of phyllosilicate nanolayers in a polymer matrix can be carried out with one of the three main techniques: (1) solution intercalation, (2) *in situ* intercalative polymerization, and (3) melt intercalation.[102–104]

Polysaccharide nanofillers represent the second popular group of nanofillers used for starch nanocomposites. A series of polysaccharides with similar chemical structures, such as cellulose, starch, chitin and chitosan, can be produced as different forms of nanofillers (nanowhiskers and nanoparticles), which can be employed for fabricating starch nanocomposites (Table 10.3). Cellulose nanowhiskers (CNWs) and starch nanoparticles (SNPs) have already been widely used, whereas the use of chitin nanoparticles has been less investigated. One of the advantages in using them is the similar polysaccharide chemical structure of the nanofiller and the matrix, which could benefit the nanofiller–matrix interactions.[8]

Another group of nanofillers is carbonaceous nanofillers. This group is highly interesting for incorporation into starch-based materials although

**Table 10.3** Summary of polysaccharide nanofillers for starch-based nanocomposites. Reproduced from Xie *et al.*[8] with permission of Elsevier.

| Source | Preparation method | Morphology | Dimensional characteristics[a] |
|---|---|---|---|
| *Plant cellulose* | | | |
| Flax | Acid hydrolysis by $H_2SO_4$ | Slender rods | *L*: 100 nm–500 nm<br>*D*: 10 nm–30 nm |
| Hemp | Acid hydrolysis by $H_2SO_4$ | – | *D*: 30 ± 10 nm |
| Ramie | Acid hydrolysis by $H_2SO_4$ | Spindle shape | *L*: 350 nm–700 nm<br>*D*: 70 nm–120 nm;<br>*L*/*D*: 6 |
| Wood | Mechanical shearing, enzymatic treatment, mechanical shearing and high-pressure homogenization | Fibres | *L*: several μm<br>*D*: 30 ± 10 nm |
| Pea hull | Acid hydrolysis by $H_2SO_4$ | Needles or rod shape with one or two sharpened ends | *L*: 400 ± 200 nm<br>*D*: 12 ± 6 nm |
| Cassava bagasse | Acid hydrolysis by $H_2SO_4$ | Long and curved elongated particles | *L*: 360–1700 nm<br>*D*: 2–11 nm |
| Wheat straw | Steam explosion (with NaOH), acidic treatment with HCl and high shear mechanical treatment | Fibre | *D*: 10–60 nm (most in the range 30–40 nm) |
| Microcrystalline cellulose (MC, commercially available) | Acid hydrolysis by $H_2SO_4$ | Needle shape | *L*: ~200 nm<br>*D*: ~5 nm |
| | Coagulated from an NaOH–urea–$H_2O$ solution of MC by ethanol–HCl aqueous solution | Particles | *D*: 50–100 nm |
| Tunicin (animal cellulose) | Acid hydrolysis by $H_2SO_4$ | Slender parallelepiped rods | *L*: 500 nm–2 μm<br>*D*: 10 nm<br>*L*/*D*: 50–200 |
| Bacterial cellulose | Bioinspired bottom-up process by *Acetobacter* sp. bacteria during the gelatinization of starch | Coherent network of interconnected nanofibrils | *D*: 100–150 nm |
| | Enzymatic hydrolysis of *Acetobacter xylinum* bacterial cellulose by *Trichoderma reesei* endoglucanase | Fibre | *D*: 90 nm |
| *Starch* | | | |
| Waxy maize | Acid hydrolysis by $H_2SO_4$ | Platelet-like particles | *L*: 40–60 nm<br>*W*: 15–30 nm<br>*T*: 6–8 nm |
| | Acid hydrolysis by $H_2SO_4$ | Particle | *D*: 50 nm |

**Table 10.3** (*Continued*)

| Source | Preparation method | Morphology | Dimensional characteristics[a] |
|---|---|---|---|
| Regular maize | Treatments of gelatinized starch paste with ethanol and citric acid | Particles | *D*: 50–100 nm |
| Chitin | Acid hydrolysis by HCl; repeated sonication processes | Particles | *D*: 50–100 nm |
| Chitosan | Physical crosslinking between tripolyphosphate and proteinized chitosan | Particles | *D*: 50–100 nm |

[a] *L*, length; *D*, diameter; *W*, width; *T*, thickness; *L*/*D*, aspect ratio.

they have not been extensively studied so far. Carbonaceous nanofillers such as CNTs, graphite and carbon black (CB) represent a promising group of nanofillers that not only generate performance improvements but also new functionalities, especially electrical conductivity and electroactivity. CNT-reinforced nanocomposites have already shown great potential in biomedical applications such as sensors and stimulators of bone cells, although the toxicity of CNTs still appears controversial.[106–111] There have already been some exciting reports on starch–CNT nanocomposites, whereas those reinforced by other carbon nanofillers scarcely exist. CNTs can be either single-walled (SWCNTs) or multi-walled (MWCNTs). Despite the wide use of CNTs in other polymer nanocomposite systems, the use of CNTs as nanofillers for starch-based materials has only more recently been initiated and mostly MWCNTs were involved.[97,112–115] This may due to the lower price and greater abundance of MWCNTs than SWCNTs. Further, MWCNTs exhibit high aspect ratios ($\sim$1000) and excellent mechanical ($E \approx 1$ TPa), thermal and electrical properties.[116–118]

Graphite and graphite oxide combine the lower price and layered structure of phyllosilicates with the superior thermal and electrical properties of CNTs.[119–121] As-prepared graphite cannot be dispersed in water or organic solvents, which makes the fabrication of nanocomposites difficult; graphite oxide is hydrophilic and can form strong physical interactions with a polymer such as starch due to its various oxygen functional groups including hydroxyl, epoxide, carbonyl and carboxyl groups.[122]

Moreover, nanocomposites can be reinforced by metalloid oxides, metal oxides and metal chalcogenides. Metalloid oxides (*e.g.* $SiO_2$, $Sb_2O_3$), metal oxides (*e.g.* ZnO, $TiO_2$, $ZrO_2 \cdot nH_2O$) and metal chalcogenides (*e.g.* CdS, CdSe) are grouped together here because of their similar chemical categories, preparation methods and nanofiller reinforcement mechanisms. Novel applications are expected for nanocomposites reinforced by this type of nanofillers. For example, metal oxides and chalcogenides are normally semiconductor materials. Incorporation of such nanofillers into a polymer can result in nanocomposites that can be used as components for photovoltaic solar cells, light-emitting diodes, photodiodes and gas sensors. Only solution methods have been used to fabricate the starch-based nanocomposites.[8,123]

In conclusion, a wide variety of nanofillers have been examined with starch. Phyllosilicates (especially montmorillonite of the smectite group) have mostly been utilized owing to their advantages such as wide availability, low cost and high aspect ratio and thus vast exposed surface area (and also the swelling nature). Polysaccharide nanofillers represent the second most popular group owing to their abundance in nature, the biological sources and the chemical similarity to starch. Nevertheless, the preparation of these bio-nanoparticles is time consuming and involves acid hydrolysis in multiple steps, which is not eco-friendly. Many other nanofillers, such as carbonaceous nanofillers, metalloid oxides, metal oxides and metal chalcogenides, have also been used.[8]

Following the incorporation of a nanofiller, starch-based materials generally show improvements in some of their properties such as mechanical properties (typically tensile strength ($\sigma$), Young's modulus ($E$) and storage modulus ($E'$), glass transition temperature ($T_g$), thermal stability, moisture resistance, oxygen barrier property and biodegradation rate. The improvements can be fundamentally ascribed to the homogeneous dispersion of the nanofiller in the matrix and the strong interface adhesion, which can contribute to the formation of a rigid nanofiller network and influence the molecular and crystalline structures in the matrix. To realize these, the key point to be addressed is the nanofiller–matrix compatibility, which mainly depends on the surface chemistry of the nanofiller and is usually achieved by hydrogen bonding, although other factors, such as the plasticizer(s)/additive(s), the starch type and chemical modification, the presence of other polymer(s) and the processing and annealing conditions, also have strong influences. However, above all, the major role is played by the nanofiller itself, of which the aspect ratio/surface area, chemistry and mechanical properties could be influenced by its preparation and modification. Nevertheless, how the nanofiller affects the crystalline structure and crystallinity of the starch matrix has not been investigated much recently. These properties can be greatly affected by the formulation (*e.g.* the amylose content of the starch and the type and content of the plasticizer), the processing conditions (*e.g.* temperature, pressure, shearing and orientation) and the storage conditions (*e.g.* time, temperature and relative humidity). Further, phase separation of the plasticizer, the starch and/or the nanofiller may occur in the system, with the different domains showing different recrystallization/anti-crystallization behaviours. These factors may account for the discrepancies in some of the results such as $T_g$ and moisture resistance. With improved properties that are comparable to those of traditional petroleum-based polymers such as polyethylene and polypropylene, the current applications of starch-based materials can be greatly enhanced and widened. The renewable resource and inherent environmental friendliness of such materials can justify their wide use for a sustainable future. In particular, the use of starch nanocomposites as new packaging materials would be based on their biodegradation and improved barrier and mechanical properties.[8]

In the future, it is still very important to carry out research to test new nanofillers to be incorporated into starch for developing promising nanocomposites with excellent performance and new functionalities to be competitive in the materials world. In addition, the heterogeneous dispersion of the nanofiller and the phase separation issue that existed in some of the past studies should also be addressed. While chemical manipulations might help to some extent, the future research should also emphasize the importance of using processing techniques such as extrusion which are more aligned with efficient industrial production. Hence research is also needed on how thermomechanical treatment in this kind of processing can assist in achieving a well dispersive structure without adding a detrimental effect to the final properties due to molecular degradation of the starch.[8,124]

### 10.3.2 Starch Nanoparticles (SNPs)

As mentioned earlier, starch is environmentally friendly, hence SNPs can be suggested as one of the promising biomaterials for novel utilization in foods, cosmetics and medicines, and also various composites.[82]

Most starches in the native form present limitations such as high viscosity, susceptibility to retrogradation, limited digestibility for some and limited solubility for others. For these reasons, most starches used in foods or industrial applications are first modified. Starches are chemically, enzymatically and/or physically modified to accentuate their positive characteristics, diminish their undesirable qualities (such as high viscosity, susceptibility to retrogradation and lack of process tolerance) or add new attributes (retention, film formation, digestibility, solubility, *etc.*). Indeed, mild acid hydrolysis has been used for a long time to modify starch and its properties. In industry, starch slurries are treated with dilute HCl or $H_2SO_4$ at 25–55 °C for various periods to produce "acid-modified" starch used as sizing agents, in the production of gum candies and in paper and paper board production. In recent studies either of these two acids has been used for preparing starch nanocrystals.[125]

In addition to its renewable nature, relatively low cost and worldwide availability, starch is also perceived as an attractive filler to replace fossil-based materials such as carbon black in nanocomposite applications. Carbon black is manufactured by burning oil or natural gas under controlled conditions, and is the most important reinforcing agent used in the rubber industry. Because of its origin from petroleum, carbon black causes pollution and gives to rubber a black colour. SNPs, in contrast, are fully renewable and reactive nanoparticles, which have shown good reinforcing and barrier properties associated with white or transparent colours.[126]

Research has revealed that the starch granules consist of numerous nano-sized semicrystalline blocklets. Through mild hydrolysis using acids and/or enzymes, the nanoblocklets could be isolated from starch.[127–132] Physical treatments may also disintegrate the starch granules and thus release the nanoblocklets.[133,134] These SNPs have crystalline moieties with the advantages inherent from starch granules, which include renewability and biodegradability.[82]

Along with recent interest in nanomaterials, some researchers have utilized SNPs as fillers in composites and found that their incorporation improved not only the mechanical properties but also the biodegradability of the composites. In addition, SNPs have been reported to be applicable in other areas, such as foods, cosmetics and pharmaceuticals.[135–138] However, the industrial utilization of SNPs has so far been limited, and only two applications (BioTRED and EcoSphere) have been reported. BioTRED (Novamont, Italy) is a tyre filler developed in collaboration with Goodyear (USA): nanoparticles derived from maize starch were used to replace part of the carbon black and silica in tyres. The incorporation of the SNPs provided an improvement in the relining resistance of the tyre with environmental advantages. EcoSphere is a starch-based biolatex that replaces oil-based latex, and has been used as a replacement for petroleum-based coatings and binders for paper and paper board.[82] Although, they have not yet been commercially applied in foods, they show high potential.

SNPs are often referred to as starch nanocrystals. However, some workers have distinguished starch nanocrystals from SNPs because SNPs may include amorphous matrices.[139] However, it becomes almost impossible to distinguish the terms "starch nanocrystals" and "SNPs", as both terms have been used to refer to the crystalline parts of starch remaining after hydrolysis or other physical treatments.[82]

### *10.3.2.1 Preparation of SNPs*

Acid hydrolysis has been widely used for the preparation of nanoparticles from polysaccharides because it is a simple process and easy to control. The crystalline regions in starch granules are more resistant to acid hydrolysis than the amorphous regions, hence crystalline moieties can be isolated by mild acid hydrolysis, typically using HCl or $H_2SO_4$. Because the mild acid hydrolysis may selectively erode the amorphous regions, the SNPs in the reaction solution have high crystallinity. However, this selective hydrolysis using acids generally requires long periods and the recovery yield is therefore relatively low. Another problem also arises from the presence of a large number of hydroxyl groups, which tend to re-form the supramolecular interactions that are characteristic of starch, yielding aggregates.[139] This effect hinders the industrial application of SNPs. For these reasons, many researchers have been attempting to find other procedures using physical treatments or a combination of different methods.

The preparation of SNPs may be classified into "top-down" and "bottom-up" processes. In a top-down process, nanoparticles can be produced from structure and size refinement through breakdown of larger particles. In a bottom-up process, nanoparticles can be prepared from a build-up of atoms or molecules in a controlled manner that is regulated by thermodynamic means such as self-assembly. Although a few studies have been reported on the basis of the bottom-up approach, such as nanoprecipitation, most studies on the preparation of SNPs have employed top-down methods.[139,140]

The hydrolysis of starch granules is commonly monitored in two ways, by measuring either the content of soluble sugar in solution (phenol–sulfuric acid) or the recovery of insoluble starch residues. Most starches exhibit a two-stage hydrolysis pattern, with fast initial hydrolysis followed by slow subsequent hydrolysis.[141,142] However, some workers have distinguished three stages of the acid hydrolysis: rapid, slow and very slow.[143] The initial stage of acid hydrolysis is thought to involve the hydrolysis of the amorphous parts within starch granules, whereas the slow stage is attributed to the erosion of the crystalline regions.[144] Two hypotheses have been proposed to account for the slow hydrolysis of the crystalline domains in starch granules. One is that the dense packing in the crystalline regions retards the penetration of hydronium ($H_3O^+$) ions and the other is that the hydrolysis of the glucosidic bonds in crystalline domains requires a change from a chair to a half-chair conformation.[145] If the crystalline structure immobilizes the sugar conformation due to its rigidity, then this transition (chair→half-chair) would be sterically impossible.[82]

The crystalline residues remaining after prolonged acid hydrolysis consist of agglomerated particles a few tens of nanometres in diameter. When observed at the microscopic scale by transmission electron microscopy (TEM), internal information regarding these microcrystals could be obtained. It has been observed under TEM that an edge-on view of the lamellae was formed by the association of amylopectin side-branches. By subjecting native waxy maize starch to hydrochloric acid hydrolysis (2.2 M HCl, 2 weeks), it was found that the lamellar arrangement had been disrupted to some extent. In other words, a certain number of $\alpha$-(1 → 6) bonds located in the amorphous regions between the crystalline lamellae were hydrolysed. Because the branching points located in the interlamellar areas were more readily hydrolysed, the insoluble residues were fairly well individualized. Indeed, the individual residues could be observed by microscopy after 6 weeks of hydrolysis. The hydrolysis residues were crystalline nanoplatelets of approximately 6–8 nm thickness, 20–40 nm length and 15–30 nm width.[131] However, many studies reported that the SNPs had spherical or polygonal shapes that often existed as microscale aggregates.[132,146] The preparation of these starch nanoplatelets required a long duration of hydrolysis (40 days of treatment) and the recovery yields from the starch granules were relatively low. Moreover, the nanoplatelets tended to aggregate to form microscale particles.[82]

Hydrolysis using $H_2SO_4$ instead of HCl for the preparation of SNPs has been reported. Compared with HCl hydrolysis, $H_2SO_4$ hydrolysis shortened the preparation time and increased the yield of SNPs. It was claimed that the formation of sulfate–ester linkages on the surface of nanoparticles during the hydrolysis should limit the flocculation of nanoparticles and thus produce a nanosuspension with increased stability.[143] This effect was confirmed by comparing the sedimentation rate of the suspensions obtained by using each acid. The stability of nanosuspensions also depends on the dimensions of the dispersed particles and polydispersity.[133] As the particles obtained are smaller and show greater uniformity, the suspension of the particles that remains has greater stability. The sulfate–ester bonds on the starch surface may decrease the thermal stability of the nanoparticles, which could negatively affect their use in composites. By adding a small amount of ammonia, the thermal stability of the SNP suspensions could be enhanced.[147]

It is obvious that acid hydrolysis is the procedure most commonly used in preparing nanoparticles from starch. Most studies of acid hydrolysis were carried out with waxy maize starch. However, other sources of starch having different amylose contents and crystallinity have also been tested.[148] Nevertheless, acid hydrolysis is an energy-intensive process for the production of SNPs and the recovery of SNPs is low. The salts produced by neutralizing soluble sugars should be carefully removed from the reaction medium, with no or little loss of SNPs. Obtaining uniform SNPs with a high yield is still a difficult task in the process using acid hydrolysis.[82]

Another approach for starch hydrolysis is enzymatic treatment, but not much research has been carried out on the enzymatic process for the preparation of SNPs compared with acid hydrolysis. Enzymatic hydrolysis of waxy rice starch with $\alpha$-amylase for 3 h induced the fragmentation of starch granules with selective dissolution of the amorphous regions. After the enzymatic hydrolysis, the volume-based size distribution of the starch exhibited two major peaks at 0.5 and 3.6 μm and a shoulder at 0.1 μm. However, the average diameter of the particles remained fairly large ($\sim$500 nm), which was greater than that of the blocklets (20–500 nm). It was assumed that the enzymatic hydrolysis of the amorphous regions induced the fragmentation of the granules and a portion of the blocklets could be separated.[132]

Combination of both enzymatic and acid hydrolysis has been used for the preparation of SNPs, and it was reported that SNP preparation could be performed faster by using this approach. Pretreatment of starch with glucoamylase for 2 h could effectively decrease the time required for acid hydrolysis. This pretreatment created pathways for the acids to diffuse inside the granules, which made the amorphous regions more readily hydrolysable. At the same extent of hydrolysis ($\sim$70%), the SNPs from the pretreated starch could be obtained in a higher yield but with a larger size ($\sim$145 nm) compared with those obtained from the non-treated starch (usually between 50 and 100 nm). The extent of acid hydrolysis that would normally have required 5 days was obtained after only 45 h (yields of $\sim$15%). Atomic force microscopy (AFM) and X-ray diffraction (XRD) confirmed that the starch particles obtained were nano-sized.[149]

In terms of physical treatments, these are promising techniques as they are environmentally friendly. Several such techniques have been established.

A high-pressure homogenization method that is simple, versatile and environmentally friendly has been developed for the isolation of SNPs.[133] High-pressure homogenization through a microfluidizer is based on the manipulation of a continuous flow of liquid through microfabricated channels. Actuation of the liquid flow is implemented by external pressure sources, external mechanical pumps, integrated mechanical micropumps or electrokinetic mechanisms. In a system that has an electronic–hydraulic intensifier pump, the product stream accelerates to high velocities, creating high shearing within the product stream. When 5% starch slurry (high-amylose maize starch) was passed through a specially designed microfluidizer 20 times under a pressure of 207 MPa, the particle size of starch granules decreased from 3–6 μm to 10–20 nm. This size reduction was attributed to the breakage of the hydrogen bonding inside the large particles by the mechanical shear forces. However, concurrently, it was also observed that partial or complete destruction of the crystalline structure occurred. In addition, because only low concentrations of starch slurry could be processed for the homogenization, the recovery yield for each process was low.[150–152]

Ultrasound defines the sound waves at a frequency that is above the normal human hearing range (>15–20 kHz). It is generated with either piezoelectric or magnetostrictive transducers that create high-energy vibrations. These vibrations are amplified and transferred to a sonotrode or probe, which is in direct contact with the fluid. The effect of ultrasound on starch depends on many parameters, such as sonication power and frequency, temperature and time of the treatment and properties of starch dispersion (*e.g.* solid concentration and botanical origin). Another important factor is the formation of bubbles of gases in the suspension medium, which bombards starch granules before they collapse. This process is called cavitation. Rapidly collapsing bubbles could also cause the shear forces that may break the polymer chains.[153,154]

Recently, ultrasonication of starch suspensions was introduced as a process for the preparation of SNPs.[134] In this study, an aqueous suspension of waxy maize starch (1.5% solids) was treated by ultrasonication in a water-bath (8 °C) for 75 min. Based on SEM observations, it was found that mechanical collision along with high shear forces brought about progressive erosion of the starch particles starting from the surface, which underwent further fragmentation until a limiting size was reached (between 30 and 100 nm). Furthermore, at the low temperature during the ultrasonication, water molecules could not diffuse inside the amylopectin chains and no plasticization of the amylopectin phase was likely to take place. This was the first approach to produce nano-sized starch particles by using the physical method of high-intensity ultrasonication. Compared with the common acid hydrolysis, this process offers the advantages of being rapid and easy to implement without the need to undertake repeated washing treatment because no chemical reagent is used. Because no further purification is needed after the SNPs have been produced in solution, the recovery yield may be

100%. However, based on wide-angle XRD analysis, ultrasonication seriously disrupted the crystalline structure in the starch and appeared to lead to nanoparticles that had low crystallinity or an amorphous structure.[134]

Nanoparticle formation *via* reactive extrusion has also been investigated. Reactive extrusion has been defined as a concurrent reaction in extrusion processing of polymers. Basically, the premixed starch and plasticizer were loaded in a twin-screw extruder and reversible crosslinkers, such as glyoxal, were added. The SNPs could be produced through crosslinking during reactive extrusion and are considered to be "regenerated" SNPs.[139,152]

During the extrusion, starch is subjected to relatively high pressure (up to 103 psi), heat and mechanical shear forces. The hydrogen bonds between starch chains, which provide integrity to the starch granules, can be broken under high shear force and temperature. As a result, significant structural changes, including gelatinization, melting and fragmentation, might occur. Starch particles smaller than 400 nm could be produced by using reactive extrusion. Owing to the limited water content, complete gelatinization did not occur during the extrusion. However, at a high extrusion temperatures, starch granules soften and partially melt and become mobile. The softened and melted starch granules were physically torn apart by the shear force to allow water to transfer into the interior of the starch granules. Similarly to other physical treatments, a high energy level of extrusion caused mechanical damage to starch crystals. The SNPs prepared by extrusion had a very low viscosity but were stable for more than 6 months.[155]

More recently, SNPs were prepared by gamma irradiation and were compared with the nanoparticles obtained from acid hydrolysis.[156] Gamma irradiation has been suggested as a rapid and convenient modification technique that fragments large molecules by cleaving the glycosidic linkages.[157] Gamma irradiation can generate free radicals that are capable of hydrolysing chemical bonds, thereby producing smaller fragments of starch called dextrin.[158] SNPs with sizes of approximately 20 and 30 nm were obtained from cassava and waxy maize starches, respectively, by applying a dose of 20 kGy with an irradiation rate of 14 kGy $h^{-1}$.[156] Thermal characterization by a simultaneous instrumental analysis revealed that the SNPs obtained from irradiation were more susceptible to thermal degradation than the parent native starch. This suggests that SNPs have a large number of hydroxyl groups on their surface where the thermal degradation starts, similarly to the thermal properties for SNPs obtained by acid hydrolysis.[159] However, similarly to the SNPs obtained from other physical treatments, the nanoparticles obtained by gamma irradiation of starch displayed amorphous XRD patterns. The explanation given for this was that the transformation to an amorphous structure was not related to the gamma irradiation. Instead, the authors suggested that it was caused by the heat treatment (85 °C) prior to the gamma irradiation, which was necessary to obtain stable dispersions.[82]

Moreover, the preparation of SNPs from waxy maize starches by a combined process of acid hydrolysis (2 days at 40 °C) and subsequent facile ultrasonication (3 min) was reported.[160] Using acid hydrolysis alone, the SNP preparation took several days. However, acid hydrolysis for 2 days was sufficient to remove most of the amorphous regions in starch (50–70% of starch). The relative crystallinity of waxy maize starch was reported to range from 30 to 50%.[142,161] Subsequently, ultrasonic treatment (60% vibration amplitude, 3 min) was applied to the redispersed suspension of the large microparticles of starch hydrolysates, which had been recovered by mild centrifugation (500 rpm, 10 min). By this facile process, the microparticles could be completely transformed to nanoparticles. However, the ultrasonication often disrupted the crystalline structure of starch and the recovery yield of the crystalline SNPs then became low (less than 30%). A procedure for SNP preparation without disrupting the crystallinity by using acid hydrolysis and ultrasonication was suggested. A low temperature (4 °C) was applied to facilitate the association of the starch chains, minimizing the disruption of crystallinity and formation of a new crystalline structure in the SNPs by chain association. The starch hydrolysates, obtained after 6 days of acid hydrolysis, were more resistant to the subsequent ultrasonication than those obtained after 2 or 4 days, regardless of the hydrolysis temperature. Compared with the ordinary process with acid hydrolysis alone, the combination of acid hydrolysis and ultrasonication increased the recovery yield of SNPs from 15 to 78%. In addition, the low-temperature acid hydrolysis in the combined process was an effective procedure for the mass production of crystalline SNPs.[162,163]

As described previously, small starch particles (~500 nm) could be prepared by hydrolysing waxy rice starch using α-amylase. However, ultrasonication after this enzymatic treatment increased the mean diameter of the starch hydrolysates, indicating that the ultrasonication induced their aggregation. Under the size distribution profile, the population density of the two main fractions (peaks of 0.5 and 3.6 μm) was decreased by the ultrasonication. Additionally, when the degree of the enzymatic hydrolysis was greater, the size increase resulting from ultrasonication was more significant. This finding indicates that the starch hydrolysates became more susceptible to the sonication treatment when the hydrolysis level was higher. Therefore, to use the combined process of enzymatic hydrolysis and ultrasonication for producing SNPs, more precise control of the hydrolysis and ultrasonication is needed.[82,132]

Another approach is a nanoprecipitation process that involves the successive addition of a dilute solution of polymer to a solvent, which leads to polymer precipitation on the nanoscale. This method is essentially based on the interfacial deposition of polymers following the displacement of a semipolar solvent that is miscible with water from a lipophilic solution. This approach has several advantages: large amounts of toxic solvents and external energy sources are avoided and submicron particle sizes with narrow size distributions can be obtained.[164] Nanospheres of acetylated waxy maize starch were produced by a nanoprecipitation process. Distilled water was added to a dispersion of waxy maize starch acetate in acetone to form nanospheres and the acetone was vaporized from the aqueous suspension. The size of the nanospheres depended on the polymer concentration in the acetone. The mean diameter increased from 249 to 720 nm as the concentration of the starch in the acetone increased from 1 to 20 mg mL$^{-1}$.[164]

Similarly, SNPs were produced by precipitation of starch paste solution with ethanol as the precipitant and then modified using citric acid (CASN).[165] When ethanol was added dropwise to a starch solution, gelatinized SNPs were gradually precipitated. Therefore, the gelatinization destroyed the A-type crystalline structure of maize starch and the SNPs exhibited the VH-type crystalline structure. However, most of the crystals of CASN had disappeared. The CASN ranged in size from approximately 50 to 100 nm and was not gelatinized during the processing of the nanocomposites because of the crosslinking by citric acid.[165]

Subsequently, a process to obtain SNPs based on complex formation between amylomaize starch and *n*-butanol and subsequent enzymatic hydrolysis was developed. First, a dilute starch solution was placed in the upper compartment of a glass filtration apparatus and allowed to pass gravimetrically through a membrane filter (PTFE, 10 μm pore size, 47 mm diameter) into the bottom compartment, which was filled with *n*-butanol. The filtration apparatus was kept at 70 °C in a convection oven for 6 days. The starch–butanol complex was isolated as a precipitate and subjected to α-amylolysis (porcine pancreatic α-amylase, EC 3.2.1.1, activity 1122 units/mg). Because the amylose–butanol complex that formed in the butanol layer contained a large portion of amorphous matrices, selective removal of these was required in order to isolate the nanoparticles. The subsequent enzymatic treatment made the SNPs spherical or oval in shape with diameters of 10–20 nm. However, there was a significant loss of the starch by hydrolysis (85–90%), hence the overall yield of the nanoparticles was very low.[166]

A facile green method for obtaining nanoparticles from proso millet starch using enzymatic hydrolysis and recrystallization was reported. Cooked starch was treated with pullulanase and the solution was stored at 4 °C for the starch to recrystallize. The recrystallized starch was isolated by centrifugation as SNPs with a size range of 20–100 nm. The study showed that debranching with pullulanase facilitated the recrystallization of the starch chains. Compared with the conventional acid hydrolysis process, this approach has the advantages of being fairly rapid and giving a higher yield (~55%). This process also offers the advantage of that no chemical reagent is added during the preparation process.[167]

The synthesis of SNPs by an emulsion–crosslinking technique was reported, involving the dispersion of the aqueous phase containing hydrophilic natural materials such as starch and crosslinkers in the oil phase in the presence of emulsifiers. The emulsion can generate small particles through a crosslinking reaction. However, the particles obtained from this emulsion–crosslinking approach were relatively large, on the microscale. The average diameter of the starch microspheres was 19 μm with a range from 2.5 to 25 μm. To reduce the size of these particles, the emulsion should contain nanoscale droplets, which is generally referred to as a miniemulsion, submicron emulsion or nanoemulsion. This is because that the emulsion droplets maintain their shape and size within the dispersed phase. An attempt was made to reduce the particle size by using an ionic liquid-in-oil (IL/O) microemulsion system instead of traditional water-in-oil (W/O) emulsion approach. By substituting the water phase using 1-octyl-3-methylimidazolium acetate, SNPs could be obtained with an average diameter of 96.9 nm.[168,169] Another possibility of producing SNPs was through a water-in-ionic liquid (W/IL) microemulsion system. This was an efficient and environmentally friendly approach because no toxic organic reagents such as cyclohexane were used as the oil phase.[170]

A miniemulsion may be produced by specially designed devices that provide sufficient mechanical energy (*e.g.* rotor–stator, sonicator, high-pressure homogenizer and membrane systems). The use of these mechanical devices increases the emulsification efficiency and reduces the cost with a smaller amount of surfactant. A high-pressure homogenizer at pressures ranging from 10 to 60 MPa was used to produce a miniemulsion in which nanoparticles displayed good sphericity of shape and a comparatively uniform size distribution. The size of the particles varied depending on the process parameters, including surfactant content, water-to-oil ratio, starch concentration, homogenization pressure and number of cycles. On applying more energy, the size of the emulsion might be increased owing to the recoalescence of droplets.[168,171]

The various methods for producing crystalline and amorphous SNPs are summarized schematically in Figure 10.5.

### *10.3.2.2 Characterization of SNPs*

In terms of morphology, Figure 10.6 illustrates micrographs of SNPs obtained by different preparation methods and Figure 10.7 shows those of SNPs of different botanical origins. The morphology of SNPs appeared to be related to the crystalline structure of the original starches. Overall, the nanocrystals produced from A-type starches (*e.g.* waxy maize, normal maize, wheat starch) showed square-like particles, whereas those from B-type starches (*e.g.* high-amylose maize, potato) produced round-shaped particles. Assuming that the nanoparticles represent the blocklets in starch granules, this difference indicates that the blocklets in the A- and B-type starches have different morphologies. The differences in the arrangement of amylopectin double helices in crystal lattices might induce the formation of different blocklets.[146]

From the TEM observations illustrated in Figure 10.7, all the SNPs had round or oval shapes, regardless of the origin of the starch. The nanoparticles from B-type starch (*e.g.* 69.7 nm for high-amylose maize) and C-type starch (*e.g.* 53.7 nm for mungbean) were larger than those from the A-type starches (*e.g.* 41.4 nm for waxy maize and 41.0 nm for normal maize). The larger nanoparticles from B- and C-type starches might be partly related to

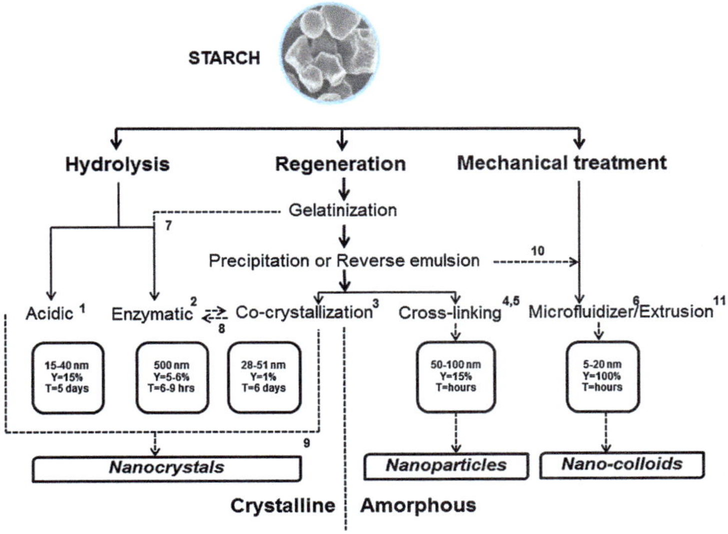


**Figure 10.5** Different ways of producing crystalline and amorphous SNPs. Hydrolysis leads to nanocrystals, whereas regeneration and mechanical treatment lead to both amorphous and crystalline particles in the final batch. (1) Acid hydrolysis; (2) enzymatic hydrolysis (may produce blocklets not nanoparticles); (3) complex formation and (4) and (5) precipitation of gelatinized starch in a non-solvent followed by cross-linking reaction; (6) analogy with microfibrillated cellulose; (7) and (8) combined enzymolysis followed by crystallization; (9) classical path of enzymatic hydrolysis; (10) emulsion crosslinking technology using a high-pressure homogenizer; (11) extrusion.
Reproduced from Le Corre and Angellier-Coussy[126] with permission of Elsevier.

their greater resistance to acid hydrolysis compared with A-type starches. It was also notable that the size of the SNPs appeared to be influenced by the amylose content. The more amylose present, the larger were the particles produced. Starches with similar amylose contents, such as maize, potato and wheat starches, produced SNPs of similar sizes.[146,148] However, in contrast, it has been reported that the SNPs had a plate-like structure.[131] The morphology and size of the SNPs might not be exactly the same as those of starch granules because they could be changed depending on the physical and chemical parameters used for the preparation.[82]

Although there are some differences in the morphology of SNPs, it is obvious that the SNPs obtained from the hydrolysis of starch granules tend to self-aggregate, forming microscale agglomerates. The aggregation behaviour of SNPs can be explained by the presence of a large number of hydroxyl groups on their surface, which readily participate in the formation of hydrogen bonds or van der Waals attraction between SNPs.[156,172] This aggregation behaviour of SNPs greatly limits their industrial applications. In the fabrication of nanocomposites with SNPs, blending of an aqueous SNP

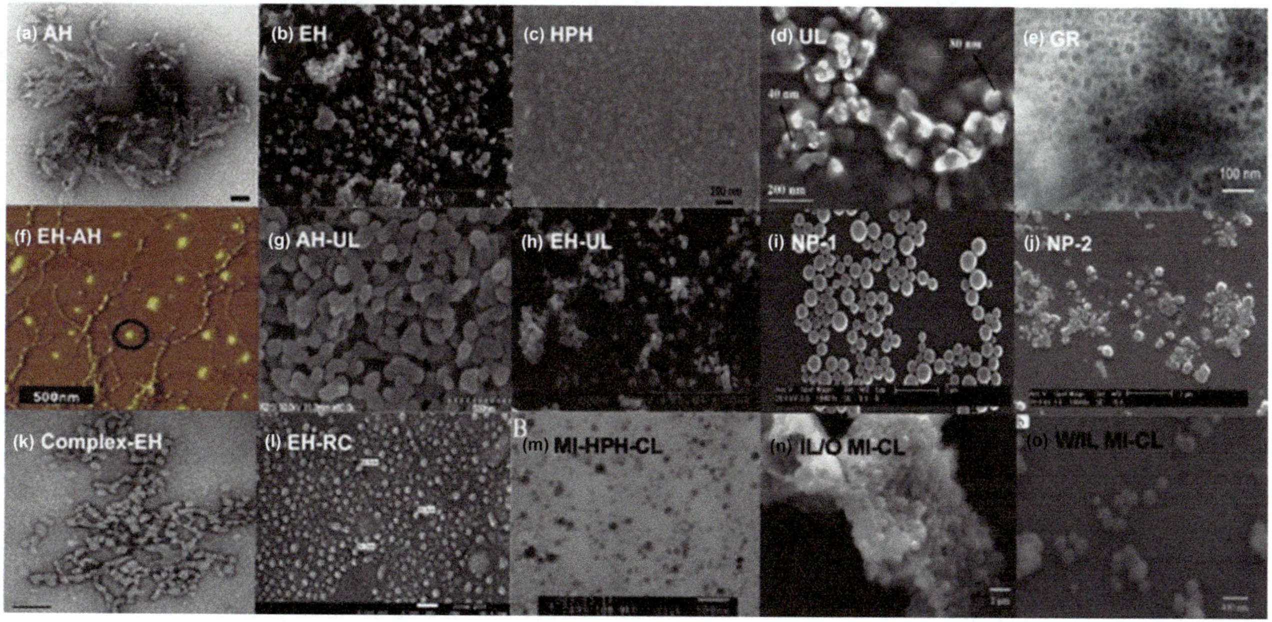


**Figure 10.6** Morphology of SNPs obtained by various preparation methods. (a) TEM image of waxy maize SNPs obtained by acid hydrolysis; (b) SEM image of waxy rice SNPs obtained by enzymatic hydrolysis; (c) SEM image of high-amylose maize SNPs obtained by high-pressure homogenization; (d) SEM image of waxy maize SNPs obtained by ultrasonication; (e) TEM image of waxy maize SNPs obtained by gamma irradiation; (f) AFM image of waxy maize SNPs obtained by enzymatic hydrolysis and acid hydrolysis; (g) SEM image of waxy maize SNPs obtained by a combination of acid hydrolysis and ultrasonication; (h) SEM image of waxy rice SNPs obtained by enzymatic hydrolysis followed by ultrasonication; (i and j) SEM images of waxy maize starch acetate (SA) nanospheres and maize SNPs obtained by nanoprecipitation, respectively; (k) TEM image of amylomaize starch–butanol complexes isolated by enzymatic hydrolysis; (l) SEM image of proso millet SNPs obtained by a combination of enzymatic hydrolysis and recrystallization; (m) TEM image of soluble SNPs obtained by high-pressure homogenization of water-in-oil (W/O) microemulsion followed by crosslinking through sodium trimetaphosphate (STMP); (n) SEM image of normal maize SNPs obtained by a combination of ionic liquid-in-oil (IL/O) microemulsion preparation and its crosslinking through epichlorohydrin; (o) SEM image of SNPs obtained by a combination of water-in-ionic liquid (W/IL) microemulsion preparation using an acid-treated starch followed by crosslinking through epichlorohydrin.
Reproduced from Kim *et al.*[82] with permission of Elsevier.

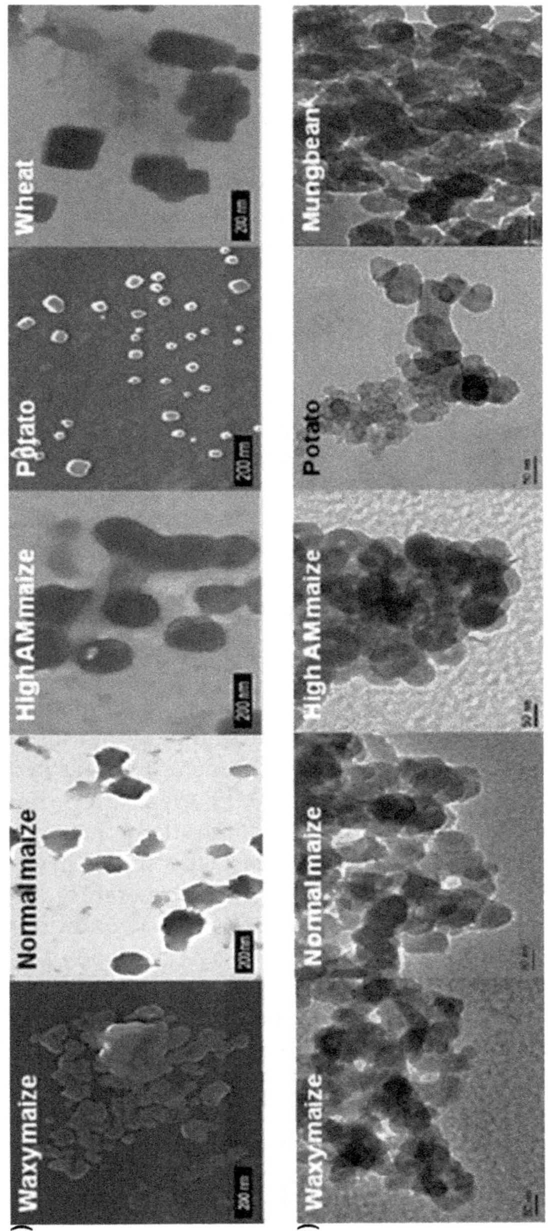


**Figure 10.7** Morphology of SNPs of different botanical origins. (a) SEM images after acid hydrolysis using dilute sulfuric acid (3.16 M, 5 days); (b) TEM images after acid hydrolysis using dilute sulfuric acid (3.16 M, 5 days). Reproduced from Kim *et al.*[82] with permission of Elsevier.

suspension with a matrix solution is required.[173] Homogeneity with nano-sized SNPs is critical to achieve a desirable mechanical performance of the nanocomposites. To avoid the aggregation, facile physical treatments such as sonication or homogenization were performed before mixing with a polymeric matrix. More recently, it was found that aggregated parallelepiped nanoplatelets (1.5 μm) changed to monodispersed spherical-like nanoparticles (50 nm) on increasing the dispersion pH from 2.07 to 11.96. This finding was attributed to the retardation of the aggregation tendency induced as a result from the increased repulsion forces among the SNPs that had become electronegative under alkaline conditions.[172]

Regarding the crystallinity of SNPs, most of the studies on SNPs were focused on the changes in the crystalline structure of starch during the production process of SNPs. The relative crystallinity (RC) of the SNPs with respect to the original crystallinity of starch was assumed to be positively related to the recovery yield of the SNPs. Compared with native starches, the RC of the SNPs increased because the amorphous regions in starch were selectively removed by the acid hydrolysis. From the maize starches containing different amounts of amylose, the RC of the SNPs obtained increased as the amylose content decreased. On comparison of starches having similar contents of amylose (21–28%) but different botanical origins, no significant differences were observed in the degree of crystallinity of their SNPs. This demonstrated that the most important parameter in determining the degree of crystallinity of SNPs was the amylose content in starch.[146,174,175]

An attempt was made to isolate crystalline SNPs, which were resistant to subsequent ultrasonic treatment, by using different temperature profiles: isothermally at 4 or 40 °C and under cycled temperatures of 4/40 °C. The starch hydrolysates obtained from the low temperature hydrolysis (4 °C) exhibited exceptionally high crystallinity (33%) even after the ultrasonic treatment, although their RC was slightly lower than that of native waxy maize starch (36.37% *versus* 33%). In contrast, the starch hydrolysates obtained by the same acid hydrolysis at a higher temperature had less crystallinity under either isothermal (27.68%) or cycled conditions (26.42%), indicating that the disruptions of the starch crystals depended on the hydrolysis temperature. Maintaining a lower temperature during hydrolysis helped the starch granules remain more rigid and intact during the hydrolysis although the hydrolysis occurred more slowly. A mild hydrolysis prior to the physical fragmentation is needed to protect the starch crystals against the physical treatment and to assist the fragmentation. In contrast to cellulose nanoparticles, which are recovered in almost pure crystal form, SNPs may be produced in a crystalline shape. The crystallinity of SNPs is influenced by the inherent properties of starch, such as botanical source, amylose content and crystalline structure, and also the processing parameters for SNP preparation such as hydrolysis temperature and ultrasonic power and level. The crystallinity of SNPs is an advantage because the SNPs remain intact and are readily isolated from the suspension.[174]

With regard to thermal transition properties, the thermal transition behaviour of SNPs has been characterized by differential scanning calorimetry (DSC). Compared with native starch, the endothermic melting transition peak of SNPs prepared by acid hydrolysis was observed in wider ranges. This trend was observed regardless of the processing methods for SNPs.[148,156,160,167,176] However, the melting temperatures and enthalpy depended on the parameters for the SNP preparation and the botanical sources of starch.

The broad melting range of SNPs could be attributed to the heterogeneity of the SNPs in which amorphous and crystalline phases were in a mixture.[156] It was reported that some A-type starches, such as waxy and normal maize starches, displayed broad melting endotherms over wider temperature ranges than the corresponding native starches.[148,176] In contrast, no endothermic peak was found for the SNPs prepared from B-type potato and high-amylose maize starches and C-type mungbean starch. It was hypothesized that the crystals in the SNPs from the B-and C-type starches might have melting temperature below room temperature and therefore were readily disrupted in water added for DSC analysis.[148] These cold-water-soluble SNPs could have similar properties to pregelatinized starch.[177] As shown in the XRD patterns, the B-type crystals were more susceptible to acid hydrolysis than the A-type crystals.[148] In native starch, the B-type crystalline packing is less dense than the A-type crystalline packing and hence more mobile and more prone to disruption.[178]

It was observed that the SNPs from B-type starches (*e.g.* high-amylose maize, potato) showed slightly higher melting temperatures than those of A-type SNPs.[148] The higher thermal stability of the B-type SNPs could be attributed to the longer chains in the amylopectin helices of the B-type starches, which could form more stable crystals.[179]

By changing the preparation conditions, such as hydrolysis temperature and time, SNPs could be obtained with different melting characteristics. The melting temperatures for the SNPs were also slightly increased by either removal of the amorphous regions or minor rearrangement of the starch chains that may occur during the ultrasonic treatment. The physical treatments that apply high levels of energy could disrupt the crystallinity of SNPs.[163]

The thermal transition characteristics of the SNPs are important in their applications. As an example, the endothermic transition of SNPs should be carefully considered when they are used in the production of composites with other polymers where thermal melting of the polymers typically occurs. When the SNPs should remain as a reinforcing filler, thermal resistance of the polymeric blends against the composite preparation conditions is required. The crystals in SNPs originate from those in native starches, which are generally stable when the moisture content is limited. By changing the physical parameters and starch sources for the SNP preparation, the thermal transition characteristics and resistance of the SNPs might be controlled to some extent. Therefore, when SNPs are used in composites, the preparation process of the SNPs should be carefully designed to render suitable thermal properties. Based on the reported data, in a case where thermal resistance is essential for the utilization of SNPs such as in composites, acid hydrolysis at a high temperature such as 40 °C and A-type starches might be favourable.[82,156]

Concerning molecular composition, the molecular features of SNPs have rarely been studied and only a few published reports are available. Waxy maize SNPs (Figure 10.8), which consisted of double helices containing $\alpha$-(1→6) linkages between the parallel strands, were progressively degraded into low molecular weight oligosaccharides, which were subjected to debranching enzymes (isoamylase and pullulanase) followed by $\beta$-amylase. Eight different dextrins were found in the SNPs prepared by the enzymatic treatment and classified into two major groups: populations A and B. Population A consisted of multi-branched dextrins with an average degree of polymerization (DP) of 31.7. Population B, which was approximately one-third of the total dextrins, contained at least one branch and its average DP was 14.2. However, the molar percentages of branched chains in populations A and B were similar. This indicates that the SNPs could be considered to possess principally homogeneous and regular molecular structures.[148,180]

The chain length and chain distribution of starch depend on the botanical origins, hence the chain conformation after debranching the SNPs was influenced by the botanical sources of starch.[148] The amylopectin chains isolated by debranching were designated as long B ($B \geq 2$), short B (B1) and very short A chains.[181] The chain length distribution of the waxy maize SNPs showed that the proportion of short chains (B1 and A chains) were higher than that of native waxy maize starch. This is attributed that the long amylopectin chains such as $B \geq 2$ were more susceptible to acid hydrolysis. However, other starches that contained amylose (normal maize, high-amylose maize, potato and mungbean starches) exhibited increased proportions of both $B \geq 2$ and B1 plus A chains because the amylose was readily disintegrated by the acid hydrolysis. The chain length distributions of the SNPs revealed that the amylose and long amylopectin chains were highly susceptible to acid hydrolysis, possibly because these chains are located mostly in the amorphous regions in starch granules.[181]

Regarding the rheological properties of SNPs, the pasting behaviour of SNPs was different from that of native starch. When heated in excess water, native maize starch granules swelled and the ordered granule structure was disrupted in the gelatinization temperature range, which resulted in an increase in viscosity. The SNPs, however, readily dissolved in water, even at temperatures lower than the gelatinization temperature of native starch. The pasting viscosity of SNPs was much lower than that of native starch, because the SNPs had been produced by acid hydrolysis. However, when the amount of the SNPs increased, the pasting viscosity increased and the pasting pattern appeared somewhat different from that of native starch. The viscosity of the SNPs increased continuously with increase in temperature (*e.g.* from 50 to 95 °C) and then reached a plateau at 95 °C. There was no

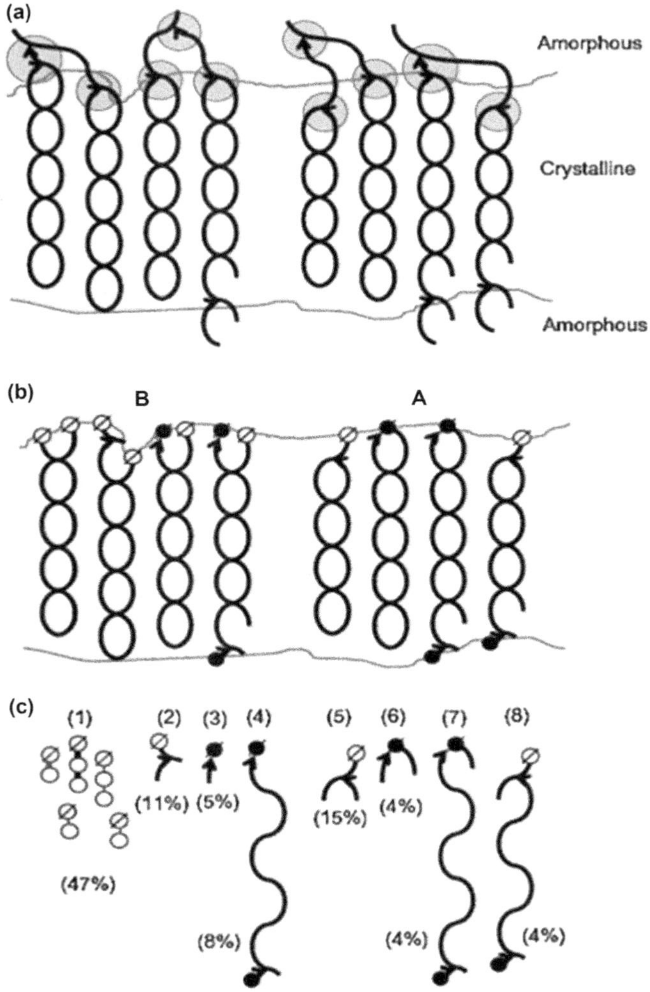


**Figure 10.8** (a) Hypothetical model of branched building blocks of clusters, as found in the amorphous and crystalline lamellae of starch granules. (b) Molecular composition of nanocrystals formed after acid treatment of the granules. Examples of possible structures found in populations A and B are shown. Symbols Ø represent reducing ends and the corresponding filled symbols represent reducing ends involved in α-(1→6) linkage, being resistant to debranching enzymes; ●, single residues in a branch also resistant to debranching. (c) Dextrins of different types (1–8) formed from the structures in (b) by β-amylolysis.
Reproduced from Angellier-Coussy *et al.*[180] with permission of Elsevier.

breakdown during the stage of hot shearing. When the temperature decreased from 95 to 50 °C, the viscosity of the SNPs paste increased again. Overall viscograms revealed that the SNPs do not show the rapid viscosity increase induced by the granule swelling such as is shown by native starch, possibly because of the reduced size. However, the heating induced the melting of crystals in SNPs, forming a paste with a viscosity increase. As nanofillers, the SNPs should not be destroyed during the process of composite preparation. In most composites, water is rarely used and polymers and SNPs were typically blended in organic solvents. Hence the SNPs remain as crystalline solids during processing such as hot pressing or casting–evaporation.[165]

At a given shear rate, the viscosity of starch particle suspensions increased significantly with decrease in particle size, especially when the size was reduced to 600 nm or smaller. This relationship could be attributed to the fact that, at a given mass concentration, the smaller the particles resulted in more particles present in the solution. Thus the interaction between the particles was increased so that the viscosity of the suspension increased. Compared with the traditionally cooked starch paste at a low concentration (3%), the SNP suspension at a much higher concentration (9%) showed much lower viscosity at the same shear rate of 1000 $s^{-1}$. This indicates that the use of SNPs allows suspensions with much higher solid contents with lower viscosity, which could be desirable in paper making and coating. The high solid content may increase the bonding strength when the suspension is used as an adhesive. The low viscosity of SNP suspensions could increase the absorption of starch on cellulose substrates. The SNPs were also suggested as a candidate for industrial applications such as thickening agents.[82,152]

#### *10.3.2.3 Utilization of SNPs*

The major application of SNP is in composites and they have been suggested as reinforcement fillers in polymeric composites. A wide range of polymeric matrices, including both natural and synthetic polymers, have been suggested for composites with SNPs. Synthetic polymers are versatile materials for many industrial applications because of their excellent physical properties and chemical resistance. However, most synthetic polymers are not biodegradable or biocompatible. For these reasons, recent research has been focused on the use of environmentally-friendly polymers, including a variety of natural polymers such as starch, pullulan, polylactide and soybean protein. By incorporating the SNPs in the synthetic polymer matrix, not only the physical properties but also the biodegradability of the composite are enhanced.[82,137,138]

Several studies emphasized the importance of having a uniform mixing of the reinforcing filler within the polymeric matrix of a composite formulation. Especially when the content of fillers was increased (more than 40%), it was considered that the nanocrystals tended to self-aggregate, which decreased the surface area for mutual interactions with polymers, resulting in decreases in strength and modulus.[164,182,183]

Composite films can be prepared from blends of SNPs and polymers using different methods with various techniques such as casting and solvent evaporation, hot pressing, compression moulding and extrusion.[184] Except in the casting–evaporation method, the composite films are subjected to a relatively high temperature and/or pressure.[138] Therefore, substantial melting of the SNPs may occur during the preparation of composites, especially when water is present in the blends. For this reason, the casting–evaporation method was suggested as a better choice than other thermal methods for film preparation.[176]

For mechanical properties such as tensile strength ($\sigma$, MPa), elasticity or Young's modulus ($E$, MPa) and elongation at break ($\varepsilon_b$, %) of the nanocomposites containing SNPs, the reinforcing effect of SNPs was evident from the tensile tests of their composite products. In most cases, the incorporation of SNPs resulted in increases in both the tensile strength and elastic modulus of the composites but a decrease in elongation at break.[137,138,185–187] For instance, nanocomposites of natural rubber with waxy maize SNPs exhibited reinforcing effects from the SNPs. Up to the SNP content of ~20 wt%, elongation at break values decreased slightly from 1960 to 1500% and then decreased more rapidly to 920% for the natural rubber film filled with 30 wt% SNPs. In contrast, Young's modulus increased nearly exponentially with increase in SNP content, from 0.64 MPa for the unfilled natural rubber matrix to 77.8 MPa for the composite film filled with 30 wt% SNPs. The SNPs appeared to function as a good substitute for carbon black since the addition of only 10 wt% of SNPs to natural rubber induced a reinforcing effect similar to that observed with 26.6 wt% addition of carbon black.[188] However, SNPs may be less competitive than cellulose nanoparticles, so that a larger amount of the SNPs is necessary to achieve a similar reinforcing effect. One of the major reasons for this is the difference in geometry of the nanoparticles. Cellulose nanocrystals have a higher aspect ratio (50–500 nm in length and 3–5 nm in diameter) than SNPs, which have a length of 20–40 nm and a width of 15–30 nm.[180,184] It has been reported that the geometric aspect ratio, which is defined as the ratio of length to diameter ($L/d$), is an important factor of composite fillers that controls the mechanical properties of the nanocomposites. The nanoparticles that have a higher aspect ratio give a better reinforcing effect.[189–191]

Various kinds of particles, *e.g.* hydrophobic silica, carbon nanotubes, latex and microfibrillated cellulose and bacterial cellulose nanocrystals, have been used as emulsion stabilizers.[192–201] In a recent study, it was suggested that SNPs can be used as a stabilizer in oil-in-water emulsions.[202] An emulsion prepared with the equivalent volume ratio of water and paraffin could be stabilized by adding an aliquot of SNP dispersion. It was found that the addition of >0.02 wt% of SNPs stabilized the emulsion during storage for more than 2 months without coalescence of oil droplets.[202] However, the emulsion became phase separated when the SNPs melted on heating at 80 °C for 2 h. The SNPs may be used in various emulsions not only for foods but also cosmetics and pharmaceuticals. However, additional studies are needed on the utilization of SNPs in these areas.

Another specific application of SNPs is as fat replacers in foods. Fat replacers are substances that imitate the organoleptic or physical properties of triglycerides but cannot replace fat on a gram-for-gram basis.[203,204] Native and modified starches can sometimes be used to replace fat.[204] The particle size of the starches plays an important role in determining both the fat-like taste and mouthfeel.[125,205] The most common modifications of starch when used as a fat replacer include chemical depolymerization (*e.g.* acid or enzymatic hydrolysis) and mechanical disintegration of starch granules. SNPs may be a promising candidate as fat replacers because of their small size. It is expected that when SNPs are blended with other components, the mixture will form a smooth cream-like material that has properties similar to those of fat. Additionally, the use of SNPs may result in a decrease in calorie content by replacing high-calorie fat with carbohydrate. Research is still needed on this aspect of applications of SNPs.[125]

With regard to packaging components, SNPs have attracted interest as a material in barrier films for food packaging. The barrier properties of packaging films are related primarily to water vapour transmission and oxygen permeability. Several studies reported a decrease in water vapour permeability when maize SNPs were incorporated. It was found that adding 30–40% of waxy maize SNPs led to a significant decrease in the water vapour permeability of sorbitol-plasticized pullulan film.[137] However, a 40% decrease in permeability for a cassava starch film on reinforcing with 2.5% of SNPs has also been reported.[206] The SNPs may also improve the barrier properties with respect to oxygen permeability. It was found that the SNPs from waxy maize starch could reduce the oxygen diffusion and permeability of a nanocomposite film prepared with natural rubber.[207] This finding was attributed to the platelet-like SNPs, which might block the migration of oxygen molecules through the film.

Moreover, SNPs can be used as drug carriers and implant materials. Nanoparticles have the ability to deliver a wide range of molecules to different locations in the body for sustained periods of time. A higher intracellular uptake of nanoparticles has been reported compared with micron-sized particles. Nanoparticles can access a variety of biological tissues because of their tiny size and mobility.[208] Indeed, nanoparticles larger than 230 nm could congregate in the organs, especially in the spleen owing to the capillary size of this organ. Starch has been used as a delivery carrier for tumour-targeted drugs and transdermal drugs.[209,210] It was used as a carrier for phenethylamines, acetylsalicylic acid (aspirin) and estrone.[211–213] Chemically modified starches or SNPs have also been used in sustained drug delivery systems. For example, a crosslinked high-amylose starch was used as a matrix for the controlled release of Contramid.[214] Propylated SNPs that had been loaded with different types of drug (flufenamic acid, testosterone and caffeine) showed enhanced effectiveness in permeation through human skin.[210] Dialdehyde SNPs (DASNPs) that were conjugated with 5-fluorouracil were found to have enhanced inhibition *in vitro* against breast cancer cells (MF-7).[215] However, some studies have mentioned that the nanoparticles used in drug delivery may have a toxic effect in the human body.[216]

In addition, the biodegradability and biocompatibility of SNPs make them an excellent candidate for implant materials. Biodegradable implants offer clear advantages over traditional metal implants, which often cause stress shielding.[217,218] SNPs can retain their strength long enough to support a healing bone and then they can gradually and harmlessly disintegrate in the patient's body. More research should be undertaken on the practical application of SNPs in this area.[82]

Concerning other non-food applications, chemically modified SNPs can be used as adsorbents for the removal of aromatic organic pollutants from water. The chemical modification of the nanoparticles (*e.g.* grafting, cross-linking) further enhanced the adsorption capacity. Another application is as a thermoresponsive conducting material. The highly crosslinked and nanofilled polymeric systems are likely to exhibit good electrical conductivity with high rigidity.[219]

Moreover, SNPs can be used as a binder in paper making and paper coating. Cooked starch has been widely used as a paper-making additive. The retention of the cooked starch on the paper matrix is based on the absorption of starch. Thus, the amount of starch absorbed is limited by the saturation of absorption on cellulose substrates. Another problem is the high viscosity of the starch paste after cooking of raw starch, which might cause operational problems. With the addition of SNPs, the viscosity of the paste can be substantially reduced, whereas the binding capability can be increased.[220]

## 10.4 Conclusion

Starch as an agro-sourced polymer has attracted much attention in recent years owing to its strong advantages such as low cost, wide availability and total biodegradability. However, the fundamental properties (mechanical properties, moisture sensitivity, *etc.*) of native starches need to be enhanced to enable such materials to be truly competitive with conventional petroleum-based polymers over a wider range of applications. One of the most promising technical advances has been the development of starch nanocomposites. Starch is mostly converted into starch nanocomposites to enhance its properties and to overcome its drawbacks. Starch can be used as the matrix either alone or in combination with other polymers and with the incorporation of different types of nanofillers. In another approach, starch granules are transformed into nanoparticles that are introduced as fillers in various polymer matrices or used directly as the ingredient.

For the past few decades, the preparation and utilization of nanoparticles originating from polysaccharides (*e.g.* starch, cellulose, chitin) have been extensively studied. Although SNPs could be obtained by the classical acid hydrolysis with the inherent crystallinity, some drawbacks remain to be overcome, such as an extended period (longer than 5 days) and low yield (less than 20%) of preparation for practical utilization. Therefore, recent research has been focused on the production of SNPs by alternative processes such as physical treatments (*e.g.* high-pressure homogenization, extrusion and ultrasonication). Compared with acid hydrolysis, these physical treatments generate SNPs faster and in higher yields. However, it is difficult to control the physical parameters to minimize the destruction of starch crystals.

Despite the great potential of SNPs for industrial applications, more research should be carried out on efficient nanoparticle production on a large industrial scale. A simple and economic process and improvement of the recovery yield from hydrolysis still remain to be achieved. Additionally, SNPs have a strong tendency to aggregate, so the recovery of powder products is difficult. However, the extensive hydrophilic nature of the SNPs provides significantly high reactivity, which is one of their advantages. The incorporation of SNPs resulted in physical properties superior to those of conventional microscale composites. Even when a small amount of nanoparticles was added to a matrix, the resulting nanocomposites had a strong reinforcing effect. In addition, positive impacts such as an improvement in barrier properties were also found. It is obvious that starch-based nanoparticles are potential fillers with respect not only to their mechanical properties but also to their renewability and biodegradability. Although most studies on the application of SNPs have been focused on nanocomposites, their utilization as carriers for biofunctional ingredients and drugs should also be considered. In the food industries, SNPs may be used as a novel ingredient for the control of rheological properties and texture, and a potential application is the replacement of fat.

# References

1. M. Rossi, F. Cubadda, L. Dini, M. L. Terranova, F. Aureli, A. Sorbo and D. Passeri, *Trends Food Sci. Technol.*, 2014, **40**, 127.
2. European Food Safety Authority, *EFSA J.*, 2009, **958**, 1.
3. L. Rashidi and K. Khosravi-Darani, *Crit. Rev. Food Sci. Nutr.*, 2011, **51**, 723.
4. J. Weiss, P. Takhistov and D. J. McClements, *J. Food Sci.*, 2006, **71**, R107.
5. FAO/WHO, *FAO/WHO Expert Meeting on the Application of Nanotechnologies in the Food and Agriculture Sectors: Potential Food Safety Implications. Meeting Report,* 2010, http://www.fao.org/docrep/012/i1434e/i1434e00.pdf.
6. M. Cushen, J. Kerry, M. Morris, M. Cruz-Romero and E. Cummins, *Trends Food Sci. Technol.*, 2012, **24**, 30.
7. C. E. Handford, M. Dean, M. Henchion, M. Spence, C. T. Elliott and K. Campbell, *Trends Food Sci. Technol.*, 2014, **40**, 226.
8. F. Xie, E. Pollet, P. J. Halley and L. Avérous, *Prog. Polym. Sci.*, 2013, **38**, 1590.

9. D. Ray and S. Maiti, in *Natural Polymers, Nanocomposites*, ed. M. J. John and T. Sabu, Royal Society of Chemistry, Cambridge, 2012, vol. 2, p. 185.
10. C. Silvestre, D. Duraccio and S. Cimmino, *Prog. Polym. Sci.*, 2011, **36**, 1766.
11. S. Tinkle, S. E. McNeil, S. Mühlebach, R. Bawa, G. Borchard, Y. C. Barenholz, L. Tamarkin and N. Desai, *Ann. N. Y. Acad. Sci.*, 2014, **1313**, 35.
12. Q. Chaudhry and L. Castle, *Trends Food Sci. Technol.*, 2011, **22**, 595.
13. Q. Chaudhry, M. Scotter, J. Blackburn, B. Ross, A. Boxall, L. Castle, R. Aitken and R. Watkins, *Food Addit. Contam.*, 2008, **25**, 241.
14. N. Duran and P. D. Marcato, *Int. J. Food Sci. Technol.*, 2013, **48**, 1127.
15. P. Ezhilarasi, P. Karthik, N. Chhanwal and C. Anandharamakrishnan, *Food Bioprocess Technol.*, 2013, **6**, 628.
16. A. Grobe and M. E. Rissanen, *Recent Pat. Food Nutr. Agric.*, 2012, **4**, 176.
17. J. K. Momin, C. Jayakumar and J. B. Prajapati, *Emir. J. Food Agric.*, 2013, **25**, 10.
18. S. Ranjan, N. Dasgupta, A. R. Chakraborty, S. M. Samuel, C. Ramalingam, R. Shanker and A. Kumar, *J. Nanopart. Res.*, 2014, **16**, 1.
19. B. S. Sekhon, *Nanotechnol., Sci. Appl.*, 2010, **3**, 1.
20. I. J. Joye, G. Davidov-Pardo and D. J. McClements, *Trends Food Sci. Technol.*, 2014, **40**, 168.
21. A. Madene, M. Jacquot, J. Scher and S. Desobry, *Int. J. Food Sci. Technol.*, 2006, **41**, 1.
22. D. J. McClements, E. A. Decker, Y. Park and J. Weiss, *Crit. Rev. Food Sci. Nutr.*, 2009, **49**, 577.
23. D. J. McClements and Y. Li, *Adv. Colloid Interface Sci.*, 2010, **159**, 213.
24. L. Sagalowicz and M. E. Leser, *Curr. Opin. Colloid Interface Sci.*, 2010, **15**, 61.
25. G. P. Gruère, *Food Policy*, 2012, **37**, 191.
26. S. L. Gorbach, *Am. J. Gastroenterol.*, 2000, **95**, S2.
27. S. Alfadul and A. Elneshwy, *Afr. J. Food, Agric., Nutr. Dev.*, 2010, **10**, 2719.
28. E. L. Lien, *Am. J. Clin. Nutr.*, 2003, 77, 1555S.
29. R. Apan, A. M. Cozmuta, A. Peter, C. Nicula and L. M. Cozmuta, *Amfiteatru Economic.*, 2014, **36**, 483.
30. W. Han, Y. Yu, N. Li and L. Wang, *Chin. Sci. Bull.*, 2011, **56**, 1216.
31. D. Restuccia, U. G. Spizzirri, O. I. Parisi, G. Cirillo, M. Curcio, F. Iemma, F. Puoci, G. Vinci and N. Picci, *Food Control*, 2010, **21**, 1425.
32. J.-W. Rhim, H.-M. Park and C.-S. Ha, *Prog. Polym. Sci.*, 2013, **38**, 1629.
33. N. Sozer and J. L. Kokini, *Trends Biotechnol.*, 2009, **27**, 82–89.
34. S. W. Wijnhoven, W. J. Peijnenburg, C. A. Herberts, W. I. Hagens, A. G. Oomen, E. H. Heugens, B. Roszek, J. Bisschops, I. Gosens and D. Van De Meent, *Nanotoxicology*, 2009, **3**, 109–138.
35. S. D. F. Mihindukulasuriya and L. T. Lim, *Trends Food Sci. Technol.*, 2014, **40**, 149.

36. K. L. Yam, P. T. Takhistov and J. Miltz, *J. Food Sci.*, 2005, **70**, R1.
37. Q. Chaudhry, L. Castle and R. Watkins, *Nanotechnologies in Food*, Royal Society of Chemistry, Cambridge, 2010.
38. S. Bhattacharya, J. Jang, L. Yang, D. Akin and R. Bashir, *J. Rapid Methods Autom. Microbiol.*, 2007, **15**, 1.
39. D. Lange, C. Hagleitner, A. Hierlemann, O. Brand and H. Baltes, *Anal. Chem.*, 2002, **74**, 3084.
40. M. García, M. Aleixandre, J. Gutiérrez and M. Horrillo, *Sens. Actuators, B*, 2006, **113**, 911.
41. A. Baeumner, *Food Technol.*, 2004, **58**, 51.
42. T. Vo-Dinh, B. M. Cullum and D. L. Stokes, *Sens. Actuators, B*, 2001, **74**, 2.
43. J. T. Mabeck and G. G. Malliaras, *Anal. Bioanal. Chem.*, 2006, **384**, 343.
44. EFSA Scientific Committee, *EFSA J.*, 2011, **9**, 21.
45. M. A. Hamburg, *Science*, 2012, **336**, 299.
46. C.-F. Chau, S.-H. Wu and G.-C. Yen, *Trends Food Sci. Technol.*, 2007, **18**, 269.
47. H. Bouwmeester, P. Brandhoff, H. J. P. Marvin, S. Weigel and R. J. B. Peters, *Trends Food Sci. Technol.*, 2014, **40**, 200.
48. S. Dekkers, H. Bouwmeester, P. M. Bos, R. J. Peters, A. G. Rietveld and A. G. Oomen, *Nanotoxicology*, 2013, 7, 367.
49. S. Dekkers, P. Krystek, R. J. Peters, D. P. Lankveld, B. G. Bokkers, P. H. van Hoeven-Arentzen, H. Bouwmeester and A. G. Oomen, *Nanotoxicology*, 2011, **5**, 393.
50. G. Oberdörster, E. Oberdörster and J. Oberdörster, *Environ. Health Perspect.*, 2005, 823.
51. K. W. Powers, S. C. Brown, V. B. Krishna, S. C. Wasdo, B. M. Moudgil and S. M. Roberts, *Toxicol. Sci.*, 2006, **90**, 296.
52. K. Tiede, A. B. Boxall, S. P. Tear, J. Lewis, H. David and M. Hassellöv, *Food Addit. Contam.*, 2008, **25**, 795.
53. C. Szakal, S. M. Roberts, P. Westerhoff, A. Bartholomaeus, N. Buck, I. Illuminato, R. Canady and M. Rogers, *ACS Nano*, 2014, **8**, 3128.
54. M. Hassellöv, J. W. Readman, J. F. Ranville and K. Tiede, *Ecotoxicology*, 2008, **17**, 344.
55. F. von der Kammer, S. Legros, T. Hofmann, E. H. Larsen and K. Loeschner, *TrAC, Trends Anal. Chem.*, 2011, **30**, 425.
56. R. Peters, G. ten Dam, H. Bouwmeester, H. Helsper, G. Allmaier, R. Ramsch, C. Solans, M. Tomaniová, J. Hajslova and S. Weigel, *TrAC, Trends Anal. Chem.*, 2011, **30**, 100.
57. J. P. Helsper, R. J. Peters, L. Brouwer and S. Weigel, *Anal. Bioanal. Chem.*, 2013, **405**, 1181.
58. L. Calzolai, D. Gilliland and F. Rossi, *Food Addit. Contam.*, 2012, **29**, 1183.
59. Z. Zhang, F. Kong, B. Vardhanabhuti, A. Mustapha and M. Lin, *J. Agric. Food Chem.*, 2012, **60**, 10762.

60. H. Bouwmeester, I. Lynch, H. J. Marvin, K. A. Dawson, M. Berges, D. Braguer, H. J. Byrne, A. Casey, G. Chambers and M. J. Clift, *Nanotoxicology*, 2011, **5**, 1.
61. S. K. Mwilu, A. M. El Badawy, K. Bradham, C. Nelson, D. Thomas, K. G. Scheckel, T. Tolaymat, L. Ma and K. R. Rogers, *Sci. Total Environ.*, 2013, **447**, 90.
62. R. Peters, E. Kramer, A. G. Oomen, Z. E. Herrera Rivera, G. Oegema, P. C. Tromp, R. Fokkink, A. Rietveld, H. J. Marvin and S. Weigel, *ACS Nano*, 2012, **6**, 2441.
63. A. P. Walczak, R. Fokkink, R. Peters, P. Tromp, Z. E. Herrera Rivera, I. M. Rietjens, P. J. Hendriksen and H. Bouwmeester, *Nanotoxicology*, 2012, **7**, 1198.
64. A. Cockburn, R. Bradford, N. Buck, A. Constable, G. Edwards, B. Haber, P. Hepburn, J. Howlett, F. Kampers and C. Klein, *Food Chem. Toxicol.*, 2012, **50**, 2224.
65. D. E. Lefebvre, K. Venema, L. Gombau, L. G. Valerio, J. Raju, G. S. Bondy, H. Bouwmeester, R. P. Singh, A. J. Clippinger, E.-M. Collnot, R. Mehta and V. Stone, *Nanotoxicology*, 2015, **9**, 523.
66. K. Loeschner, N. Hadrup, K. Qvortrup, A. Larsen, X. Gao, U. Vogel, A. Mortensen, H. R. Lam and E. H. Larsen, *Part. Fibre Toxicol.*, 2011, **8**, 18.
67. M. van der Zande, R. J. Vandebriel, E. Van Doren, E. Kramer, Z. Herrera Rivera, C. S. Serrano-Rojero, E. R. Gremmer, J. Mast, R. J. Peters and P. C. Hollman, *ACS Nano*, 2012, **6**, 7427.
68. D. K. Kim and J. Dobson, *J. Mater. Chem.*, 2009, **19**, 6294.
69. A. Kumari, S. K. Yadav and S. C. Yadav, *Colloids Surf., B*, 2010, **75**, 1.
70. S. Saranya and K. V. Radha, *Polym.-Plast. Technol. Eng.*, 2014, **53**, 1636.
71. D. M. Zalk, S. Y. Paik and P. Swuste, *J. Nanopart. Res.*, 2009, **11**, 1685.
72. J. Höck, T. Epprecht, E. Furrer, M. Gautschi, H. Hofmann, K. Höhener, K. Knauer, H. Krug, L. Limbach, P. Gehr, B. Nowack, M. Riediker, K. Schirmer, K. Schmid, C. Som, W. Stark, C. Studer, A. Ulrich, N. von Götz, A. Weber, S. Wengert and P. Wick, *Guidelines on the Precautionary Matrix for Synthetic Nanomaterials, Version 3.0*, Federal Office of Public Health and Federal Office for the Environment, Berne, 2013.
73. B. Van Duuren-Stuurman, S. R. Vink, K. J. Verbist, H. G. Heussen, D. H. Brouwer, D. E. Kroese, E. Tielemans and W. Fransman, *Ann. Occup. Hyg.*, 2012, 525.
74. J. M. Davis, *J. Nanosci. Nanotechnol.*, 2007, 7, 402.
75. D. R. Hristozov, S. Gottardo, M. Cinelli, P. Isigonis, A. Zabeo, A. Critto, M. Van Tongeren, L. Tran and A. Marcomini, *Nanotoxicology*, 2014, **8**, 117.
76. D. R. Hristozov, A. Zabeo, C. Foran, P. Isigonis, A. Critto, A. Marcomini and I. Linkov, *Nanotoxicology*, 2014, **8**, 72.
77. I. Linkov, F. K. Satterstrom, J. Steevens, E. Ferguson and R. C. Pleus, *J. Nanopart. Res.*, 2007, **9**, 543.

78. T. Tervonen, I. Linkov, J. R. Figueira, J. Steevens, M. Chappell and M. Merad, *J. Nanopart. Res.*, 2009, **11**, 757.
79. I. Linkov, M. E. Bates, L. J. Canis, T. P. Seager and J. M. Keisler, *Nat. Nanotechnol.*, 2011, **6**, 784.
80. S. F. Hansen, K. A. Jensen and A. Baun, *J. Nanopart. Res.*, 2014, **16**, 1.
81. H. J. Marvin, H. Bouwmeester, M. Bakker, E. D. Kroese, D. van de Meent, F. Bourgeois, R. Lokers, H. van der Ham and L. Verhelst, *J. Nanopart. Res.*, 2013, **15**, 1.
82. H.-Y. Kim, S. S. Park and S.-T. Lim, *Colloids Surf. B*, 2015, **126**, 607.
83. K. Oksman, A. P. Mathew, D. Bondeson and I. Kvien, *Compos. Sci. Technol.*, 2006, **66**, 2776.
84. A. Sorrentino, G. Gorrasi and V. Vittoria, *Trends Food Sci. Technol.*, 2007, **18**, 84.
85. X. Ma, J. Yu and J. F. Kennedy, *Carbohydr. Polym.*, 2005, **62**, 19.
86. X. Z. Mo, Y. X. Zhong, C. Q. Liang and S. J. Yu, *Adv. Mater. Res.*, 2010, **87**, 439.
87. K. G. Satyanarayana, G. G. Arizaga and F. Wypych, *Prog. Polym. Sci.*, 2009, **34**, 982.
88. M. G. Lomelí-Ramírez, S. G. Kestur, R. Manríquez-González, S. Iwakiri, G. B. de Muniz and T. S. Flores-Sahagun, *Carbohydr. Polym.*, 2014, **102**, 576.
89. W. S. Ratnayake and D. S. Jackson, in *Advances in Food and Nutrition Research*, ed. L. T. Steve, Academic Press, London, 2008, vol. 55, p. 221.
90. J. W. Donovan, *Biopolymers*, 1979, **18**, 263.
91. L. S. Lai and J. L. Kokini, *Biotechnol. Prog.*, 1991, **7**, 251.
92. F. Xie, H. Liu, P. Chen, T. Xue, L. Chen, L. Yu and P. Corrigan, *Int. J. Food Eng.*, 2007, **2**, 1.
93. L. Yu, T. Kealy and P. Chen, *Int. Polym. Process.*, 2006, **21**, 283.
94. P. Chen, L. Yu, T. Kealy, L. Chen and L. Li, *Carbohydr. Polym.*, 2007, **68**, 495.
95. F. Xie, L. Yu, L. Chen and L. Li, *Carbohydr. Polym.*, 2008, **72**, 229.
96. H. Liu, F. Xie, L. Yu, L. Chen and L. Li, *Prog. Polym. Sci.*, 2009, **34**, 1348.
97. A. P. Kumar, D. Depan, N. Singh Tomer and R. P. Singh, *Prog. Polym. Sci.*, 2009, **34**, 479.
98. F. Annabi-Bergaya, *Microporous Mesoporous Mater.*, 2008, **107**, 141.
99. E. Duquesne, S. Moins, M. Alexandre and P. Dubois, *Macromol. Chem. Phys.*, 2007, **208**, 2542.
100. S. Sinha Ray, K. Okamoto and M. Okamoto, *Macromolecules*, 2003, **36**, 2355.
101. L. Avérous and P. J. Halley, *Biofuels, Bioprod. Biorefin.*, 2009, **3**, 329.
102. M. Alexandre and P. Dubois, *Mater. Sci. Eng., R*, 2000, **28**, 1.
103. S. Sinha Ray and M. Okamoto, *Prog. Polym. Sci.*, 2003, **28**, 1539.
104. S. Sinha Ray and M. Bousmina, *Prog. Mater. Sci.*, 2005, **50**, 962.
105. D. R. Paul and L. M. Robeson, *Polymer*, 2008, **49**, 3187.
106. B. S. Harrison and A. Atala, *Biomaterials*, 2007, **28**, 344.

107. J. Wang, *Electroanalysis*, 2005, **17**, 7.
108. E. Lahiff, C. Lynam, N. Gilmartin, R. O'Kennedy and D. Diamond, *Anal. Bioanal. Chem.*, 2010, **398**, 1575.
109. Y. Xiao and C. M. Li, *Electroanalysis*, 2008, **20**, 648.
110. L. M. Famá, V. Pettarin, S. N. Goyanes and C. R. Bernal, *Carbohydr. Polym.*, 2011, **83**, 1226.
111. M. G. Valdés, A. C. V. González, J. A. G. Calzón and M. E. Díaz-García, *Microchim. Acta*, 2009, **166**, 1.
112. Z. Spitalsky, D. Tasis, K. Papagelis and C. Galiotis, *Prog. Polym. Sci.*, 2010, **35**, 357.
113. X. Cao, Y. Chen, P. R. Chang and M. A. Huneault, *J. Appl. Polym. Sci.*, 2007, **106**, 1431.
114. X. Ma, J. Yu and N. Wang, *Compos. Sci. Technol.*, 2008, **68**, 268.
115. Z. Liu, L. Zhao, M. Chen and J. Yu, *Carbohydr. Polym.*, 2011, **83**, 447.
116. S. A. Curran, P. M. Ajayan, W. J. Blau, D. L. Carroll, J. N. Coleman, A. B. Dalton, A. P. Davey, A. Drury, B. McCarthy and S. Maier, *Adv. Mater.*, 1998, **10**, 1091.
117. P. Ajayan, *Chem. Rev.*, 1999, **99**, 1787.
118. H. Qi, K. Teo, K. Lau, M. Boyce, W. Milne, J. Robertson and K. Gleason, *J. Mech. Phys. Solids*, 2003, **51**, 2213.
119. H. Kim and C. W. Macosko, *Polymer*, 2009, **50**, 3797.
120. T. Wei, G. Luo, Z. Fan, C. Zheng, J. Yan, C. Yao, W. Li and C. Zhang, *Carbon*, 2009, **47**, 2296.
121. L. Jiang, X. P. Shen, J. L. Wu and K. C. Shen, *J. Appl. Polym. Sci.*, 2010, **118**, 275.
122. R. Li, C. Liu and J. Ma, *Carbohydr. Polym.*, 2011, **84**, 631.
123. D. Y. Godovsky, in *Biopolymers: PVA Hydrogels, Anionic Polymerisation, Nanocomposites*, Advances in Polymer Science, Springer, Berlin, 2000, vol. 153, pp. 163–206.
124. E. Ruiz-Hitzky and F. M. Fernandes, *Prog. Polym. Sci.*, 2013, **38**, 1391.
125. B. Kaur, F. Ariffin, R. Bhat and A. A. Karim, *Food Hydrocolloids*, 2012, **26**, 398.
126. D. Le Corre and H. Angellier-Coussy, *React. Funct. Polym.*, 2014, **85**, 97.
127. D. J. Gallant, B. Bouchet and P. M. Baldwin, *Carbohydr. Polym.*, 1997, **32**, 177.
128. H. Tang, T. Mitsunaga and Y. Kawamura, *Carbohydr. Polym.*, 2006, **63**, 555.
129. N. L. García, L. Ribba, A. Dufresne, M. Aranguren and S. Goyanes, *Carbohydr. Polym.*, 2011, **84**, 203.
130. H. Namazi and A. Dadkhah, *Carbohydr. Polym.*, 2010, **79**, 731.
131. J.-L. Putaux, S. Molina-Boisseau, T. Momaur and A. Dufresne, *Biomacromolecules*, 2003, **4**, 1198.
132. J.-Y. Kim, D.-J. Park and S.-T. Lim, *Cereal Chem.*, 2008, **85**, 182.
133. D. Liu, Q. Wu, H. Chen and P. R. Chang, *J. Colloid Interface Sci.*, 2009, **339**, 117.
134. S. Bel Haaj, A. Magnin, C. Pétrier and S. Boufi, *Carbohydr. Polym.*, 2013, **92**, 1625.

135. N. Lin, J. Huang, P. R. Chang, L. Feng and J. Yu, *Colloids Surf., B*, 2011, **85**, 270.
136. P. Wongpanit, N. Sanchavanakit, P. Pavasant, T. Bunaprasert, Y. Tabata and R. Rujiravanit, *Eur. Polym. J.*, 2007, **43**, 4123.
137. E. Kristo and C. G. Biliaderis, *Carbohydr. Polym.*, 2007, **68**, 146.
138. H. Angellier, S. Molina-Boisseau, L. Lebrun and A. Dufresne, *Macromolecules*, 2005, **38**, 3783.
139. D. b. Le Corre, J. Bras and A. Dufresne, *Biomacromolecules*, 2010, **11**, 1139.
140. Y. Tan, K. Xu, L. Li, C. Liu, C. Song and P. Wang, *ACS Appl. Mater. Interfaces*, 2009, **1**, 956.
141. R. Hoover, *Food Rev. Int.*, 2000, **16**, 369.
142. L. Jayakody and R. Hoover, *Food Res. Int.*, 2002, **35**, 665.
143. H. Angellier, J.-L. Putaux, S. Molina-Boisseau, D. Dupeyre and A. Dufresne, *Macromol. Symp.*, 2005, **221**, 95.
144. A. Imberty, H. Chanzy, S. Perez, A. Buleon and V. Tran, *Macromolecules*, 1987, **20**, 2634.
145. K. Kainuma and D. French, *Biopolymers*, 1971, **10**, 1673.
146. D. LeCorre, J. Bras and A. Dufresne, *J. Nanopart. Res.*, 2011, **13**, 7193.
147. N. Lin, J. Huang, P. R. Chang, D. P. Anderson and J. Yu, *J. Nanomater.*, 2011, 20.
148. H.-Y. Kim, J. H. Lee, J.-Y. Kim, W.-J. Lim and S.-T. Lim, *Starch Stärke*, 2012, **64**, 367.
149. D. LeCorre, E. Vahanian, A. Dufresne and J. Bras, *Biomacromolecules*, 2011, **13**, 132.
150. L. Martín-Banderas, M. Flores-Mosquera, P. Riesco-Chueca, A. Rodríguez-Gil, Á. Cebolla, S. Chávez and A. M. Gañán-Calvo, *Small*, 2005, **1**, 688.
151. M. S. Hale and J. G. Mitchell, *Nano Lett.*, 2001, **1**, 617.
152. D. Song, Y. S. Thio and Y. Deng, *Carbohydr. Polym.*, 2011, **85**, 208.
153. J. Y. Zuo, K. Knoerzer, R. Mawson, S. Kentish and M. Ashokkumar, *Ultrason. Sonochem.*, 2009, **16**, 462.
154. P. Tomasik and M. F. Zaranyika, in *Advances in Carbohydrate Chemistry and Biochemistry*, ed. D. Horton, Academic Press, San Diego, CA, 1995, vol. 51, pp. 243–318.
155. L. S. Lai and J. L. Kokini, *J. Rheol.*, 1990, **34**, 1245.
156. M. Lamanna, N. J. Morales, N. L. García and S. Goyanes, *Carbohydr. Polym.*, 2013, **97**, 90.
157. S. Singh, N. Singh, R. Ezekiel and A. Kaur, *Carbohydr. Polym.*, 2011, **83**, 1521.
158. Y. Yu and J. Wang, *Food Res, Int.*, 2007, **40**, 297.
159. N. L. García, M. Lamanna, N. D'Accorso, A. Dufresne, M. Aranguren and S. Goyanes, *Polym. Degrad. Stab.*, 2012, **97**, 2021.
160. H. T. Kim, E. J. Yun, D. Wang, J. H. Chung, I.-G. Choi and K. H. Kim, *Bioresour. Technol.*, 2013, **136**, 582.
161. D. Cooke and M. J. Gidley, *Carbohydr. Res.*, 1992, **227**, 103.

162. H. Angellier, L. Choisnard, S. Molina-Boisseau, P. Ozil and A. Dufresne, *Biomacromolecules*, 2004, **5**, 1545.
163. H.-Y. Kim, D. J. Park, J.-Y. Kim and S.-T. Lim, *Carbohydr. Polym.*, 2013, **98**, 295.
164. A. P. Mathew and A. Dufresne, *Biomacromolecules*, 2002, **3**, 609.
165. X. Ma, R. Jian, P. R. Chang and J. Yu, *Biomacromolecules*, 2008, **9**, 3314.
166. J.-Y. Kim and S.-T. Lim, *Carbohydr. Polym.*, 2009, **76**, 110.
167. Q. Sun, M. Gong, Y. Li and L. Xiong, *Carbohydr. Polym.*, 2014, **111**, 133.
168. A.-m. Shi, D. Li, L.-j. Wang, B.-z. Li and B. Adhikari, *Carbohydr. Polym.*, 2011, **83**, 1604.
169. G. Zhou, Z. Luo and X. Fu, *Ind. Crops Prod.*, 2014, **52**, 105.
170. G. Zhou, Z. Luo and X. Fu, *J. Agric. Food Chem.*, 2014, **62**, 8214.
171. S. M. Jafari, E. Assadpoor, Y. He and B. Bhandari, *Food Hydrocolloids*, 2008, **22**, 1191.
172. B. Wei, X. Hu, H. Li, C. Wu, X. Xu, Z. Jin and Y. Tian, *Food Hydrocolloids*, 2014, **36**, 369.
173. L. Ren, M. Jiang, L. Wang, J. Zhou and J. Tong, *Carbohydr. Polym.*, 2012, **87**, 1874.
174. J. Canselier, H. Delmas, A. Wilhelm and B. Abismail, *J. Dispersion Sci. Technol.*, 2002, **23**, 333.
175. B. Duan, P. Sun, X. Wang and C. Yang, *Starch Stärke*, 2011, **63**, 528.
176. D. LeCorre, J. Bras and A. Dufresne, *Carbohydr. Polym.*, 2012, **87**, 658.
177. H.-M. Lai, *Food Chem.*, 2001, **72**, 455.
178. Z. G. Luo, X. Fu, Q. Y. Gao and S. J. Yu, *Int. J. Food Sci. Technol.*, 2011, **46**, 429.
179. T. A. Waigh, M. J. Gidley, B. U. Komanshek and A. M. Donald, *Carbohydr. Res.*, 2000, **328**, 165.
180. H. Angellier-Coussy, J.-L. Putaux, S. Molina-Boisseau, A. Dufresne, E. Bertoft and S. Perez, *Carbohydr. Res.*, 2009, **344**, 1558.
181. S. Hizukuri, *Carbohydr. Res.*, 1986, **147**, 342.
182. H. Zheng, F. Ai, P. R. Chang, J. Huang and A. Dufresne, *Polym. Compos.*, 2009, **30**, 474.
183. D. W. Schaefer and R. S. Justice, *Macromolecules*, 2007, **40**, 8501.
184. R. J. Moon, A. Martini, J. Nairn, J. Simonsen and J. Youngblood, *Chem. Soc. Rev.*, 2011, **40**, 3941.
185. G. Chen, M. Wei, J. Chen, J. Huang, A. Dufresne and P. R. Chang, *Polymer*, 2008, **49**, 1860.
186. H. Angellier, S. Molina-Boisseau, P. Dole and A. Dufresne, *Biomacromolecules*, 2006, **7**, 531.
187. J. Yu, F. Ai, A. Dufresne, S. Gao, J. Huang and P. R. Chang, *Macromol. Mater. Eng.*, 2008, **293**, 763.
188. M. Arroyo, M. Lopez-Manchado and B. Herrero, *Polymer*, 2003, **44**, 2447.
189. D. Dubief, E. Samain and A. Dufresne, *Macromolecules*, 1999, **32**, 5765.

190. B. Jiang, X. Chen, Z. Huang, Z. Zhang, S. He, S. Wang, Z. Wu and F. Qiu, *Front. Med. China*, 2007, **1**, 24–29.
191. K. K. H. Wong, M. Zinke-Allmang, J. L. Hutter, S. Hrapovic, J. H. Luong and W. Wan, *Carbon*, 2009, **47**, 2571.
192. B. P. Binks and S. O. Lumsdon, *Langmuir*, 2000, **16**, 8622.
193. B. P. Binks and C. P. Whitby, *Langmuir*, 2004, **20**, 1130.
194. B. Midmore, *Colloids Surf., A*, 1998, **132**, 257.
195. N. P. Ashby and B. P. Binks, *Phys. Chem. Chem. Phys.*, 2000, **2**, 5640.
196. Y. Nonomura and N. Kobayashi, *J. Colloid Interface Sci.*, 2009, **330**, 463.
197. M. Shen and D. E. Resasco, *Langmuir*, 2009, **25**, 10843.
198. N. P. Ashby, B. P. Binks and V. N. Paunov, *Chem. Commun.*, 2004, 436.
199. B. P. Binks and S. O. Lumsdon, *Langmuir*, 2001, **17**, 4540.
200. M. Andresen and P. Stenius, *J. Dispersion Sci. Technol.*, 2007, **28**, 837.
201. I. Kalashnikova, H. Bizot, B. Cathala and I. Capron, *Langmuir*, 2011, **27**, 7471.
202. C. Li, P. Sun and C. Yang, *Starch Stärke*, 2012, **64**, 497.
203. P. A. Lucca and B. J. Tepper, *Trends Food Sci. Technol.*, 1994, **5**, 12.
204. J. Jane, L. Shen, L. Wang and C. Maningat, *Cereal Chem.*, 1992, **69**, 280.
205. J. Daniel and R. Whistler, *Cereal Foods World*, 1990, **35**, 321.
206. N. L. García, L. Ribba, A. Dufresne, M. I. Aranguren and S. Goyanes, *Macromol. Mater. Eng.*, 2009, **294**, 169.
207. H. Angellier, S. Molina-Boisseau and A. Dufresne, *Macromolecules*, 2005, **38**, 9161.
208. J. Kreuter, *J. Controlled Release*, 1991, **16**, 169.
209. S. Xiao, C. Tong, X. Liu, D. Yu, Q. Liu, C. Xue, D. Tang and L. Zhao, *Chin. Sci. Bull.*, 2006, **51**, 1693.
210. M. Santander-Ortega, T. Stauner, B. Loretz, J. Ortega-Vinuesa, D. Bastos-González, G. Wenz, U. Schaefer and C. Lehr, *J. Controlled Release*, 2010, **141**, 85.
211. B.-Z. Weiner, M. Tahan and A. Zilkha, *J. Med. Chem.*, 1972, **15**, 410.
212. P. Stjärnkvist, T. Laakso and I. Sjöholm, *J. Pharm. Sci.*, 1989, **78**, 52.
213. C.-Y. Won, C.-C. Chu and T.-J. Yu, *Carbohydr. Polym.*, 1997, **32**, 239.
214. V. Lenaerts, I. Moussa, Y. Dumoulin, F. Mebsout, F. Chouinard, P. Szabo, M. Mateescu, L. Cartilier and R. Marchessault, *J. Controlled Release*, 1998, **53**, 225.
215. S. Xiao, X. Liu, C. Tong, L. Zhao, X. Liu, A. Zhou and Y. Cao, *Chin. Sci. Bull.*, 2012, **57**, 3226.
216. A. Nel, T. Xia, L. Mädler and N. Li, *Science*, 2006, **311**, 622.
217. D. S. Brodke, S. Gollogly, R. Alexander Mohr, B.-K. Nguyen, A. T. Dailey and K. N. Bachus, *Spine*, 2001, **26**, 1324.
218. M. C. Kennady, M. R. Tucker, G. E. Lester and M. J. Buckley, *Int. J. Oral Surg.*, 1989, **18**, 170.
219. M. Valodkar and S. Thakore, *Carbohydr. Res.*, 2010, **345**, 2354.
220. V. Prasad, A. Shaikh, A. Kathe, D. Bisoyi, A. Verma and N. Vigneshwaran, *J. Mater. Process. Technol.*, 2010, **210**, 1962.

CHAPTER 11

# Effects of Additives on the Properties of Starch

WEI WANG,[a] HONG YANG*[a,b,c,d] AND MIN CUI*[e,f]

[a] College of Food Science and Technology, Huazhong Agricultural University, Wuhan, Hubei 430070, China; [b] Key Laboratory of Environment Correlative Dietology, Huazhong Agricultural University, Ministry of Education, Wuhan, Hubei 430070, China; [c] National R&D Branch Center for Conventional Freshwater Fish Processing (Wuhan), Wuhan, Hubei 430070, China; [d] Aquatic Product Engineering and Technology Research Center of Hubei Province, Wuhan, Hubei 430070, China; [e] State Key Laboratory of Agricultural Microbiology, Huazhong Agricultural University, Wuhan, Hubei 430070, China; [f] Laboratory of Animal Virology, College of Veterinary Medicine, Huazhong Agricultural University, Wuhan, Hubei 430070, China
*Email: yangh@mail.hzau.edu.cn; cuimin@mail.hzau.edu.cn

## 11.1 Introduction

Native starches suffer from problems such as shear-thinning and syneresis, and generally do not have ideal properties for the preparation of processed food products. Usually, native starches are chemically modified to improve their tolerance to processing conditions (*e.g.* heat, shear and acidic), enhance their texture, provide cold storage and freeze–thaw stability, control water mobility, *etc.*[1–7] However, addition of a suitable additive can also overcome the shortcomings of native starches. For example, hydrocolloids can protect starch granules against shearing during heating or cooking, improve product texture, hold moisture and protect against syneresis.[8]

RSC Green Chemistry No. 37
Starch-based Blends, Composites and Nanocomposites
Edited by Visakh P. M. and Long Yu

Published by the Royal Society of Chemistry, www.rsc.org

Starch–additive combinations are also considered to be relatively clean and safe and more readily accepted by consumers compared with modified starches.

Proteins, lipids and starches, three macromolecules, have been widely studied in the past decade, including their combinations and blends. Recently, there has been some research on the interaction of starch and small molecules, such as salts, sugars and amino acids. This chapter provides an overview of recent progress in the interactions between starch and its additives, including the effects of additives on the properties of starch, the mechanisms of the interactions between starch and additives and the recent applications of additives in starch-based products.

## 11.2 Starch–Hydrocolloid Blends

### 11.2.1 Introduction

Non-starch polysaccharides are usually called hydrocolloids or food gums.[8] Food hydrocolloids have been widely applied in the food industry as thickeners, gelling agents, coating agents, stabilizers and emulsifiers. The amount of food hydrocolloids added to foods is usually very small compared with other ingredients, but they can improve the performance of the food system effectively.[9,10] Gums can compensate the shortcomings of native starch, hence reducing the need for chemically modified starch.[8] The sources of hydrocolloids can be roughly divided into four categories: plant, animal, microbial and semisynthetic. For example, gum arabic is made of hardened sap taken from the acacia tree, and guar gum is primarily the ground endosperm of guar beans.[11] Pectin is extracted from the peel and pomace of fruits,[12] and agar and carrageenan are extracted from seaweed.[13] Gelatin is produced by hydrolysis of proteins of bovine and fish origins. Xanthan gum is a kind of acidic extracellular heteropolysaccharide produced by *Xanthomonas campestris.*[14] Other important hydrocolloids are cellulose derivatives, such as carboxymethylcellulose (CMC).[15] Since both starch and hydrocolloids are polysaccharide molecules with a certain degree of similarity, hydrocolloids can affect the gelatinization and retrogradation behaviors of starch and improve the textural properties and freeze–thaw stability of starchy foods.[8] However, different hydrocolloids with diverse structures, molecular weights and ionic charges would have different effects on the properties of starch.[16,17]

### 11.2.2 Characteristics of Starch–Hydrocolloid Blends

#### *11.2.2.1 Effects of Hydrocolloids on the Pasting Properties of Starch*

Hydrocolloids, in most cases, inhibit the swelling of starch granules and leaching of starch polymers, primarily amylose.[8] With the addition of okra gum, wheat and corn starch granules exhibited a smaller size during heating.[18] Xanthan gum also inhibited the swelling of potato and waxy maize starch granules and maintained their integrity as a physical barrier.[14,16,19] Pullulan, which is produced by *Aureobasidium pullulans*, has the same effect as xanthan gum.[20] There was a significant delay of starch granule swelling when CMC and sodium alginate were added to water chestnut starch.[21] The swelling inhibitory effect of CMC was greater than that that of sodium alginate on potato, sweet potato and taro starches.[22] With the increase in the amount of corn fiber gum, there was a reduction in the swelling power of and the amount of amylose leached from maize starch gels.[23] The swollen granules in rice starch gel with konjac glucomannan (KGM) were smaller than those without KGM.[24] The viscous phase surrounding the swollen starch granules might probably be due to KGM interacting with leached amylose. Carrageenans were both excluded and adsorbed on crosslinked waxy maize starch granules during gelatinization, and the percentage of trapped κ-carrageenan was higher than that of ι-carrageenan.[25] The addition of locust bean gum had little influence on the swelling of modified waxy maize starch granules.[26]

Generally, swelling power has a negative relationship with peak viscosity and breakdown of starch paste, which are related to the stability of the granules to retain their integrity.[27] However, the addition of hydrocolloids could make such a trend more complex. A decrease in peak viscosity of native potato starch or anionic modified potato starch with negatively charged xanthan gum was observed, but an increase in that of cationic modified potato starch with xanthan gum.[19] Heyman *et al.* also reported a decreasing trend in the viscosity of waxy potato, waxy maize and pea starches in the presence of xanthan gum,[14] because xanthan gum inhibited granule swelling in most cases.[8] The addition of CMC and sodium alginate to water chestnut starch caused a significant decrease in the peak viscosity.[21] The viscosity of rice flour was greatly reduced by β-glucan.[28] Corn fiber gum lowered the peak viscosity and breakdown of maize starch, but increased the final viscosity.[23]

However, there have also been a few of reports of hydrocolloids enhancing granule swelling. Rosell *et al.* found that xanthan gum induced a steady increase in the swelling power of rice starch, likely due to the shear force exerted by the hydrocolloid on the surface of the swollen starch granules or the osmotic pressure exerted by the continuous hydrocolloid phase on the starch phase.[27] Mandala and Bayas also observed a similar effect of xanthan gun on wheat starch.[29] Higher concentrations of KGM resulted in higher values of peak viscosity, breakdown, final viscosity and setback in rice starch paste.[24] There were large increases in pasting peak viscosity, final viscosity and breakdown values of tapioca starch with the increase in xanthan gum.[30] Because guar gum did not have the ability to cover the surface of starch granules like xanthan gum, the majority of the guar gum was present in the continuous phase, which led to an increase in the viscosity of the native waxy maize starch.[6] After adding guar gum, the cooked potato matrix maintained its viscosity similarly to an undigested cooked potato sample throughout *in vitro* digestion, due to lower starch hydrolysis.[31,32] Guar and xanthan gums induced an increase in the viscosity of rice starch, and such an increase in the viscosity of guar– or xanthan gum–rice starch mixtures started much earlier than that of starch alone.[27] There was no change in the paste viscosity during holding at 95 °C in the xanthan gum–rice starch mixture, which indicated that xanthan gum stabilized the integrity of the starch granules during heating.[27] Hydroxypropylmethylcellulose (HPMC) did not modify the pasting properties of rice starch during heating, but increased the viscosity during cooling.[27] Sometimes, the viscosity behavior of starch solution is also associated with hydrocolloid concentration. The viscosity of wheat and corn starches increased with addition of low levels of okra gum to improve the thickening ability, but decreased at higher levels.[18]

The pasting temperature of potato and tapioca starches increased on adding xanthan gum,[19,33] but the addition of xanthan gum did not change the gelatinization enthalpy and transition pattern of starch according to the differential scanning calorimetry (DSC) thermogram.[33] Xanthan gum did not affect the gelatinization of waxy starch.[14] Wheat and corn starches showed higher peak temperatures in the presence of okra gum, indicating slowing of their gelatinization.[18] There were significant increases in gelatinization temperature and enthalpy of sorghum starch with the addition of cactus (*Opuntia ficus-indica*) mucilage.[23] There was delayed gelatinization on adding CMC and sodium alginate to water chestnut starch.[21] In the presence of KGM, the enthalpies of potato and broad bean starches were decreased, whereas their gelatinization peaks were shifted to higher temperatures.[34] Nonetheless, the gelatinization temperature of rice flour with β-glucan was decreased, even though β-glucan reduced starch gelatinization.[28]

### *11.2.2.2 Effects of Hydrocolloids on Gel Texture and Freeze–Thaw Stability of Blends*

Hydrocolloids are often added to the starch to improve the taste of foods due to their great performance with respect to gelatinization and transparency. However, as polysaccharide macromolecules, mixtures of starch and hydrocolloids can suffer thermodynamic incompatibility and phase separation phenomenon, especially in the gel system.[8] Galactomannan can weaken the network structure because galactomannan and amylose molecules are incompatible.[35] In all binary and tertiary component gels formed by mixing gelatin or gellan with maize starch or egg white, antagonistic effects were observed, and the composite gels were weaker and more brittle than the single-component gels.[36] Gellan composite gels were comparatively weaker and showed more viscous behavior compared with gelatin-based gels.

The texture profiles of starch–hydrocolloid mixed gels depend not only on the varieties of starch and hydrocolloids but also on the concentration of hydrocolloids. Only a suitable hydrocolloid at a proper concentration can endow the gel with excellent qualities.[8] Five non-starch polysaccharides (guar gum, xanthan gum, CMC, tapioca fiber and tamarind seed fiber) contributed to the textural modification of rice starch gels, resulting in higher gel hardness and adhesiveness values, especially for the gels with guar gum and CMC.[37] There was an improvement in the mechanical strength of corn starch film with increase in CMC content.[15] However, CMC had a negative impact on the inner structure of noodles and reduced the cohesiveness of cooked instant noodles.[38] Guar and xanthan gums resulted in decreased elasticity and lowered resistance to deformation of gels. The incorporation of HPMC resulted in a more elastic network but a low resistance to deformation. These hydrocolloids led to weaker structure with less gel-like character than rice starch alone.[27] Kappa-type carrageenan and sodium ions greatly affected the rigidity of crosslinked waxy maize starch gel.[39] Low concentrations ($<0.2\%$ w/w) of carrageenan and high concentrations of gellan ($>0.3\%$ w/w) could significantly increase the hardness and adhesiveness of rice starch gels.[40] KGM could not improve the adhesiveness, chewiness and hardness of rice starch gel.[41] The rice starch gel with KGM was softer than the starch gel alone, which clearly indicated that KGM retarded the aggregation of swollen granules of starch.[24] A small amount of $Na_2CO_3$ (0.1–0.2 wt% based on starch) resulted in increased elasticity of KGM-induced wheat starch gel but did not affect the structural order during retrogradation.[42] This indicated that $Na_2CO_3$ could promote strong interchain association between KGM and starch. When cactus mucilage was mixed with sorghum starch to prepare gels, it resulted in enhancement of the gel structure, probably due to hydrogen bonds in starch molecules, followed by a higher mucilage–starch interaction with water.[23] There was an increase in the plasticity on addition of either κ-carrageenan, guar gum or xanthan gum to potato starch gel.[43] However, the plasticity of the mixed gels was lost to a significant extent within the first day of storage, resulting in an elastically characteristic gel, which then rheologically and texturally stabilized for subsequent 30-day storage.

Hydrocolloids decrease or prevent the instability of gels during storage or freeze–thaw cycles. KGM retards the rice starch gel retrogradation induced by freeze–thaw treatment.[24] The effect of hydrocolloids on the stabilization of potato starch gel depended on the gel concentration during long-term storage. For example, hydrocolloids had such an effect only in the case of 3% starch gels. They essentially increased the gel strength, which decreased during storage. However, such an effect became insignificant in the case of 4 and 5% gels.[43] The zeta potential of xanthan gum and potato starch combination was significantly higher than that of potato starch alone, which implied that the combination had greater electronegativity, stability and resistance to retrogradation.[19] During room temperature storage, potato starch gel with xanthan gum was more stable than the gel with guar gum or κ-carrageenan.[44] Xanthan gum enhanced the retrogradation of tapioca starch during the initial storage but retarded the process on further storage at 5 °C. Tapioca starch–xanthan gum gel kept for 14 days showed lower

Young's modulus than starch gel alone, confirming the retardation of retrogradation by xanthan gum.[33] In general, hydrocolloids promoted short-term retrogradation but retarded long-term retrogradation. The addition of pullulan (*Aureobasidium pullulans*) to rice starch not only inhibited the short-term retrogradation of amylose, but also retarded and decelerated the long-term retrogradation of amylopectin.[45] Hypothetically, the effect of hydrocolloids on the reduction of syneresis is due to the retardation of amylose leaching, the increase in viscosity of the continuous phase and the interaction between certain leached amylose and gums.[8,24] In general, hydrocolloids can extend the shelf-life of starch-based products.

#### *11.2.2.3 Effects of Hydrocolloids on the Rheological Properties of Blends*

In mixed systems of starch, hydrocolloid and water, the rheological properties of starch-based paste are strongly determined by the integrity of the granules within the hydrocolloid solution.[8] When the granules are more disrupted, the rheological properties depend on the phase behavior of the hydrocolloid and the dissolved starch molecules. The presence of xanthan gum resulted in increases in the consistency index and vane yield stress of sweet potato starch.[46] With increase in xanthan gum concentration, the dynamic moduli of sweet potato starch–xanthan gum mixture increased but tan $\delta$ decreased. The addition of xanthan gum appeared to contribute to the elasticity of the weak network of the starch paste. The presence of xanthan gum strongly influenced the structure development rate constant ($k$) of gels during aging, which suggests that the phase separation process caused by the incompatibility phenomena between the amylose in starch and xanthan gum could increase the elastic characteristics of the starch–xanthan gum mixture.

Xanthan gum increased the storage modulus ($G'$) and loss modulus ($G''$) of tapioca starch dispersion during gelatinization.[30] The maximum $G'$ increased with increase in xanthan gum content but the maximum of $G''$ was not so dependent on xanthan gum. After heating at 95 °C, the paste of tapioca starch–xanthan gum tended to be more solid-like with increasing xanthan gum concentration. The $G'$ of tapioca starch–xanthan paste increased at 10 °C, indicating network formation. Furthermore, xanthan gum increased the thermostability of the steady shear viscosity of tapioca starch gels with and without sucrose.[47]

Crosslinked tapioca starch–xanthan gum mixture was used as a thickener in commercial chili sauce.[48] The sauce with xanthan gum showed enhanced flow characteristics, with an undetectable yield stress, the lowest consistency coefficient and the highest flow behavior index. Guar gum primarily modified the rheological properties of waxy maize starch paste by its presence in the continuous phase, whereas xanthan gum also induced indirect effects by better preserving the granular structure.[16] Association of xanthan gum with the granule surface led to a stabilizing effect during shearing.[16] Compared with starch alone, the addition of guar gum reduced the yield stress, but xanthan gum significantly increased it. Xanthan gum solution led to a weak gel owing to its rigid conformation and molecular entanglements. Moreover, xanthan gum could enhance the granule interaction of waxy maize starch.[49] An increase in guar gum in the continuous phase resulted in encumbering granule–granule interactions and possibly lowering of the yield stress. The consistency index was strongly increased by guar gum, but only moderately by xanthan gum.[49,50]

Regarding the dynamic viscoelasticity of mixed gels, xanthan gum and agar increased the $G'$ values of rice starch by blending, whereas KGM decreased $G'$.[51] Okra gum–wheat or –corn starch blends were pseudoplastic materials irrespective of the blend composition.[18] The presence of okra gum led to lower viscosity and instigated more pseudoplasticity in the starch-based blend. In another words, a greater addition of okra gum caused more shear-thinning behavior. Flow curves for all okra gum–starch blends showed narrow hysteresis loops, indicating small changes in the structure of the gels in the applied shear range. The $G'$ and $G''$ values of maize starch were increased when corn fiber gum was added.[23]

There was prolonged softness of zein–maize starch dough that was supplemented with HPMC or oat bran with a high content of β-glucan. Those doughs exhibited a reduction in age-related stiffening, highly extensional viscosity and strain hardening, suggesting appropriate rheological properties for bread making performance.[52] The greater resistance of β-glucan-incorporated rice dough might be useful in minimizing the mechanical damage of products during transport and handling.[28]

#### 11.2.2.4 *Effects of Hydrocolloids on the Functional Properties of Blends*

Starch is an important ingredient of staple foods, such as pasta and noodles. The mechanism of the glycemic response, in fact, involves enzymatic hydrolysis of starch in the human digestive process. Hydrocolloids are able to influence the oil-binding capacity, water-binding capacity and digestibility of starch. Sodium alginate and CMC enhanced the lipophilic character of water chestnut starch.[21] Addition of sodium alginate increased the oil-binding capacity of starch to a greater extent compared with starch with CMC. The functional properties of starch, such as oil- and water-binding capacity, could be increased significantly with addition of hydrocolloids, such as sodium alginate and CMC.[22]

The addition of three non-starch polysaccharides (NSPs) (agar, xanthan gum and KGM) suppressed rice starch hydrolysis in the gels compared with starch alone. At the same concentration, agar had a lower suppressing effect on starch hydrolysis than xanthan gum and KGM.[51] The suppressing effect of NSPs on starch hydrolysis in starch gels was due not only to the rigidity of the gel, but also to the interaction between starch and non-starch polysaccharides.[51] Guar gum decreased the enzymatic hydrolysis and glycemic index of potato starch.[32] Addition of guar gum to cooked potato reduced its hydrolysis during *in vitro* digestion.[31] When guar gum, xanthan gum, CMC, tapioca fiber and tamarind seed fiber were added to waxy and non-waxy rice starches, all non-starch polysaccharides had little or no effect on starch digestibility at the concentrations applied.[37]

### 11.2.3 Mechanisms of Starch–Hydrocolloid Interactions

The effect of hydrocolloids on starch is dependent on both the origin and amount of starch and hydrocolloid. The amount of amylose in the starch makes a difference to the properties of the composite system.[53] The effects of guar and xanthan gums on starch have frequently been compared because of the nature of their respective differences.[16,54] Okra gum increased the viscosity of wheat starch or corn starch at the 5 and 10% levels, but lowered the viscosity at 15%.[18] In addition, coexisting ions and the pH value can also affect the starch–hydrocolloid system.[55] Hydrocolloid molecules might penetrate, adsorb on or be excluded by granules, restricting their swelling and leaching.[13] Further, hydrocolloids could interact with the amylose leaching from the starch granules to produce a more complex matrix surrounding the gelatinized granules, thus leading to further restricted swelling and pasting.[18,27]

Hydrocolloids can affect the properties of starch in various ways. In general, water is one of the most important factors in starch gelatinization and retrogradation. Owing to their hydrophilicity and thickening, hydrocolloids obviously have effects on the amount of available water and its mobilization, which in turn change the properties of starch.[18] Apart from the role of water, the electrostatic interactions between starch and hydrocolloids, such as hydrogen bonding[23] and strong electrostatic repulsion,[19,21] play important roles in the properties of the mixture. The electrostatic interaction is prominent in potato starch owing to its negatively charged phosphate groups and the negatively charged hydrocolloids.[56]

### 11.2.4 Applications of Starch–Hydrocolloid Blends

In starchy foods, such as noodles and jelly products, starch with hydrocolloids usually has a better texture and appearance. The gel is prepared by combination of wheat starch and hsian-tsao leaf gum, called grass jelly, which is consumed in a beverage.[8] Onion pasta with HPMC can have better quality and sensory characteristics.[57] Hydrocolloids are widely used in starch noodle processing.[58] The addition of a small amount of gum can improve the rehydration characteristics of noodles during cooking and modify the texture and overall mouthfeel of the finished product.[41] Hydrocolloids also contribute to reducing cooking losses and the swelling index of starch noodles.[28,59]

Food processing usually involves high temperatures, high shearing or an acidic environment. Native starch does not generally have ideal properties for the preparation of food products, and hydrocolloids are used to improve the tolerance to processing conditions. Xanthan-crosslinked tapioca starch mixture could interact synergistically in acidic conditions to enhance the gel-like characteristics and shear-thinning behavior of chili sauces.[48] In frozen starchy foods, hydrocolloids are able to reduce the syneresis of starch in freeze–thaw cycles.[60] The addition of hydrocolloids to frozen starchy foods can increase the viscosity of the non-freezing phase, reduce the free volume of the solute molecules and increase the number of microcrystals and low-temperature stability in frozen foods.[61] Hydrocolloids can control or slow the growth rate of ice crystals to improve the quality of frozen foods and extend their shelf lives.[8] The addition of alginate into frozen dumplings could improve the stability and resistance to aging of starch.[62]

Hydrocolloids are often used as a dietary fiber and low-calorie food, owing to their own anti-digestibility that can reduce starch digestibility in the system. The addition of guar gum to cooked potato could delay and decrease the extent of starch hydrolysis during *in vitro* digestion. This might be of help in the food industry with the use of guar gum for the development of potato-based foods with slow digestion.[31] Noodles with 10% of oat β-glucan incorporated showed a large increase in soluble fiber content and significantly reduced glycemic index and carbohydrate digestibility rate.[63] Hydrocolloids can enhance the viscosity of starch at low fat contents. Locust bean gum could be used to replace the fat content with the desirably textural attributes of emulsion-based products,[26,64] which may be useful in the development of reduced-fat foods with desirably sensory properties. A mixture of κ-carrageen and pregelatinized crosslinked waxy maize starch can be used as a more effective thickening agent in fat replacement products.[65]

In general, appropriate hydrocolloids can improve the characteristics of starch in terms of rheology, gelatinization properties, gel characteristics, freeze–thaw stability and thermodynamics. They provide greater accessibility to raw materials and further reduce food production costs. Nowadays, research is mainly focused on various properties of starch and hydrocolloid blends, and not enough on their application in food processing. There is still a need for more applications of hydrocolloid–starch blends in the food industry.

## 11.3 Starch–Protein Blends

### 11.3.1 Introduction

Both starches and proteins are biological macromolecules that possess versatile functionalities, such as emulsification, foaming, thickening and gelation properties. Proteins have either hydrophobic or hydrophilic molecules that can migrate spontaneously to an oil–water interface. Once at the interface, the protein has the ability to interact with neighboring molecules and form a strongly cohesive, viscoelastic foam that can withstand morphological and mechanical motion.[66] Starch–protein blends are widely used in food products, such as sausages, bioplastic edible film,[67,68] low-fat foods[69] and gluten-free products.[70] The binary or even ternary system contributes new functionality and performance to the products.[66] Further, low-fat foods and gluten-free products are gradually acquiring an indispensable position in the food industry with regard to the requirements of obese and celiac disease patients.[69,70]

### 11.3.2 Characteristics of Starch–Protein Blends

#### *11.3.2.1 Effects of Proteins on the Pasting Properties of Blends*

As early as 1994, Doublier *et al.* reported the effect of adding sodium caseinate on the pasting behaviors of wheat, potato and tapioca starches.[71] This addition resulted in decreases in swelling and solubility for all starches. Noisuwan *et al.* investigated the adsorption of sodium caseinate and whey protein isolate by both normal and waxy rice starch granules using sodium dodecyl sulfate poyacrylamide gel electrophoresis (SDS-PAGE).[72] In the case of sodium caseinate, $\alpha_s$-casein was adsorbed more preferentially and to a greater extent than β-casein by both normal and waxy rice starches. The amount of adsorbed β-lactoglobulin was higher than that of α-lactalbumin, but no preferential adsorption was observed in the case of whey protein isolate. The waxy maize starch granule structure was reinforced by both α- and β-caseinate, but not by whey proteins (α-lactalbumin, β-lactoglobulin).[73] Confocal laser scanning microscopy (CLSM) showed that the reinforcement was caused by caseinate bound to the indigenous surface protein of starch granules by hydrophobic association. This might restrict the diffusion of water into the starch granules and make the gelatinized granules smaller and tougher.[73,74]

Addition of strong, medium and weak gluten significantly decreased the peak viscosity of wheat starch, but had no significant effect on pasting temperature.[75] Strong gluten with a higher gluten index had a more obvious effect on the pasting properties, such as peak viscosity. Addition of both globulin and glutelin extracted from rice to rice starch caused significant decreases in its pasting, breakdown and final viscosities.[76] In lentil starch–lentil protein composites, there was an increase in the pasting temperature but a decrease in the paste viscosity with increase in the amount of lentil protein.[77] There was an absence of breakdown at higher proportions of protein in the mixtures due to the resistance of lentil starch to mechanical shearing under hot conditions. The viscosity of sago starch decreased significantly with increase in the amount of fish gelatin.[67] However, pea protein isolate increased the viscosity of both corn and cassava starches, due to the crosslinking between the hydrophilic groups of protein and starch molecules.[78]

In the case of rice starch, the presence of glutelin caused an increase in pasting temperature of the starch paste.[79] The presence of globulin, on the other hand, resulted in a decrease in all pasting parameters of rice starch.[76]

Noisuwan *et al.* systematically studied the effect of four different milk proteins [milk protein concentrate (MPC), skim milk powder (SMP), whey protein isolate (WPI) and sodium caseinate] on the pasting behaviors of both normal and waxy rice starches.[80] Normal rice starch with SMP and sodium caseinate showed an increase in the temperature of peak viscosity, but a decrease with MPC and WPI. The effect of lactose and free cations in milk proteins on the properties of dairy protein–starch mixtures should also be considered. Addition of whey protein concentrate to cassava starch decreased the glass transition temperature ($T_g$) of the blends, probably due to the Maillard reaction of by-products.[81] There was a single $T_g$ for sago starch–fish gelatin blends, indicating their good compatibility at the concentration levels used.[67]

### *11.3.2.2 Effects of Proteins on the Gel Texture and Freeze–Thaw Stability of Blends*

Cassava starch (1.5 wt%) significantly improved the textural properties of myofibrillar protein gels.[82] The gelation behavior of the starch–protein composite gels was remarkably affected by whether starch or protein was the dominant fraction and also extrinsic factors, such as pH and ionic strength.[77] Lentil protein appeared to have an antagonistic effect on the gel network because it weakened the firmness of starch-rich composite gels. Non-covalent interactions (hydrophobic and hydrogen bonding) and covalent bonding contributed to the gel structure and firmness of these composite gels.[77] Set yoghurt is mainly composed of a network of casein, which was strengthened by adding crosslinked acetylated starch.[83] Owing to the absorption of modified starch, the particle size of the casein micelles showed an increasing trend, which prevented flocculation of the casein micelles and contributed to the stability of the set yoghurt.

Because whey protein can form gels under certain conditions, mixtures of whey protein and starch have attracted more attention than casein in terms of improving gel properties.[66] The addition of calcium promoted cold-set gelation in a WPI-modified waxy maize starch system. The presence of swollen starch granules led to discontinuous protein network formation and lower gelation kinetics in comparison with protein-only WPI gel.[84] Similar results were reported for whey protein and crosslinked waxy maize starch by Dang *et al.*[85]

The addition of fish gelatin to sago starch solution resulted in a significantly lower tensile strength of the film.[67] In such blend, gelatin was a continuous phase, while starch was a separated phase distributed in gelatin and covered with gelatin.[86] The addition of zein and gluten to cassava starch resulted in consolidated and homogeneous structural foams and increased the flexural and compressive strength.[87] The addition of glutelin to rice starch led to a higher hardness and adhesive properties of starch gel, due to its good gelation properties. The presence of globulin, on the other hand, resulted in a decrease in the textural parameters of rice starch gel.[76]

Yang *et al.* proposed a continuous distribution of whey protein interspersed with discontinuous starch inclusions.[88] This phase morphology showed an *in vitro* reduction in α-amylase digestion of the polysaccharide in the presence of whey protein.[88] After cooling, denatured whey protein formed a continuous matrix suspending filler inclusions of gelatinized starch. The addition of calcium ions greatly influenced the gelation kinetics of whey protein, leading to significant variability in the hydrolytic potential of α-amylase on wheat starch.[89] Aggregated WPI particles filled the space among the continuous starch granules. However, when the pH was increased to 9, the starch granules were much smaller and embedded in a continuous, thick WPI network.[66] Fitzsimons *et al.* conducted a systematic rheological study on co-gels of WPI–crosslinked waxy maize starch and found a continuous WPI network containing a dispersed starch phase at high WPI concentrations.[90] At low WPI concentrations there was a continuous starch network but it was fragmented by WPI when gelled.

The addition of pea protein isolate led to a change in corn starch network from an elastic-like to a more viscous-like nature with less solid-like character, whereas the opposite effect was observed in a cassava gel network.[78] In the presence of hydrolyzed (alcalase-treated) pea protein isolate, cassava and corn starch gels showed a significant increase in water release, whereas those with transglutaminase crosslinking tended to show a reduced released.[78]

Starch retrogradation is the main cause of the deterioration of the quality of starch-containing foods during storage. Soy protein hydrolyzates retarded maize starch retrogradation to a certain extent, while the polypeptide containing seven amino acids was probably the key component in significantly inhibiting maize starch retrogradation.[91] The addition of sodium caseinate to corn starch matrices reduced the degree of crystallinity of starch film and inhibited starch recrystallization during film storage.[92] Both native maize and chemically modified maize or tapioca starches can release the syneresis of reduced-fat yogurt. Modified starches exhibited a better performance than native starch because hydrophilic groups were incorporated.[69]

#### 11.3.2.3 *Effects of Proteins on the Rheological Properties of Blends*

The $G'$ value was increased with increase in the ratio of normal maize starch to protein in processed cheese, whereas $G'$ decreased with the addition of waxy maize starch.[93] The $\tan\delta$ value at high temperature (90 °C) decreased with the addition of both normal maize starch and waxy maize starch, but normal maize starch caused a greater reduction in the $\tan\delta$ value. These properties were influenced to a lesser extent in processed cheeses containing high-amylose (70%) maize starch. At high ratios of starch to protein in the system, the thermo-irreversible gel of retrograded normal starch possibly dominated the rheological properties of the processed cheese.

During heating, the rheological profiles of a binary mixture of wheat starch and whey protein exhibited a dramatic increase in $G'$ at temperatures more closely related to those observed for single whey protein rather than wheat starch.[88] With added pregelatinized potato starch, all model salad dressings prepared with dried egg yolk or sodium caseinate showed shear-thinning behavior with yield stress. With increase in the amount of pregelatinized potato starch, $G'$ and $G''$ of the dressings increased but $\tan\delta$ decreased.[94] With a small amount of starch, the dressing demonstrated liquid character ($\tan\delta > 1$). However, with more starch, it showed a weak gel-like response ($\tan\delta < 1$).

The $G'$ and $G''$ values of the set yogurt system were increased with increasing concentration of crosslinked acetylated starch.[83] The set yogurt system, which was added the crosslinked acetylated starch, exhibited shear-thinning behavior and yielded a pseudoplastic fluid. WPI–crosslinked waxy maize starch (CWMS) mixtures exhibited non-Newtonian shear-thinning with an anti-thixotropic behavior.[85] The addition of WPI into a starch suspension dramatically modified the flow behavior with a marked decrease in the yield stress. The solid-like behavior of CWMS alone disappeared, but a liquid-like behavior was observed.

Supplementation of rice starch-based dough with proteins from vegetable sources (egg albumin, calcium caseinate, pea protein and soy protein isolates) led to more structured dough matrices (higher viscoelastic moduli and steady viscosities and lower $\tan\delta$, instantaneous and retarded elastic compliances), which might be magnified with increase in protein content.[70] Chemically modified starch could induce a positive impact on flow and viscoelastic properties by contributing to the formation of more stable milk gels, compared with a full-fat yogurt.[69]

#### *11.3.2.4 Effects of Proteins on the Functional Properties of Starch*

Manipulation of polyelectrolyte addition is critical in order to provide the optimum spatial structure for maximum protection of starch against enzymatic degradation.[89] The presence of globulin initially accelerated the rate of water absorption of rice starch during cooking, but slowed it in the later stages. However, in both cases, the ultimate amount of water absorbed was significantly lower than that by starch alone.[76] Addition of pregelatinized potato starch was sufficient to stabilize dressings in term of creaming and fat-holding capacity.[94] The water-holding capacity (WHC) of myofibrillar protein from grass carp gels increased with addition of modified starches cassava, hydroxypropylated cassava and crosslinked hydroxypropylated cassava starches, while hydroxypropylated cassava starch improved the WHC most effectively.[82] Starches had a negative effect on the cooking loss of surimi-beef gels, and the gel with corn starch showed the highest cooking loss.[95]

### 11.3.3 Mechanisms of Starch–Protein Interactions

The addition of milk proteins had a greater effect on rice starch than on waxy rice starch, because the amounts of protein and fat were higher for the rice starch.[72,80] The adsorption of proteins on starch granules was mediated by indigenous lipid and protein on the starch granule surface.[74] However, milk proteins were also adsorbed on the starch granules when they washed the starch granules to remove indigenous lipids and proteins.[72] The presence of the milk proteins either at the interface of the granules or in the voids (pits, channels and holes) that were present at the hilum of starch granules would restrict the diffusion of water into the starch granules during pasting.[72] The starch was adsorbed on the surface of casein micelles and could prevent flocculation of casein micelles.[83] In general, protein molecule could directly interact with starch granules, restricting the access of water to the starch granules during gelatinization.[76] Water-soluble proteins, such as albumin, may limit the amount of free water available through the formation of hydrogen bonds with water molecules, which would break later at higher temperatures.[76]

### 11.3.4 Applications of Starch–Protein Blends

Starch is the most common ingredient used in sausages. It can not only decrease the use of meat in the products to reduce the cost, but also improve the quality of the products.[82,95] Gluten and starch are the two main ingredients of wheat flour dough. Decreasing the gluten content prolonged the time to develop and increase the dough density.[96] It is necessary to study new gluten-free foods for people with celiac disease, since some are allergic to wheat proteins. A batter of rice starch and soy protein isolate had the highest consistency, viscous and elastic moduli and adhesive force and showed the most similar rheological behavior to wheat flour batter.[97] Blends of albumen protein and starch (potato and corn starches) were suitable raw materials for bioplastics, exhibiting high transparency and improved mechanical behavior.[98] The addition of zein and gluten could improve the mechanical properties of cassava starch foam with a slightly inferior water resistance, which might make it an alternative to expanded polystyrene foam for packing less moist foods.[87]

Cold-set whey protein gels may be useful in developing reduced-calorie foods by removing fat and/or starch.[84] The addition of starch to reduced-fat yogurts contributed to the formation of a more stable dispersed, acidified, milk gel system, while chemically modified starch exhibited better performance than native starch.[69] Crosslinked acetylated starch contributed to the stability of the set yoghurt as a potential stabilizer.[83] Starch was used to replace rennet casein in processed cheese to retain its firmness and melt properties by using a suitable starch at the appropriate concentration.[93] Pregelatinized potato starch may be a suitable ingredient in a low-fat dressing.[94] Octenyl succinylated waxy maize starch improved the overall quality of reduced-fat Minas fresh cheese.[99] In general, there is a great market demand for the application of protein–starch combinations in food products.

## 11.4 Starch–Lipid Blends

### 11.4.1 Introduction

Lipids are divided into three groups: simple lipids (*e.g.* monoglycerides, diglycerides and triglycerides), compound lipids (*e.g.* phospholipids) and derived lipids [*e.g.* free fatty acids (FFAs) and long-chain alcohols].[100] Lipids in starchy foods play an important role in the caloric density, texture and flavor, which alter the physicochemical properties of starchy foods.[101]

It is well known that amylose forms single-helical complexes with lipids and other compounds, such as iodine and some fat-soluble bioactive substances.[102,103] Depending on different coexisting compounds, amylose forms six-, seven- or eightfold left-handed single helices stabilized by hydrogen bonds.[104] In general, more complicated of lipid molecules with longer chain length fatty acids showed greater formation of amylose helices with larger cavities to accommodate the guest molecules.[104,105] The dissociation temperature of amylose–lipid complexes increases with increasing length of the hydrocarbon chains of the lipids, but decreases with increasing number of olefinic bonds in the hydrocarbon chains. After heating at a temperature above the dissociation temperature, amorphous amylose–lipid complexes (form I) can further rearrange into lamellar crystallites (form II).[104] The lipid came into the helical structure of amylose and formed amylose–lipid complexes, which led to resistance to the starch digestion rate. A novel resistant starch (RS5) was developed *via* processing high-amylose corn starch with FFAs.[106]

### 11.4.2 Characteristics of Starch–Lipid Blends

#### *11.4.2.1 Effects of Lipids on the Pasting Properties of Blends*

In general, the pasting properties of starch are affected by the amylopectin, amylose and lipid contents. Amylopectin contributes to pasting and swelling of starch granules, but amylose and lipid inhibit the swelling. Amylose–lysophosphatidylcholine complex formation led to less water ingress and less swelling during pasting.[107] Complexing with lauric acid restricted the swelling of maize starch and high-amylose maize starch granules, but did not affect waxy maize starch, probably owing to a to lack of amylose.[108] Probably, lipids could also be adsorbed on the surface of starch granules, which, in turn, inhibited the swelling.[109] The addition of soy lecithin facilitated the swelling of normal corn and high-amylose corn starches, because soy lecithin with amphiphilic characteristics removed the endogenous lipids from the starch granules.[110]

The chain length of fatty acids could play a role in retarding disruption of the starch granules. For example, myristic acid (C14) did not prevent the disruption of pea starch granules, but stearic acid (C18) was able to prevent granule disruption effectively.[105] There were interesting protrusions on the surface of swollen amylomaize starch granules after heating with fatty acids.[105,111] These protrusions could be attributed to amylose molecules that partly exited from the starch granules, immediately interacting with the available myristate ions present in the milieu to form a helical complex.[105] When the heating temperature was increased, the trapped amylose molecules were released from the granules.

FFAs (palmitic, stearic, oleic and linoleic acids) decreased the peak viscosities of corn and tapioca starches during cooking, and the addition of corn oil to both starches also decreased their viscosities.[110] There was a negative correlation between final viscosity and setback in barley flour with lipids.[112] Lauric acid and the incubation temperature showed pronounced effects on the viscosity profile of the maize starch complex, but had little influence on waxy maize and high-amylose maize starches. High-amylose maize starch could not be fully gelatinized below the Brabender temperature, while waxy maize starch had a small amount of starch–lauric acid complex that changed the viscosity. However, maize starch with lauric acid showed a higher pasting temperature, lower peak viscosity and breakdown during the holding period at 95 °C.[108]

Monocaprylin (C8:0) did not induce major changes to the viscosity of wheat starch. Saturated monoglycerides with C10–C16 fatty acids increased the cool paste viscosity during the first few heat–cool cycles, which was followed by a gradual decrease to a stable cool paste viscosity.[109] The addition of unsaturated fatty acids (olive oil and sunflower oil) resulted in higher viscosity of starch (Tesco plain flour), compared with those of the control (no lipid) and saturated fats (butter and medium-chain triglycerides) added.[113] The addition of glycerol monostearate postponed pasting and decreased the breakdown of wheat starch paste during the 95 °C isothermal holding phase.[114]

The melting temperature ($T_m$) of starch–fatty acid mixtures increased with increase in the number of carbon atoms and also with a decrease in the number of double bonds in the fatty acid, but the melting enthalpy exhibited the opposite trend. A higher $T_m$ reflected greater stability of the complex, and unsaturated fatty acids led to a lower $T_m$ than saturated fatty acids.[115] Soy lecithin and FFAs (palmitic, stearic, oleic and linoleic acids) formed amylose–lipid complexes with corn, tapioca and high-amylose corn starches, and there were well-defined amylose–lipid complex dissociation peaks in the DSC thermograms.[110] The addition of corn oil led to the formation of a helical complex with amylose, but to a lesser extent, and there was no amylose–lipid complex dissociation peak in the DSC thermograms. In the starch–lauric acid complex, the temperature range of gelatinization became narrower, while the gelatinization temperatures (onset $T_o$, peak $T_p$ and completion $T_c$) increased. This suggested that the addition of lipid played a role in retarding starch gelatinization.[108] The DSC thermograms of the starch–lipid complex exhibited three peaks: the first peak (48–55 °C) was due to the melting of free lipids, the second peak presumably represented the melting of amylopectin double helices and type I starch–lipid complex and the third peak was attributed to the melting of the type II complex.[108] Type I complex was the less ordered form and melted at a lower temperature whereas type II complex was more crystalline and melted at a higher temperature. The enthalpy ($\Delta H$) value of type II complex increased when the incubation temperature was increased. This suggested that a high incubation temperature was beneficial to the formation of type II complex.[108] Starch–lipid complexes with monoglycerides containing longer chain fatty acids were more stable and displayed higher dissociation and melting temperatures than those with shorter chain lengths.[109]

### 11.4.2.2 *Effects of Lipids on the Retrogradation Properties of Blends*

Amylose–lipid complex formation can affect the progress of retrogradation. It is important to insert the lipid into the helical structure of amylose and form amylose–lipid complexes, which can determine the starch texture in the solid state. The addition of soy lecithin and FFAs (palmitic, stearic, oleic and linoleic acids) to corn starch decreased the gel strength of the blends, resulting in lamellar crystallite formation and loss of water-binding capacity, whereas corn oil enhanced the gel strength.[110] Thermoplastic starch film with soybean oil was more homogeneous and had greater elongation at break and resistance.[116] An increase in oil content resulted in significant lowering of the tensile breaking strength, but a concurrent increase in elongation at break of the starch film, which was more resistant to fracture.[117] Higher glycerol concentration resulted in a lower tensile strength of cassava starch film.[118] The addition of palm oil to cassava starch foam gave the lowest flexural strength and compressive strength, but increased the water resistance.[87] Teff starch with stearic acid was more viscous and non-gelling.[119] Mixing of corn starch with sodium palmitate was able to prevent retrogradation.[120]

### 11.4.2.3 *Effects of Lipids on the Crystallization Properties of Blends*

Amylose plays the major role in crystallite formation; a higher amylose content of the starch led to greater crystallinity, regardless of being inside or outside the starch granules.[105] The degree of crystallinity of amylomaize–fatty acid was much higher than that of the normal maize product, and the crystallization behavior of the amylose–lipid complex was affected by the crystallization conditions, such as moisture content and temperature. A higher heating temperature led to a higher degree of crystallinity. At 75 °C, only the V-type pattern was present on adding fatty acid, since the maize starch granules lost their A-type crystallinity pattern.[105]

High-amylose maize starch with lauric acid displayed a mixture of B- and V-type patterns. However, the V-type pattern became dominant and the B-type pattern decreased on incubation at 80 °C.[108] Similarly, amylomaize with fatty acids heated at 75 °C displayed a mixture of B- and V-type patterns.[105] At 85 °C, the V-type pattern became dominant but the B-type pattern was still visible. At 98 °C, the B-type pattern had totally disappeared. The peak observed at around 20.15°(2$\theta$) was attributed to the formation of a starch–lauric acid complex, whereas the peaks observed at 21.5 and 23.9°(2$\theta$) were identified as corresponding to the crystalline pattern of aggregating lauric acid.[108] The relative crystallinity of starch was inversely proportional to the amylose content, indicating that the relative crystallinity of starch was mainly attributable to amylopectin double helices.[108]

At barrel temperatures less than 90 °C, extrudates with almond flour showed a pattern of closely matching A-type wheat starch, unlike those with only wheat flour. The preservation of A-type crystallinity in almond flour was due to the lubrication of flour particles by the lipid (present in almond flour) that tended to protect the starch granules from dispersal.[121] At barrel temperatures above 90 °C, the pattern changed to the V-hydrate form. The formation of double helices (B pattern) was decreased and partially replaced by the formation of a starch–lipid complex with low long-range structural order ($V_I$ pattern) during pasting. Continuous shearing in combination with repeated heating and cooling resulted in a further increase in crystallinity and sharpening of starch–lipid complex reflections ($V_{II}$ pattern) accompanied by increased longer range structural order.[109] Monocaprylin with shorter chain fatty acids formed an unstable Type I complex with a $V_I$ pattern, whereas monoglycerides with C10–C16 fatty acids first formed Type I complexes that subsequently converted into Type II complexes.[109] An increase in the unsaturation of long-chain fatty acids could reduce the stability of inclusion complexes with the amylose in potato starch.[109]

#### *11.4.2.4 Effects of Lipids on the Functional Properties of Blends*

The complexation of amylose with lysophosphatidylcholine decreased the susceptibility of wheat starch granules by α-amylase, because of the new V-helix formation that hindered less water ingress and enzymatic attack.[107] After cooking with corn, tapioca, waxy corn and high-amylose corn starches, corn oil, soy lecithin and four other FFAs (palmitic, stearic, oleic and linoleic acids) significantly decreased the starch hydrolysis rate, except waxy corn starch, which lacks amylose.[110] The interaction between starch and lipids could be useful in producing starchy foods with a reduced glycemic index.[110] In general, the hydrolysis rate of starch can be reduced by the addition of fatty acids or lipids.[115]

However, Clegg *et al.* found that the levels of sugars in the digest could increase with increase in fat concentration.[113] This is contrary to the literature that indicates that lipids can form starch–lipid complexes that prevent enzyme hydrolysis of starch. Once the fat became emulsified by bile at the beginning of the intestinal phase, it could cause the breakdown of the starch–lipid structure and rapid release of the sugar into the digest, faster than for the control.[113] The starch–lipid complex structure was broken down earlier and the starch released was augmented in the gastric phase.

### 11.4.3 Mechanisms of Starch–Lipid Interactions

Amylose–lipid complexes can be formed during gelatinization. The first step is the formation of a lipid layer around the granules before the gelatinization temperature is reached. The aliphatic chains then diffuse into the interior of the granules and form a complex on the granule surface. Finally, the complex on the granule surface functionally acts as a natural barrier to prevent water diffusion into the granule interior and allow it to maintain its structure during hydrothermal processing.[122] Only the aliphatic moiety of the lipid can be inserted at each end of an amylose chain.[109] Monoglycerides with longer chain fatty acids have more aliphatic groups available for insertion into the amylose chain compared with monoglycerides with shorter chain fatty acids, thereby producing a more stable complex.[105,109] However, there is an opposite view that the shorter fatty acid chain leads to more aliphatic groups being available for amylose to interact with the fatty acid.[115] Vasiliadou *et al.* reported the same result.[111] A fatty acid with a long carbon chain forms a poor complex with amylose because of the poor dispersivity in the gelatinized starch. Further, the complex index also increased with increase in the number of double bonds in the fatty acid.[115]

Amylose could entangle with amylopectin, generate a strong crystalline network or complex with lipids and form a V-amylose complex to inhibit amylopectin dissociation, which in turn affects starch swelling.[108] The formation of the V-amylose complex increased the number of inter- and intramolecular hydrogen bonds and maintained the integrity of starch granules during heating and shearing.[108] It could provide stability to the gel by the formation of physical junction zones between amylose and the external chains of amylopectin molecules.[102] The ratio between intact granules and granule remnants was related to the textural properties of the starch paste. There were large numbers of swollen granules with a minimal amount of polymer leachate in the extra-granular matrix. The addition of lipids prevented gels from being formed. The amylose–lipid complex is probably a V-type single-helical amylose pattern, hence there is no formation of double helices, junction zones and gel network.[102,119]

### 11.4.4 Applications of Starch–Lipid Blends

Film with oil exhibits less moisture.[116] The addition of epoxidized soybean oil (ESO) enhanced both the Young's modulus and the tensile strength of thermoplastic starch film.[123] Soybean oil was considered to be a good compatibilizer for blends of thermoplastic starch and poly(butylene adipate-*co*-terephthalate) (PBAT).[116] A coating formulation with cassava starch, glycerol, carnauba wax and stearic acid has good barrier and mechanical properties, and it can potentially be used in minimally processed fruit.[118] The addition of lipids to the starch film mainly plays the role of a water barrier.[124,125] The water vapor permeability of the high-amylose starch–glycerol film decreased significantly with the addition of natural wax.[126]

Lipid was mixed into the starch film to obtain desirable performance characteristics, such as waterproofing.[127–129]

Corn starch film with plant essential oils, *Zataria multiflora* Boiss. (ZEO) or *Mentha pulegium* (MEO), incorporated showed a strong potential for used as active films to inhibit the growth of *Escherichia coli* and *Staphylococcus aureus.*[117] When feruloylated monoacyl- and diacylglycerols (FAGs) were encapsulated as microdroplets within a starch matrix, the FAGs retained their ultraviolet-absorbing efficacy after steam jet cooking.[130] By using monoglycerides, unsaturated oils successfully substituted for saturated fats in the production of healthier bakery products.[131]

Current techniques for producing starch–guest inclusion complexes require a relatively high processing temperature and a long time. A new method was introduced to form a starch–guest inclusion complex that involved first precipitating "empty" V-type amylose helices and then "inserting" guest molecules into the helices at a temperature well below the melting point of the V-type amylose.[104] The high-pressure homogenization process (100 MPa) could be used as an effective means to promote the complex formation of starch and fatty acids.[132]

## 11.5 Blends of Starch and Salts, Sugars and Other Additives

### 11.5.1 Effects of Acids on the Properties of Starch

Acid is the most important additive in the food industry as an acidity regulator, antimicrobial or flavoring agent. Also, it may have some effects on the functional properties of other food constituents, such as starch.[133] It is well known that starch can be hydrolyzed by acids, so-called "acid modification of starch" or "acid-thinned starch".[134] Acids primarily attack the amorphous regions within starch granules, whereas both amylose and amylopectin are hydrolyzed simultaneously by acids.[134] The formation of some cracks and spots was observed on the acid-treated granules,[133] and these cracks and spots facilitated the penetration of water into the granules, leading to an increase in water solubility. The intrinsic viscosity of starches was decreased in the presence of acids, particularly after gelatinization resulting from starch degradation by the acids.[135] Chen *et al.*[136] and Majzoobi *et al.*[137] considered that the decrease in starch viscosity on adding amino acids was primarily due to the acidic environment that they caused. The acids also led to a reduction in gelatinization temperature and enthalpy of gelatinization of starch.[138]

The starch gel becomes softer, less cohesive, elastic and gummy when acid is added. Acid-thinned potato starch produced much firmer gels than acid-thinned corn and rice starches, possibly due to the relatively higher percentage of long-branched chains in the amylopectin of potato starch.[134] Crosslinked wheat starch was more resistant to the acid. Hence in some acidic foods, manufacturers often use crosslinked starch to reduce the negative influence caused by the decrease in viscosity.[135]

### 11.5.2 Effects of Alkalis on the Properties of Starch

Alkaline treatment with agents such as sodium hydroxide is widely used in the production of many food products, including waxy rice dumplings and yellow alkaline noodles, to enhance the quality characteristics of color, flavor and texture.[139] After alkaline treatment, many starch granules were deformed with depressions evident on the surface of the granules and they appeared to adhere to each other.[140] Alkaline treatment of starch resulted in damage to the starch granule surface, leaching of amylose and formation of cavities inside the granules, which increased the *in vitro* digestibility of starch granules.[140] Alkaline treatment significantly decreased the peak viscosity, breakdown and setback of pea, sago and potato starches.[140–142] In contrast, alkali-treated corn starch showed an increase in peak viscosity, which was attributed to the removal of protein or lipid from the granule surface.[141] Retrogradation was markedly reduced by alkaline treatment, possibly due to depolymerization of starch polymer chains.[142]

### 11.5.3 Effects of Salts on the Properties of Starch

The effects of salts on starch performance have been widely studied.[143–145] However, there is considerable controversy in the literature about the influence of the salt on the physicochemical properties of the starch. Depending on their variety and concentration, salts can cause an elevation or depression of starch gelatinization and also retrogradation. Various explanations have been proposed.

Considering the effect of salt on starch, it normally involves the Hofmeister series, which is a classification of ions in order of their ability to change protein solubility or denaturation.[146,147] Except for proteins, there is a similar sequence for the effects of ions on the stability and solubility of other macromolecule colloidal solutions, such as starch. Anions appear to have a greater effect than cations and are usually ordered as $F^- < SO_4^{2-} < Cl^- < NO_3^- < Br < ClO_3 < I < ClO_4^- < SCN^-$. The order of cations is usually given as $NH_4^+ < K^+ < Na^+ < Li^+ < Mg^{2+} < Ca^{2+}$. For the center with $Cl^-$, ions at the front of the series (*e.g.* $F^-$) are of smaller diameter and stronger electric field intensity or polyvalent ions and tend to protect the links between the molecules, and therefore decrease the solubility of molecules ("salting out" or "kosmotropes"). In contrast, later ions in the series (*e.g.* $I^-$) are of larger diameter and weaker electric field intensity and tend to break the links between the molecules, and thus increase the solubility of molecules ("salting in" or "chaotropes").[143,144,148,149]

The effects of ions on the properties of potato starch follow the order of the Hofmeister series.[144] The salting-out ions are able to decrease the solubility, swelling power, transparency and particle size of potato starch significantly, whereas the salting-in ions increase these properties. Further, there is an increase in gelatinization temperature and enthalpy with the addition of salting-out ions owing to the structure-making effect, whereas there is a reverse trend with the addition of salting-in ions owing to its structure-breaking effect.[144]

Different concentrations of NaCl solution had different impacts on the gelatinization process of different starches.[143] NaCl inhibited the gelatinization of corn starch. However, for potato starch, low-concentration NaCl inhibited gelatinization whereas high-concentration NaCl significantly promoted gelatinization, which might related to the negatively charged phosphorus groups. The ionic repulsion generated by these groups weakened the association between the molecules and increased the water-binding capacity and swelling of the granules.[143,144] The effects of salts on starch gelatinization also depended on the moisture content of the system. With limited water, water availability was a limiting factor, whereas with excess water, it was more likely that the presence of the NaCl acted *via* electrostatic screening of starch molecules and inhibited the initial water ingress into the granules, thus reducing the swelling.[145]

### 11.5.4 Effects of Sugars on the Properties of Starch

In general, addition of sugars can delay starch gelatinization, and disaccharides delay gelatinization more than monosaccharides.[150] The peak viscosity and final viscosity increased with increase in sucrose content. The gelatinization onset ($T_o$), peak ($T_p$) and completion ($T_c$) temperatures increased significantly with increase in sugar. The enthalpy of tapioca starch increased at low sugar concentrations, whereas it decreased when higher concentrations of sugar were added.[150] Sucrose increased the gelatinization temperature of tapioca starch more than glucose.[150] The number of equatorial hydroxyl groups in sucrose (6.2) is greater than that in glucose (4.56), which played a key role in stabilizing the sugar–starch interaction and increased the gelatinization temperature. Li *et al.* reached a similar conclusion.[151] With increase in sugar concentration, the activation energies ($E_a$) of blends showed a gradual increase as revealed by non-isothermal kinetic analysis, indicating that sugar retarded the process of gelatinization.[150] Sugar had a protective effect on the crystalline structure of starch during gelatinization, and the gelatinization temperature increased with increase in sugar concentration.[151] Trisaccharides and tetrasaccharides were more effective in inhibiting the gelatinization than disaccharides, with the order monosaccharides < disaccharides < trisaccharides < tetrasaccharides.

The hysteretic loop area of tapioca starch gel with sugar was smaller than that of the control.[150] This indicated that sugar reduced the thixotropy, because the sugar–starch interaction caused the gel network to be less destroyed by shearing strain. All samples exhibited pseudo-plastic and shear-thinning behavior. Addition of sugar increased the consistency coefficient of starch.[150] The $G'$ value of a tapioca starch–sucrose blend was increased with increase in sucrose content but tan $\delta$ was decreased, indicating that sucrose increased the solid-like characteristics of the blend. The hydroxyl groups of sugar interacted with water and inhibited the movement of water. The available water was correspondingly decreased, and the motion of starch chains was suppressed. Additionally, sugar altered the structure of water, had easy access to the hydration layer of starch chains and produced a strong anti-plasticizing effect compared with water alone. Furthermore, the formation of sugar–starch interactions in the amorphous region made it difficult to damage the crystallization region in the starch granules.

### 11.5.5 Effects of Amino Acids on the Properties of Starch

The physical properties of starch depend not only on the variety of amino acids, but also on their addition. The addition of lysine (Lys) increased the swelling power, solubility, peak viscosity and breakdown and decreased the pasting temperature. However, glycine (Gly) had little or no obvious effect on the pasting properties, swelling power and solubility.[152] There were a decrease in $G'$ and an increase in $G''$ and tan $\delta$ after the addition of Lys or Gly. Lys led to greater increase in tan $\delta$ than Gly at the same concentration. A maize starch–amino acid mixture exhibited shear-thinning flow and liquid-like behavior.[152] Charged amino acids, such as Lys, had greater effects on the pasting and rheological properties than did neutral amino acids, such as Gly.[152]

Chen *et al.*[136] and Cui *et al.*[153] further studied the effects of charged amino acids on potato starch with respect to the unique charge characteristics. Charge-carrying amino acids (Lys, arginine (Arg), aspartic acid (Asp) and glutamic acid (Glu)) significantly decreased the swelling power, solubility, light transmittance, $L^*$ value and gel strength of potato starch, but increased syneresis during freeze–thaw treatment. Neutral amino acids (phenylalanine (Phe) and methionine (Met)) did not cause modifications in starch gels. During heating, potato starch with fortified charge-carrying amino acids showed a lower peak $G'$ compared with Phe and Met. Generally, charge-carrying amino acids could modify the physicochemical properties and improve the nutritional value of starch-based products.

## 11.6 Conclusions and Future of Starch Blends

When two or more ingredients, such as starch and protein, are mixed together, the physicochemical properties of the mixed system depend on their original properties and relative concentrations, water content and operating temperature. From the perspective of the literature during the last decade, the focus of research on starch blends from low molecular weight molecules (*e.g.* sugar, salt) shifted to biomacromolecules (*e.g.* hydrocolloids, proteins, lipids) and functional substances (*e.g.* amino acids). These additives not only fundamentally change some native properties of starch and improve the processing and storage performance of starchy foods, but also, to a certain extent, improve the nutritional values and reduce costs. Much more attention has been devoted to the functional properties of starch and additive blends, such as low-calorie, low-oil and gluten-free properties. As the emphasis on food and health has increased, people have gradually tended to accept additions of natural substances, instead of chemical modification. This research upsurge will continue in the future.

Although much progress has been made in the field of starch blends, many fundamental aspects remain unanswered. One is how to demonstrate the interaction between two or more ingredients at the molecular level and molecular simulation. Another aspect that has not been thoroughly explored is the effect of additives on the kinetics of starch gelatinization. Such an understanding of the interactions between starch and its additives will greatly help its practical application in various industries.

## Acknowledgements

The authors acknowledge financial support from the Ministry of Scientific and Technology China (Grant No. 2012BAD28B06) and Fundamental Research Funds for the Central Universities (Project 2013PY096).

## References

1. H. J. Chung, K. S. Woo and S. T. Lim, *Carbohydr. Polym.*, 2004, **55**, 9–15.
2. K. Kaewtatip and J. Thongmee, *Mater. Des.*, 2013, **45**, 586–589.
3. S. H. Koo, K. Y. Lee and H. G. Lee, *Food Hydrocolloids*, 2010, **24**, 619–625.
4. J. Liu, B. Wang, L. Lin, J. Zhang, W. Liu, J. Xie and Y. Ding, *Food Hydrocolloids*, 2014, **36**, 45–52.
5. A. V. Singh and L. K. Nath, *Int. J. Biol. Macromol.*, 2012, **50**, 14–18.
6. P. A. M. Steeneken, A. J. J. Woortman and A. A. C. M. Oudhuis, *Food Hydrocolloids*, 2011, **25**, 410–418.
7. R. Wongsagonsup, T. Pujchakarn, S. Jitrakbumrung, W. Chaiwat, A. Fuongfuchat, S. Varavinit, S. Dangtip and M. Suphantharika, *Carbohydr. Polym.*, 2014, **101**, 656–665.
8. J. N. BeMiller, *Carbohydr. Polym.*, 2011, **86**, 386–423.
9. J. A. Ramírez, R. M. Uresti, G. Velazquez and M. Vázquez, *Food Hydrocolloids*, 2011, **25**, 1842–1852.
10. P. Varela and S. M. Fiszman, *Food Hydrocolloids*, 2013, **30**, 477–484.
11. K. T. Roberts, *Food Res. Int.*, 2011, **44**, 1109–1114.
12. T. Witczak, M. Witczak and R. Ziobro, *J. Food Eng.*, 2014, **124**, 72–79.
13. A. Matignon, P. Barey, M. Desprairies, S. Mauduit, J. M. Sieffermann and C. Michon, *Food Hydrocolloids*, 2014, **35**, 597–605.
14. B. Heyman, D. De Hertogh, P. Van der Meeren, F. Depypere and K. Dewettinck, *Carbohydr. Polym.*, 2013, **96**, 568–577.
15. B. Ghanbarzadeh, H. Almasi and A. A. Entezami, *Innovative Food Sci. Emerging Technol.*, 2010, **11**, 697–702.

16. B. Heyman, W. H. De Vos, F. Depypere, P. Van der Meeren and K. Dewettinck, *Food Hydrocolloids*, 2014, **35**, 546–556.
17. H. S. Kim and J. N. BeMiller, *Carbohydr. Polym.*, 2012, **88**, 1164–1171.
18. M. S. Alamri, A. A. Mohamed and S. Hussain, *Food Hydrocolloids*, 2013, **30**, 541–551.
19. X. Cai, Y. Hong, Z. Gu and Y. Zhang, *Food Res. Int.*, 2011, **44**, 3079–3086.
20. L. Chen, Q. Tong, F. Ren and G. Zhu, *Int. J. Biol. Macromol.*, 2014, **66**, 325–331.
21. K. Gul, C. S. Riar, A. Bala and M. S. Sibian, *LWT—Food Sci. Technol.*, 2014, **59**, 348–355.
22. K. S. Pramodrao and C. S. Riar, *Food Hydrocolloids*, 2014, **35**, 613–619.
23. S. Qiu, M. P. Yadav, H. Chen, Y. Liu, E. Tatsumi and L. Yin, *Carbohydr. Polym.*, 2015, **115**, 246–252.
24. S. Charoenrein, O. Tatirat, K. Rengsutthi and M. Thongngam, *Carbohydr. Polym.*, 2011, **83**, 291–296.
25. A. Matignon, G. Moulin, P. Barey, M. Desprairies, S. Mauduit, J. M. Sieffermann and C. Michon, *Carbohydr. Polym.*, 2014, **99**, 345–355.
26. C. Chung, B. Degner and D. J. McClements, *Food Hydrocolloids*, 2014, **35**, 315–323.
27. C. M. Rosell, W. Yokoyama and C. Shoemaker, *Carbohydr. Polym.*, 2011, **84**, 373–382.
28. S. Heo, S. Jeon and S. Lee, *LWT—Food Sci. Technol.*, 2014, **55**, 627–631.
29. I. G. Mandala and E. Bayas, *Food Hydrocolloids*, 2004, **18**, 191–201.
30. P. Chantaro, R. Pongsawatmanit and K. Nishinari, *Food Hydrocolloids*, 2013, **31**, 183–194.
31. A. Bordoloi, J. Singh and L. Kaur, *Food Chem.*, 2012, **133**, 1206–1213.
32. M. A. Gularte and C. M. Rosell, *Carbohydr. Polym.*, 2011, **85**, 237–244.
33. R. Pongsawatmanit, P. Chantaro and K. Nishinari, *J. Food Eng.*, 2013, **117**, 333–341.
34. J. M. Schwartz, K. Le Bail, C. Garnier, G. Llamas, D. Queveau, B. Pontoire, G. Srzednicki and P. Le Bail, *Food Hydrocolloids*, 2014, **41**, 71–78.
35. M. Alloncle, J. Lefebvre, G. Llamas and J. L. Doublier, *Cereal Chem.*, 1989, **66**, 90–93.
36. T. C. Tan, W. T. Foo, M. T. Liong and A. M. Easa, *J. King Saud Univ., Sci.*, 2014, **26**, 311–322.
37. P. Arranz-Martínez, K. Srikaeo and A. L. González-Sánchez, *Bioact. Carbohydr. Diet. Fibre*, 2014, **4**, 6–15.
38. A. L. Choy, B. K. May and D. M. Small, *Food Hydrocolloids*, 2012, **26**, 2–8.
39. A. Espinosa-Dzib, M. Ramírez-Gilly and A. Tecante, *Food Hydrocolloids*, 2012, **28**, 248–257.
40. M. Huang, J. F. Kennedy, B. Li, X. Xu and B. J. Xie, *Carbohydr. Polym.*, 2007, **69**, 411–418.
41. B. X. Fu, *Food Res. Int.*, 2008, **41**, 888–902.
42. Y. Zhou, C. G. Winkworth-Smith, Y. Wang, J. Liang, T. J. Foster and Y. Cheng, *Carbohydr. Polym.*, 2014, **114**, 357–364.

43. M. Krystyjan, G. Adamczyk, M. Sikora and P. Tomasik, *Food Hydrocolloids*, 2013, **31**, 270–276.
44. H. M. Baranowska, M. Sikora, M. Krystyjan and P. Tomasik, *J. Food Eng.*, 2012, **109**, 685–690.
45. L. Chen, F. Ren, Z. Zhang, Q. Tong and M. M. A. Rashed, *Carbohydr. Polym.*, 2015, **115**, 415–421.
46. H. M. Choi and B. Yoo, *Food Chem.*, 2009, **116**, 638–643.
47. P. Chantaro and R. Pongsawatmanit, *J. Food Eng.*, 2010, **98**, 44–50.
48. C. Gamonpilas, W. Pongjaruvat, A. Fuongfuchat, P. Methacanon, N. Seetapan and N. Thamjedsada, *J. Food Eng.*, 2011, **105**, 233–240.
49. M. Chaisawang and M. Suphantharika, *Carbohydr. Polym.*, 2005, **61**, 288–295.
50. P. Achayuthakan and M. Suphantharika, *Carbohydr. Polym.*, 2008, **71**, 9–17.
51. T. Sasaki and K. Kohyama, *Food Chem.*, 2011, **127**, 541–546.
52. H. Andersson, C. Öhgren, D. Johansson, M. Kniola and M. Stading, *Food Hydrocolloids*, 2011, **25**, 1587–1595.
53. H. S. Kim, B. Patel and J. N. BeMiller, *Carbohydr. Polym.*, 2013, **98**, 1438–1448.
54. B. Heyman, W. H. De Vos, P. Van der Meeren and K. Dewettinck, *Food Hydrocolloids*, 2014, **39**, 85–94.
55. W. Samutsri and M. Suphantharika, *Carbohydr. Polym.*, 2012, **87**, 1559–1568.
56. N. Singh, N. Isono, S. Srichuwong, T. Noda and K. Nishinari, *Food Hydrocolloids*, 2008, **22**, 979–988.
57. G. Rajeswari, S. Susanna, P. Prabhasankar and G. V. Rao, *Food Bioscience*, 2013, **4**, 13–20.
58. H. Z. Tan, Z. G. Li and B. Tan, *Food Res. Int.*, 2009, **42**, 551–576.
59. E. Silva, M. Birkenhake, E. Scholten, L. M. C. Sagis and E. van der Linden, *Food Hydrocolloids*, 2013, **30**, 42–52.
60. Y. G. Lü, J. Chen, X. Q. Li, L. Ren, Y. Q. He and L. B. Qu, *LWT—Food Sci. Technol.*, 2014, **59**, 403–410.
61. T. Tran, K. Thitipraphunkul, K. Piyachomkwan and K. Sriroth, *Starch Stärke*, 2008, **60**, 61–69.
62. N. Wang, A. Wang and Z. Pan, *Acta Agric. Jiangxi*, 2009, **8**, 048.
63. C. L. Choo and N. A. A. Aziz, *Food Chem.*, 2010, **119**, 34–40.
64. C. Chung, B. Degner and D. J. McClements, *Food Res. Int.*, 2013, **54**, 829–836.
65. I. Gladkowska-Balewicz, I. T. Norton and I. E. Hamilton, *Food Hydrocolloids*, 2014, **42**, 355–361.
66. T. Considine, A. Noisuwan, Y. Hemar, B. Wilkinson, J. Bronlund and S. Kasapis, *Food Hydrocolloids*, 2011, **25**, 2008–2017.
67. A. A. Al-Hassan and M. H. Norziah, *Food Hydrocolloids*, 2012, **26**, 108–117.
68. Q. Sun, C. Sun and L. Xiong, *Carbohydr. Polym.*, 2013, **98**, 630–637.
69. C. Lobato-Calleros, C. Ramírez-Santiago, E. J. Vernon-Carter and J. Alvarez-Ramirez, *J. Food Eng.*, 2014, **131**, 110–115.

70. F. Ronda, M. Villanueva and C. Collar, *LWT—Food Sci. Technol.*, 2014, **59**, 12–20.
71. J. L. Doublier, C. Marzin, S. Visdeloup and J. Lefebvre, *Carbohydr. Polym.*, 1994, **25**, 228–229.
72. A. Noisuwan, Y. Hemar, B. Wilkinson and J. E. Bronlund, *Carbohydr. Polym.*, 2011, **84**, 247–254.
73. A. P. Kett, V. Chaurin, S. M. Fitzsimons, E. R. Morris, J. A. O'Mahony and M. A. Fenelon, *Food Hydrocolloids*, 2013, **30**, 661–671.
74. K. J. Ryan and M. S. Brewer, *Food Chem.*, 2007, **104**, 619–629.
75. J. S. Chen, Z. Y. Deng, P. Wu, J. C. Tian and Q. G. Xie, *Agric. Sci. China*, 2010, **9**, 1836–1844.
76. G. Baxter, C. Blanchard and J. Zhao, *J. Cereal Sci.*, 2014, **60**, 414–420.
77. M. Joshi, P. Aldred, J. F. Panozzo, S. Kasapis and B. Adhikari, *Food Hydrocolloids*, 2014, **35**, 226–237.
78. P. D. Ribotta, A. Colombo and C. M. Rosell, *Food Hydrocolloids*, 2012, **27**, 185–190.
79. J. S. BAO, *Rice Sci.*, 2008, **15**, 69–72.
80. A. Noisuwan, J. Bronlund, B. Wilkinson and Y. Hemar, *Food Hydrocolloids*, 2008, **22**, 174–183.
81. L. García, A. Cova, A. J. Sandoval, A. J. Müller and L. M. Carrasquel, *Carbohydr. Polym.*, 2012, **87**, 1375–1382.
82. F. Sun, Q. Huang, T. Hu, S. Xiong and S. Zhao, *Int. J. Biol. Macromol.*, 2014, **64**, 17–24.
83. B. Cui, Y. M. Lu, C. P. Tan, G. Q. Wang and G. H. Li, *Food Hydrocolloids*, 2014, **35**, 576–582.
84. C. Chung, B. Degner and D. J. McClements, *LWT—Food Sci. Technol.*, 2013, **54**, 336–345.
85. H. V. Dang, C. Loisel, A. Desrumaux and J. L. Doublier, *Food Hydrocolloids*, 2009, **23**, 1678–1686.
86. N. Zhang, H. Liu, L. Yu, X. Liu, L. Zhang, L. Chen and R. Shanks, *Carbohydr. Polym.*, 2013, **92**, 455–461.
87. N. Kaisangsri, O. Kerdchoechuen and N. Laohakunjit, *Carbohydr. Polym.*, 2014, **110**, 70–77.
88. N. Yang, Y. Liu, J. Ashton, E. Gorczyca and S. Kasapis, *Food Chem.*, 2013, **137**, 76–82.
89. Y. Kim, J. I. Kee, S. Lee and S. H. Yoo, *Food Chem.*, 2014, **145**, 409–416.
90. S. M. Fitzsimons, D. M. Mulvihill and E. R. Morris, *Food Hydrocolloids*, 2008, **22**, 468–484.
91. X. Lian, W. Zhu, Y. Wen, L. Li and X. Zhao, *Int. J. Biol. Macromol.*, 2013, **59**, 143–150.
92. A. Jiménez, M. J. Fabra, P. Talens and A. Chiralt, *Food Hydrocolloids*, 2012, **29**, 265–271.
93. A. Ye, S. Hewitt and S. Taylor, *Food Hydrocolloids*, 2009, **23**, 1220–1227.
94. G. Bortnowska, J. Balejko, V. Schube, G. Tokarczyk, N. Krzemińska and K. Mojka, *Carbohydr. Polym.*, 2014, **111**, 624–632.

95. F. Zhang, L. Fang, C. Wang, L. Shi, T. Chang, H. Yang and M. Cui, *Meat Sci.*, 2013, **93**, 533–537.
96. F. Koksel and M. G. Scanlon, *J. Cereal Sci.*, 2012, **56**, 445–450.
97. F. Ronda, B. Oliete, M. Gómez, P. A. Caballero and V. Pando, *J. Food Eng.*, 2011, **102**, 272–277.
98. J. González-Gutiérrez, P. Partal, M. García-Morales and C. Gallegos, *Carbohydr. Polym.*, 2011, **84**, 308–315.
99. V. R. Diamantino, F. A. Beraldo, T. N. Sunakozawa and A. L. B. Penna, *LWT—Food Sci. Technol.*, 2014, **56**, 356–362.
100. E. Fahy, D. Cotter, M. Sud and S. Subramaniam, *Biochim. Biophys. Acta, - Mol. Cell Biol. Lipids*, 2011, **1811**, 637–647.
101. C. Chung, B. Degner and D. J. McClements, *Food Res. Int.*, 2012, **48**, 641–649.
102. S. Dhillon and K. Seetharaman, *J. Cereal Sci.*, 2011, **54**, 374–379.
103. U. V. Lay, Ma, J. D. Floros and G. R. Ziegler, *Carbohydr. Polym.*, 2011, **83**, 1869–1878.
104. L. Kong and G. R. Ziegler, *Carbohydr. Polym.*, 2014, **111**, 256–263.
105. S. Exarhopoulos and S. N. Raphaelides, *J. Cereal Sci.*, 2012, **55**, 139–152.
106. E. Fuentes-Zaragoza, E. Sánchez-Zapata, E. Sendra, E. Sayas, C. Navarro, J. Fernández-López and J. A. Pérez-Alvarez, *Starch Stärke*, 2011, **63**, 406–415.
107. S. Ahmadi-Abhari, A. J. J. Woortman, A. A. C. M. Oudhuis, R. J. Hamer and K. Loos, *Carbohydr. Polym.*, 2013, **97**, 436–440.
108. F. Chang, X. He and Q. Huang, *Food Hydrocolloids*, 2013, **32**, 365–372.
109. J. Blazek, E. P. Gilbert and L. Copeland, *J. Cereal Sci.*, 2011, **54**, 151–159.
110. Y. Ai, J. Hasjim and J. L. Jane, *Carbohydr. Polym.*, 2013, **92**, 120–127.
111. E. Vasiliadou, S. N. Raphaelides and E. Papastergiadis, *LWT—Food Sci. Technol.*, 2014, **60**, 698–707.
112. D. Cozzolino, S. Roumeliotis and J. Eglinton, *Food Res. Int.*, 2013, **51**, 444–449.
113. M. E. Clegg, P. S. Thondre and C. J. K. Henry, *Food Res. Int.*, 2011, **44**, 636–641.
114. J. A. Putseys, C. J. Gommes, P. Van Puyvelde, J. A. Delcour and B. Goderis, *Carbohydr. Polym.*, 2011, **84**, 1141–1150.
115. K. Kawai, S. Takato, T. Sasaki and K. Kajiwara, *Food Hydrocolloids*, 2012, **27**, 228–234.
116. R. P. H. Brandelero, M. V. Grossmann and F. Yamashita, *Carbohydr. Polym.*, 2012, **90**, 1452–1460.
117. M. Ghasemlou, N. Aliheidari, R. Fahmi, S. Shojaee-Aliabadi, B. Keshavarz, M. J. Cran and R. Khaksar, *Carbohydr. Polym.*, 2013, **98**, 1117–1126.
118. M. Chiumarelli and M. D. Hubinger, *Food Hydrocolloids*, 2014, **38**, 20–27.
119. T. V. D'Silva, J. R. N. Taylor and M. N. Emmambux, *J. Cereal Sci.*, 2011, **53**, 192–197.
120. J. A. Byars, G. F. Fanta and J. A. Kenar, *Carbohydr. Polym.*, 2013, **95**, 171–176.

121. T. De Pilli, K. Jouppila, J. Ikonen, J. Kansikas, A. Derossi and C. Severini, *J. Food Eng.*, 2008, **87**, 495–504.
122. E. Hernández-Hernández, C. A. Ávila-Orta, B. S. Hsiao, J. Castro-Rosas, J. A. Gallegos-Infante, J. Morales-Castro, L. Araceli Ochoa-Martínez and C. A. Gómez-Aldapa, *J. Cereal Sci.*, 2011, **54**, 69–75.
123. R. Belhassen, F. Vilaseca, P. Mutjé and S. Boufi, *Ind. Crops Prod.*, 2014, **53**, 261–267.
124. V. C. R. Schmidt, L. M. Porto, J. B. Laurindo and F. C. Menegalli, *Ind. Crops Prod.*, 2013, **41**, 227–234.
125. R. Ortega-Toro, A. Jiménez, P. Talens and A. Chiralt, *Food Hydrocolloids*, 2014, **38**, 66–75.
126. D. Muscat, R. Adhikari, S. McKnight, Q. Guo and B. Adhikari, *J. Food Eng.*, 2013, **119**, 205–219.
127. A. Jiménez, M. J. Fabra, P. Talens and A. Chiralt, *Food Hydrocolloids*, 2012, **26**, 302–310.
128. A. Jiménez, M. J. Fabra, P. Talens and A. Chiralt, *J. Food Eng.*, 2013, **116**, 695–702.
129. A. Jiménez, M. J. Fabra, P. Talens and A. Chiralt, *Food Hydrocolloids*, 2013, **30**, 408–418.
130. D. L. Compton, J. A. Kenar, J. A. Laszlo and F. C. Felker, *Ind. Crops Prod.*, 2007, **25**, 17–23.
131. L. Manzocco, S. Calligaris, S. Da Pieve, S. Marzona and M. C. Nicoli, *Food Res. Int.*, 2012, **49**, 778–782.
132. S. Meng, Y. Ma, D. W. Sun, L. Wang and T. Liu, *J. Cereal Sci.*, 2014, **59**, 25–32.
133. S. Wang, J. Yu, J. Yu, J. Pang and H. Liu, *Food Hydrocolloids*, 2008, **22**, 1283–1290.
134. L. Wang and Y. J. Wang, *Starch Stärke*, 2001, **53**, 570–576.
135. M. Majzoobi and P. Beparva, *Food Chem.*, 2014, **147**, 312–317.
136. W. Chen, H. Zhou, H. Yang and M. Cui, *Food Chem.*, 2015, **167**, 180–184.
137. M. Majzoobi, A. Farahnaky, J. Jamalian and M. Radi, *Food Chem.*, 2011, **124**, 795–800.
138. T. Komiya and S. Nara, *Starch Stärke*, 1986, **38**, 9–13.
139. *Asian Noodles: Science, Technology and Processing*, ed. G. G. Hou, John Wiley & Sons, Inc., Hoboken, NJ, 2010.
140. S. Wang and L. Copeland, *Food Chem.*, 2012, **135**, 1635–1642.
141. M. Z. Nor Nadiha, A. Fazilah, R. Bhat and A. A. Karim, *Food Chem.*, 2010, **121**, 1053–1059.
142. A. A. Karim, M. Z. Nadiha, F. K. Chen, Y. P. Phuah, Y. M. Chui and A. Fazilah, *Food Hydrocolloids*, 2008, **22**, 1044–1053.
143. Q. Li, Q. Xie, S. Yu and Q. Gao, *Food Hydrocolloids*, 2014, **35**, 392–402.
144. H. Zhou, C. Wang, L. Shi, T. Chang, H. Yang and M. Cui, *Food Chem.*, 2014, **156**, 137–143.
145. L. Day, C. Fayet and S. Homer, *Carbohydr. Polym.*, 2013, **94**, 31–37.
146. D. A. Garrec and I. T. Norton, *J. Colloid Interface Sci.*, 2012, **379**, 33–40.

147. J. Lyklema, *Chem. Phys. Lett.*, 2009, **467**, 217–222.
148. F. B. Ahmad and P. A. Williams, *J. Agric. Food Chem.*, 1999, **47**, 3359–3366.
149. S. Moelbert, B. Normand and P. De Los Rios, *Biophys. Chem.*, 2004, **112**, 45–57.
150. X. Zhang, Q. Tong, W. Zhu and F. Ren, *J. Food Eng.*, 2013, **114**, 255–261.
151. Q. Li, H. Li and Q. Gao, *Food Hydrocolloids*, 2015, **43**, 803–811.
152. J. Lu, Z. Luo and Z. Xiao, *Food Res. Int.*, 2012, **49**, 612–617.
153. M. Cui, L. Fang, H. Zhou and H. Yang, *Food Chem.*, 2014, **151**, 162–167.

# Subject Index